Condensed Matter and Materials Physics Conference 2010

(CMMP10)

Journal of Physics: Conference Series Volume 286

Coventry, United Kingdom
14 – 16 December 2010

Editor:

John Inglesfield

ISBN: 978-1-61782-667-2
ISSN: 1742-6588

Printed from e-media with permission by:

Curran Associates, Inc.
57 Morehouse Lane
Red Hook, NY 12571

Some format issues inherent in the e-media version may also appear in this print version.

Printed by Curran Associates, Inc. (2011)

Institute of Physics
Dirac House, Temple Back
Bristol BS1 6BE UK

Phone: 44 1 17 929 7481
Fax: 44 1 17 920 0979

techtracking@iop.org

Additional copies of this publication are available from:

Curran Associates, Inc.
57 Morehouse Lane
Red Hook, NY 12571 USA
Phone: 845-758-0400
Fax: 845-758-2634
Email: curran@proceedings.com
Web: www.proceedings.com

TABLE OF CONTENTS

BIOLOGICAL AND SOFT MATTER PHYSICS

SUPERCONDUCTIVITY

NANOSCALE MAGNETISM AND SPINTRONICS

NANOPHYSICS AT LOW TEMPERATURE AND QUANTUM MEMORY

SURFACES AND INTERFACES

COMPUTER SIMULATION OF SURFACES AND INTERFACES

ENERGY MATERIALS, AND MATTER UNDER EXTREME CONDITIONS

PREFACE

This is the first proceedings for a CMMP Conference, and we are delighted that nearly 60 papers are presented in this volume. CMMP10 was itself a very successful conference, with 422 participants and over 300 papers spread over 23 symposia. Over the years, this conference has had its up and downs, starting off life as the Manchester Solid State Physics Conference, held annually at the beginning of January. Having been held for many years during the Easter holidays, it is now held in the week before Christmas, which seems to suit most people, and the CMMP conference is once more on an upward slope. A special effort was made this year to include symposia on 'Biological Physics', 'Soft Matter in Action', and 'Polymer Physics', and one of the highlights of the conference was the Mott lecture presented by Professor Athene Donald, on 'Self assembly of proteins'. Biological and soft matter physics is well represented in this volume, with 9 papers, and we hope that it will continue to play a large part in future CMMP conferences.

Many people contributed to the success of the conference, and the organisers would especially like to thank the team from the Institute of Physics Conference Office, led by Claire Garland. The local organisers and conference office from the University of Warwick did a splendid job, and we all enjoyed an excellent conference atmosphere, culminating in the conference buffet on the Wednesday evening. A huge amount of work falls on the symposium organisers, who choose the invited speakers (32 this year), and put together the oral and poster sessions. They were, on the whole, very patient and put up with the nagging of the conference secretary and chairman.

This year there were 7 plenary lectures – always the mainstay of the conference. As well as the annual Mott lecture, the Wohlfarth Memorial lecture – sponsored by the Magnetism Group – is presented at the CMMP conference. This year the Wohlfarth lecture was given by Guido Meier, from the University of Hamburg, with the title 'Imaging magnetization dynamics on its intrinsic time scale'.

The opening plenary lecture of the conference was given by Marshall Stoneham, on 'Where quantum physics meets biology', presented with his usual panache. Marshall, who had been President of the Institute of Physics since 1 October 2010, stayed for the whole conference, and participated with enthusiasm, interacting with young physicists and the older members of the community alike. It was with great sadness and almost disbelief that we learned of his death on 18 February 2011. We will all miss him, and these proceedings are dedicated to his memory.

John Inglesfield

Marshall Stoneham, FRS
18 May 1940 – 18 February 2011

Photograph by Philip Wade

Conference Chairman
Peter Weightman

Physics Department
University of Liverpool

Conference Secretary and Proceedings Editor

John E Inglesfield
School of Physics and Astronomy
Cardiff University

Conference Sponsors

We would like to thank the following sponsors of CMMP10:
AWE, EPSRC, IOP Publishing.

Peer review statement

All papers published in this volume of *Journal of Physics: Conference Series* have been peer reviewed through processes administered by the proceedings Editors. Reviews were conducted by expert referees to the professional and scientific standards expected of a proceedings journal published by IOP Publishing.

Using bespoke fluorescence microscopy to study the soft condensed matter of living cells at the single molecule level

Q Xue, O Harriman and M C Leake[1]
Clarendon Laboratory, Dept of Physics, Parks Road, Oxford University, Oxford OX1 3PU, UK

E-mail: m.leake1@physics.ox.ac.uk

Abstract. The use of bespoke imaging tools and analysis can offer significant insight into the living counterpart of soft condensed matter. The soft matter of biological systems consists of molecular building blocks, a staple of which is protein. Protein molecules, so small that 1 billion would fit on the full-stop at the end of this sentence, carry out most of the vital activities in living cells. Many of these processes require the assembly of multiple proteins into remarkable biological machines. Obtaining the blueprints for the architecture of these machines is essential for understanding the workings of the cell. Here, we discuss recent biological physics experiments on functional single-celled organisms in which one can apply bespoke fluorescence microscopy imaging and analysis to monitor the number and dynamics of several different proteins at the nanometre length scale to a precision of single molecules.

1. Introduction

One of the greatest discoveries in the life sciences, made in the 19th century, was that living organisms were composed of cells, and the make-up of cells is what we could describe as soft condensed matter, stuff like DNA, fats and proteins. Much life science research as we know it today is essentially concerned with understanding the mesoscopic molecular mechanisms that make this soft-matter perform the functions of life. But why study such molecular details in living cells when one can do so in the test-tube? The test-tube environment is significantly more controllable, less contaminated and comes with less measurement noise. The best answer is that cells are not test-tubes. A test-tube experiment is a much reduced version of the native biology with only the components which one thinks/hopes are important included. But we now know definitively that even the simplest cells are not just bags of chemicals, but rather have localized processes in both space and time. A related issue is that of localized effects of the physical and chemical environment in the cell which are essentially impossible to faithfully replicate in the test-tube. Also, the effective numbers of molecules involved in many processes are often low, sometimes just a few per cell, and these minimal stoichiometry conditions are not easy to reproduce in the test-tube.

A second question is why use fluorescence microscopy as a measuring tool. In many cases, some of which we illustrate here, it allows us to visualize specific molecular machines inside living cells with relatively minimal perturbation to the native physiology, so in other words it is a relatively gentle probe which allows one to see molecular detail of soft-matter without disrupting it. Here, we will show how the limits of fluorescence microscopy can be pushed towards a single-molecule

[1] To whom any correspondence should be addressed.

precision over an imaging time scale which is realistic in light of the underlying process being studied, in other words in "real time", exploring a sampling rate from Hz up to kHz; a range of three orders of magnitude for characteristic time scale which encapsulates many vital biological processes [1, 2].

2. Counting single molecule subunits in molecular machines using bespoke TIRF

It is relatively easy to see single fluorescent molecules in vitro using bespoke laser excitation total internal reflection fluorescence, or so-called TIRF (figure 1), illustrated here with green fluorescent protein (GFP) molecules binding transiently to a glass coverslip. If we draw a region of interest around each of these spots of light and sum the pixel intensity, this gives us a measure for total photon emission flux. We can observe firstly the total emission flux is relatively constant across a population of molecules, secondly each molecule photobleaches in a step-like manner, the size of the step being a function not only of local laser excitation intensity but also of the immediate physical and chemical environment. This has significance if the molecule is inside a living cell; the absolute brightness depends not only on where in that cell it is but also on what the cell happens to be doing at that time.

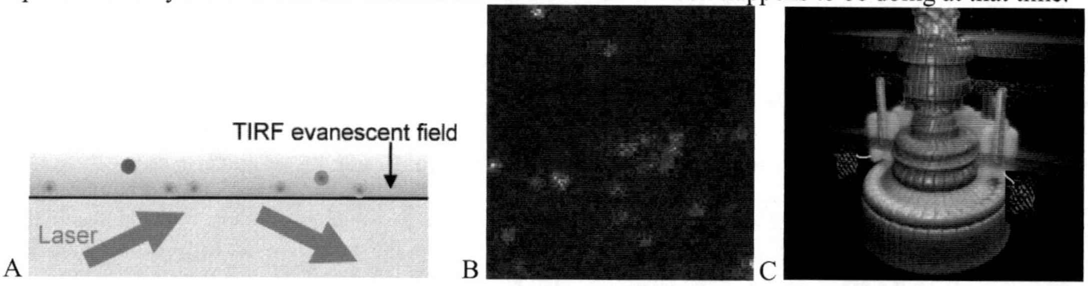

Figure 1. (A) Schematic of transient binding of GFP to a glass coverslip in a TIRF excitation field. (B) Typical experimental single frame snapshot of single GFP molecules bound to glass (false-colour). (C) Schematic of the bacterial flagellar motor molecular machine with attached GFP tags (green).

One example of live-cell imaging involves whole cell motility in bacteria. Bacteria swim by the concerted rotation of several semi stiff helical filaments rotating as one super-filament, at the base of which is an intricate and remarkable molecular machine known as the bacterial flagellar motor (figure 1), one of Nature's primordial wheels. By tinkering with the genetics of the cell so as to fuse GFP to one of the proteins implicated in force generation in this machine, so-called MotB, every molecule of MotB produced will then have a GFP fused to it as a single-molecule fluorescent tag [3]. To visualize this motor one can use a cell rotation assay in which the cell is tethered to a microscope coverslip via one of its filaments. Since the motor is still functional this then causes the whole body of the cell to rotate about this point, like holding down a car wheel and watching the whole car rotate. In TIRF, sampling at typically 1-10 Hz, one generally sees a spot of light at the centre of cell rotation.

One can then quantify fluorescence in exactly the same way as was done with the previous in vitro assay by drawing a region of interest around the spot of intensity. One sees typically exponential photobleach decays, though towards the end of the bleach there is step-like behaviour, made clearer on expanded sections (figure 2). It is more complicated than the in vitro example because the contribution of additional background fluorescence not associated with the motor itself needs to be taken into account, but the principle is the same. One can measure the underlying periodicity in the intensity by constructing a pair-wise difference distribution and then look at its associated power spectrum, which here indicates a fundamental peak at around five thousand counts on our detector. By looking more closely at less noisy data from cells in which the background component was pre-bleached one indeed sees steps of roughly integer multiples of this value. In other words this is an assay that can now pull out the brightness of a single GFP molecule which is unique not only to that single cell, but to that individual molecular machine in that single cell. Dividing the extrapolated initial unbleached intensity by this measured GFP brightness gives an estimate for stoichiometry, here of around 22 molecules.

The width on this distribution is ~6 molecules, but simulations and other experiments show that much of this is inherent variation in the cell population.

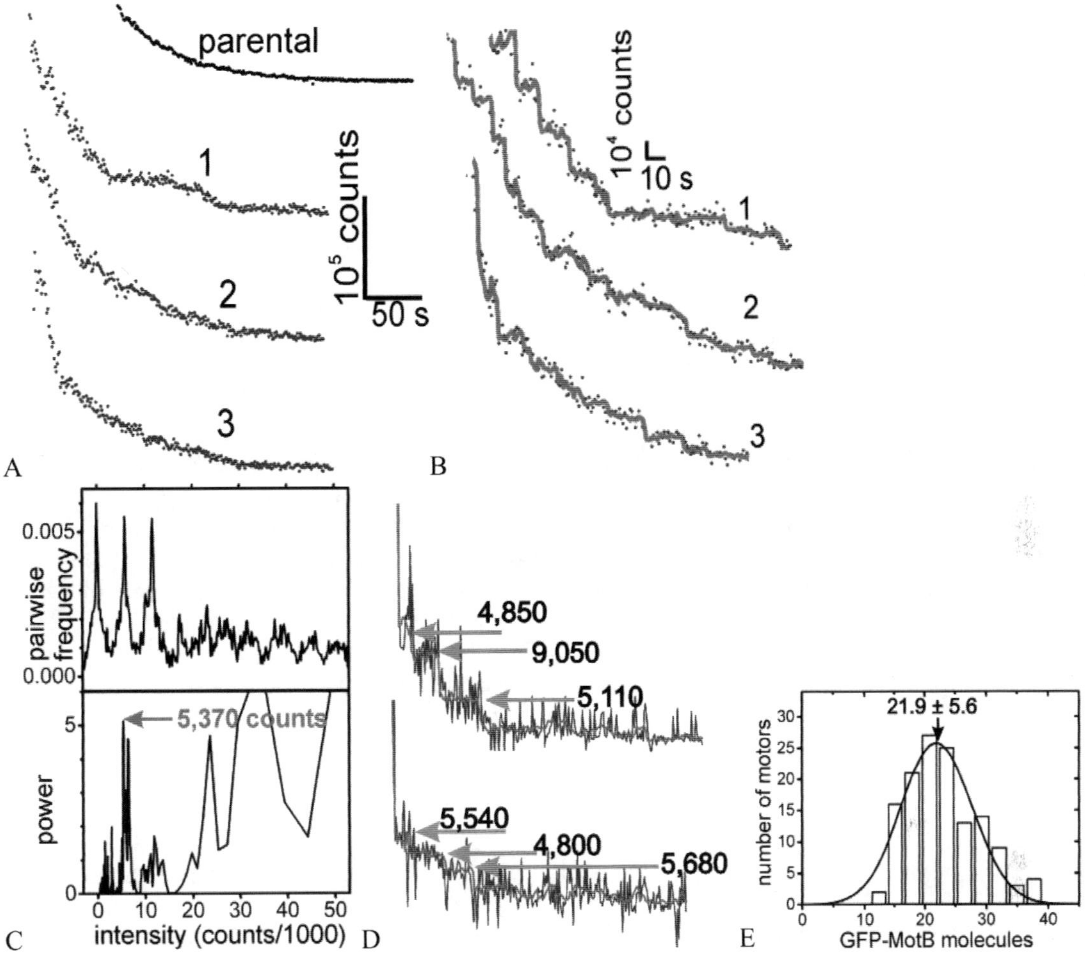

Figure 2. (A) Three typical photobleach intensity traces with (B) expanded sections, raw (blue) and filtered (red) indicated. (C) Pair-wise difference distribution and associated power spectrum. (D) Expanded photobleach sections of pre-bleached cells, step size indicated (arrows). (E) Distribution of molecular stoiochiometry for MotB molecule in the bacterial flagellar motor, Gaussian fit indicated.

2.1. Simulating photobleaching and stoichiometry measurement

To characterize the accuracy of our estimates we generate realistic simulations of the photobleach process using known values for stoichiometry and run these through the same analysis algorithms. We assume that photobleaching is characterized by a series of unitary photobleach events whose probability for occurrence is Poisson distributed. We add in realistic background camera noise in addition to independent photon emission noise from each photoactive fluorescent molecule present in the measurement region of interest (figure 3). This indicates that the overall precision in our stoichiometry estimates using this method is ~10% or less [4].

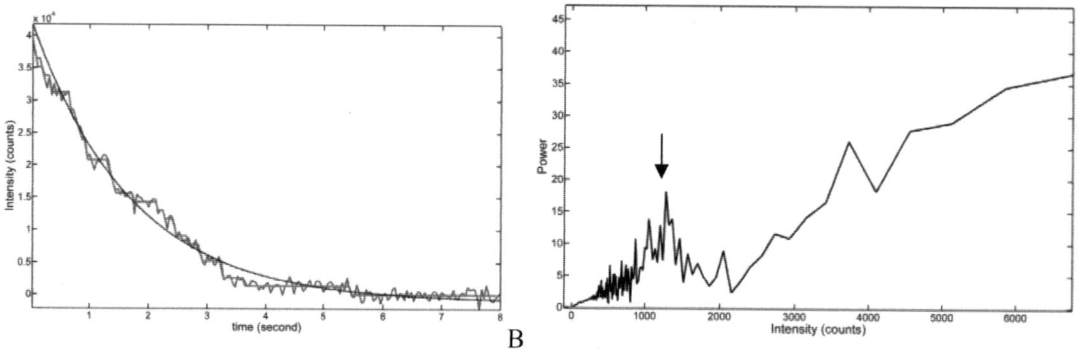

A B

Figure 3. (A) Simulated photobleach, noise-free (red), noise-added (magenta), exponential fit (black) indicated, using 30 initial photoactive proteins, unitary step here of 1,300 counts. (B) Power spectrum of pair-wise difference distribution indicating correct estimation of peak at ~1,300 counts (arrow).

3. Dynamics of single molecules in the cell membrane and turnover in molecular machines

Furthermore, one can monitor protein dynamics. Here one first delimits a high intensity volume using a bespoke focused laser for use in fluorescence recovery after photobeaching, so-called FRAP (figure 4). Looking closely at the cell indicates initially two spots of light, then a flash, then just one spot of light; photobleaching one motor but leaving another motor in the same cell but only one micron away relatively unaffected. There are non-motor components to fluorescence which include a membrane component not incorporated in the motor, and one can monitor the recovery of this component after the focussed laser bleach, showing there must be membrane proteins free to diffuse back into the bleached region, which we can model using Monte Carlo simulations utilising the known geometry of the cell and imposing different values of theoretical diffusion coefficient, and by comparing these to the experimental data we can estimate an effective diffusion coefficient (figure 4). A complementary approach is to track spots of intensity directly as they diffuse on a faster time scale, and both approaches broadly agree to give an estimate which is similar to those of other free membrane proteins of similar mass, so in other words when MotB is not incorporated into a motor it is relatively mobile.

For a motor inside the bleach zone half of the intensity recovers after a period of five minutes. Similarly, for a motor outside the bleach zone its intensity falls by a similar amount over the same time. This is fluorescence loss in photobleaching (FLIP), and what this means is that turnover of the MotB protein has occurred, either involving photobleached or photoactive GFP depending upon its position relative to the centre of the focused laser bleach. Modelling this using a modified Monte Carlo with a diffusion-to-capture element, indicates that the dwell time for a force-generating unit in the motor is ~30 seconds. This is rapid, like driving a car down a motorway but changing one of the wheels at the same time. One unresolved question is whether this turnover is unique to the flagellar motor or is a more generic feature of many different molecular machines.

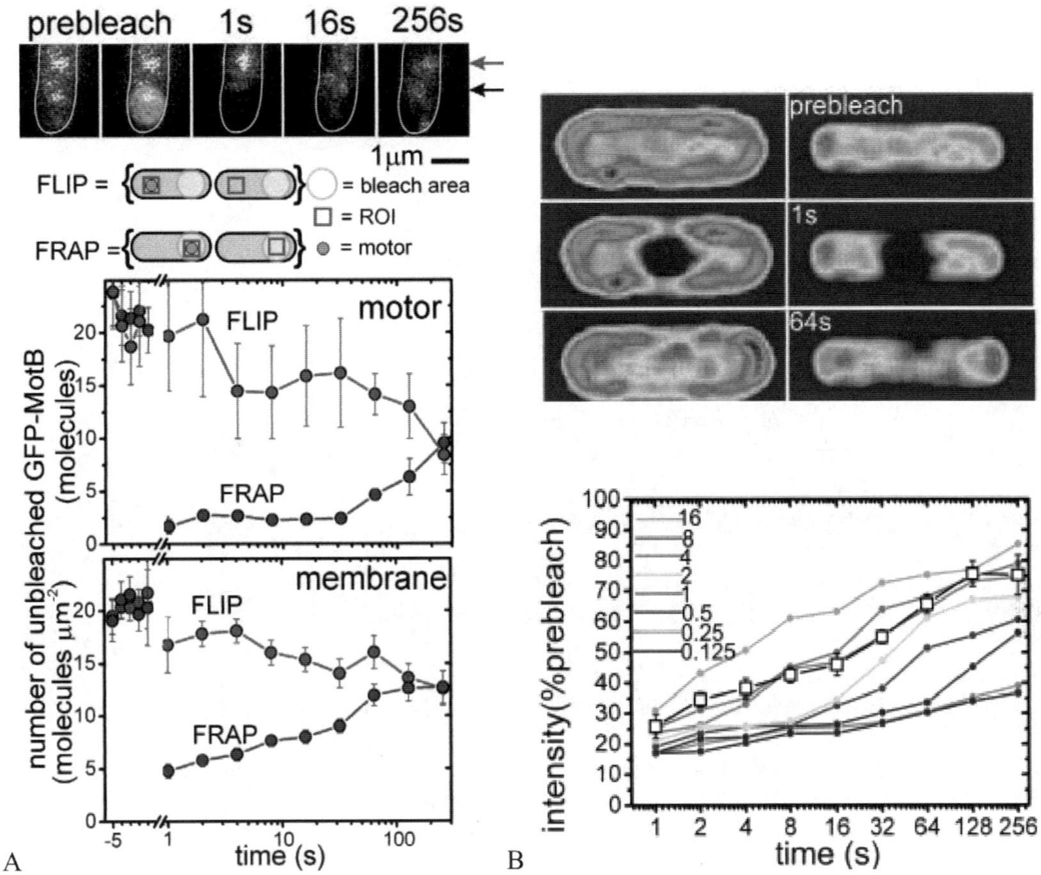

Figure 4. (A) FRAP/FLIP to measure molecular turnover, bleach zone (circle) centred over one motor in a bacterial cell (blue arrow), with another motor in same cell outside the bleach zone (red arrow), changes to intensity of different components indicated. (B) Monte Carlo simulations of bleaching and diffusion in a single cell (upper panel), intensity recovery of membrane component indicated for simulations as a function of time post bleach, effective diffusion coefficients indicated in units of $\mu m^2 s^{-1}$ (coloured lines), compared against experimental values (black squares).

3.1. Tracking molecular machines

To apply similar stoichiometry assays to diffusing membrane proteins requires robust methods for tracking them. To do so one can first automatically segment the outlines of cells, and use this to both generate image masks and cell vital statistics which can be used in coordinate transformations to map between the camera and the curved surface of each cell [5, 6]. Particles can be automatically detected in each fluorescence image and tracked, typically sampling at 10-100 Hz, using an iterative Gaussian fitting process [7]. One can beat the optical resolution limit of roughly half a wavelength by increasing the number of photons sampled, and this is the basis for Fluorescence Imaging at the one nanometre level (FIONA). Having robust tracking implies pulling out the right numbers, and here simulation is everything. One can generate realistic point spread functions using either full 3D Bessel formulations or the experimentally measured point spread function directly, adding noise at realistic levels. A bacterial cell membrane is not a 2D plane, so one can add in a more realistic curved geometry, modelling the cell surface as a cylinder capped by hemispheres, and set simulated particles to diffuse on that surface, figure 5 shows four such particles which start in a row at the bottom of the cell, and two focal planes have been simulated, one through the midline which might be used during epifluorescence, and one at the bottom which might be used in TIRF. These data are run through the same automated detection and tracking algorithms to predict how well they perform, and the conclusion is that under video-rate conditions the sensitivity is equivalent to around 1 or 2 fluorescent protein molecules with an associated localization precision of 30-40 nm. If a typical diffusing complex has a stoichiometry of the order of 10s of molecules then this localization precision drops to 5 nm or less, the width of a cell membrane.

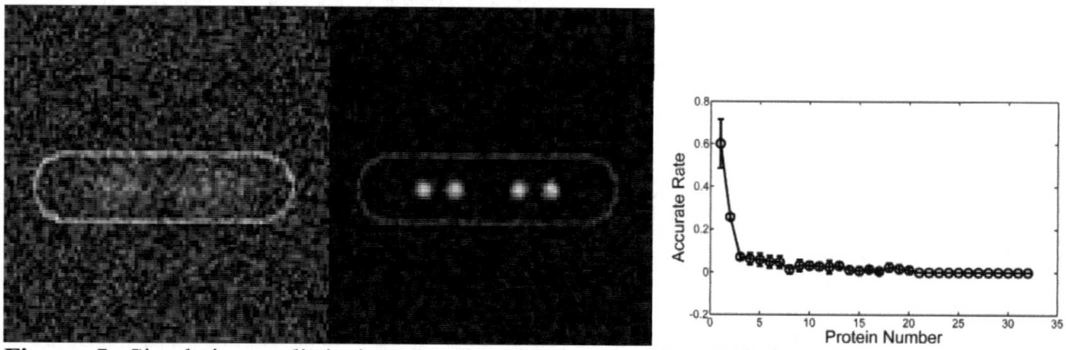

Figure 5. Simulating realistic images of molecular machine diffusion of over cell membranes, showing epifluorescence (left panel), TIRF (middle panel) and the localization precision ("accurate rate") in units of pixels as a function of number of fluorescent proteins per molecular machine.

4. Visualizing molecular machines inside the cell using fast bespoke "slimfield" imaging

The live-cell examples discussed thus far involved membrane proteins at relatively low concentration. There are advantages to this, one is that molecular machines can be imaged using TIRF since they are relatively close to the coverslip surface, second that since they are at low concentration the mean separation is generally greater than the point spread function width and so distinct complexes can be detected. Even in systems having higher concentrations bespoke optical tricks can be applied to effectively reduce the number of photoactive proteins present, either using stochastic photoactivation or photobleaching of a sub-population [8, 9]. A third advantage is that complexes diffuse relatively slowly, so are easier to track. Using simple diffusion theory one can calculate that mobile membrane proteins need to be imaged faster than around 100 ms to be seen unblurred, whereas the limit for proteins inside the cell, in the cytoplasm, is more like just a few milliseconds, typically 3 or less. This means that conventional video-rate imaging is too slow, and simply reducing the image time per frame results in the signal being swamped by noise. The test cytoplasmic system we studied was that of DNA replication, again using a model bacterial organism. DNA replication is performed by a multiprotein machine in the cytoplasm called the replisome. Previously, several functional cell strains were made which had all 10 or so of the different components of this machine individually labelled with a fluorescent tag, this time with a yellow fluorescent protein called YPet, typical epifluorescence snapshots showing either one or two spots per cell depending upon the stage in the cell growth cycle.

To probe this machine with a better "real time" precision, we used a simple optical trick to generate bespoke "slimfield" illumination [10]. This essentially involves underfilling the back aperture of an objective lens to produce a laterally expanded confocal volume wider than a diffraction-limited focal waist, which can be catered to be large enough to encompass single bacterial cells with relatively little correction needed for non-uniformity in intensity. The effect is to produce a high intensity in the vicinity of single cells allowing ~kHz sampling rates whilst still being above the camera noise. By doing this we could see distinct spots frame by frame on the millisecond time scale (figure 6) and were able to generate step-like photobleach traces to a single-molecule precision [11].

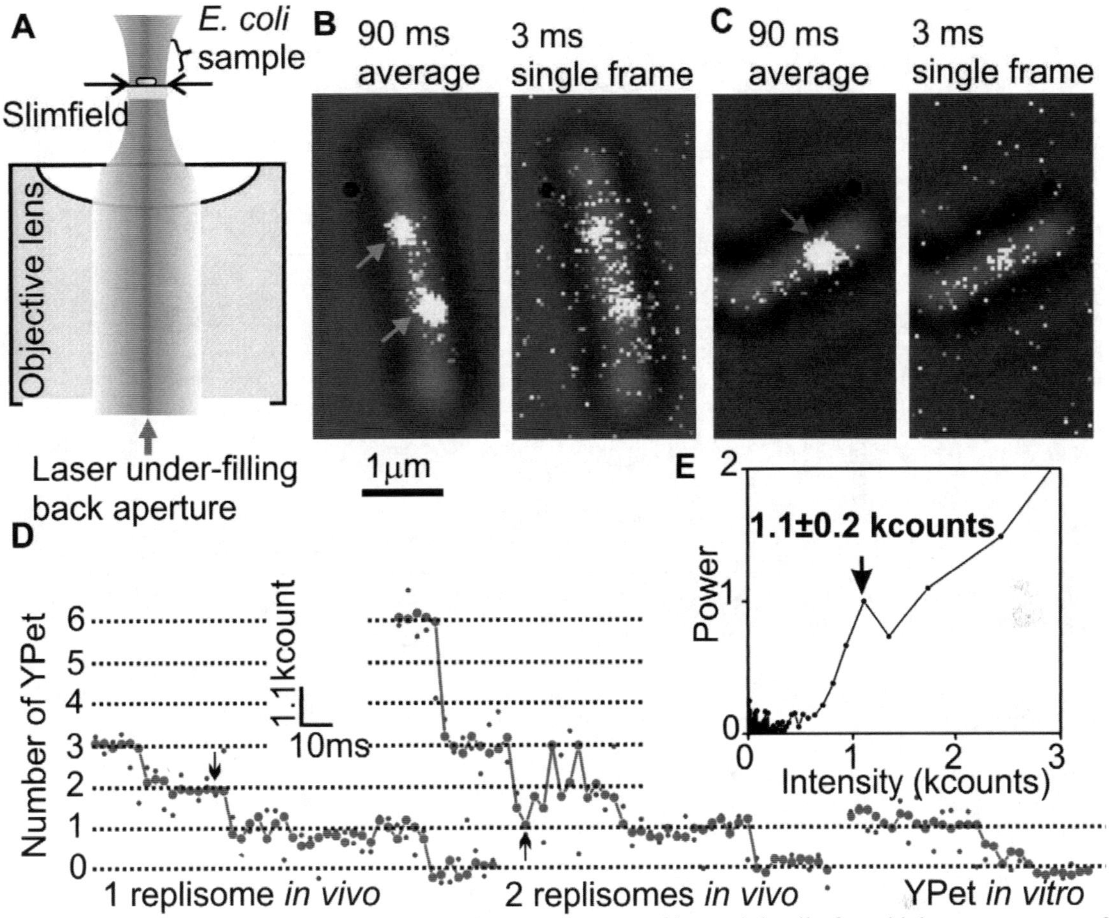

Figure 6. (A) Schematic of slimfield illumination. Images of bacterial cells for which a component of the DNA polymerase enzyme DnaQ has been fused to YPet, either having (B) 2 spots or (C) 1 spot per cell. (D) Typical photobleach trace from spots in cells which had two spots per cell (left panel) and in cells which had one spot per cell (right panel) compared against a single photobleach step for purified surface immobilized YPet in vitro, and (E) power spectrum of pair-wise difference distribution for photobleach trace showing unitary peak brightness.

5. Conclusions

Here we show how using a combination of bespoke optical microscopy and image analysis the unitary photobleach step of fluorescent proteins can be measured in a variety of living cells, in order to apply this as a measuring tool for estimating the stoichiometry of components within functional molecular machines. Furthermore, one can extend the bespoke microscopy and imaging to permit tracking of mobile machines in the cell membrane, and can push the sampling limit up the kHz level to permit imaging of biological machines in the cytoplasm of living cells. The immediate future development of this work lies in extending these methodologies into the multi-colour regime (figure 7), which should allow measurement of two or more different protein components in the same cell at the same time, which may go some way to addressing several questions of living soft condensed matter complexity.

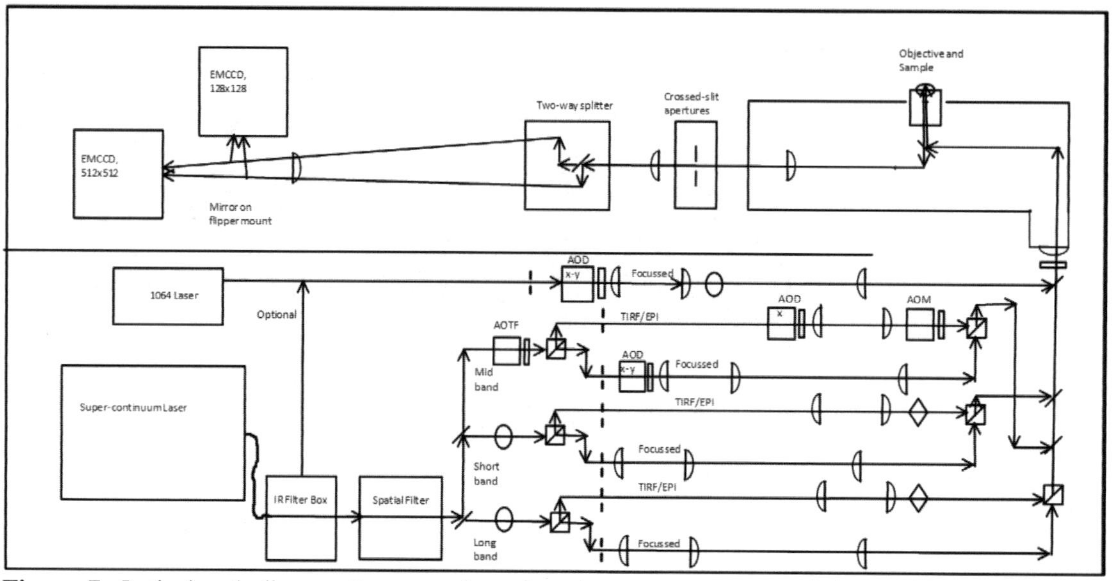

Figure 7. Optical path diagram for a new bespoke microscope setup permitting multi-colour imaging over "real time" scales spanning several orders of magnitude (work in progress, O. Harriman).

6. References

[1] Leake MC 2010 *Commun Integr Biol.* **3** 415

[2] Dobbie IM, Robson A, Delalez N and Leake MC 2009 *J Vis Exp.* **31** 1508

[3] Leake MC, Chandler JH, Wadhams GH, Bai F, Berry RM and Armitage JP 2006 *Nature* **443** 355

[4] Delalez NJ, Wadhams GH, Rosser G, Xue Q, Brown MT, Dobbie IM, Berry RM, Leake MC, and Armitage JP 2010. *Proc Natl Acad Sci U S A* **107** 11347

[5] Leake MC, Greene NP, Godun RM, Granjon T, Buchanan G, Chen S, Berry RM, Palmer T and Berks BC 2008 *Proc Natl Acad Sci U S A* **105** 15376

[6] Xue Q, Jones NS and Leake MC 2010 *Proc. IEEE Internat. Symp. Biomed. Imaging (ISBI): From Nano to Macro* 161

[7] Xue Q and Leake MC 2009 *Proc. IEEE Internat. Symp. Biomed. Imaging (ISBI): From Nano to Macro* 1158

[8] Lenn T, Leake MC and Mullineaux CW 2008 *Mol Microbiol.* **70** 1397

[9] Lenn T, Leake MC and Mullineaux CW 2008 *Biochem Soc Trans.* **36** 1032

[10] Plank M, Wadhams GH and Leake MC 2009 *Integr Biol (Camb)* **10** 602

[11] Reyes-Lamothe R, Sherratt DJ and Leake MC 2010 *Science* **328** 498

Acknowledgments

This research was supported by a Royal Society University Research Fellowship (M.C.L.), BBSRC via the Oxford Centre for Integrative Systems Biology (Q.X.), EPSRC (O.H. and M.C.L.), the Oxford University Press (M.C.L) and a Hertford College Oxford Research Fellowship (M.C.L.)

Journal of Physics: Conference Series **286** (2011) 012002 doi:10.1088/1742-6596/286/1/012002

Integration of FIRST, FRODA and NMM in a coarse grained method to study Protein Disulphide Isomerase conformational change.

J. Emilio Jimenez-Roldan[1,2] **S. A. Wells**[1] **R. B. Freedman**[2] **and R. A. Roemer**[1]

1 Department of Physics and Centre for Scientific Computing, University of Warwick, Coventry, CV4 7AL, UK

2 School of Life Sciences, University of Warwick, Coventry, CV4 7AL, UK

E-mail: `j.e.jimenez@warwick.ac.uk`

Abstract. Our simulations of protein motion are based on a rigidity-based coarse graining criteria which is able to identify flexible and constraint regions and on Normal Modes of Motion (NMM) which map out the directions of motion for a given network. The NMM and rigidity constraints are integrated together by the FRODA module to identify large scale conformational changes at low computational cost. In this context we identify the limits for conformational change of a large multi domain protein -Protein Disulphide Isomerase, by re-assessing the principal NMM at intermediate points along a trajectory as the protein moves over the trajectory defined by the initial principal NMM.

1. Introduction

Although computationally expensive, protein simulations to explore dynamics, fluctuations and motions via molecular dynamics (MD) have proved to be a valuable tool for investigating molecular motions up to a few hundred nanoseconds. However, there is a growing interest in studying processes involving larger proteins and processes of the order of milliseconds or longer to complete. Coarse grained (CG) models are developed with the motivation of simulating large systems for longer time scales at a lower computational cost in order to have a dynamic rapport with experimental observables. Hence, CG models average out atomic degrees of freedom based on individual CG constraints, so that it is possible to investigate large mesoscopic phenomena which are usually beyond the scope of atomistic detailed models. Ideally a good CG model will reproduce the physical behaviour of an experiment or a more computationally costly simulation. Hence, the value of a CG model is defined by how well it captures the essential features of a molecular network and simplifies the handling of the network so that it brings both time and performance advantage. Since CG models allow one to explore larger systems and larger time scales, they have been used most often to study issues on protein folding, conformational change and processes that involve macromolecular interactions involving large time scales.

There is a growing research focus on the role of CG models in interpreting experimental data [1], protein rigidity [2, 3], protein folding [4], multi-scale methods [5], and membrane proteins [6]. In particular there has been a development of CG models which specialise in protein flexibility and in exploring their accessible conformational space [7, 8]. Such methods

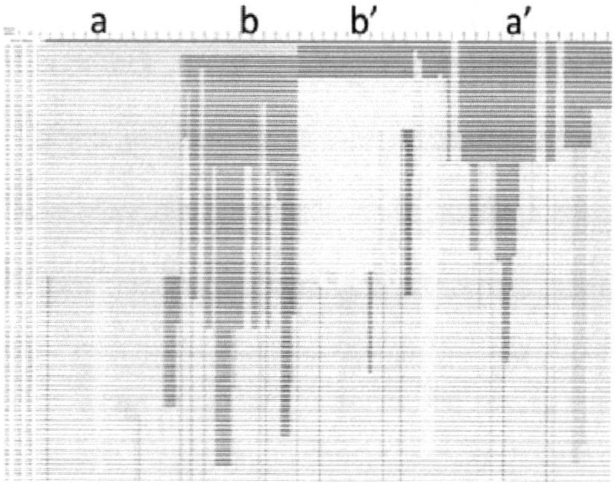

Figure 1. PDI's Rigid Cluster Decomposition graph defined for its four domains a-b-b'-a'. Each colour defines a rigid cluster. The energy cut off increases downwards, so each horizontal line is created when there is a change in the protein rigidity as we remove the weakest hydrogen bond and test the network rigidity. The rigid residues are defined by a colour thick line and the flexible ones by black thin line. Hence the lines at the top have low energy cut off and more bond density than the ones further down.

have been used to simulate conformational changes either by using stereochemically acceptable pathways [9] or normal modes of motion [10–13].

However, there is no method that will identify the limits of the accessible conformational space for proteins. In this paper we present the preliminary results of our CG simulation strategy for the principal Normal Mode (pNM) which is defined as the lowest frequency non-trivial NMM. We investigate the limits of Protein Disulphide Isomerase (PDI) conformational change for its principal mode while using our integrated coarse grained method. Further, we investigate the distance that PDI's domains can move freely without steric clashes or network bond constraints along the trajectory defined by the pNM and finally we show that it is possible to to study large conformational changes at a computational cost of minutes using our CG strategy.

Protein Disulphide Isomerase is a 522 amino acid protein with four domains `a-b-b'-a'`plus a linker region between `b'` and `a'` and a C-terminal extension by a domain [14]. PDI catalytic activity is essential in the formation of disulphide bonds in secretory proteins and it is required to be a conformationally flexible molecule to perform its catalytic activity [15].

2. Methods

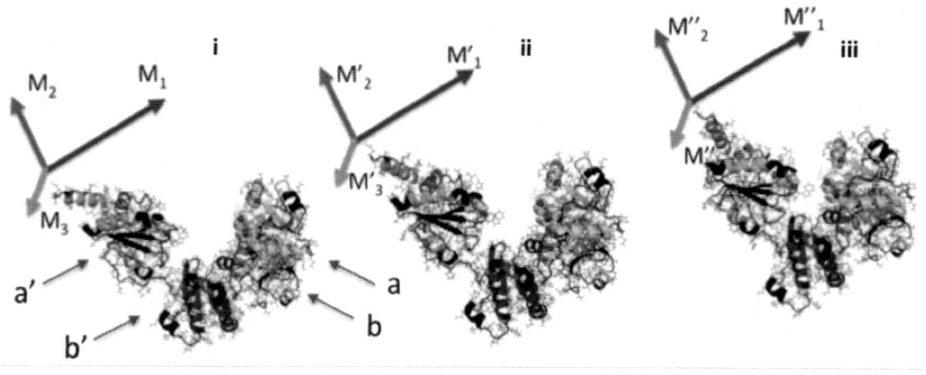

Figure 2. PDI's RCD graph defined for its four domains a-b-b'-a'. Within each domain, a colour defines a rigid cluster. The energy cut off increases downwards, so each horizontal line is created when there is a change in the protein rigidity as we remove the weakest hydrogen bond and test the network rigidity.The rigid residues are defined by a colour thick line and the flexible ones by black thin line.

We obtain the Rigid Cluster Decomposition (RCD) shown in Figure 1 using FIRST [7], a software that identifies rigid and flexible regions or clusters in a protein for a given energy cut off. Such clusters are used to coarse grain the simulation and are considered to move as a single unit. The set of NMM are obtained using ElNemo software [16, 17]. The directed motion for PDI is simulated using the software FRODA [18] which integrates the NMM and the rigidity constraints from the RCD to provide a CG simulation at a very low computational cost. As the protein structure moves along the conformational path defined by a given initial NMM the new conformational structure is recorded at given intervals or frames.

Each frame represents a computer cycle in which the protein is exploring the conformational space in the direction given by a NMM. We record a new structure for every 100 frames and up to 2500 frames. For each of these structures we calculate the new set of NMM particular to that new conformation which I refer to as intermediate NMM. This process is illustrated in Figure 2 where a set of three structures show PDI closing (from left to the right) and the intermediate NMM (M_i-M_i'-M_i'') calculated for each frame. The positive and negative direction of motions are chosen arbitrarily.

Thereafter, we calculate how each intermediate set of NMM have changed with respect the initial NMMs by obtaining the dot product between the vector that defines each of the intermediate NMM and the initial NMM. Even though in this paper we present the assessment of the pNM only, we show how it is possible to identify changes between the initial pNM and the intermediate pNM calculated as the protein moves along the trajectory defined by the initial pNM.

3. Results

The dot product between the initial pNM and the intermediates pNM shown in Figure 3-Left reveal that there is a number of frames for which the initial pNM holds good correlation with the intermediates pNM. This is true for all the energy cut off selected. Hence we can identify the range of conformational change that the protein is free from steric effects and for which it moves freely over the initial pNM. Using the intermediate structural conformations obtained

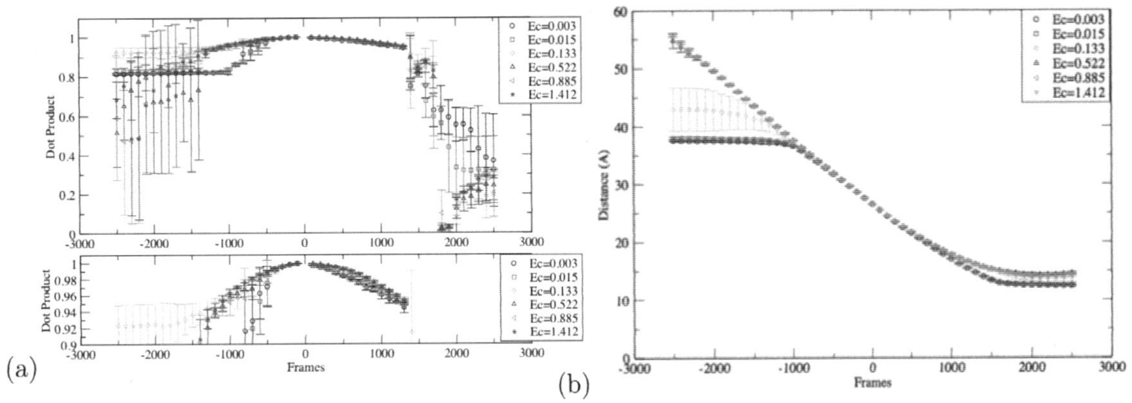

Figure 3. (a) Dot product between the pNM calculated initially and the pNM calculated at each of the 2500 frames, for the positive and negative direction of motion. We include calculations for six different energy cut off each of which include a different number of bonds and therefore a different rigidity patter. (b) Calculated distance between the two active sites, within the a and a' domain versus frames over the two directions of motion for the pNM. The minimum distance remains similar for different energy cut off but the maximum distance increases for higher energy cut off.

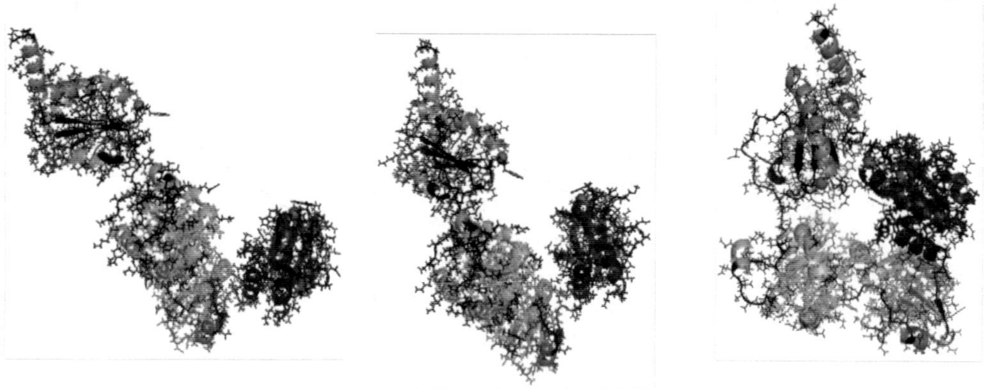

Figure 4. PDI pNMM conformational change. In red, the active sites of the protein linked by a dotted yellow line identifying the distance between them. Other colours represent rigid clusters in different domains.

using the pNM we show in Figure 3-Right the variation in distance between the two PDI active sites in domains a and a' as the motion over the initial pNM progresses over each frame. The 3D representation of PDI's conformational change and its active sites relative motion is show in Figure 4. The simulated conformational change of PDI over the pNM trajectory reveals that the distance between PDI's active sites spans from approximately 14Å to over 55Å depending on the energy cut off, Figure 3-Right. The difference in the accessible conformational space between different energy cut off may be due to the large number of hydrogen bonds which are present at low energy cut off which would make the bond network less flexible to the point of restricting protein mobility for a lower energy cut off.

4. Discussion

The simulation of conformational changes for large proteins are out of reach for most atomistic models. Our coarse grained simulations however are able to map out PDI's conformational change along the pNM trajectory but also, equally important, the simulation time scale is of a few hours. Furthermore, the results here presented are in agreement with previous experimental work stating that PDI is required to be a conformationally flexible molecule to perform its catalytic activity [15].

Acknowledgements

We gratefully acknowledge support by EPSRC and BBSRC.

References

[1] Russel D, Lasker K, Phillips J Schneidman-Duhovny D, Velazquez-Muriel J A and Sali A 2009 Curr Opin Cell Biol, 2197-108
[2] Wells S A, Jimenez-Roldan J E, Roemer R A 2009 Phys. Biol. 6 046005
[3] Clementi C 2008 Curr Opin Struct Biol 2008 18 10-15
[4] Shervood P, Brooks B R and Sansom M S P 2008 Curr Opin Struct Biol 18 630-640
[5] Yaliraki S N and Barahona M 2007 Philos Trans R Soc A Math Phys Eng Sci 365 2921-2934
[6] Bahar I, Lezon T R, Bakan A and Shrivastava I H 2010 Chem Rev 110 1463-1497
[7] Jacobs D J, Rader A J, Thorpe M F and Kuhn L A 2001 Proteins: Struct Funct Genet 44 150-165
[8] Raveh B, Enosh A, Schueler-Furman O and Halperin D 2009 PLoS Comput Biol 5 e1000295
[9] Farteel D W, Speranskiy K and Thorpe M F 2010 Proteins 78 2908-2921
[10] Moritsugu K, Njunda B M and Smith J C 2010 J Phys Chem B 114 1479-1485

[11] Rueda M, Chacn P and Orozco M 2007 Structure 15 565-575
[12] Nakasako M, Maeno A, Kurimoto E, Harada T, Yamaguchi Y, Oka T, Takayama Y, Iwata and Kato K 2010 Bichemistry 49 6953-6962
[13] Dykeman E C and Sankey O F 2010 J. Phys. Condens. Matter 22 423202
[14] Freedman RB,Klappa P andRuddock LW 2002 EMBO reports 3 2 136-140
[15] Tian G, Kober F, Lewandrowski U, Sickmann A, Lennarz W J and Schindelin H 2008 The J. of Biological Chemistry. 283 33630-33640
[16] Suhre K and Sanejouand Y H 2004 Nucleic Acids Research 32 W610-W614
[17] Suhre K and Sanejouand Y H 2004 Acta Cryst D 60 796-799
[18] Wells S, Menor S, Hespenheide B and Thorpe M F 2005 Phys. Biol. 2 S127

AFM Nanotools for Surgery of Biological Cells

J.D. Beard[1], **S.N. Gordeev**[1] **and R.H. Guy**[2]

[1]Department of Physics, Claverton Down, University of Bath, Bath, BA2 7AY, UK
[2]Department of Pharmacy and Pharmacology, Claverton Down, University of Bath, Bath, BA2 7AY, UK
E-mail: jdb28@bath.ac.uk

Abstract. Using a method of electron-beam induced deposition, we have been able to fabricate specialized AFM probes with application as "nanotools" for the manipulation of biological structures ("nanosurgery"). We describe several such tools, including a "nanoscalpel", "nanoneedles" for probing intracellular structures, and a "nanotome" which can separate surface layers from a biological structure. These applications are demonstrated by performing nanomanipulation on corneocyte cells from the outer layer of human skin.

1. Introduction

There is increasing interest among biologists in the ability to manipulate both single cells and smaller intracellular components at the micro- and nano-scale. This manipulation is often described as "nanosurgery". In recent years, technologies for the manipulation of nanoscale structures, including optical tweezers [1], laser microscalpels [2], and the atomic force microscope [3], have allowed nanosurgery to be applied in a variety of different ways. Nanosurgery has been used to isolate intracellular organelles [1] or single cells [4], for the dissection of viruses [5], collagen fibrils [6], plasmids [7], chromosomes [3], and for in situ dissection of bacterial cell walls [8] to image their internal structure. It has also been used to cut single cytoskeleton filaments [2] and to create incisions in cell membranes [4].

Of the techniques available to biologists, the atomic force microscope (AFM) can manipulate structures with the highest precision. AFM probe tips are typically ~5-10 nm in radius and can be positioned with nanometre accuracy, allowing for very precise control of the manipulation combined with high resolution 3D imaging of the sample. [9] However, standard AFM probes are not ideal for manipulation. They are typically pyramidal in shape with a low aspect ratio, decreasing the precision of nanomanipulation as the tip penetrates deeper into a sample, and are usually fabricated from crystalline silicon or silicon nitride, which are relatively brittle materials that can easily be damaged during manipulation. High lateral precision is desirable in nanosurgery; just as in macroscopic surgery and dissection, low precision can cause great damage to the tissue. Several approaches have been used to improve the aspect ratio of AFM tips, including the attachment of carbon nanotubes to the AFM probe tip [10,11,12,13] and the fabrication of a needle-like structure on the tip apex using focused ion beam milling. Needles fabricated by this technique have been used for microinjection of biomolecules (including DNA) into cells and specific organelles. [14,15,16,17,18,19]

Another technique which has been widely applied for the modification of AFM probes is electron beam induced deposition (EBID). [20,21,22,23,24,25] EBID is a highly versatile technique for the

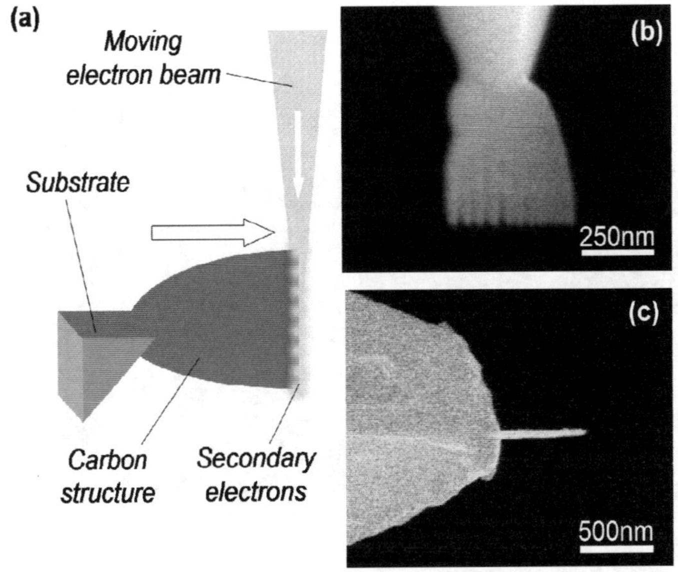

Figure 1: (a) Fabrication of nanotool structures using electron beam induced deposition. (b) and (c) Nanoscalpel blades imaged from the side (b) and from the top (c) using SEM.

fabrication of two-dimensional and three-dimensional nanostructures. [26,27,28,29,30,31,32,33] To fabricate a structure using EBID, a high-energy focused electron beam in an electron microscope is used to "write" the structure. The electron beam causes the emission of lower-energy secondary electrons from the atoms of the substrate, which are able to decompose hydrocarbon precursor molecules adsorbed onto the substrate surface. The sources of these hydrocarbon molecules are the trace gases and normal contaminants present in the vacuum system of the electron microscope [23,25,34], although some groups have performed EBID by deliberately introducing precursor molecules using a gas source. The decomposition of these molecules creates a deposit of carbon atoms on the substrate surface, creating a nanoscale structure from amorphous carbon.

We have used EBID to fabricate a variety of different structures on standard AFM probes for different nanomanipulation applications, including "nanoscalpel" blades which can be used to cut and dissect biological structures as well as for mechanical nanolithography of inorganic materials. [35] In this paper, we describe a number of different "nanotools" which can be applied to different nanosurgical applications, including high-aspect ratio "nanoneedles" for microinjection into cells, extraction of material from target sites on the surface of a cell, and indentation on cells to determine their mechanical properties. Other nanotools which we have fabricated include a "nanotome" for removing material from the surface of cells to image the underlying structures *in situ*. In future, these tools could be used for sophisticated manipulation of biological cells, offering a new approach to the investigation of the properties of cells.

We have also investigated the mechanical properties of these probes by bending the EBID carbon blades using AFM manipulation. In the process we discovered that the deposited structures are extremely flexible, capable of recovering elastically from deformations of up to ~75% of their total length. [36] This indicates that they are highly robust, and able to withstand and recover from large applied forces during nanomanipulation.

2. Fabrication of Nanotools

Fabrication of a nanostructure on an AFM probe was performed using a scanning electron microscope (Hitachi SE-4300) equipped with a Raith electron beam lithography (EBL) system. Typically, a beam energy of 20kV and a beam current of 0.1nA were used. In our experiments, the pressure within the SEM system was ~10^{-3} Pa. The electron beam was moved laterally from the AFM tip apex, depositing a line of carbon which extends beyond the tip as a self-supporting structure (see Figure 1a). [37] The dimensions and geometry of the nanostructure depend on the velocity at which the beam spot is moved over the surface. At low beam spot velocities (1-6nm/s), the deep penetration of the electron beam into the substrate and deposited structure means that it grows as a plate-like or blade-like structure in the plane of the electron beam (see Figure 1 b-c). The blade-like shape of this structure makes it ideal as a "nanoscalpel" for making fine incisions in the surface of a sample such as a cell, thereby acting as a dissection tool.

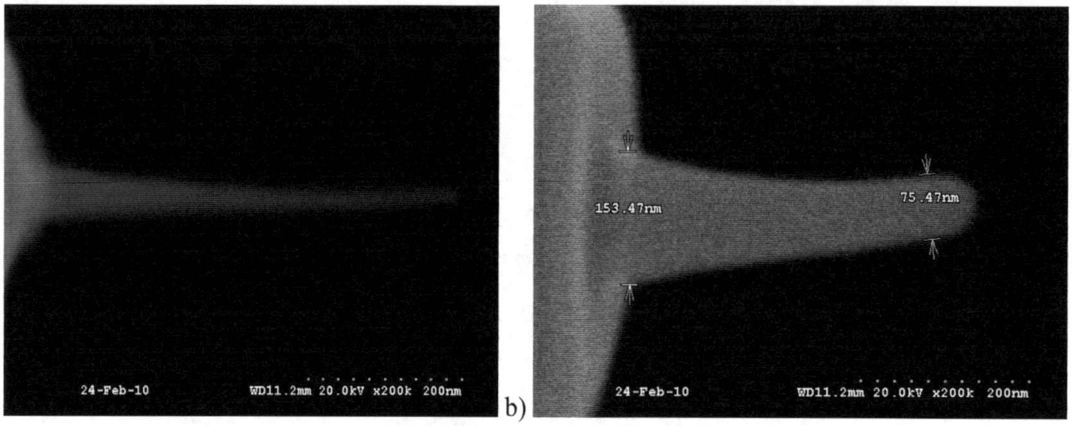

a) b)

Figure 2: (a) Thin nanoneedle probe for high-resolution imaging. (b) Needle after exposure for ~26 mins in SEM, depositing additional carbon to stabilize needle against bending.

The growth of the structure in the lateral direction is restricted by the much smaller penetration of the secondary electrons through the deposited material. The width of this blade-like structure decreases with increasing beam spot velocity. For higher velocities (6-9 nm/s) the blade width and thickness are approximately equal (~20-30nm), resulting in a cylindrical, needle-like structure.

The thin needles fabricated by this technique can be used as high-aspect ratio AFM probes for the profiling of samples with steep, high surface features or deep trenches. An example of a nanoneedle is shown in Figure 2a. However, these thin structures are flexible and this limits the force which can be applied. [36] To stabilize the needle, it can be imaged at high magnification (concentrating the electron beam into a smaller area and so increasing the beam flux), depositing additional carbon onto the needle and increasing its diameter. Typically, imaging for ~30 minutes at a magnification of 250,000x can increase the diameter of a needle from about 20 to 90 nm. During this process, the growth rate is even on all sides of the nanoneedle; it therefore retains a well-defined geometry with a circular cross-section and a smooth, hemispherical tip. An example of a tip stabilized by thickening the needle using this process is shown in Figure 2b. Control of the geometry is important if the needle is to be used as an indenter to determine the mechanical properties of a biological structure because its shape determines the dynamics of its interaction with the sample during indentation. The simultaneous SEM imaging of the nanoneedle during the stabilization process allows control of the diameter and shape of the needle.

The EBID process is highly versatile and can be used to create complex structures. This allows the fabrication of highly specialized nanotools on AFM probes for manipulation of samples. For example, a "nanotome" device was fabricated which can be used to remove thin layers from the surface of a biological structure using a single pass of the AFM probe, in a manner similar to a vegetable peeler. This consists of two parallel nanoscalpel blades with a suspended filament between the blade tips. Examples of these structures are shown in Figure 3a-c. To fabricate this structure, the parallel blades were tilted after EBID fabrication to expose only the blade tips to the beam. This prevents the growth of multiple filaments, which often occur in depositions extending from the edge of thick substrates. Since the primary electron beam can penetrate deep into the substrate, the generated secondary electrons can cause deposition from multiple sites at the edge of a thick substrate, producing multiple separate filaments. [38] In the case of the nanotome fabrication, tilting the blades to deposit a filament across their tips allowed the successful deposition of a single carbon filament supported by the nanoscalpel blades. The growth process is illustrated in Figure 3d.

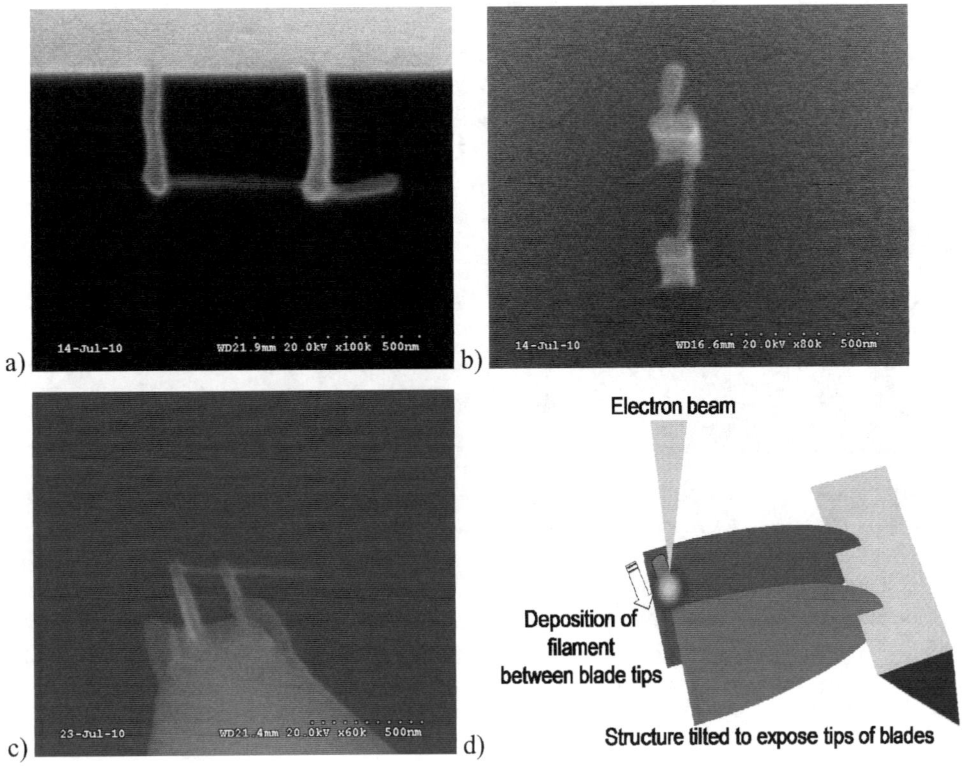

Figure 3: (a) Suspended filament between nanoscalpel blades. (b) Image from above showing single filament.(c) Similar structure deposited on an AFM tip, creating a "nanotome". (d) Process of fabrication of a single filament across tips of parallel blades.

3. Cell Sample Preparation

To test the capabilities of these probes, we performed investigations on corneocytes. These cells form the outermost layer of mammalian skin, the stratum corneum, which forms a barrier against water loss from the body. The cells were isolated using an adhesive stripping technique from the upper skin layers of healthy human subjects. Commercial hair removal wax, heated to 65°C on glass slides, was used as an adhesive to remove the cells from the skin surface. This method is an adaptation of a commonly used tape-stripping technique. [39]

Corneocytes were chosen because they are easily harvested and prepared, and have a comparatively rigid structure compared to other cell types (making them easier to manipulate). They can also be imaged using AFM in air without the need for dehydration, fixation or other complex sample preparation techniques, unlike most other mammalian cell types which must be imaged in liquid if not fixed. We were able to image the corneocytes in air using AFM; examples of AFM images of corneocytes are shown in Figure 4.

4. Manipulation of Cells

We successfully showed that nanoscalpel probes could be used to make incisions of controlled depth on the surface of the corneocytes. To control the force on the AFM probe during an incision, a feedback loop was implemented to hold the AFM cantilever deflection at a fixed setpoint value during the cut. The force exerted on the sample is proportionate to the cantilever deflection and so this feedback loop holds the applied force at a constant value. This allows the applied force and hence the cut depth to be controlled. It also prevents a large tip-sample force from breaking the nanoscalpel or detaching it from the AFM tip. Incisions using a nanoscalpel probe on cells and on inorganic structures have recently been described in detail [35].

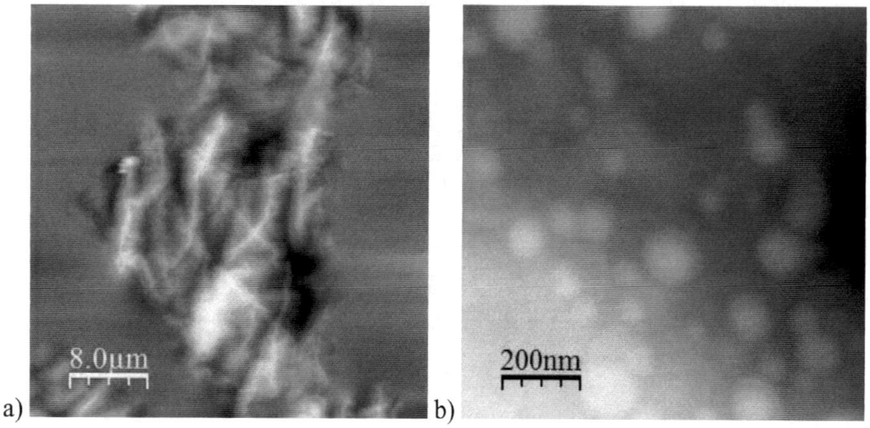

Figure 4: (a) AFM image of corneocytes. (b) Image of a small area of a corneocyte surface, showing many small, round structures which are probably protein clusters.

Manipulation and imaging using the nanotool probes was carried out using a Veeco Multimode IIIa AFM running Nanoscope software. The manipulation program was written such that the AFM probe would first profile the sample surface along the cut path in tapping mode. The probe would then return to the start of the cut path, and the tip-sample force would be increased until the cantilever deflection reached the chosen setpoint, then moved along the cut path to create a cut. An example of an incision on the surface of a corneocyte is shown in Figure 5.

The high aspect ratio of the nanoscalpel allows penetration into the cell to large depths without loss of lateral precision, creating a narrow incision in the surface. The penetration into the cell can be calculated by comparing the topographic profile taken before the cut with the path taken by the nanoscalpel tip during the cut. The depth of the cut was shown by this method to increase in proportion to the applied force; this could be used to control the depth of penetration of the scalpel blade. This point is illustrated in Figure 6.

EBID nanoneedle probes could be used for the accurate imaging of the corneocyte surface, and for indentation on the surface to probe mechanical properties of the cell. The penetration of the needle can be measured by examination of the AFM force curve; the deformation of the cell can be measured by comparing the deflection of the AFM cantilever recorded during an indent on the cell surface to that on a rigid sample (e.g., a clean silicon wafer) which the nanoneedle cannot penetrate. This process is illustrated in Figure 7a. The high aspect ratio cylindrical needle shape allows it to penetrate the sample to a large depth without loss of lateral precision. An example of the indent remaining in the corneocyte surface is shown in Figure 7b.

Nanoneedles thickened to ~75 nm diameter were able to exert forces of up to ~1.5 μN on a sample without detectable damage or deformation of the needle structure. The thickened nanoneedles can be used for imaging cell surfaces and for indentation studies. During these experiments, we found that indenting on a protein cluster on the surface of a corneocyte cell apparently led to its removal, leaving a pore in the surface of the cell. Indentation a second time appeared to replace an object at approximately the same site; it is likely that this object was the same protein cluster. This indicates that the structure had adhered to the nanoneedle, and could be transplanted to a desired site using the nanoneedle tip. This shows that AFM nanotools may be used for the extraction of samples from cells, and for the manipulation of subcellular structures. This process is shown in Figure 8.

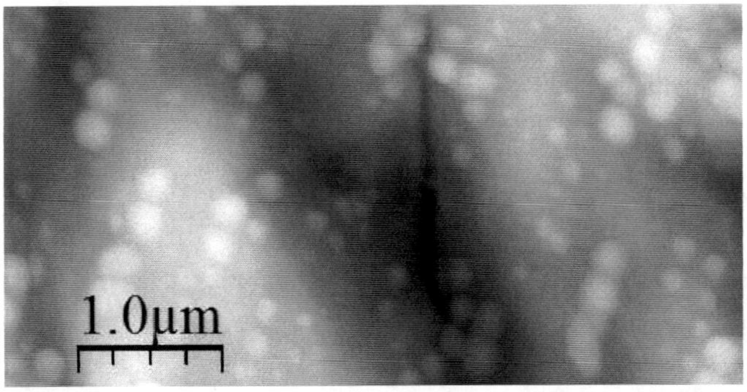

Figure 5: AFM image of the surface of a corneocyte showing an incision created using a nanoscalpel probe.

It was also shown that nanotome probes can be used to separate the outer layers of a biological structure from the underlying layers. The two nanoscalpel blades create parallel cuts in the cell surface, while the suspended filament cuts beneath the surface layer to separate it from the underlying material. This is illustrated diagrammatically in Figure 9a.

A slight asymmetry in the length of the parallel nanoscalpel blades meant that only one blade tip touched the sample surface during normal imaging, allowing the nanotome probe to be used to capture reasonably accurate images of the sample surface. This also permitted the alignment of the nanotome over a target site on the surface. Increasing the tip-sample force and making an incision using the same manipulation program as was used for the nanoscalpel brought both blades into contact, creating two parallel cuts. After this manipulation, the probe was passed across the cuts in a perpendicular direction. This moved aside the surface layer between the parallel cuts, indicating that it had been separated from the underlying structures by the suspended filament. After displacing the upper layer in this way, an unmodified sharp AFM probe could be used to image the structures which had previously been concealed beneath the outer layer of the corneocyte, as shown in Figure 9b.

5. Conclusions

A number of different AFM tools for the manipulation of biological structures have been developed using electron-beam induced deposition (EBID). These include a "nanoscalpel" capable of making incisions to a controlled depth in the surface of biological objects, and "nanoneedle" probes which can be used as high-aspect ratio probes for the imaging of samples with steep surface gradients. Thickening of these nanoneedles by deposition of additional carbon onto the needle creates a stiffer structure which can exert large forces on a sample; these stabilized needles can be used to indent on the surface of a biological object, penetrating it to a large depth without loss of lateral precision. This could be used for nanoindentation experiments to determine the mechanical properties of intracellular structures, and for the microinjection of materials (e.g., DNA) into the cell. The needle can also be used for the extraction of a sample from the cell, as shown by the experiment in which a protein cluster was removed from the surface of a cell, and then replaced, by indentation using a nanoneedle.

Suspended filaments were successfully fabricated between the tips of parallel nanoscalpel blades using EBID. This created a "nanotome" tool which can be used to remove part of the outer layer of corneocyte cells by AFM manipulation, exposing structures beneath for AFM imaging *in situ*.

These nanotools provide a new route to the specialized nanomanipulation and nanodissection of biological structures. Combined with the precision and force sensitivity of AFM, they can be used to manipulate cells at the scale of individual organelles and subcellular structures. In future, these tools could open up novel methods of investigation of the structure and function of biological systems at the nanoscale.

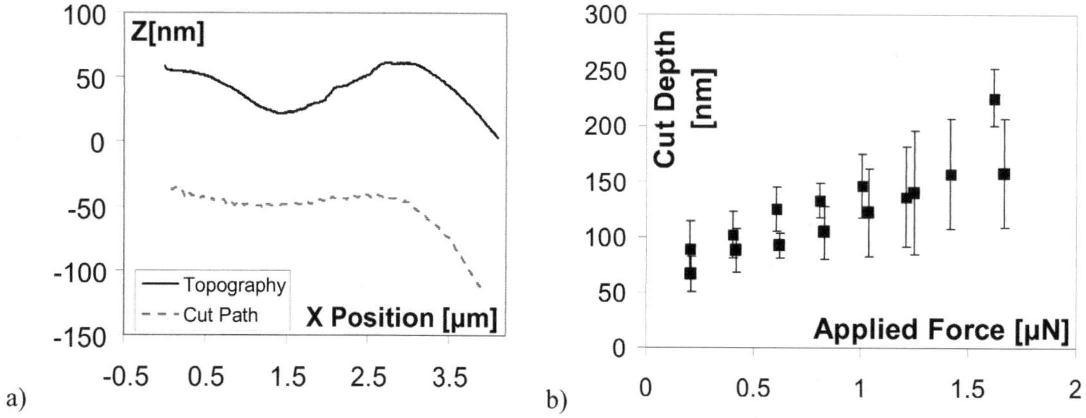

Figure 6: (a) Comparison of undamaged surface topography of a corneocyte with the path of a nanoscalpel during cut. (b) Depth of cut as a function of applied cutting force; data compiled from incisions on 2 different corneocyte cells.

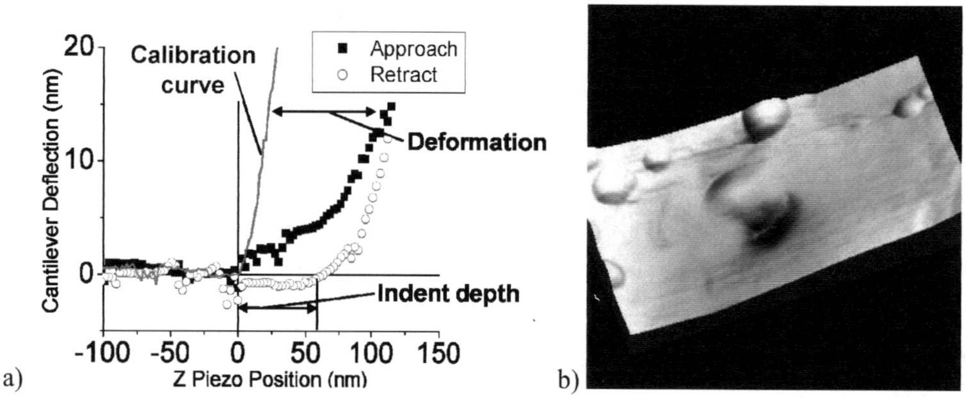

Figure 7: (a) Force curve recorded during an indent using a nanoneedle; the deformation of the cell and the remaining indentation depth can be determined from the curve by comparison to a calibration curve recorded on a rigid substrate. (b) AFM image of a nanoneedle indent on the corneocyte surface. (1.0x0.5μm area).

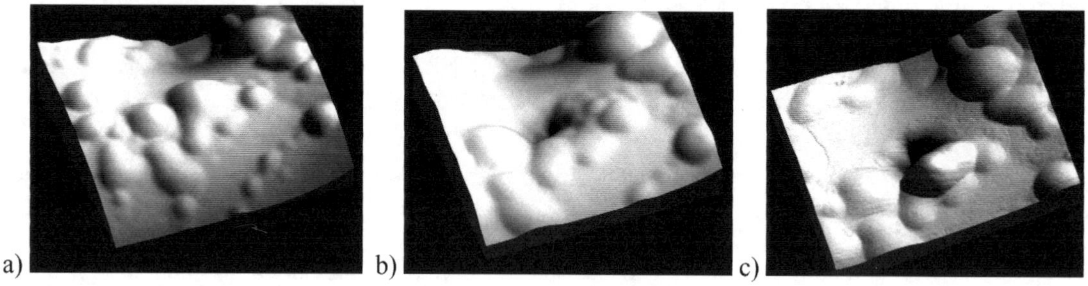

Figure 8: Removal of a protein cluster using a nanoneedle. (a) AFM image of corneocyte surface before indent. (b) Surface after indent showing pore created in surface by nanoneedle tip. (c) Image after indentation was performed a second time at the same site. The protein cluster has been deposited back onto the surface of the cell after adhering to the needle during the first indent. Imaged area is 1.0x0.75μm.

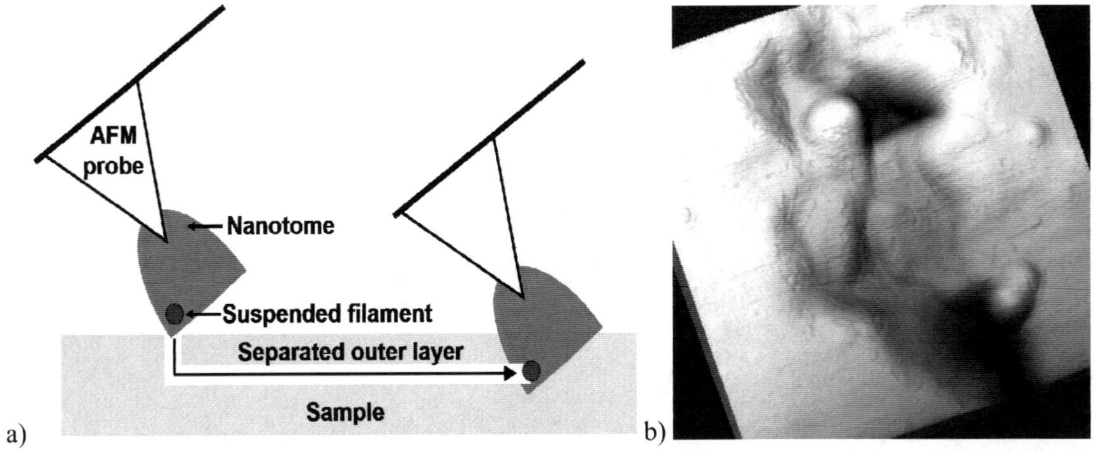

Figure 9: (a) Use of a nanotome to remove an area of the cell sample surface. (b) Area of corneocyte surface which has had a section of surface removed using the nanotome. Imaged area is 1.0x1.0μm.

References

[1] Shelby JP Edgar JS and Chiu DT 2005 *Photochem. Photobiol.* **81** 994-1001
[2] Shen N Datta D Schaffer C B Donald P L Ingber E and Mazur E 2005 *Mech. Chem. Biosystems* **2** 17–25
[3] Stark R W Rubio-Sierra F J Thalhammer S and Heckl W M 2003 *Eur. Biophys. J.* **32** 33–9
[4] Kohli V Elezzabi AY and Acker JP 2005 *Las. Surg. Med.* **37** 227-230
[5] Falvo MR Washburn S Superfine R Finch M Brooks FP Chi V and Taylor RM 1997 *Biophys. J.* **72** 1396-1403
[6] Wen C and Goh M 2004 *Nano Lett.* **4** 129-132
[7] Henderson E 1992 *Nucleic Acids Res.* **20** 445
[8] Firtel M Henderson G and Sokolov I 2004 *Ultramicroscopy* **101** 105-109
[9] Binnig G Quate C and Gerber C 1986 *Phys. Rev. Lett.* **56** 930-933
[10] Dai H Hafner J Rinzler A Colbert D and Smalley R 1996 *Nature* **384** 147-150
[11] Hafner J Cheung C Woolley and Lieber C 2001 *Prog. Biophys. Mol. Bio.* 77 73-110
[12] Ye Q Cassell A Liu H Chao K Han J and Meyyappan M 2004 *Nano Lett.* **4** 1301-1308
[13] Yenilmez E Wang Q Chen R Wang D and Dai H 2002 *Appl. Phys. Lett.* **80** 2225
[14] Obataya I Nakamura C Han S Nakamura N and Miyake J 2005 *Nano Lett.* **5** 27-30
[15] Han S Nakamura C Obataya I Nakamura N and Miyake J 2005 *Biochem. Bioph. Res. Co.* **332** 633-639
[16] Obataya I Nakamura C Han S Nakamura N and Miyake J 2005 *Nanobiotechnology* **1** 347-352
[17] Han S Nakamura C Obataya I Nakamura N and Miyake J 2005 *Biosens. Bioelectron.* **20** 2120-2125
[18] Vakarelski I Brown S Higashitani K and Moudgil B 2007 *Langmuir* **23** 10893-10896
[19] Patil A Sippel J Martin G and Rinzler A 2004 *Nano Lett.* **4** 303-308
[20] Akama Y Nishimura E Sakai A and Hurakami H 1990 *J. Vac. Sci. Tech.* **8** 429.
[21] Keller D and Chih-Chung C 1992 *Surf. Sci.* **268** 333-9
[22] Yamaki M Miwa T Yoshimura H and Nagayama K 1992 *J. Vac. Sci. Tech. B* **10** 2447-50
[23] Schiffmann K 1993 *Nanotech.* **4** 163
[24] Zenhausen F Adrian M Heggeler-Bordier B Ardizzoni F and Descouts P 1993 *J. App. Phys.* **73** 7232
[25] Chen I-C Chen L-H Orme C Quist A Lal R and Jin S 2006 *Nanotech.* **17** 4322-6
[26] van Dorp W F and Hagen C W 2008 *J. App. Phys.* **104** 081301

[27] Christy R 1960 *J. App. Phys.* **31** 1680-3
[28] Baker A and Morris W 1961 *Rev. Sci. Instrum.* **32** 458
[29] Broers A Molzen W Cuomo and J Wittels N 1976 *App. Phys. Lett.* **29** 596
[30] Randolph S J Fowlkes J D and Rack P D 2006 *CRC Cr. Rev. Sol. State.* **31** 55
[31] Gazzadi G C Frabboni S and Menozzi C 2007 *Nanotech.* **18** 445709
[32] Botman A Mulders J J L Weemaes R and Mentink S 2006 *Nanotech.* **17** 3779-85
[33] Cho C C and Bernasek S L 1989 *J Appl. Phys.* **65** 3035-43
[34] Wendel M Lorenz H and Kotthaus J 1995 *Appl. Phys. Lett.* **67** 3732
[35] Beard J D Burbridge D J Moskalenko A V Dudko O Yarova P L Smirnov S V and Gordeev S N 2009 *Nanotech.* **20** 445302
[36] Beard J D and Gordeev S N 2010 *Nanotech.* **21** 475702.
[37] Mollenstedt G 1988 *Optik* **78** 132-43
[38] Gazzadi G C Frabboni S and Menozzi C 2007 *Nanotech.* **18** 445709
[39] Gorzelanny C Goerge T Schnaeker E-M Thomas K Luger TA and Schneider SW 2006 *Exp. Dermatol.* **15** 387-391

Journal of Physics: Conference Series **286** (2011) 012004 doi:10.1088/1742-6596/286/1/012004

Fabrication and optical properties of gold nanowire arrays

C L McClatchey, A Murphy, J McPhillips and R Pollard

School of Mathematics and Physics, Queen's University of Belfast, University Road, Belfast, BT7 1NN

cmcclatchey01@qub.ac.uk

Abstract. In this work the optical properties of arrays of gold nanowires in transmission and attenuated total reflectance (ATR) geometry has been examined further, with a view to maximising the sensitivity by finding optimum nanowire and array dimensions. Gold nanowire arrays were grown by electrochemical deposition into a nano-porous alumina template. The effect of changing the nanowire aspect ratio on the optical properties was investigated.

1. Introduction

The potential of noble metal nanostructures in the field of plasmonic sensing is well known [1-3] due to the high sensitivity of the plasmonic resonance to small changes in the surrounding dielectric environment. This sensitivity has been exploited recently to perform label-free biosensing measurements, in both transmission [3] and attenuated total reflectance (ATR) geometries [4], on substrate-bound nanowire arrays. ATR measurements on a non-optimised array of gold nanowires immersed in a range of aqueous solutions of glycerine of different concentrations showed a sensitivity of 30,000nm/RIU [4]. This is two orders of magnitude higher than that currently offered by conventional SPR-based sensors [5].

2. Fabrication and Results

Gold nanowire arrays with differing aspect ratio and inter-wire spacing were grown by electrochemical deposition into a nano-porous alumina template. The diameter and spacing of the nanowires formed was varied by altering the anodisation and deposition parameters. The length of the nanowires was controlled by monitoring in-situ optical extinction measurements during the gold deposition.

When illuminated with plane polarized light above the critical angle, a plasmonic mode is observed as a resonance in the near-IR region in the reflectance spectrum. This mode is non-localised and is in fact guided in nature; it exists in the thin sub-wavelength slabs of the nanowires. Figure 1 shows the change in resonant position as the angle of incidence is changed. The effect of changing the nanowire array geometry was studied experimentally by creating two nanowire samples with the same height but different diameters. It was found that decreasing the aspect ratio of the nanowires led to a blue-shift in the ATR resonance position, as shown in Figure 2.

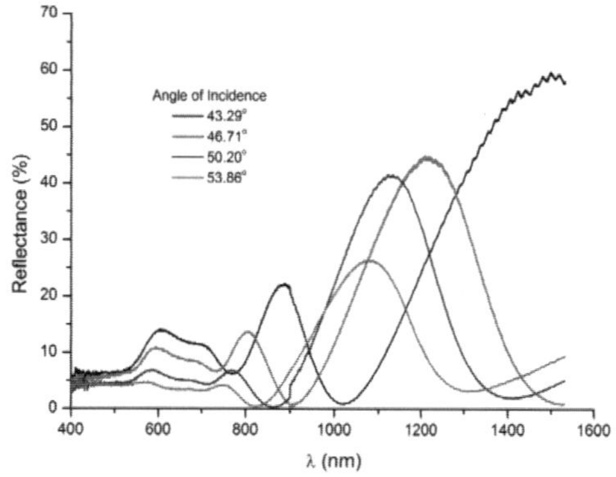

Figure 1 Graph showing reflectance of an array of gold nanowires embedded in alumina over a number of angles of incidence. The resonant mode is highly sensitive to changes in the refractive index between the wires.

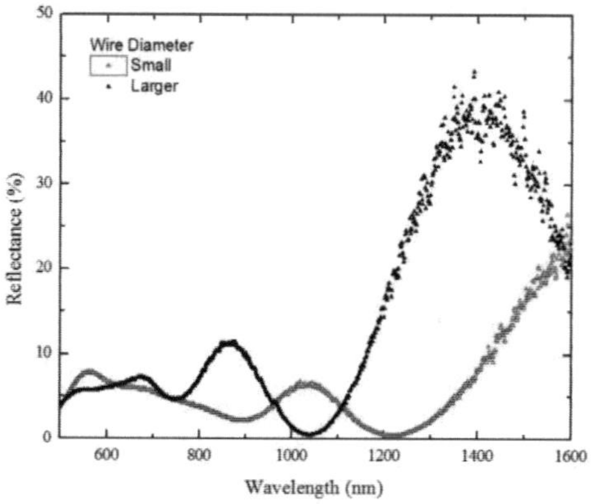

Figure 2 Graph showing shift in ATR spectrum with changing aspect ratio. The wires with larger diameter have a more blue-shifted resonance than those with smaller diameter.

3. Outlook

Since the plasmonic resonance observed in nanowire arrays illuminated in ATR geometry is highly sensitive to small changes in the refractive index of the material surrounding the nanowires means they have a large potential for biological sensing applications. We have already demonstrated the tunability of this resonance by fabricating nanowire arrays with differing aspect ratio; we now wish to find use finite element modelling based on Maxwell's equations to a quantitative explanation of the experimental results.

References

[1] Liedberg B, Nylander C and Lundstrom I, *Sensors and Actuators* **4** (1983)
[2] Jonsson M P, Dahlin A B, Jönsson P, Höök F, *Biointerphases* **3** (2008)
[3] McPhillips J, Murphy A, Jonsson M P, Hendren W R, Atkinson R, Höök F, Zayats A V and Pollard R J, A*CS Nano* **4** (2010)
[4] Kabashin A V, Evans P, Pastkovsky S, Hendren W, Wurtz G A, Atkinson R, Pollard R, Podolskiy V A and Zayats A V, *Nature Materials* **8** (2009)
[5] Anker J N et al., Biosensing with Plasmonic Nanosensors, Nature *Materials* **7** (2008)

Condensed Matter and Materials Physics Conference (CMMP10) IOP Publishing
Journal of Physics: Conference Series **286** (2011) 012005 doi:10.1088/1742-6596/286/1/012005

Complex Permittivity of Pure Water Measured by Vector Network Analysis at W-Band

Bin Yang*, Kastriot Shala, Xiaoming Liu, Hansheng Su and Robert S. Donnan*
School of Electronic Engineering and Computer Science, Queen Mary University of London, Mile End Road, London, E1 4NS, UK.

E-mail*: bin.yang@elec.qmul.ac.uk; robert.donnan@elec.qmul.ac.uk

Abstract. Preliminary pure-water transmittance measurements over the W-band (75 – 110 GHz) are performed using a well-designed quasi-optical bench and vector network analyser (VNA/QO). The measurements are an initial trial to explore vector network analysis in studies of micro-biological systems.

1. Introduction

Water, the commonest and most studied liquid on earth, continues to warrant detailed metrology in view of its varied complex influences in the function of micro-biological systems – e.g. proteins. Terahertz (THz) spectrometric investigations of condensed-phase *biological* systems are relatively new [1, 2 and 3] and research interests cover: simple crystalline forms of amino acids; carbohydrates and polypeptides to the more complex aqueous forms of small proteins; DNA and RNA. However, the role of water in bio-molecular dynamics is still unclear and so careful acquisition of reference data from pure water is warranted. In this paper therefore, preliminary pure-water transmittance measurements over the W-band (75 – 110 GHz) are performed using a *vector network analyser* and *quasi-optical bench* (VNA+QO).

2. Measurement System

A well-designed, quasi-optical transmissometer [4] driven by a HP N5244A Vector Network Analyzer (VNA) from 75 to 110 GHz has been used for this work. Compared with FTIR and THz-TDS, the Agilent VNA (spanning DC to 1 THz), is able to offer a greater dynamic range and signal-to-noise ratio of around 100 to 140 dB. The VNA also provides greater spectral resolution down to 1 MHz. This is superior to a typical 3 GHz resolution for TDS and FTIR interferometers [5]. The plane-wave angular-spectrum of a beam having a Gaussian amplitude distribution at its beam-waist of width 50mm is of Gaussian form with an angular-width near to 1 degree at 60 GHz and proportionally smaller than this for higher frequencies [6]; furthermore, the beam's angular-width narrows appreciably as the beam refracts into the sample. This small angular-width ensures that the transmissometer essentially measures the sample-plate's *plane-wave* transmittances. A large sample-plate is required (width 100 mm) in order that the beam is negligibly truncated at the edges of the plate. It is also quite easy and simple to extend the spectrometer to THz energies by using the same quasi-optical circuit, by replacing only the corrugated feed-horns with a higher frequency band pair.

Conventional methods such as FTIR and THz-TDS have been greatly restricted to using powdered or lyophilized samples due to strong attenuation of THz frequencies by water. Our study has therefore developed a new bespoke cuvette system enabling analysis of bio-molecules and water in aqueous solution. The cuvette system consists of two 1.1 mm thick boron glass plates (a material generally transparent over millimeter and sub-millimeter wavelengths) with a 0.28 mm spacer (four metallic corner-located pads creating a constant gap between the two plates). Inlet and outlet valves are placed at opposite ends of the plates for flushing without the need to dismantle the cuvette.

3. Theoretical Debye Model for Water Analysis

The structure of a 3-layer system i.e. glass-water-glass can be represented in a corresponding 3-term product ABCD matrix. Based on the materials' dielectric properties and thicknesses, the transmittance matrix is:

$$\begin{pmatrix} A & B \\ C & D \end{pmatrix} = \begin{pmatrix} A_g & B_g \\ C_g & D_g \end{pmatrix} \cdot \begin{pmatrix} A_w & B_w \\ C_w & D_w \end{pmatrix} \cdot \begin{pmatrix} A_g & B_g \\ C_g & D_g \end{pmatrix}. \tag{1}$$

Where, subscript 'g' stands for boron glass and 'w' stands for water. The detailed equations of A, B, C and D for each layer are referenced from [7]. Therefore, the transmittance (T) and reflection (R) of the structure are

$$T = \frac{2}{A+B+C+D}; \; R = \frac{A+B-C-D}{A+B+C+D}. \tag{2}$$

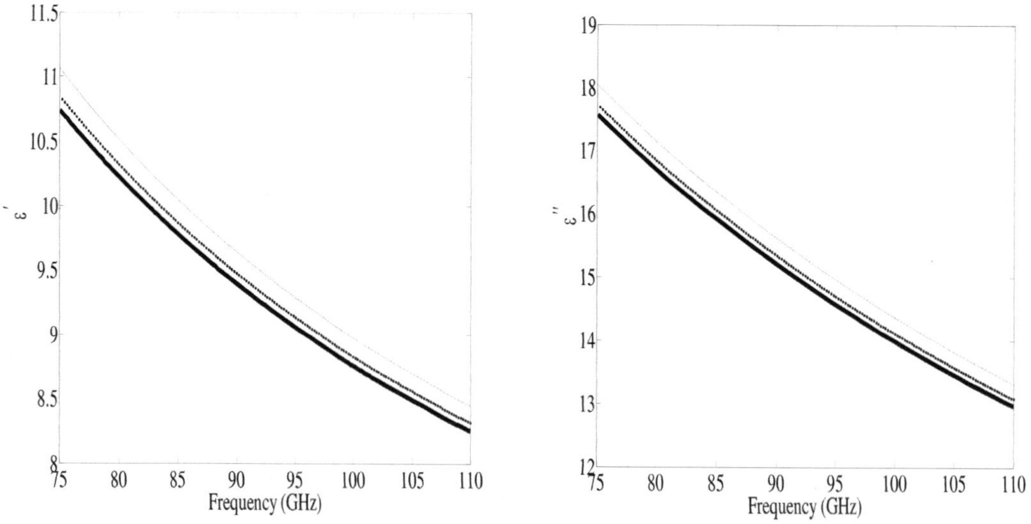

Fig.1: Temperature-dependent dielectric constant of pure water at W-band based on the Double Debye Model (suitable for frequencies below 1 THz): (left-plot) real part of the complex permittivity and (right-plot) imaginary part. The bold line is for 24.4^{0}C, the dashed for 25^{0}C and the light for 26.2^{0}C.

The permittivity of boron glass is non-dispersive over the W-band at 4.463+i0.066, and the transmittance loss is very low with $\tan(\delta) \cong 10^{-3}$. In 1991, Liebe et al [8] summarised a large volume of published experimental permittivity data of liquid water, applied them, and expressed a standard dispersion formula for the permittivity of water from 0.1 to 1 THz. The model is made from two relaxation terms, and the theoretical permittivity of pure water with temperature dependence is plotted in Fig.1. From Fig.1, the permittivity variations are small but clear when the temperature is slightly changed. Clear differences are therefore expected in experimental transmittance spectra.

4. Results and Discussion

Separate pure water tests were repeated on three consecutive days, with ambient temperature varying slightly on each. In Fig.2 (left), the results from the three days are plotted together. Small differences are observed as expected. Based on the Debye model and the free-space 3-layer ABCD model, the theoretical transmittance spectra is compared with the experimental data as shown in Fig. 2(right). The observed difference could arise for two reasons: one, the theoretical permittivity, according to the Debye model holds from 100 GHz and up, but our horns are designed to operate from 75 GHz. The second may be due to mechanical effects on the water by the close-lying plates.

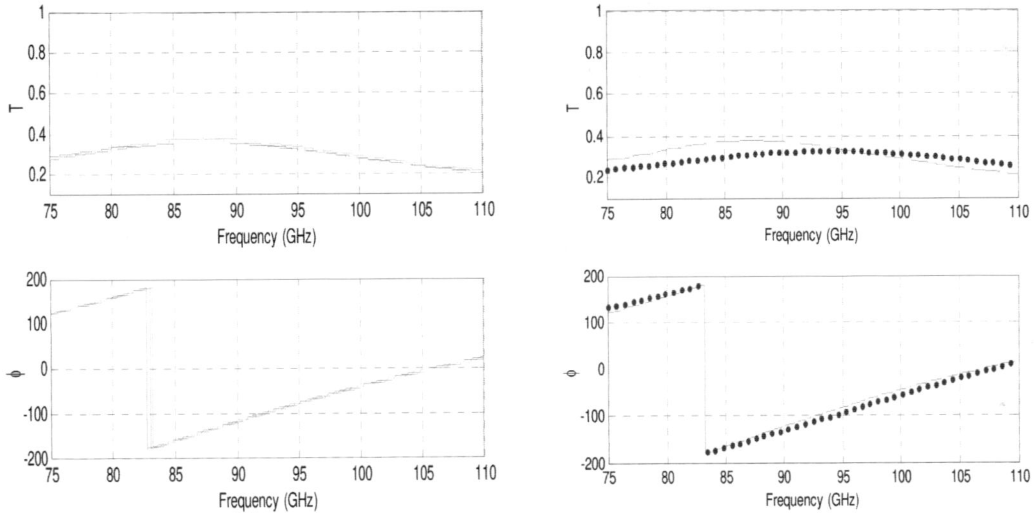

Fig.2: (left) Measured glass/water/glass transmittance spectra (upper plot is the transmittance amplitudes and lower is the phase): black is at 24.4^{0}C, blue 25^{0}C and red 26.2^{0}C; (right) theoretical 3-layers transmittance at 25^{0}C based on the ABCD model (blue dotted line) compared with experimental data.

The measurements are part of an initial on-going trial to explore the capabilities of vector network analysis for the study of soft condensed systems. More refined temperature- and pressure-controlled sample holders are required and such are being developed. The VNA/QO apparatus and methods is soon to be extended to higher bands of the sub-THz domain for analysis of amino acids.

5. Conclusions

The apparatus of VNA/QO and ABCD signal processing is seen in trials to-date to enable high measurement stability and repeatability from bulk pure-water tests. The aim is to apply these further in studies of biological water and other biological samples in sub-THz bands where response is expected but as yet not measured.

The errors shown in Fig.2 between experiments and model are seen to be less at the cut-on frequency of validity for the Debye model; however, further model explorations and measurement system characterisation are needed to expose systematic errors.

References

[1]D.M.Leitner, M.Gruebele and M.Havenith, HFSP Journal, vol.2, No.6, Dec 2008, 314-323.
[2]D.F.Plusquellic,K.Siegrist, E.J.heilweil and O.Esenturk, ChemPhysChem, 2007,8,2412-2431.
[3]A.G.Markelz, IEEE Journal of Selected Topics in Quantum Electronics, vol.14, No.1, Jan/Feb 2008.
[4]B. Yang, R. J. Wylde, D. H. Martin, P. Goy, R .S. Donnan and S. Caroopen, IEEE Trans.on Microw. Theory Tech, Vol 58, No.12, Dec, 3587-3597, 2010.
[5]P.Y.Han, M.Tani, M.Usami,S.Kono,R.Kersting and X-C.Zhang, J.Applied Physics, vol.89, 4,2001,2357-2359.
[6]D.H.Martin and J.W.Bowen, IEEE Trans.Microw.Theory.Tech, vol.41, no.10, pp.1676-1690, Oct.1993.
[7]D.M.Pozar, Microwave Engineering, 3rd edition, Wiley, Chapter 4.
[8]H.J.Liebe, International Journal of Infrared and Millimeter Waves, vol.12, No.7, 1991, page 659-675.

Rigidity analysis of HIV-1 protease

J W Heal[1], S A Wells[2], E Jimenez-Roldan[2], R F Freedman[3] and R A Römer[2]

[1] MOAC Doctoral Training Centre, University of Warwick, Coventry, CV4 7AL, United Kingdom
[2] Department of Physics and Centre for Scientific Computing, University of Warwick, Coventry, CV4 7AL, United Kingdom
[3] School of Life Sciences, University of Warwick, Coventry, CV4 7AL, United Kingdom

E-mail: jack.heal@warwick.ac.uk

Abstract. We present a rigidity analysis on a large number of X-ray crystal structures of the enzyme HIV-1 protease using the 'pebble game' algorithm of the software FIRST. We find that although the rigidity profile remains similar across a comprehensive set of high resolution structures, the profile changes significantly in the presence of an inhibitor. Our study shows that the action of the inhibitors is to restrict the flexibility of the β-hairpin flaps which allow access to the active site. The results are discussed in the context of full molecular dynamics simulations as well as data from NMR experiments.

1. Introduction

The relationships between protein structure, rigidity, and function are important in biophysics. In this study the software FIRST is used to create rigidity profiles of a set of 280 X-ray crystal structures of HIV-1 protease that have been solved to a resolution of at least 2.0 Å and made available via the protein data bank (PDB). FIRST runs the pebble game algorithm [1] to identify rigid clusters and flexible regions within a protein by matching atomic degrees of freedom with the constraints present due to bonding, and can carry out the rigidity analysis without high computational cost. By rapidly extracting information from all of the available high resolution HIV-1 protease structures we use FIRST to build an overview of the information available in the PDB. In a recent study [2], FIRST is applied to a set of individual structures from each of six different proteins, one of which is HIV-1 protease. Here we examine the enzyme in more detail by making a comprehensive study of the high resolution structures available.

Molecular dynamics (MD) studies on structures of HIV-1 protease [3] are informative yet computationally intensive and as such cannot offer an overview of the available structures. Experimentally, NMR [4]and fluorescence spectroscopy [5] have been used to study the flexibility of the protein. HIV-1 protease is a well-documented drug target due to its role in the life-cycle of HIV-1 [6] and many inhibitors have been designed in an attempt to block HIV-1 replication [7]. The structures have a range of crystallisation conditions, inhibitors and mutations. Initial findings from the analysis are presented, and work is ongoing to determine other underlying trends in the rigidity data of the 280 structures.

2. Method

In order to prepare the structures for rigidity analysis, all the water molecules are removed and the REDUCE software is used to add the hydrogen atoms that are not present in the X-ray crystal structure and to flip side chains where necessary. Uninhibited protease structures are created in PYMOL by manually by deletion of the inhibitor. A new PDB file is created representing the protein in a conformation which is part of the ensemble explored by the protein in its natural flexible motion [8]. The MD study in [9] also used this method to get structures of the apoenzyme and the holoenzyme. Covalent bonds, hydrogen bonds, salt bridges and hydrophobic interactions are included as constraints that restrict the freedom of the atoms in a molecule. Weaker, longer-range electrostatic forces and van der Waals forces are omitted [10]. FIRST builds a rigidity profile of the protein consisting of rigid substructures linked together by flexible sections [11]. The profile is calculated at a particular energy cutoff value $E_{\rm cut}$, and whilst other constraints are maintained, only hydrogen bonds whose energy lie below $E_{\rm cut}$ are included. Constraints are removed as $E_{\rm cut}$ is lowered and the rigidity profile of the protein is modified. Rigid cluster decomposition (RCD) plots show the rigidity profile of the protein at each $E_{\rm cut}$ value which resulted in a change to the flexibility of a C_α atom in the protein backbone. The horizontal axis of an RCD shows the C_α atoms in the primary structure of the protein with flexible regions indicated by thin black lines and rigid regions by thick coloured blocks. The colour associates each residue with a particular cluster and the vertical axis shows the $E_{\rm cut}$ value used to determine each rigidity profile. Examples of RCD plots can be seen in Figure 2.

As in [2] overall rigidity of the protein is expressed by the function f_5, defined in (1). It is determined by how many of the C_α atoms are included in the five largest rigid clusters. The total number of C_α atoms in the protein is N_{C_α}. Rigid clusters are numbered in order of decreasing size so that $n_i(E_{\rm cut})$ is the number of C_α atoms in rigid cluster i at cutoff value $E_{\rm cut}$.

$$f_5(E_{\rm cut}) = \frac{1}{N_{C_\alpha}} \sum_{i=1}^{5} n_i(E_{\rm cut}). \tag{1}$$

Rigidity dilution (RD) graphs plot f_5 against $E_{\rm cut}$ and show how the rigidity of the main chain of the protein is lost as the cutoff value is lowered.

3. Results and discussion

Figure 1 shows two representations of the protein corresponding to $E_{\rm cut}$ values at the beginning 1(a) and towards the end 1(b) of the bond dilution. Rigidity is maintained in two α-helices towards the base of the protein and at the centre where the active site is located. However, at an $E_{\rm cut}$ of -1.305 kcal/mol, the protein is almost entirely flexible. The same is observed in the structure 1B6J, for which two RCD plots with and without the inhibitor are shown in Figure 2. The regions surrounding residues 50 and 150 of the protein correspond to the β-hairpin flaps which close over the active site of the enzyme. The flaps become flexible much earlier in the bond-breaking process when there is no active-site inhibitor present. This overall trend was observed in all of the structures studied. Figure 3 shows an RD plot of the structure 1B6J with and without an inhibitor attached. In the RD plot, data is recorded in the graph whenever the rigidity of the C_α atoms changes. Only when a C_α atom ceases to be part of one of the five largest rigid clusters in the protein is f_5 reduced. This results in the stepwise nature of the plots. With no inhibitor, the f_5 value of 1B6J drops faster with decreasing $E_{\rm cut}$, although both structures exhibit rapid rigidity loss in agreement with [2].

4. Conclusion and outlook

The overall trend in the rigidity analysis of the 280 crystal structures reveals that the main effect of the inhibitors is to rigidify the flaps of the enzyme which move to permit access to the active

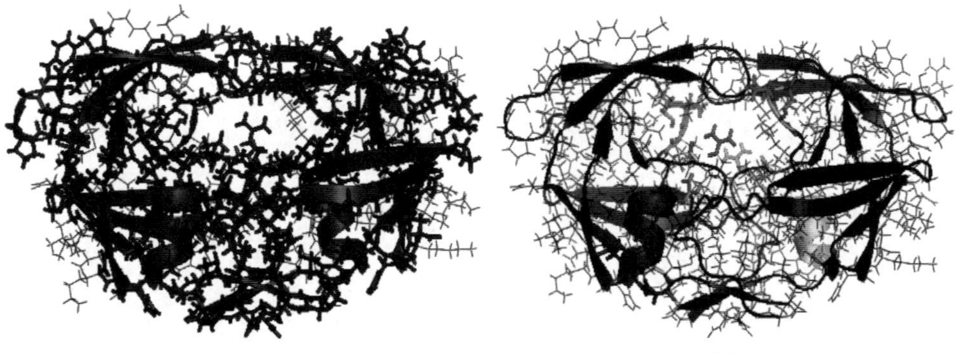

(a) (b)

Figure 1. PYMOL is used to show the structure 1EC0, its inhibitor having been removed. The protein is largely rigid in (a) with $E_{cut} = 0$ kcal/mol, and flexible in (b), where $E_{cut} = -1.305$ kcal/mol. The secondary structure is shown in a black 'cartoon' representation, and the rigid clusters are shown in different colours for clarity.

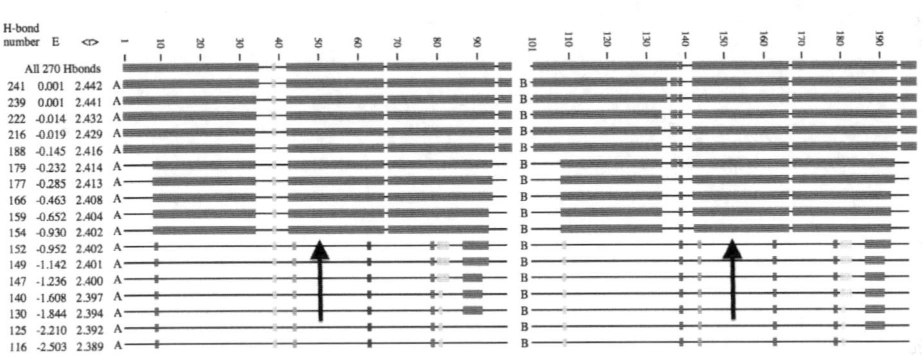

(a) Inhibited.

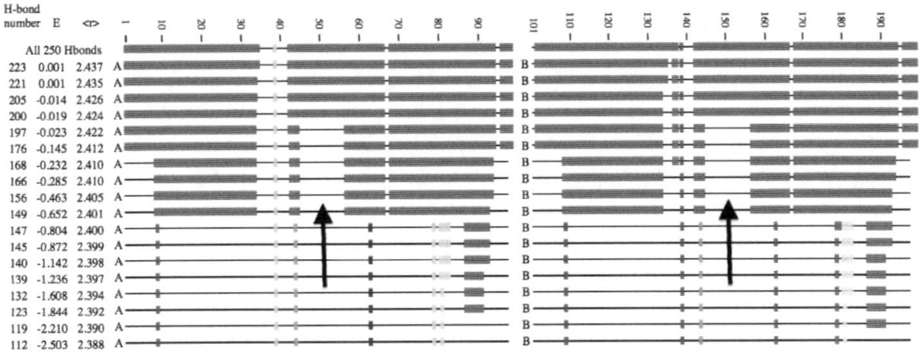

(b) Uninhibited.

Figure 2. The RCD plot for the structure1B6J is shown in (a), and with its inhibitor having been removed in (b). The thin black lines represent flexible regions of the protein and the thick coloured lines show rigid clusters. The columns on the left hand side of the plot show the number of remaining hydrogen bonds, E_{cut} and the mean coordination value <r>, which is an indication of the average number of interactions per atom. The rigidity loss patterns are similar, however the regions indicated by arrows become flexible at higher E_{cut} in (b) than in (a).

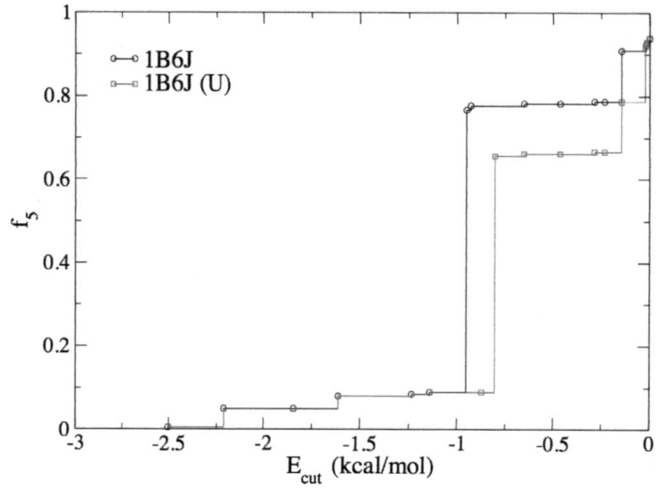

Figure 3. The RD plot for the protein with PDB code 1B6J. The suffix (U) signifies the uninhibited copy of the protein. As E_{cut} is lowered, rigidity is lost. Main chain rigidity is lost more rapidly in the uninhibited copy.

site. That the structures studied vary in resolution, inhibitors, mutations, and crystallisation conditions suggests that this trend is robust against structural variation. The results tie in with MD simulations such as [9] and experimental evidence [12], where the protein flaps are reported as being mobile in the native state but restricted by the presence of an active-site inhibitor. To expand on the analysis using FIRST, we are currently exploring the use of the mobility software FRODA in combination with elastic network modelling to study predicted protein mobility.

References

[1] Jacobs D J and Thorpe M F 1995 *Physical Review Letters* **75** 4051

[2] Wells S, Jimenez-Roldan J and Römer R 2009 *Physical Biology* **6**

[3] Collins J R, Burt S K and Erickson J W 1995 *Nat Struct Mol Biol* **2** 334–338

[4] Freedberg D I, Ishima R, Jacob J, Wang Y, Kustanovich I, Louis J M and Torchia D A 2009 *Protein Science* **11** 221–232 ISSN 09618368

[5] Rodríguez E J, Debouck C, Deckman I C, Abu-Soud H, Raushel F M and Meek T D 1993 *Biochemistry* **32** 3557–3563 ISSN 0006-2960 PMID: 8466899

[6] Erickson J W 1995 *Nat Struct Mol Biol* **2** 523–529

[7] Flexner C 1998 *The New England Journal of Medecine* **338** 1281–1293

[8] Henzler-Wildman K and Kern D 2007 *Nature* **450**

[9] Hornak V, Okur A, Rizzo R C and Simmerling C 2006 *Proceedings of the National Academy of Sciences of the United States of America* **103** 915 –920

[10] Jacobs D, Rader A, Kuhn L and Thorpe M 2001 *Proteins: Structure, Function and Genetics* **44** 150–165

[11] Thorpe M, Jacobs D, Chubynsky N and Rader A 1999 *Rigidity theory and applications* (Kluwer Academic, New York)

[12] Ishima R, Freedberg D I, Wang Y, Louis J M and Torchia D A 1999 *Structure* **7** 1047–1055, S1–S12 ISSN 0969-2126

Dynamic model of gene regulation for the lac operon

Maia Angelova and Asma Ben-Halim

Intelligent Modelling Lab, School of Computing, Engineering and Information Sciences, Northumbria University, Newcastle upon Tyne NE2 1XE, UK

E-mail: maia.angelova@northumbria.ac.uk, asma.benhalim@northumbria.ac.uk

Abstract. Gene regulatory network is a collection of DNA which interact with each other and with other matter in the cell. The *lac* operon is an example of a relatively simple genetic network and is one of the best-studied structures in the Escherichia coli bacteria. In this work we consider a deterministic model of the *lac* operon with a noise term, representing the stochastic nature of the regulation. The model is written in terms of a system of simultaneous first order differential equations with delays. We investigate an analytical and numerical solution and analyse the range of values for the parameters corresponding to a stable solution.

1. Introduction

Bacterial genome plays a central role in controlling the cellular processes, such as the response of a cell to environmental signals, the differentiation of cells and groups of cells in the unfolding of developmental programmes, the replication of the DNA preceding the cell division and many others [1, 2, 3, 4]. Proteins synthesized from genes may function as transcription factors binding to regulatory sites of other genes, as enzymes catalyzing metabolic reactions, or as components of signal transduction pathways. The degradation of proteins and the immediate DNA products can also be regulated in the cell. The proteins involved in the regulatory functions are produced by other genes. This gives rise to genetic regulatory network consisting of regulatory interactions between DNA, RNA, proteins and small molecules. A simple network consists of one or more input gene, metabolic, and signaling pathways, regulatory proteins that integrate the input signals, several target genes, and the RNA and proteins produced from target genes.

The *lac* operon [1] is an example of a relatively simple genetic network and is one of the well-studied and best-understood structures in the Escherichia coli (*e.coli*). It consists of a promoter, and operator region and three larger structural genes, *lacZ*, *lacY* and *lacA*, with a preceding regulatory operon responsible for producing a repressor, R, protein. In the absence of glucose available for cellular metabolism, but in the presence of external lactose, lactose is transported into the cell by a permease. Intracellular lactose, L, is then broken down into allolactose, A, first and then glucose and galactose by the enzyme β-galactosidase, B. The allolactose feeds back to bind with the lactose repressor and enable the transcription process to proceed [5, 6].

A number of mathematical models of the *lac* operon have been developed (see for example [3, 4, 5, 7, 8, 11, 12]). In this work, we consider a deterministic model of the lac operon with a noise term representing the stochastic nature of the regulation. The model is written in terms of a system of simultaneous first order differential equations with delay. We have investigated the analytical and numerical solutions and analysed the range of values for the parameters leading to stable solutions. The effects of the noise term on the concentrations levels are also presented.

2. Deterministic model

Deterministic models have been widely used to analyse genetic regulatory systems [2, 3, 4]. In many cases they are given by a system of ordinary differential equations (ODE). The ODEs formalism models the concentration of RNAs, proteins and other molecules by time dependent variables. The interactions in the network have a form of functional and differential relations between concentrations. The network dynamics can be described by Michaelis-Menten enzyme kinetics [4]. However, the deterministic model based on ODEs, describes an average response of the system. It assumes that the concentration varies continuously and deterministically, both of which assumptions may be questionable in case of gene regulation particularly of systems with small number of gene responses. For large systems, there is a large number of gene responses and the gene regulatory network can be realistically described with a set of deterministic ODEs. Time delays are common and substantial in biochemical processes. They are essential for the system and can protect it against transient loss of input signal, improve the accuracy of reading the information and filter non-beneficial pulses. However, time delays are not always beneficial as they may play a negative role in the stability of the gene network.

First, we will consider a deterministic model (without noise) following the works of Yildirim and Mackey [8, 9, 10, 11]. The model is written in terms of a system of simultaneous first order ODEs with time delays (DDEs) and is based on Michaelis-Menten enzyme kinetics. The full model [8] consists of five DDEs describing the lactose system in *e.coli* of positive feedback. This model is simplified [9], assuming a constant quantity of lactose L inside the cell (equilibrium of internal and external cellular lactose) and ignoring the dynamics of permease (assumed as a constant permease),

$$\frac{dM}{dt} = \alpha_M f(A_{\tau_M}) - \tilde{\gamma}_M M, \tag{1}$$

$$\frac{dB}{dt} = \alpha_B e^{-\mu\tau_B} M_{\tau_B} - \tilde{\gamma}_B B, \tag{2}$$

$$\frac{dA}{dt} = \alpha_A B h(L) - \beta_A B g(A) - \tilde{\gamma}_A A. \tag{3}$$

Here M is the mRNA concentration, n is the number of molecules of allolactose required to inactivate the repressor R. The model takes into account the delays in the response of mRNA, β-galactosidase and allolactose. The factor $e^{-\mu\tau_M}$ accounts for the dilution of A through growth during the transcriptional period, where μ is the bacterial growth rate, $A_{\tau_M} = A(t-\tau_M)$. The description of the same system without time delays is achieved with $\tau_Y = 0, Y = A, B, M$. The rate of change of M is a balance between a production term $\alpha_M f$,

$$f(A_{\tau_M}) = \frac{1 + K_1(e^{-\mu\tau_M} A_{\tau_M})^n}{K + K_1(e^{-\mu\tau_M} A_{\tau_M})^n}. \tag{4}$$

The functions $h(L)$ and $g(A)$ do not depend on time delays,

$$h(L) = \frac{L}{K_L + L}, \quad g(A) = \frac{A}{K_A + A}. \tag{5}$$

The parameters of the system, $K_i, i = 1, 2, L$, $\alpha_Y, Y = A, B, M$ and β_A, are deterministic mass-action kinetic rate constants. K_1 is the equilibrium constant for the repressor-allolactose reaction, $K = 1 + K_2 R_{tot}$, and R_{tot} is the total amount of repressor R. The loss of rate of B is given by $\tilde{\gamma}_B B$. Similarly, $\tilde{\gamma}_M$ and $\tilde{\gamma}_A$ represent the loss of rate of M and A respectively. $\tilde{\gamma}_Y Y = (\gamma_Y + \mu) Y$ is made up of degradation term $\gamma_Y Y$ and effective loss due to dilution μY for $Y = M, B, A$. The values of these parameters are estimated experimentally in [12].

The model was tested using the estimates for the parameters in [12]. Details of the numerical and analytical solution are given in [9]. The solutions indicate that the steady states depend on the intracellular lactose concentration L and growth rate μ. We have further investigated the analytical and numerical solution of this simplified model and analysed the range of values for the parameters corresponding to a stable solution using Routh-Hurwitz criterium [4]. The numerical solutions are obtained with Matlab (using Euler and Rounge-Kutta methods, dde23 routine and symbolic toolbox).

The steady state point (M_*, B_*, A_*), is depicted when the derivatives of M, B and A are equal to zero, $\frac{dM}{dt} = 0$, $\frac{dB}{dt} = 0$, $\frac{dA}{dt} = 0$. Applied to equations (1)-(3), this gives a condition for the steady state,

$$f(A_*) = \theta \frac{A_*}{\left[h(L) - \frac{\beta_A}{\alpha_A} g(A_*)\right]}, \qquad \theta = \frac{\widetilde{\gamma}_A \widetilde{\gamma}_B \widetilde{\gamma}_M e^{\mu \tau_B}}{\alpha_A \alpha_B \alpha_M}. \tag{6}$$

A graphical solution for $n = 2$ is illustrated on Figure 1 with initial conditions $A = M = 0, B = 1$. The dashed curve represents the left hand side, $f(A_*)$, of (6), while the right hand side, which is a monotonically increasing function of L, is plotted for three values of L. Steady state exists for intracellular lactose in the range $40\mu\text{M} \leq L \leq 55.4\mu\text{M}$. The location of the steady state is given by the intersection of the curves representing each side of (6). Depending on the value of L, one, two or three steady states are possible for $n = 2$. For example, three steady states exist for $L = 40\mu\text{M}$. The range of allolactose concentration is always positive, thus $h(L) - \frac{\beta_A}{\alpha_A} g(A_*) > 0$, which bounds the region where no steady states exist.

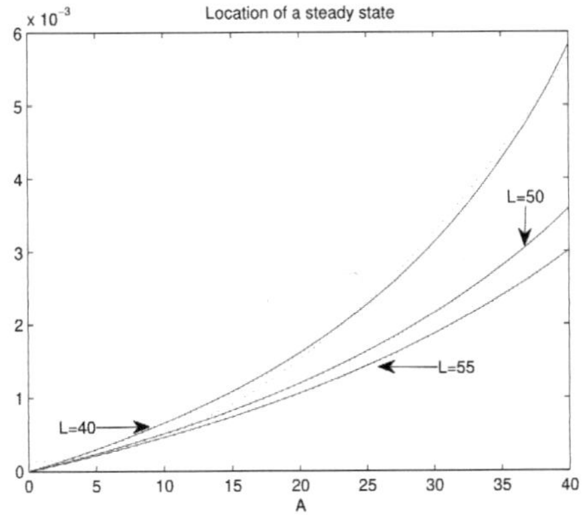

Figure 1. Steady states obtained by graphical solution of (6), A (μM).

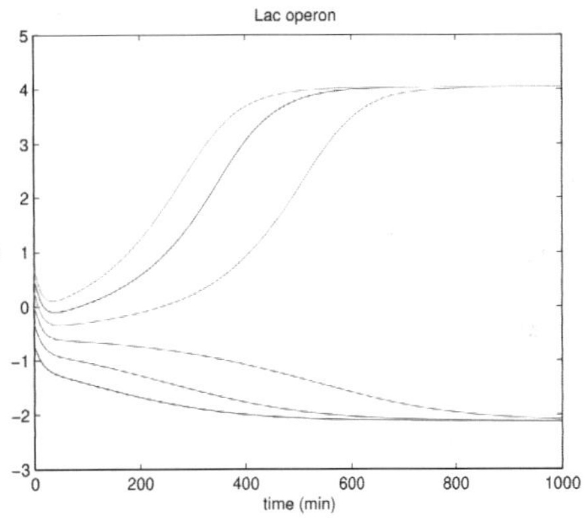

Figure 2. Bistability of β-galactosidase (μM) as a function of time (min) for various initial values of B.

β-galactosidase regulatory pathway is the most essential of the regulatory mechanisms in the lactose operon [1, 5, 6]. The concentration of β-galactosidase as a function of time was simulated and shows (Figure 2) that the steady states display bistability, in agreement with [8, 9].

The concentrations of mRNA and β-galactosidase are plotted on Figures 3 and 4 without time delay and with time delay respectively with initial values $M = 0, B = 1$. mRNA levels reach saturation faster then β-galactosidase levels. B is more sensitive to time delays, which also reduce the levels of concentration.

3. Stochastic correction

Gene expression is a complex process regulated at several stages in the synthesis of proteins. Apart from the regulation of DNA transcription, RNA translation and post-translational

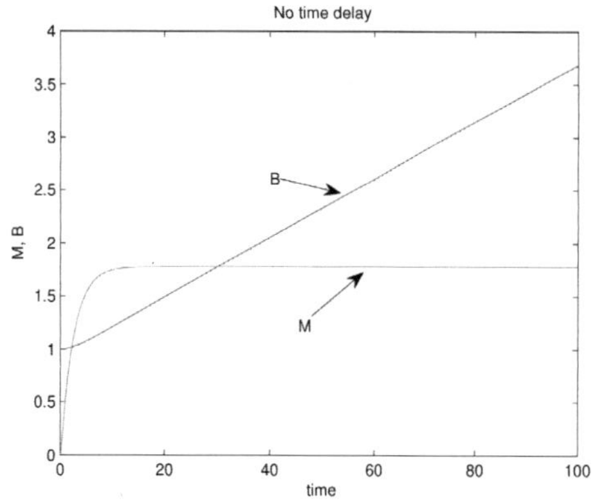

Figure 3. M (μM) and B (μM) versus time (min) without time delay.

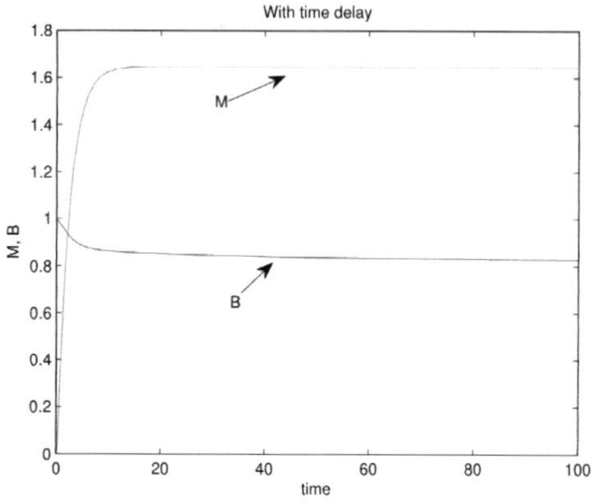

Figure 4. M (μM) and B (μM) versus time (min) with time delay.

modification of proteins, gene expression is also part of the regulatory process. Gene regulation is a stochastic process as it happens in living organisms and there are many fluctuations due to delay of response and other processes taking place at the same time. For small systems such as the *lac*, a stochastic model is more appropriate, as the gene responses are few and their nature is probabilistic. We have adjusted the deterministic model using the Langevin approach [3, 7] by adding a noise term to equations (1)-(3) to account for the stochastic behavior of the system.

Let $Y = M, B, A$ and F is the function representing the right hand side of the equations (1)-(3),

$$\frac{dY}{dt} = F(Y). \tag{7}$$

We add a noise term,

$$\frac{dY}{dt} = F(Y) + \sigma\frac{dW}{dt} \tag{8}$$

where W is independent random variable of Wiener process, which satisfies the conditions of this process, namely $W_t - W_0 \sim \mathcal{N}(\mu\sigma^2)$ [14]. Here $\mathcal{N}(\mu\sigma^2)$ is a normal distribution with expected value (mean) μ and variation σ^2. μ and σ are also known as drift and volatility respectively. Thus, the stochastic differential equation (SDE) is

$$dY = F(Y)dt + \sigma dW \tag{9}$$

We have used Euler-Maruyama method, which is a technique for approximate numerical solution of SDE, and is a simple generalisation of the Euler method [15]. Let $h = t_{i+1} - t_i$, where $i = 0, ..., N$,is the iterations step and N is the number of iterations. Then

$$Y_{i+t} = Y_i + F(Y_i)h + \sigma Z_t\sqrt{h} \tag{10}$$

where Z_t are normally distributed random numbers (derivatives of a Wiener process) with mean 0 and variance 1. Applying to the model,

$$\frac{dM}{dt} = \alpha_M f(A_{\tau_M}) - \widetilde{\gamma}_M M + \lambda, \tag{11}$$

$$\frac{dB}{dt} = \alpha_B M - \widetilde{\gamma}_B B + \lambda, \tag{12}$$

$$\frac{dA}{dt} = \alpha_A Bh(L) - \beta_A Bg(A) - \widetilde{\gamma}_A A + \lambda. \tag{13}$$

where $\lambda = \sigma Z_t \sqrt{h}$. These equations were solved with and without time delays. The solutions for M and B, with initial conditions $M = A = 0$ and $B = 1$, and different noise terms are illustrated in Figures 5 and 6 respectively. The fluctuations of M and B are clearly present, as expected from the stochastic nature of the process. The levels of concentration are sensitive to the noise. The fluctuations of mRNA concentration are above and below mRNA levels without noise, while β-gallactosidase concentration mainly decreases with time in the presence of the noise term.

The effect of the stochasticity was investigated in [13], where different noise terms were investigated experimentally and by simulation and the behaviour of the deterministic and stochastic models compared. The conclusion of this work is that it is still unclear whether the stochastic gene expression in the *lac* operon is detrimental or beneficial for the cells.

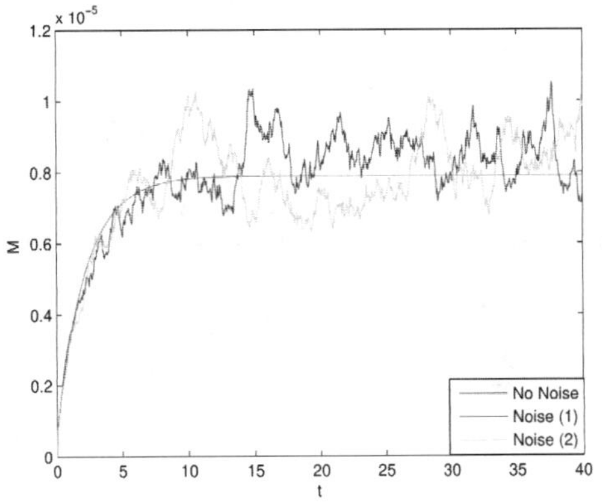

Figure 5. mRNA with different noise, M (μM).

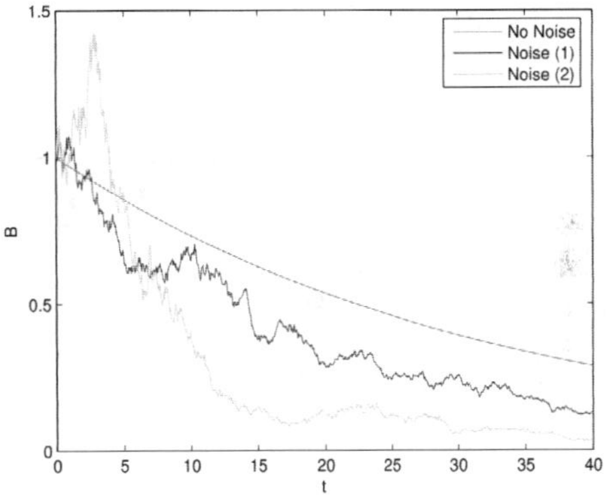

Figure 6. β-gallactosidase with different noise, B (μM).

References

[1] Ptashne M 2004 *A genetic switch* Cold Spring Harbor Laboratory Press: Cold Spring Harbor New York.
[2] Bower J and Bolouri H Editors 2001 *Computational Modelling of Genetic and Biochemical Networks* MIT Press:Boston Massachusets.
[3] De Jong H 2002 J Comp Biology 2002, 9, 67-103.
[4] Murray J D 2002 *Mathematical Biology* 3rd Edition Volume 1, IAM **17** Springer:NY Berlin.
[5] Goodwin BC 1969 *Eur J Biochem* **10** 515–522.
[6] Yagil G and Yagil E 1971 *Biophys J* **11** 11–27.
[7] Mettetal J. T, D. Muzzey J. M., Pedraza E. M., Ozbudak and van Oudenaarden A. 2006. PNAS 103: 7304-7309.
[8] Yildirim N and Mackey M C(2003) *Biophys J* **84** 2841–2851.
[9] Yildirim N, Santillan M and Mackey M C 2004 /it Chaos **14** 279–291.
[10] Mackey M C,Santillan M and Yildirim N 2004 *CR Biologies***327** 211–224.
[11] Santillan M, Mackey M.C. and Zeron E.S. 2007 . Biophysical J. 92: 3830-3842.
[12] Ackers G, Johnson A D and Shea M A 1982 *PNAS* **79** 1129–1133.
[13] van Hoek M and Hogeweg P 2007 *PLoS Comput Biol* **3**(6) doi:10.1371/journal.pcbi.0030111.
[14] Stark H and Woods J W 2002 *Probability and Random Processes with Applications to Signal Processing* Prentice Hall: New Jersey.
[15] Kloeden P E and Platen E 1999 Numerical Solutions of Stochastic Differential Equations Springer:Berlin,

Journal of Physics: Conference Series **286** (2011) 012008 doi:10.1088/1742-6596/286/1/012008

Creep test observation of viscoelastic failure of edible fats

C R Vithanage[1], M J Grimson[1], B G Smith[2] and P R Wills[1]

[1]Department of Physics
[2]Food Science Programmes, Department of Chemistry,
The University of Auckland, Private Bag 92019, New Zealand
E-mail cvit002@aucklanduni.ac.nz

Abstract A rheological creep test was used to investigate the viscoelastic failure of five edible fats. Butter, spreadable blend and spread were selected as edible fats because they belong to three different groups according to the Codex Alimentarius. Creep curves were analysed according to the Burger model. Results were fitted to a Weibull distribution representing the strain-dependent lifetime of putative fibres in the material. The Weibull shape and scale (life-time) parameters were estimated for each substance. A comparison of the rheometric measurements of edible fats demonstrated a clear difference between the three different groups. Taken together the results indicate that butter has a lower threshold for mechanical failure than spreadable blend and spread. The observed behaviour of edible fats can be interpreted using a model in which there are two types of bonds between fat crystals; primary bonds that are strong and break irreversibly, and secondary bonds, which are weaker but break and reform reversibly.

1. Introduction

The creep-and-recovery test is an important tool for investigating the viscoelastic failure of soft solids. In the creep test the strain is measured under a constant stress as a function of time at different temperatures. The test consists of two phases: the strain is first measured as a function of time at constant shear stress in the creep phase; any change in strain that occurs when the stress is removed is then measured in the recovery phase. The Maxwell model of a viscoelastic material comprises a (purely viscous) Newtonian dashpot and a (purely elastic) Hookean spring connected in series, whereas the Kelvin–Voigt (or Voigt) model comprises a Newtonian dashpot and Hookean spring connected in parallel. The complete creep curve can be accounted for by a Burger model comprising one or more Kelvin–Voigt units in a series with the Maxwell unit [1]. The creep and recovery behaviour of a Burger system [2] has been described by

$$\frac{\varepsilon(t)}{\sigma}=\frac{1}{G_0}+\frac{t}{\eta_0}+\sum_{i=1}^{n}\frac{1}{G_i}\left[1-e^{-t/\tau_i}\right] \tag{1}$$

where σ is the applied shear stress, $\varepsilon(t)$ is the strain at time t, G_0 is the instantaneous elastic modulus, η_0 is the viscosity characteristic of the Newtonian dashpot and G_i are elements of the spectrum of the retarded modulus associated with time constants τ_i. The time constant is related to the viscosity η_i of the ith Voigt unit: $\eta_i = G_i\tau_i$.

A Weibull distribution is often used to model the behaviour of composite solid substances and assess their reliability [2]. The Weibull density function [3]-[6] is given by

$$f(x_i)=\frac{\beta}{\alpha}\left(\frac{x_i-\gamma}{\alpha}\right)^{\beta-1}e^{-\left(\frac{x_i-\gamma}{\alpha}\right)^{\beta}} \tag{2}$$

where x_i is the breaking strength of the ith element, $\beta > 0$ is the shape factor, $\alpha > 0$ and $x \geq \gamma \geq 0$ are scale parameters; γ is the location parameter or x-threshold, below which we take $f(x_i) = 0$.

Fancey [7] showed that equations based on the Weibull distribution functions can be used accurately to represent the creep and recovery behaviour of polymeric materials. The Weibull shape parameter β and scale parameter α can both be estimated from creep test data represented by the equation

$$\varepsilon(t) = \varepsilon_i + \varepsilon_c \left[1 - e^{-(t/\alpha)^{\beta}} \right] \quad (3)$$

where ε_c is a constant and ε_i is the strain at time zero. The recovery of the sample following the removal of the load be used to estimate the same parameters can also be estimated based on the representation

$$\varepsilon(t) = \varepsilon_r e^{-(t/\alpha)^{\beta}} + \varepsilon_f \quad (4)$$

where ε_r and ε_f are constants [7].

Our present results have been obtained by conducting creep and recovery tests at refrigeration temperature (5 °C) and at room temperature (20 °C).

2. Materials and methods

2.1. Materials

Mainland butter and Anchor Spreadable blend (a butter and sunflower oil blend) (Fonterra Brands, Auckland, New Zealand), Pams butter (Pam's Products Ltd, Auckland, New Zealand), Sunrise table spread and MeadowLea Original spread (Goodman Fielder New Zealand Ltd, Auckland, New Zealand) (both spreads contained 70% fat) were purchased from a supermarket and transferred chilled to a 5 °C store.

2.2. Sample Preparation

Cylindrical samples (25 mm diameter) from each material were prepared using a cork borer. Using a stainless steel wire, the cylinders were sliced into discs approximately 3 mm in height for use on the rheometer and stored at 5 °C until required.

2.3. Creep Test

A rheometer (Paar Physica UDS 200, Physica Messtechnik GmbH, Stuttgart, Germany) was used with parallel plate fixtures. The lower plate was fitted with a Peltier element that kept the temperature within 0.1 °C of the set value. To prevent the samples slipping, emery paper was fixed to the upper and lower plates. The instrument was set to the zero gap position and the lower plate set to 20 °C. The upper plate was then raised and the gap set to 3 mm. A disc was placed on the lower plate and the top plate lowered to 3mm. Excess material was carefully removed and the sample allowed to equilibrate for 30 min after which a run was initiated. A digital laser thermometer (HLP Controls Pty Ltd, model 8859, Riverstone, NSW, Australia) was used to check the temperature of the upper plate prior to commencement of each run. For the samples tested at 5 °C, the Peltier plate was set at 5 °C and the run initiated immediately without an equilibration period. Stress of 250 Pa was applied to the samples in all cases.

3. Results and discussion

Figure 1 shows the creep behaviour for the five chosen edible fats, taken at two different temperatures where this was possible, fitted to the Burger model (Eq. 1). In all cases a satisfactory fit was obtained by using only one Voigt unit and use of two units did not significantly improve fit to the data. The values obtained for the rheological parameters are shown in Table 1. Results from three replicate experiments were used to estimate the value of each parameter and its precision (quantified as the standard error of the mean).

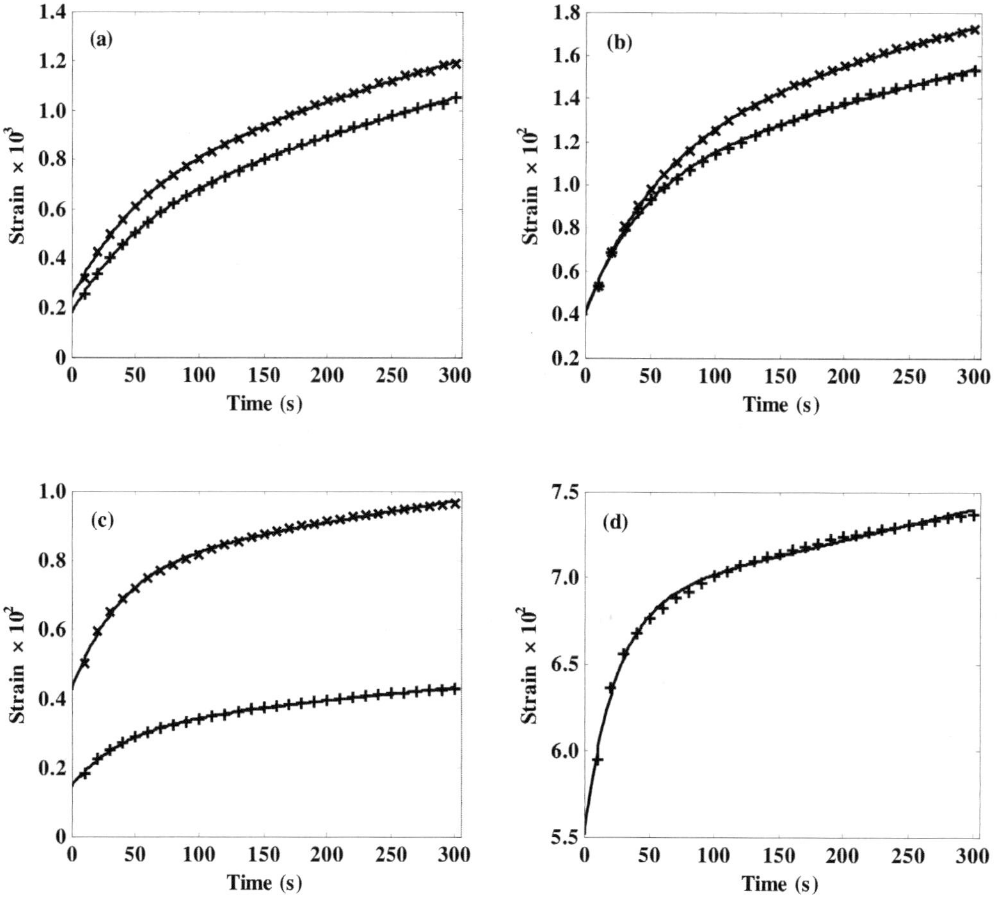

Figure 1. Creep behaviour for data fitted with one Voigt-unit for (a) Pams (×) and Mainland (+) at 5 °C, (b) Pams (×) and Mainland (+) at 20 °C, (c) Anchor (+) and MeadowLea (×) at 5 °C (d) Sunrise (+).

Table 1. Creep parameters of edible fat.

Sample	G_0 (Pa)	G_1 (Pa)	τ_1 (s)	η_1 (Pa s)	η_0 (Pa s)
Pams 5 °C	1.3×10^6	4.7×10^5	46.4	2.2×10^7	1.5×10^8
Mainland 5 °C	1.7×10^6	5.6×10^5	46.6	2.6×10^7	1.6×10^8
Anchor 5 °C	6.7×10^4	6.7×10^4	42.7	2.9×10^6	4.5×10^7
Sunrise 5 °C	4.6×10^3	2.4×10^4	38.6	9.2×10^5	1.2×10^7
Meadowlea 5 °C	1.9×10^5	1.4×10^5	42.5	5.8×10^6	7.0×10^7
Pams 20 °C	5.4×10^4	2.6×10^4	46.9	1.2×10^6	1.2×10^7
Mainland 20 °C	7.3×10^4	3.7×10^4	47.4	1.8×10^6	1.7×10^7

The recovery data was not robust enough to allow meaningful fitting of the 4-parameter Burger model.

The Burger model gives a good representation of the creep behaviour of viscoelastic materials. When stress is applied, elastic forces (represented by a series-connected Hookean spring element, elastic modulus G_0) give rise to the virtually instantaneous appearance of moderate strain due to simple molecular bond stretching. The appearance of further strain, beyond the elastic limit of these bonds and somewhat delayed, arises due to the operation of internal viscous forces (represented by Newtonian dashpot elements) as polymer molecules in close contact move relative to one another. Non-recoverable strain due to such effects is represented by the single series-connected dashpot (viscosity η_0). The Voigt unit represents that part of the process of structural change in which the secondary intermolecular bonds break and reform with a characteristic retardation time (τ_1).

At 5 °C the butters have significantly higher intrinsic mechanical strength and viscosity. These higher values are also reflected in the parameters associated with the Voigt unit. It is notable that the viscosities (η_1) associated with the (parallel connected) Voigt unit are in all cases lower than those (η_0) obtained for the Newtonian dashpot in the (series connected) Maxwellian unit, whereas the opposite is true concerning the mechanical strength: $G_0 > G_1$. This is consistent with the interpretation in which the Maxwellian unit represents the behaviour of stronger primary bonds which break irreversibly whereas the Voigt unit represents the behaviour of weaker secondary bonds that break and reform as tcontinuous stress is applied and the sample is increasingly strained.

It is noteworthy that there are no significant differences between the retardation times τ_1 for the different samples, except perhaps for one of the spreads (Sunrise). This could be interpreted as an indication that the character of the secondary bonds is similar for the all of the materials tested whether they are butters, spreads or blends.

Non-linear curve fitting was applied to both the creep (not shown) and recovery test data (Figure 2) to obtain estimates of the relevant Weibull parameters (Eqs. 3 & 4). The goodness of fit for this curve-fitting procedure was very similar to that for the Burger model procedure (Figure 1). Results of the Weibull parameters obtained for the five samples are listed in Tables 2 and 3.

Table 2. Estimated Weibull parameters for the creep test.

Samples	ε_i	ε_c	α (s)	β
Pams 5 °C	$(6.2 \pm 1.3) \times 10^{-2}$	2.1 ± 0.2	25.0 ± 4.4	$(5.1 \pm 0.9) \times 10^{-1}$
Mainland 5 °C	$(11.3 \pm 2.6) \times 10^{-2}$	1.9 ± 0.2	55.5 ± 24.7	$(6.9 \pm 0.8) \times 10^{-1}$
Anchor 5 °C	$(2.1 \pm 1.4) \times 10^{-1}$	0.9 ± 0.2	7.0 ± 1.1	$(4.6 \pm 1.3) \times 10^{-1}$
Sunrise 5 °C	3.3 ± 0.5	3.2 ± 0.5	3.2 ± 0.4	$(3.5 \pm 0.1) \times 10^{-1}$
MeadowLea 5 °C	$(0.8 \pm 0.2) \times 10^{-1}$	0.5 ± 0.2	9.9 ± 2.2	$(4.8 \pm 0.5) \times 10^{-1}$
Pams 20 °C	$(3.0 \pm 1.1) \times 10^{-1}$	2.4 ± 0.2	17.5 ± 1.5	$(5.4 \pm 0.3) \times 10^{-1}$
Mainland 20 °C	$(1.5 \pm 0.6) \times 10^{-1}$	2.3 ± 0.6	12.7 ± 6.3	$(5.0 \pm 1.5) \times 10^{-1}$

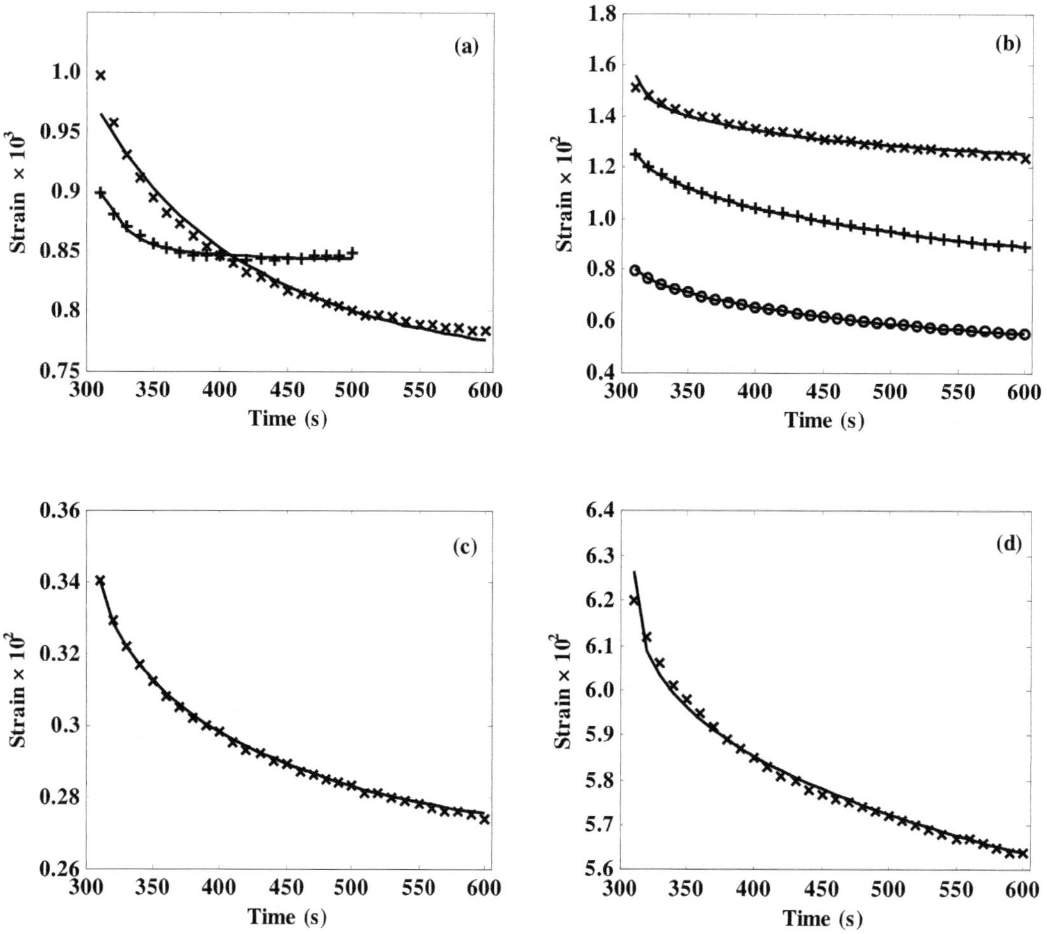

Figure 2. Weibull parameter plots using strain values for recovery test of (a) Pams (×) and Mainland (+) at 5 °C, (b) Pams (×) and Mainland (+) at 20 °C and Anchor (O) and (c) MeadowLea (×) at 5 °C (d) Sunrise (×) at 5 °C.

Table 3. Estimated Weibull parameters for the recovery test.

Samples	ε_r	ε_f	α (s)	β
Pams 5 °C	$(3.1 \pm 0.2) \times 10^{-4}$	$(6.9 \pm 0.5) \times 10^{-4}$	181.6 ± 38.1	$(5.4 \pm 0.3) \times 10^{-1}$
Mainland 5 °C	$(5.2 \pm 0.3) \times 10^{-5}$	$(8.5 \pm 0.7) \times 10^{-4}$	268.5 ± 4.2	$(7.5 \pm 0.1) \times 10^{-1}$
Anchor 5 °C	$(3.2 \pm 0.2) \times 10^{-3}$	$(4.8 \pm 0.3) \times 10^{-3}$	201.4 ± 1.4	$(7.6 \pm 0.1) \times 10^{-1}$
Sunrise 5 °C	$(6.9 \pm 0.0) \times 10^{-3}$	$(4.4 \pm 1.2) \times 10^{-2}$	183.4 ± 40.2	$(6.9 \pm 0.4) \times 10^{-1}$
MeadowLea 5 °C	$(6.6 \pm 0.3) \times 10^{-3}$	$(2.8 \pm 0.2) \times 10^{-2}$	100.5 ± 1.6	$(9.8 \pm 0.2) \times 10^{-2}$
Pams 20 °C	$(3.0 \pm 0.5) \times 10^{-3}$	$(1.2 \pm 0.1) \times 10^{-2}$	137.6 ± 24.7	$(8.2 \pm 0.6) \times 10^{-1}$
Mainland 20 °C	$(3.7 \pm 0.5) \times 10^{-3}$	$(1.0 \pm 0.1) \times 10^{-2}$	133.5 ± 6.8	$(8.5 \pm 0.4) \times 10^{-1}$

The value of the scale (life-time) parameter (α) estimated from creep test data was highest for two brands of butter followed by MeadowLea Original spread, Anchor Spreadable blend and Sunrise spread. The life-time parameter was lower for Anchor Spreadable blend and Sunrise spread than for Pams butter and Mainland butter at 20 °C and for MeadowLea Original spread. These results indicate that mechanical failure is less likely in MeadowLea Original spread, Pams butter and Mainland butter than in Anchor Spreadable blend and Sunrise spread.

The Weibull shape parameter (β) indicates whether the failure rate of internal elements of the material is increasing ($\beta > 1$), constant ($\beta = 1$) or decreasing ($\beta < 1$) over time. The experimentally obtained value of β was indistinguishable from unity for both brands of butter, indicating that the mechanical composition of these butters is, within the temperature range 5–20 °C, quite homogeneous. The other materials have values of β slightly less than unity, indicating that the distribution of strengths of individual mechanical elements is skewed toward those which are weaker. This effect is much more marked in the "spreads" as opposed to the "blend".

The results of Weibull fitting to the recovery test data indicate that the recovery process does not occur through exactly the same mechanism as the creep process. In the first place, all of the values of the shape parameter β were found to be smaller than unity for the recovery data, indicating a decreasing rate of strain diminution over time; greater numbers of weaker bonds reforming at a slower rate. However, even more striking is the large increase (typically between one and two orders of magnitude) in the life-time parameter when the substances switch from creep to recovery behaviour. The rate of recovery under zero applied stress is very much slower than the rate of deformation under the constant applied stress of 250 Pa.

4. Conclusion

A comparison of the rheometric measurements of three different types of edible fat samples showed clear differences among the products. Overall the data suggest that the microstructure of Mainland butter at 5 °C is the most robust, while Anchor and Sunrise are the least robust. Furthermore, the microstructure of the Anchor and Sunrise products is more unstable at 5 °C than the microstructure of the Pams and Mainland products at 20 °C. These differences in failure characteristics are probably due to differences in the composition and manufacturing procedure of the all products, questions which are now under investigation.

5. References

[1] Steffe J E 1992 *Yield stress:* In *phenomena and measurement Advances in Food Engineering* RP Singh and A Wirakartakusumah (ed) (Boca Raton FL: CRC Press) p 285-332.
[2] Shama, F and Sherman, P 1970 *J. Texture Studies* **1** 196-205.
[3] Mann N R Schafer R E and Singpurwalla N D 1974 *Methods for statistical analysis of reliability and life data.* (New York; USA; John Wiley and Sons) p 184-266.
[4] Bergman B 1984 *J. Materials Sci. Letters* **3** 689-692.
[5] Langlois R 1991 *J. Materials Sci. Letters* **10** 1049-1051.
[6] Mahesh S and Phoenix S L. 2004 *Int. J. Fracture* **127** 303-360.
[7] Fancey K S 2005 *J. Material Sci.* **40** 4827-4831.

In-situ conductivity and UV-VIS absorption monitoring of iodine doping-dedoping processes in poly(3-hexylthiophene) (P3HT)

V V Kislyuk[1,5], O P Dimitriev[2], A A Pud[3], J Lautru[4], I Ledoux-Rak[4]

[1] Taras Shevchenko National University of Kyiv, 64 Volodymyrs'ka str. 64, 01601 Kyiv, Ukraine
[2] Institute of Semiconductor Physics, pr. Nauki 45, 03028 Kyiv, Ukraine
[3] Institute of Bioorganic Chemistry and Petrochemistry, 50 Kharkivske shose, 02160 Kyiv, Ukraine
[4] Laboratoire de Photonique Quantique et Moleculaire, UMR CNRS 8537, Institut d'Alembert, Ecole Normale Superieure de Cachan, 61 avenue du President Wilson, 94 235 Cachan, France

E-mail: viktor.kislyuk@gmail.com

Abstract. Poly(3-hexylthiophene) (P3HT) belongs to the family of conducting polymers whose conductivity can be controlled via doping. This paper is intended to study iodine doping-dedoping processes in P3HT films via in-situ measurements of the conductivity and light absorption. The film was spin-coated on ITO (indium tin oxide) patterned in the form of four-probe configuration. The sample was doped by iodine via exposure to iodine vapours in chamber and then de-doped by spontaneous release of the dopant under ambient conditions. The doping process was found to be several orders of magnitude faster than dedoping, which is attributed to the different mechanisms of diffusion. The theory of diffusion was applied to approximate the conductivity relaxation as measured experimentally. The best-fit parameters for the approximation were used to determine the dopant diffusivity D and β parameter of the semi-empirical relation between conductivity and dopant concentration ($\sigma \propto N_D^{\beta}$). Their values were found to be (3.9±0.9) 10^{-13} cm^2/s and 4.28±0.04 respectively. The latter agrees with the value know from the literature. The change of the polaron band in the absorption spectrum correlated with the conductivity changes in a complex manner and cannot be used as a measure of the concentration of free carriers contributing into the conductivity.

1. Introduction

Since the first evidence of the possibility to produce conducting polymers [1] they have become (along with organic molecular semiconductors) the basis of the fast developing field of organic electronics. Nowadays there are many applications based on conducting polymers such as photovoltaic cells [2-4], organic transistors [5,6], and light emitting diodes [7] etc.

One of the most attractive (for electronics) features of this class of materials is the ability to control their conductivity via doping as in the case of solid state electronics. The dopant molecules play the

[5] To whom any correspondence should be addressed.

role of either electron donors or electron acceptors with respect to the π-conjugation chain of the polymer. The removal of the electron from the chain is a result of the polymer oxidation. This oxidized state is called positive polaron or polaronic cation. Sometimes it is falsely called "hole" in analogy to the electronic states formed in classical semiconductors. In contrast to a hole in solid state semiconductors, a polaron, once it is generated, induces an energy change due to the distortion of the polymer chain. Thus, the LUMO-HOMO (LUMO – lowest unoccupied molecular orbital; HOMO – highest occupied molecular orbital) gap of the chain reduces when the polaron is formed on it. The deeper oxidation with removal of the second electron from HOMO induces bipolaron formation, which further reduces the gap (figure 1). There are many evidences of doping related changes in the absorption spectrum for various types of conducting polymers and oligomers [8-11], which is indicative of the presence of polarons and bipolarons. On the other hand, when it concerns the electronics applications, it is more appropriate to distinguish the states which contribute into the net electrical conductivity. Bipolarons are associated not only to the presence of polaronic intra chain states but also and rather to interchain interactions and charge transfer processes via these localized states (hopping transport [12,13]). The comparison of light absorption spectra with electrical measurements allows one to derive quantitative characteristics of the states responsible for the conductivity.

This paper is intended to control synchronously the absorption features and the conductivity in poly(3-hexylthiophene) (P3HT) with respect to doping level changes. Our interest to this material is manifold. First, it is widely used in numerous applications [2-7]. Second, there are many experimental data on either its polaronic band behavior or its electrical characteristics. However there is a lack of the data on their correlation and quantitative estimates of the relation between polaronic features and conductivity. Third, halogens are known to be effective dopants for p-type conductivity in P3HT. In particular, iodine doping is a quite simple and straightforward process [14] and can induce a strong conductivity changes over a wide range. At the same time the doped state of P3HT in this case is rather unstable [8,9] and degrades with time via spontaneous withdrawal and evaporation of the dopant molecules, which is apparently a drawback. Nevertheless the iodine volatility is of great interest for this study, since there is a possibility to make a set of doping-dedoping cycles on the same sample in a repetitive manner. In addition, the dedoping-related time dependence of the conductivity contains some information about dopant diffusion, which is used here to derive the iodine diffusivity value. For this reason some model of the diffusion is considered and its relation to the conductivity relaxation is proposed.

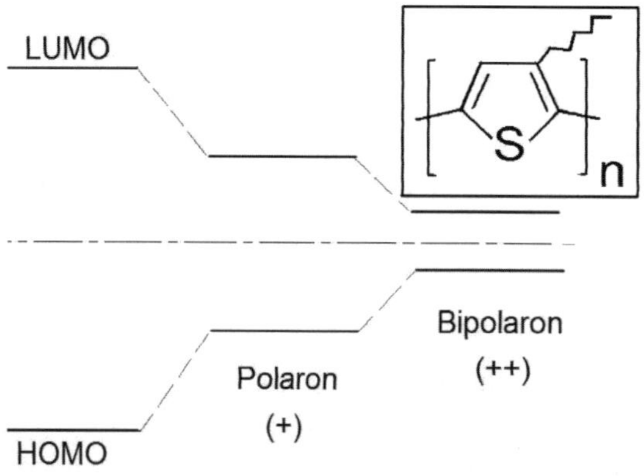

Figure 1. Illustration of gap reduction for the polaron and bipolaron states. The dot-and-dash line represents schematically Fermi level position. **Inset.** P3HT molecule

2. Experiment

The experimental study includes: i) the photolithographic preparation of the special drawing on the indium tin oxide (ITO) coverage; ii) electrical measurement as the dopant escapes; iii) measurement of the optical absorption spectra synchronously with electrical measurements.

2.1. Sample preparation

ITO coated glass substrate was patterned via standard lithographic route. The substrate was rinsed in acetone and propanol and dried in nitrogen flow. A S1805 resist was spin coated for 1 min at 1000 and dried at 115 °C for 1 min. Then it was exposed through the mask (a transparency film with the drawing made by laser printer) for 6 sec under UV radiation. The exposed fragments of the resist were removed while rinsing in a water solution of Developer 35 with a 1(developer):5(H_2O) volume ratio. ITO was etched off from the open area by concentrated HCl acid for 1 min at room temperature. Then the substrate was washed in acetone to remove the residual resist and rinsed in propanol.

Before the sample preparation the patterned ITO substrate was successively washed under ulrasonication in acetone and then in 2-propanol, then in pure water followed by oxidative treatment in boiling H_2O_2 and dried in air. **Regioregular P3HT** (Rieke Metals) film on the freshly treated ITO was formed by spin-coating of its 2 wt % solution in chlorobenzene at 2000 rpm for 20 s followed by baking at 120 °C for 30 min under argon. The film thickness measured by atomic force microscopy was 87±5 nm.

2.2. Iodine doping-dedoping

The sample was immersed in a chamber saturated with iodine vapor. To control in-situ the process, the wires from the sample contacts were brought out and kept connected to the measurement circuit. After the doping was saturated (which was monitored by the stabilization of the conductivity value) the sample was mounted into the spectrometer for less than 10s. Two different doping/dedoping cycles were made. The first one was made after long relaxation time (more than one day), and the other one followed immediately after the 1 h relaxation of the first cycle.

2.2. Conductivity measurements

A serial connection of the power supply and the amperemeter was made to 1,4 contacts (figure 2, inset), whereas the voltmeter was connected to the 2,3 inner contacts. This is a general principle widely used to avoid the contact influence, which results in an additional voltage drop when the current flows through a contact. Since the voltmeter has a high input resistance, the current is negligible in its circuit, and the voltage between contacts 2-3 relates directly to the part of the sample length where they are connected. The geometry of the contacts is chosen so that the distances between contacts are: L_{1-2}:L_{2-3}:L_{3-4}=1:1:1 (indices indicates the numbers of the respective contacts). Thus the distance between contacts 1 and 4 is three times higher than that of L_{2-3}. Prior to synchronous electrical and optical measurements, the contact resistance influence was checked at various stages of the doping level.

The experimental data relate to the reduced value of conductivity σ/σ_0 (used in further analysis) as:

$$\frac{\sigma}{\sigma_0}=\frac{I(t)}{U_{2-3}(t)}\cdot\frac{U_{2-3}(0)}{I(0)}, \qquad (1)$$

where σ_0, $U_{2-3}(0)$, $I(0)$ are initial values of conductivity, voltage between contacts 2,3 and current measured at the saturation of the doping process prior to the extraction of the sample from the chamber; and σ, $U_{2-3}(t)$, $I(t)$ are the corresponding values as a function of the time t after extraction from the iodine chamber (actually, t is the duration of the dedoping process).

2.3. Optical absorption

The optical absorption spectra were measured using an AvaSpec-2048 spectrometer, with the bare ITO coated glass substrate as a reference sample. The integration time to take a single spectrum was 5 ms with the final spectrum resulted from averaging over 100 scans; thus, the total time of the measurement was far less than the characteristic relaxation time of the samples which was of the order of several minutes. Normally, the spectral changes were recorded every 30-60 seconds after the sample was exposed to the iodine vapors, and the rate of the spectral changes was controlled every 0.5 s visually on a monitor.

Spectra were modelled by the sum of four Gaussian components via mean square fit procedure. The spectra profiles can be expressed in the following forms:

$$A(x)=\sum_{i=1}^{4}\frac{A_i}{w_i\cdot\sqrt{\pi/2}}\cdot\exp\left[-2\cdot\left(\frac{x-x_i}{w_i}\right)^2\right], \qquad (2)$$

where $x=h\nu$ is variable and the free parameters are (three for each component): $x_i=h\nu_i$ (current value of the photon energy (in eV) at its i-th peak value); A_i is the integrated area of the i-th component; w_i is approximately the full width at half maximum (FWHM), i.e.:

$$FWHM_i=\sqrt{\ln 4}\cdot w_i\approx 1.177\cdot w_i\,. \qquad (3)$$

To estimate the influence of the iodine spectrum another glass plate was immersed in the iodine chamber under the same exposition conditions, and the spectrum of the adsorbed iodine molecules was measured in the same manner as described above.

3. Results and discussions

The first and the most unexpected result was that the doping rate is much faster than the dedoping process. The conductivity rose to the saturation in about one second after the sample was immersed into the iodine vapour chamber. The reverse process (after the sample was removed from the chamber and kept under ambient) took hours. A possible interpretation of this asymmetry is presented at the end of this section. To check whether this process arises from bulk or contact phenomena, special 4-probe ITO geometry was used. Figure 2 shows that contact resistance becomes essential after 15 min of dedoping process. $R_{1\text{-}4}$ is a drastically superlinear function of time whereas the resistance between internal contacts tends to be sublinear. After 1 h of dedoping their values differ by one order of magnitude. This means that only a high level of doping can guarantee good ITO contacts to P3HT film.

The experimental time dependence of the conductivity as calculated from (1) fits well with a linear behaviour (figure 3). The best fit parameters for curve 1 (table 1) results in a value of $\beta=4.28$ (A.10) which is very close to the value $\beta\approx 4.5$ known from other experimental studies [15] and explained theoretically on the basis of the variable range hopping [13]. The repeated measurement of the conductivity relaxation (in 1 h after the first measurement) yields slightly lower values, which can be attributed to the reconstruction of the P3HT molecules, since this parameter is known to be sensitive to the dimensionality of hopping. In contrary the value of the dopant diffusivity derived from the second relaxation is a little bit higher. Probably the reason of both transformations is the same and associated with the reconstruction in the polymer film, which suppress the diffusion. To establish relations between the conductivity and polaronic states in the polymer the absorption spectra were analyzed at different time values.

The analysis of absorption spectra (figure 4) reveals that they are well described by (2). The parameters of the best mean-square fit agree with experimental results of other authors [16]. In particular the initial spectrum displays components with spectral features at x_1=1.37 eV, x_2=1.64 eV x_3=2.05 eV, whereas the peaks identified in [16] as two polaronic states and interchain interaction have positions at 1.302 eV, 1.657 eV and 2.034 eV respectively. For this reason we attribute x_1 and x_2 to polaronic transitions and x_3 to interchain feature. Thus the sum of integrals A_1 and A_2 is taken as a measure of the polaron concentration. Since the light absorption monitors the film thickness it would be more appropriate to consider number of dopant molecules per unit area of the film, which is called in electronics as Gummel Number (GN) [17]. It is the integral characteristic of the layer which is defined as integral of the concentration over the thickness of the layer. Integrating (A.3) and taking approximation (A.4) yield:

$$GN \sim \int_0^L N_D(x,t)dx \sim t^{-1/2}. \tag{4}$$

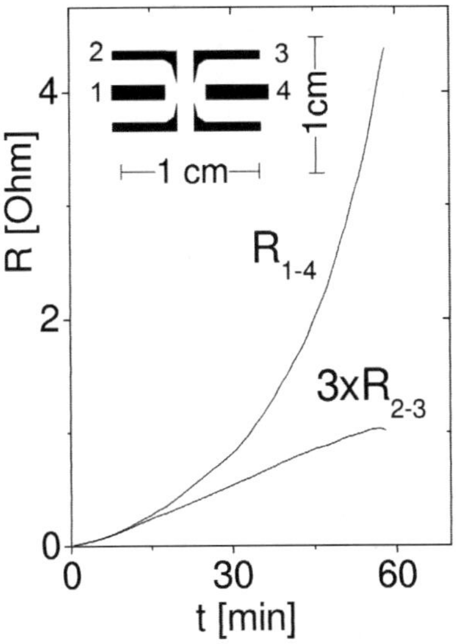

Figure 2. Time-dependence of the resistance on 1-4 and 2-3 contacts calculated as $R_{1-4} = U_{1-4}/I$, $R_{2-3} = U_{2-3}/I$, where U_{1-4}, U_{2-3} are values of the potential differences between electrodes 1-4 and 2-3 respectively, I is current. To normalize the resistances to the same length the value of R_{2-3} is multiplied by three. **Inset.** Photolithographic drawing of the ITO coverage.

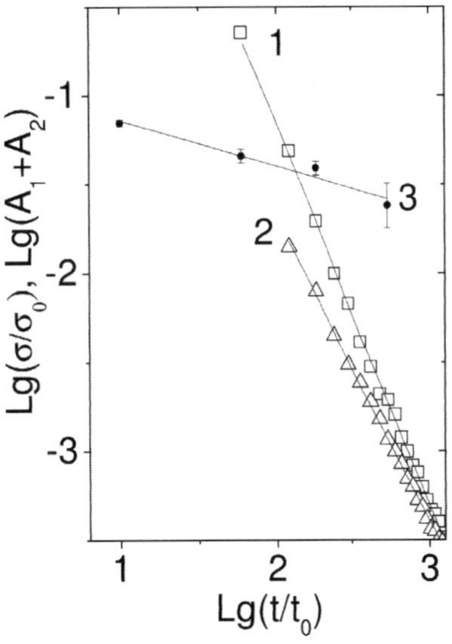

Figure 3. Time dependence in lg-lg plot ($\lg = \log_{10}$) of relative conductivity (1 – first measurement, 2 – second measurement) and 3 – overall integral intensity A_1+A_2 (A_1, A_2 are parameters of approximation) of the polaronic band in absorption spectrum. Time is reduced to $t_0 = 1$ s (see comments to (A.9)). The solid lines are the linear mean square fit approximations (parameters are presented in table 1).

Table 1. Parameters of the linear mean square fit approximations expressed by (2) of the data presented in figure 3.

Plot	Slope	Intercept	D [10^{13} cm^2/s]	β
1	-2.14 ± 0.02	3.12 ± 0.05	3.9 ± 0.9	4.28 ± 0.04
2	-1.70 ± 0.02	1.72 ± 0.04	9.8 ± 2.4	3.40 ± 0.04
3	-0.25 ± 0.04	-0.89 ± 0.07	-	-

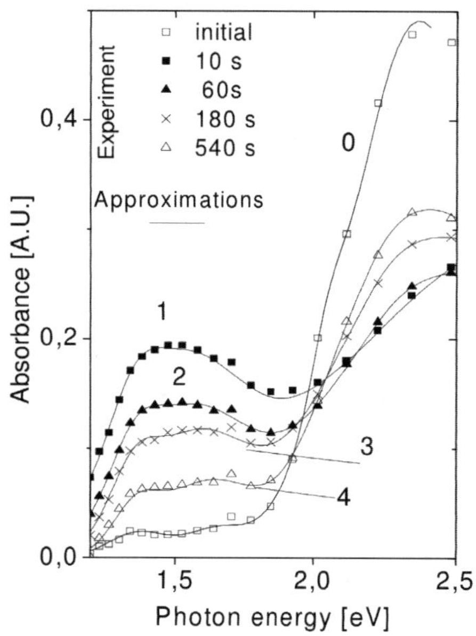

Figure 4. Absorption spectra (each 50 points are skipped). Solid lines are approximations calculated from (1) with parameters calculated by mean square fit and presented in table 2.

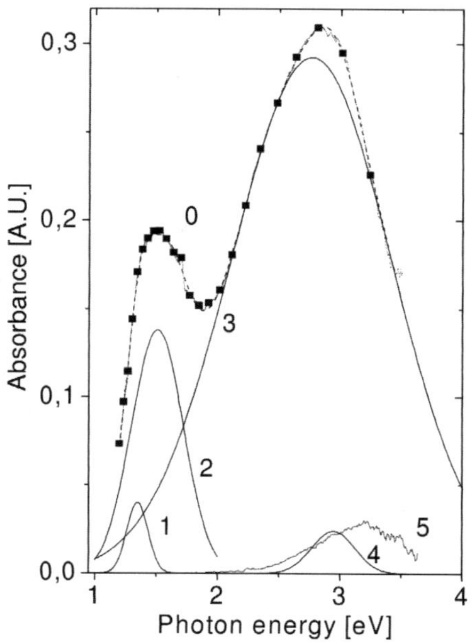

Figure 5. Absorption spectrum of the sample 10s after the extraction from the chamber with its Gaussian components. The numbers 1 – 4 correspond to the numbers of the components presented in table 2; 5 – iodine spectrum.

Thus the slope of the lg(A_1+A_2) vs lg(t/t_0) dependence is expected to be ~ – 0.5, whereas it turns out to be half this value (table 1, 3^d row). Hence the polaronic band intensity relates to the concentration of free charges contributing to the conductivity in a complicated manner. In general, the polaronic absorption is a bad approximation for quantitative estimates of the concentration of conductivity-related charges. Similarly, it is reasonable to assume that the charges responsible for conductivity belong to bipolaron species. Since the conductivity is determined by the hopping between states near the Fermi level, this assumption seems to be rather appropriate. Indeed Fermi level must be located near the middle of the gap. Bipolaronic states locate closer to the midgap than polaronic ones (figure 1). To be accurate the only tails of the bipolaronic density of states cross the Fermi level. Probably, to study this discrepancy (between time dependences of polaron band and conductivity, figure 3) in details, one should consider polaronic/boipolaronic concentration ratio, which is out of the scope of our paper. In addition, the conductivity-related states must have one more specific peculiarity. They belong to the percolative net of current paths. The dopant diffusion along these paths can be faster (presumably) than diffusion from the rest of the film.

The presence of dopant molecules has influence on electron transitions related to both inter- and intrachain interactions. Changes of interchain interactions are expected to be analyzed from x_3 component. Indeed, x_3 position is very sensitive to the doping level. Its value changes gradually from 2.3 eV after 60 s of dedoping to 2.19 eV for 540 s monotonously approaching its initial value of 2.05 eV. But it is not properly clear whether x_3 at rather high level of doping can be assigned only to interchain interaction. An alternative suggestion requires considering at least two types of neutral P3HT chain. Prior to doping, there is only one of them with maximum at 2.37 eV (x_4) which is consistent with the value for regioregular P3HT. Dopant molecules cause shortening of the polymer conjugation lebgth and the intrachain peak position blue shifts. Actually, both types of polymer chains (with reduced length and pristine) are present. Thus the values 2.3 eV and 2.19 eV derived for x_3 can be the result of superposition of interchain interaction transition and band of pristine P3HT chain.
The spectrum at 10 s is deformed in a so drastic way that seeking interchain interaction feature, which is screened by more powerful component with peak at 2.77 eV, is meaningless (figure 5).

Table 2. The best-fit values of the free parameters for the approximation of the experimental spectra presented as a sum of four Gaussians (1)

Parameter	Initial	10s	60s	180s	540s
A_1	0.006±0.0003	0.009±0.001	0.010±0.001	0.009±0.001	0.007±0.001
x_1 (eV)	1.37±0.01	1.35±0.001	1.350±0.001	1.35±0.001	1.36±0.001
w_1 (eV)	0.21±0.01	0.17±0.01	0.173±0.005	0.162±0.004	0.160±0.01
A_2	0.006±0.0004	0.073±0.001	0.048±0.001	0.042±0.001	0.028±0.004
x_2 (eV)	1.64±0.005	1.51±0.003	1.54±0.004	1.557±0.003	1.61±0.01
w_2 (eV)	0.21±0.01	0.423±0.003	0.376±0.006	0.359±0.008	0.34±0.02
A_3	0.005±0.0002	0.48±0.01	0.015±0.001	0.017±0.002	0.04±0.03
x_3 (eV)	2.05±0.001	2.77±0.001	2.30±0.002	2.25±0.004	2.19±0.005
w_3 (eV)	0.104±0.003	1.310±0.003	0.337±0.008	0.32±0.01	0.36±0.04
A_4	0.297±0.001	0.011±0.001	0.380±0.001	0.314±0.002	0.23±0.03
x_4 (eV)	2.37±0.001	2.95 ±0.004	2.71±0.002	2.56±0.01	2.51±0.04
w_4 (eV)	0.482±0.002	0.32±0.01	1.165±0.004	0.886±0.009	0.63±0.05

The spectrum measured at the highest level of doping (after 10s doping) shows many differences as compared with the other spectra (figure 5). Firstly, its x_3 and x_4 components are essentially blue shifted. As discussed above x_3 can be hardly attributed to the interchain interactions like in other spectra. It rather relates to processes similar to those associated with x_4 features on the other spectra. Instead the spectrum at the highest doping level contains its own x_4 component (figure 5, curve 4) which differs substantially from the other x_4 components. As this component is close to the spectrum of iodine (curve 5) in terms of intensity and spectral position, it is probably associated with iodine absorption. The red shift of its energetic position can be explained by a more extended localization of its exciton due to the conjugated polymer surrounding as compared to the situation when iodine molecules are separately adsorbed onto the glass surface. This behavior of the spectrum measured after a so short period of desorption is indicative of an initially very high level of dopant concentration along the whole thickness of the film. It can be assumed that just after the saturation was reached the doping level was even higher. The doping time was extremely short. How to explain such asymmetry in the doping-dedoping duration? We can propose two possible scenarios. They are complementary rather than alternative, since the first explains the enhancement of the doping process whereas the other concerns suppression of the dedoping.

First. The doping process is enhanced by the ambipolar diffusion. The dopant diffusivity is much lower than that of free carriers. Once dopant molecules enter the film, some number of positively charged mobile carriers is generated at the interface. They diffuse almost immediately to the rear surface due to a huge gradient of concentration. At the same time they involve dopant molecules. Actually, both species move with a velocity which can be characterized by some averaged diffusivity which is called ambipolar diffusivity [18]. A similar increase of the rate is observed for thermal oxidation of silicon at the beginning when the oxide layer is extremely thin [19]. The reverse process is quasi steady state while no gradients are formed or at least they are not so huge.

Second. **The dedoping process** is associated with the dissociation of the charge transfer complexes formed by the dopant molecules and conjugation chains. Electrostatic forces between polaron and negative iodine ion suppress disruption of this complex. This mechanism of dedoping is described in [9].

4. Conclusions

P3HT can be reversibly doped-dedoped with iodine vapour. The doping process is much faster than the dedoping one, which is controlled by in-situ measurements of the conductivity. The model proposed to describe the conductivity relaxation during the dedoping process is the basis of the experimental technique measuring the dopant diffusivity and the β exponent corresponding to the semi empirical power law between the conductivity and doping concentration ($\sigma \sim N_D^{\beta}$). The procedure consists in the determination of the best fitting parameters for alinear approximation of the experimental $\lg(\sigma(t)/\sigma(0))$ vs $\lg(t/t_0)$ dependence (t_0 =1s). The parameters found from (A.10) and (A.11) are β=4.28±0.04 and D=(3.9±0.9)·10^{-13} cm^2/s.

The absorption spectrum measured at different durations of the dedoping process (synchronously with the conductivity measurements) contains two polaronic features at 1.35 → 1.37 eV and 1.51→ 1.64 eV. The ranges indicate the extent of their variation with dedoping. These values are close to those measured by other authors. The analysis of the polaron band intensity demonstrates that this parameter, although correlated with the conductivity, cannot represent a determination of the free carrier concentration contributing into the conductivity. The slope of the polaronic band intensity – time dependence in a lg-lg plot is –0.25 instead of –0.5 as expected if the polaronic intensity proportionally follows the change of the free carrier concentration. It is assumed that this discrepancy is explained by the nature of the conductivity related charges which relate more to bipolaronic states.

These results are expected to contribute not only to a better understanding of the electronic processes in P3HT material but also to the general knowledge about conductivity formation in conjugated polymers. The volatility of the iodine dopant in P3HT makes this system to be a very suitable and flexible model object to study electronic processes responsible for the conductivity.

Acknowledgments

The authors gratefully acknowledge Dr Lytvyn P M and Dr Lytvyn O S for the film thickness measurements and Dr Miomandre F for providing ITO coated substrates.

5. Appendix (model of diffusion and its relation to the conductivity change while dedoping)

One-dimensional concentration distribution of a diffusing species is described by Fick's law:

$$\frac{\partial N_D(x,t)}{\partial t} = D \cdot \frac{\partial^2 N_D(x,t)}{\partial x^2}, \tag{A.1}$$

where D is the dopant diffusivity (the unknown quantity in our study); x is coordinate along the diffusion flow direction, t is time. The boundary conditions corresponding to the experiment are based on the following assumptions: i) initial distribution is uniform since the doping reaches the saturation value; ii) once the sample is extracted from the iodine chamber, the dopant concentration at the surface relaxes till zero very quickly (it is evidenced experimentally for halogen dopants [20]). Hence, the boundary conditions are:

$$\begin{cases} N_D(x,t) = N_{D0} = \underset{x}{const}, & t<0 \\ N_D(0,t) = 0, & t>0 \end{cases}. \tag{A.2}$$

The solution of (A.1), (A.2) is [21]:

$$N_D(x,t) = N_{D0} \cdot erf\left(\frac{x}{2\sqrt{D \cdot t}}\right), \tag{A.3}$$

where the abbreviation erf means the error function defined as:

$$erf(z) = \frac{2}{\sqrt{\pi}} \int_0^z e^{-z^2} dz \approx \frac{2}{\sqrt{\pi}} z. \tag{A.4}$$

The conductivity analysis based on the hopping model and evidenced experimentally demonstrated power law dependence of the conductivity on the dopant concentration [13]:

$$\sigma \sim N_D^{\beta}. \tag{A.5}$$

If N_D is a function of coordinate at any instant t, then contribution of some layer dx into the net conductivity can be presented as:

$$d\sigma(x,t) \sim \frac{1}{L} N^{\beta}(x,t) dx, \tag{A.6}$$

and the integral value reduced to the initial conductivity $\sigma_0 \sim N_{D0}^{\beta}$ is

$$\frac{\sigma}{\sigma_0} = \frac{2}{L\sqrt{\pi}} \int_0^L \left(\frac{x}{2\sqrt{D \cdot t}}\right)^{\beta} dx, \tag{A.7}$$

where approximate value of the error function is used (A.4). Taking the logarithms of both sides one yields:

$$\lg\frac{\sigma}{\sigma_0} = \lg\frac{L^{\beta}}{(\beta+1)\cdot(\pi D t)^{\beta/2}}, \tag{A.7}$$

which allows one to present the experimental data as linear dependence:

$$\lg\frac{\sigma}{\sigma_0} = a \cdot \left(\lg\frac{t}{t_0}\right) + b, \tag{A.8}$$

where slope and intercept are:

$$a = -\frac{\beta}{2},\ b = \lg\left(\frac{L^{\beta}}{(\beta+1)\cdot(\pi D t_0)^{\beta/2}}\right), \text{ respectively,} \tag{A.9}$$

and t_0 is taken as normalization factor to keep a unitless value under logarithm. We can take it as t_0=1s. Thus the best fit parameters of the linear approximation for the experimental data presented as lg($\sigma(t)/\sigma_0$) vs lg(t/t_0) dependence relates to the desired quantities:

$$\beta = -2a\,, \tag{A.10}$$

$$D = \frac{L^2}{\pi \cdot t_0} \cdot \left(\frac{10^{-b}}{\beta + 1} \right)^{2/\beta}. \tag{A.11}$$

The estimate of the relative error is derived by differentiating (A.11):

$$\delta D = 2 \cdot \delta L + \frac{2}{\beta} \cdot \ln 10 \cdot \left(\Delta b + b \cdot \delta\beta \right) + 2 \cdot \left(\frac{1}{1+\beta} + \frac{\ln(1+\beta)}{\beta} \right) \cdot \delta\beta\,, \tag{A.12}$$

where $\delta D = \dfrac{\Delta D}{D}$, $\delta\beta = \dfrac{\Delta\beta}{\beta}$, $\delta L = \dfrac{\Delta L}{L}$ are relative errors, $\Delta D, \Delta\beta, \Delta L, \Delta b$ are absolute errors.

References

[1] Chiang C K, Fincher C R, Park Y W, Heeger A J, Shirakawa H, Louis E J, Gau S C, MacDiarmid A G 1977 *Phys. Rev. Lett.* **39** 1098
[2] Sakai J, Kawano K, Yamanari T, Taima T, Yoshida Y, Fujii A, Ozaki M 2010 *Sol. En. Mater. & Sol. Cells* **94** 376
[3] Szmytkowski J 2010 *Semicond. Sci. Technol.* **25** 015009
[4] Cai W, Gong X, Cao Y 2010 *Sol. En. Mater. & Sol. Cells* **94** 114
[5] Chan C K, Richter L J, Dinardo B, Jaye C, Conrad B R, Ro H W, Germack D S, Fischer D A, Delongchamp D M, Gundlach D J 2010 *Appl. Phys. Lett.* **96** 133304
[6] Chao Y C, Chen C Y, Zan H W, Meng H F 2010 *J. Phys. D: Appl. Phys.* **43** 205101
[7] Chong L W, Chou Y N, Lee Y L, Wen T C, Guo T F 2009 *Org. Electron.* **10** 1141
[8] Schopf G and Koβmehl G 1997 Polythiophenes – Electrically Conductive Polymers *Advances in Polymer Science* vol 129 (Berlin: Springer)
[9] Koizumi H, Dougauchi H, Ichikawa T 2005 *J. Phys. Chem. B* **109** 15288
[10] Furukawa Yu 1997 Do Bipolarons Exist in Doped or Photoirradiated Conjugated Polymers ? – An Analysis Based on Studies of Model Compounds *Primary Photoexcitations in Conjugated Polymers: Molecular Exciton versus Semiconductor Band Model* ed N S Sariciftci (Singapour: World Scientific) chapter 17
[11] Chen T A and Rieke D 1993 *Synth. Met.* **60** 175
[12] Ambegaokar V, Halperin B I, Langer J S 1971 *Phys. Rev. B* **4** 2612
[13] Paasch G, Lindner T, Scheinert S 2002 *Synth. Met.* 97
[14] Winokur M J, Wamsley P, Moulton J, Smith P, Heeger A J 1991 *Macromolecules* **24** 3812
[15] Brown A R, De Leeuw D M, Havinga E E, Pomp A 1994 *Synth. Met.* **68** 65
[16] Brown J P, Thomas D S, Köhler, Wilson J S, Kim J S, Ramsdale C M, Sirringhaus H, Friend R H, *Phys. Rev. B* **67** 064203
[17] Muller S and Kamins T I 1989 *Device Electronics for Integrated Circuits* (New York: John Wiley & Sons) chapter 6
[18] Bonch-Bruyevich V L and Kalashnikov S G 1977 *Physics of Semiconductors* (Moscow: Nauka, in Russian) chapter 7
[19] Brodie I and Muray J J 1982 *The Physics of Microfabrication* (New York and London: Plenum Press) chapter 5
[20] Wu C G, Yeh Y R, Chien L N 2000 *Polymer* **41** 5839
[21] Raychenko A I *Mathematic Theory of Diffusion in Applications* (Kiev: Naukova Dumka, in Russian) chapter 2

The reexamination of thermal expansion of ferromagnetic superconductors and the pressure differential of its superconducting transition temperature-possible application to UGe_2

Rikio Konno, Nobukuni Hatayama

Kinki University Technical College, 2800 Arima-cho, Kumano-shi, Mie 519-4395, Japan

E-mail: r-konno@ktc.ac.jp

Abstract. The temperature dependence of thermal expansion of ferromagnetic superconductors below the superconducting transition temperature $T_{sc\uparrow}$ of a majority spin conduction band is reexamined. In the previous study [to be published in J. M. Phys. B] the volume differential of the kinetic energy of conduction electrons is constant. However, in this study the volume differential of the kinetic energy of conduction electrons is inconstant. The superconducting gap of the majority spin conduction band used in this study has a line node. It is appropriate to UGe_2. The pressure differential of its superconducting transition temperature is also investigated. We find that the thermal expansion coefficient has the divergence at the superconducting transition temperature. The thermodynamic Grüneisen's relation is satisfied.

1. Introduction

Many researchers have paid attention to the ferromagnnetic superconductors [1, 2] again since ferromagnetic superconductors UGe_2 [3], UCoGe [4] and URhGe [5, 6] was discovered. Recently, Hatayama and Konno [7, 8] investigated the temperature dependence of thermal expansion based on the free energy derived by Linder and Sudbo [9, 10] by Takahashi's method [11]. However, they assumed that the volume differential of the kinetic energy of conduction electrons is constant. In this study, thermal expansion behaviour is reexamined by our keeping the volume differential of the kinetic energy of conduction electrons inconstant.

Secondly, the pressure differential of the superconducting transition temperature and that of the Curie temperature are studied. Pfleiderer et al. investigated critical behaviour at the transition as a function of hydrostatic pressure [12]. Recently, Shopova and Uzunov [13] looked into the temperature-pressure phase diagram based on the Landau expansion of the free energy. Aso et al. estimated the pressure dependence of Stoner gap based on the Stoner model from the neutron intensities [14]. It seems to us that they do not satisfy the thermodynamic Grüneisen's relation. The pressure differential of the superconducting transition temperature and that of the Curie temperature based on the free energy derived by Linder and Subo [10] is unanswered. We will obtain the analytical expression of the pressure differential of the superconducting transition temperature of the majority spin conduction band and that of the Curie temperature by the mean field approximation. The thermodynamic Grüneisen's relation will be satisfied in this study.

This paper is organised as follows. In the next section thermal expansion of the ferromagnetic superconductors will be derived. In section 3, the numerical results will be provided. In section 4, the pressure differential of the superconducting transition temperature and that of the Curie temperature will be derived. Section 5 will be devoted to conclusions.

2. The derivation of thermal expansion of ferromagnetic superconductors

We begin with the following free energy [8, 10]:

$$F_{coexist}/N = F_0/N + F_T/N, \tag{1}$$

$$F_0/N = \frac{IM^2}{2} + \frac{1}{2\pi}\int_0^{2\pi} d\theta \sum_\sigma \frac{\Delta_\sigma^2(\theta)}{2g} - \frac{1}{2\pi}\sum_\sigma \int_0^{2\pi} d\theta \int_0^{E_F} d\varepsilon N(\varepsilon)\frac{E_\sigma(\varepsilon,\theta)}{2}, \tag{2}$$

$$F_T/N = -\frac{T}{2\pi}\sum_\sigma \int_0^{2\pi} d\theta \int_0^{\infty} d\varepsilon N(\varepsilon)\ln(1+e^{-E_\sigma(\varepsilon,\theta)/T}) \tag{3}$$

where F_0 is the ground state energy and F_T is the thermal part of the free energy. E_F is the Fermi energy. N is the number of magnetic atoms, $N(\epsilon)$ is the density of state, and $\Delta_\sigma(\theta)$ is the superconducting gap of the spin σ conduction band. θ is an azimuthal angle in the k_{Fx}-k_{Fy} plane and k_F is the Fermi wave number. g is the effective attractive pairing coupling constant. I is the on-site Coulomb coupling constant. M is the magnetisation. ϵ is the kinetic energy of electrons. $E_\sigma(\varepsilon,\theta)$ is given by

$$E_\sigma(\varepsilon,\theta) = \sqrt{(\varepsilon - \sigma IM - E_F)^2 + \Delta_\sigma^2(\theta)}. \tag{4}$$

Thermal expansion of the ferromagnetic superconductors is obtained as follows:

$$\omega = -K\frac{\partial F_{coexist}}{\partial V} \tag{5}$$

where K is the compressibility. The thermal expansion is

$$\omega/(NE_F) = \omega_0/(NE_F) + \omega_T/(NE_F). \tag{6}$$

ω_0 comes from the part of the ground state energy. ω_T originates from the thermal part of the free energy. ω_T is much smaller than ω_T because of $T_{sc} << T_F$ where T_F is the Fermi temperature. ω_T is negligible. By our noting $\frac{\partial\epsilon}{\partial V} = \frac{\partial \ln t}{\partial V}\epsilon$ where t is a transfer integral of electrons, ω_0 is

$$\begin{aligned}\omega_0 = -K\{&\frac{1}{2}\frac{E_F}{I}(\frac{\partial \ln I}{\partial V})\tilde{M}^2 + \frac{1}{2}E_F\frac{\partial}{\partial g}(\frac{1}{g})\sum_\sigma \frac{1}{2\pi}\int_0^{2\pi} d\theta\tilde{\Delta}_\sigma^2(\theta)\\ &-N(0)E_F[\frac{\partial \ln N(0)}{\partial V}\sum_\sigma \frac{1}{2\pi}\int_0^{2\pi} d\theta \int_0^1 dx\frac{\tilde{E}_\sigma^2(x,\theta)}{2}\\ &+\sum_\sigma \frac{\partial \ln t}{\partial V}\int_0^{2\pi}\frac{d\theta}{2\pi}\int_0^1 dx x\frac{x-1-\sigma\tilde{M}}{2\sqrt{(x-\sigma\tilde{M}-1)^2+\tilde{\Delta}_\sigma^2(\theta)}}\\ &+\frac{1}{2}\sum_\sigma A_\sigma\frac{1}{2\pi}\int_0^{2\pi} d\theta(\sqrt{(-\sigma\tilde{M})^2} - \sqrt{(-\sigma\tilde{M}-1)^2+\tilde{\Delta}_\sigma^2(\theta)})]\}\end{aligned} \tag{7}$$

with

$$A_\sigma = -\sigma \frac{\partial \ln I}{\partial V} \tilde{M} - \frac{\partial \ln E_F}{\partial V} \tag{8}$$

where $\tilde{M} = IM/E_F$, $N(0)$ is the density of states at the Fermi energy, $\tilde{\Delta}_\sigma(\theta) = \Delta_\sigma(\theta)/E_F$, and $x = \epsilon/E_F$. $\tilde{E}_\sigma(x,\theta)$ is

$$\tilde{E}_\sigma(x,\theta) = \sqrt{(x - \sigma\tilde{M} - 1)^2 + \Delta_\sigma^2(\theta)}. \tag{9}$$

The corresponding thermal expansion coefficient is given by

$$\alpha = \frac{\partial \omega}{\partial T} \tag{10}$$

where $\tilde{M} = IM/E_F$ and E_F is the Fermi energy.

We assume that the Curie temperature T_C is much higher than the superconducting transition temperature and that T_C is much lower than the Fermi temperature T_F throughout this paper. Correspondingly, the magnetisation is constant. This assumption is valid in UGe_2.

Next, we mention the superconducting gap in order to study the thermal expansion and its coefficient. Harada et al. displayed that the superconducting gap of the up-spin conduction band with the line node is in UGe_2 experimentally [15]. Therefore, we consider the following superconducting gap:

$$\Delta_\sigma(\theta) = \begin{cases} \Delta_0 \cos\theta (\sigma = \uparrow) \\ 0 (\sigma = \downarrow). \end{cases} \tag{11}$$

The superconducting order parameter at $T = 0$[K] is

$$\Delta_0(0) = 2.426 E_0 \exp(-1/c\sqrt{1 + \tilde{M}(0)}). \tag{12}$$

E_0 is the cutoff energy. E_0/E_F is set to 0.01. The weak-coupling constant $c = gN(0)/2$ is set to 0.2. The temperature dependence of the superconducting order parameter $\Delta_0(T)$ is obtained

$$\Delta_0(T) = \Delta_0(0) \tanh(1.70\sqrt{T_{sc}/T - 1}) \tag{13}$$

with

$$T_{sc} = 1.134 E_0 \exp(-2/c\sqrt{1 + \tilde{M}(\tilde{T}_{sc})}) \tag{14}$$

where T_{sc} is the superconducting transition temperature. In the next section the numerical results will be presented.

3. Results

The temperature dependence of the thermal expansion coefficient of the ferromagnetic superconductors is investigated with Eqs.(6), (7), (8), (9) and (10). Fig.1 shows the temperature dependence of the thermal expansion coefficient. The divergence of the thermal expansion coefficient appears at T_{sc}. The thermal expansion coefficient behaves in the similar way to the previous study [8] even though we keep the volume dependence of the kinetic energy of the conduction electrons inconstant. In addition, the thermodynamic Grüneisen's relation between the temperature dependence of the thermal expansion coefficient and that of the magnetic specific heat is satisfied because we use the same free energy as the free energy when the expression of the specific heat is derived.

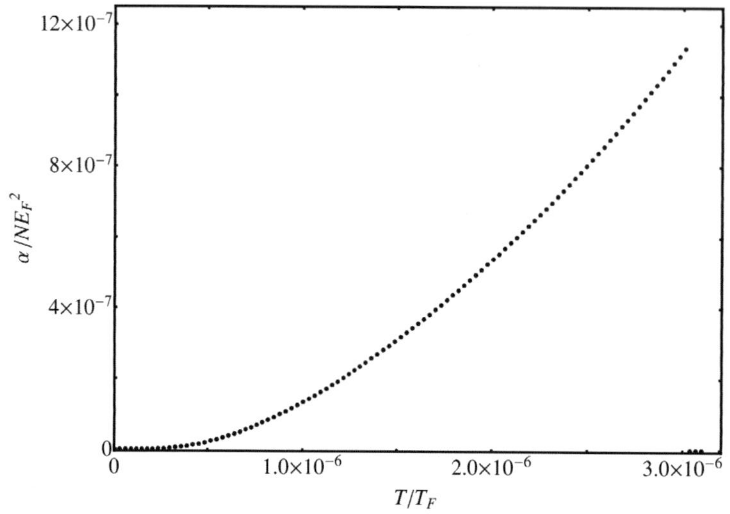

Figure 1. The reduced temperature T/T_F dependence of the thermal expansion coefficient of ferromagnetic superconductors when $E_F/I = 0.1$, $\frac{\partial \ln I}{\partial V} = 0.1$, $E_F\frac{\partial}{\partial V}(\frac{1}{g}) = 0.1$, $N(0)E_F\frac{\partial t}{\partial V} = 0.1$, $N(0)E_F = 0.1$, and $N(0)E_F\frac{\partial \ln N(0)}{\partial V} = 0.1$.

4. The pressure differential of the superconducting transition temperature and that of the Curie temperature

The pressure differential of the superconducting transition temperature and that of the Curie temperature are investigated in this section. From Eq.(14), we get the pressure differential of the superconducting transition temperature

$$\frac{\partial T_{sc}}{\partial P} = T_{sc}[\frac{\partial}{\partial P}(\ln 1.134E_0) + \frac{2}{c^2\sqrt{1+\tilde{M}(T_{sc})}}\frac{\partial c}{\partial P} + \frac{1}{c(1+\tilde{M}(T_{sc}))^{3/2}}\frac{\partial \tilde{M}(T_{sc})}{\partial P}]. \quad (15)$$

The pressure differential of the cutoff energy E_0 and that of the magnetisation will be estimated if the pressure differential of the superconducting transition temperature is observed.

We proceed to the pressure differential of the Curie temperature. The magnetisation is

$$M = IN(0)M + TN(0)\ln\frac{\cosh\frac{E_F+IM}{2T}}{\cosh\frac{E_F-IM}{2T}}. \quad (16)$$

In $T \to T_C$, this equation is expanded around the small M. After that, M is equal to zero at $T = T_C$. We obtain the Curie temperature

$$T_C = \frac{E_F}{\ln(2IN(0)-1)}. \quad (17)$$

T_C differentiate about the pressure P. We obtain the pressure differential of T_C

$$\frac{\partial T_C}{\partial P} = \frac{\frac{\partial E_F}{\partial P}}{\ln(2IN(0)-1)} - \frac{2E_F\frac{\partial(IN(0))}{\partial P}}{(2IN(0)-1)(\ln(2IN(0)-1))^2}. \quad (18)$$

If the pressure differential of T_C between experiments and the theory is analysed, the pressure differential of E_F and that of $IN(0)$ will be estimated. Eq.(16) is expanded around small $e^{-(E_F-IM)/T}$ and $e^{-(E_F+IM)/T}$ when $T_{sc} < T << T_C << T_F$. We obtain

$$M = 2IN(0)M - TN(0)e^{-E_F/T}(e^{IM/T} - e^{-IM/T}). \tag{19}$$

The magnetisation is exponential in the similar way to the result of Aso et al. [14].

5. Conclusions

We have reexamined the temperature dependence of thermal expansion of the ferromagnetic superconductors. The superconducting gap of the up-spin conduction band with the line node are assumed. We find that the divergence of the thermal expansion coefficient exists at the superconducting transition temperature like the previous study. The thermodynamic Grüneisen's relation between the temperature dependence of the thermal expansion coefficient and that of the magnetic specific heat is satisfied.

We have obtained both the analytical expressions of the pressure differential of the superconducting transition temperature and the Curie temperature. If they are compared with experimental data, the pressure differential of E_F and that of $IN(0)$ will be determined experimentally. In $T_{sc} < T << T_C << T_F$, the magnetisation is exponential. This result is consistent with that of Aso et al. [14].

5.1. Acknowledgments

We would like to thank Y. Takahashi, O. Stockert, M. Kanno , and M. Nakamori for stimulating conversations. One of the authors (R. K.) would like to also thank K. Grube, H. v. Löhneysen, F. Steglich, A. de Visser, M. B. Maple for stimulating conversations. This work is supported by the Kinki University Technical College Research Fund.

References

[1] Tachiki M and Maekawa S 1984 *Phys. Rev.* B **29** 2497
[2] Matsumoto H, Teshima R, Umezawa H, Tachiki M 1983 *Phys. Rev.* B **27** 158 and references therein.
[3] Saxena S S, Agarwal P, Ahilan K, Grosch F M, Haselwimmer R K W, Steiner M J, Pugh E, Walker I R, Julian S R, Monthoux P, Lonzarich G G, Huxley A, Sheikin I, Braithwaite D, and Flouque J 2000 *Nature* **406** 587
[4] Huy N T, Gasparini A, Nijs D E de, Huang Y K, Klaasse J C P, Gortenmulder T, Visser A de, Hamann A, Görlach T, and Löhneysen v H 2007 *Phys. Rev. Lett* **99** 067006, and references therein.
[5] Aoki D, Huxley A, Ressouche E, Braithwaite D, Flouque J, Brison J P, Lhotel E, Paulsen C, 2001 *Nature* **413**, 613
[6] Levy F, Sheikin I, Grenier B, Marcenat C, and Huxley A 2009 *J. Phys.:Condens. Matter* **21** 164211, and references therein.
[7] Hatayama N and Konno R J. Phys. Conf. Ser. **200** 012056 (2010)
[8] Hatayama N and Konno R to be published in *J.Mod.Phys.B*
[9] Linder J and Sudbo 2007 *Phys. Rev.* B **76** 054511
[10] Linder J, Sperstad I B, Nevidomskyy A H, Cuoco M, Sudbo A 2008 *Phys. Rev.* B **77**, 184511
[11] Takahashi Y and Nakano H 2006 *J. Phys. Condens. Matter* **18**, 521
[12] Pfleiderer C, MeMullan G J, Lonzarich G G 1994 *Physica* **B 199-200**, 634
[13] Shopova D V and Uzunov D I 2009 *Phys. Rev.* B **79** 064501
[14] Aso N, Motoyama G, Uwatoko Y, Ban S, Nakamura S, Nishioka T, Homma Y, Shiokawa Y, Hirota K, and N. K. Sato 2006 *Phys. Rev.* B **29** 054512
[15] Harada A, Kawasaki S, Mukuda H, Kitaoka Y, Haga Y, Yamamoto E, Itoh K M, Haller E E, Harima H 2007 *Phys. Rev.* B **75** 140502(R)

Journal of Physics: Conference Series **286** (2011) 012012 doi:10.1088/1742-6596/286/1/012012

Strong-coupling superconductivity beyond BCS and the key pairing interaction in cuprate superconductors

A.S. Alexandrov

Department of Physics, Loughborough University, Loughborough LE11 3TU, United Kingdom

E-mail: a.s.alexandrov@lboro.ac.uk

Abstract. It has been now over 20 years since the discovery of the first high temperature superconductor by Georg Bednorz and Alex Müller in 1986 and yet, despite intensive effort, no universally accepted theory exists about the origin of high-temperature superconductivity. A controversial issue on whether the electron-phonon interaction (EPI) is crucial for high-temperature superconductivity or weak and inessential has been one of the most challenging problems of contemporary condensed matter physics. I briefly review our recent theoretical results, which in conjunction with a great number of experimental observations including isotope effects, angle-resolved photoemission (ARPES), pump-probe and tunnelling spectroscopies, normal state diamagnetism and magnetic quantum oscillations provide the definite answer to this fundamental question. The true origin of high-temperature superconductivity is found in a significant finite-range Fröhlich EPI of *nonadibatic* polaronic carriers which is beyond the conventional BCS-Migdal-Eliashberg approximation.

After we have shown [1]— unexpectedly for many researchers— that the BCS-Migdal-Eliashberg (BCS-ME) theory breaks down already at the EPI coupling $\lambda > 0.5$ for any adiabatic ratio $\hbar\omega_0/E_F$, the multi-polaron physics has gained particular attention [2]. The parameter $\lambda\hbar\omega_0/E_F$, which is supposed to be small in the BCS-ME theory becomes in fact large at $\lambda > 0.5$ since the electron band narrows and the Fermi energy, E_F turns out below the characteristic phonon energy, $\hbar\omega_0$. Nevertheless, as noted in the unbiased comment by Jorge Hirsch [3], in order to explain the increasingly higher T_cs found in supposedly conventional materials, values of the electron-phonon coupling constant λ larger than 1 have been used in the conventional BCS-ME formalism. This formalism completely ignores the polaronic collapse of the bandwidth, but regrettably continues to be used by some researchers (irrespective of whether λ is small or large) failing to explain high-T_c. As a result, many researchers maintain that a repulsive electron-electron interaction is responsible for pairing providing high transition temperatures, T_c, without phonons in high-temperature superconductors.

Properly extending the BCS theory towards the strong interaction between electrons and ion vibrations, a *Bose liquid* of tightly bound electron pairs surrounded by the lattice deformation (i.e. of *small bipolarons*) was predicted [4]. Further prediction was that high temperature superconductivity should exist in the crossover region of the electron-phonon interaction strength from the BCS-like to bipolaronic superconductivity [1], Fig.1. The strong enhancement of T_c in the crossover region from BCS-like polaronic to BEC-like bipolaronic superconductivity is entirely due to a sharp increase of the density of states in a narrow polaronic band [1], which is

Journal of Physics: Conference Series **286** (2011) 012012 doi:10.1088/1742-6596/286/1/012012

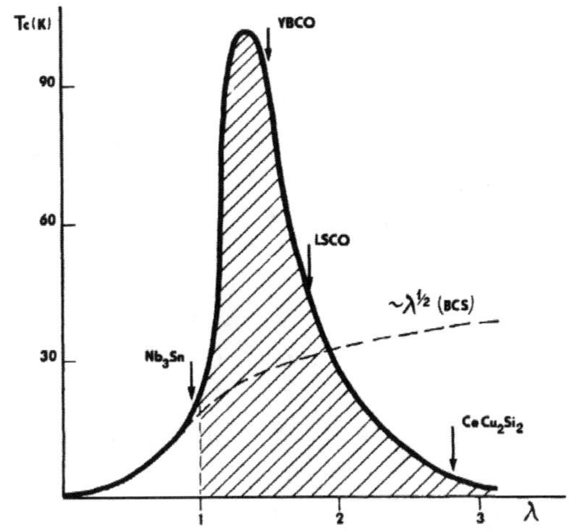

Figure 1. Breakdown of the BCS-ME approximation due to the polaron bandwidth collapse at some critical coupling λ of the order of 1. The dependence of T_c versus λ is shown in the polaron-bipolaron crossover region compared with the BCS-ME dependence (dotted line). (Reproduced from Alexandrov A S 1988 *Phys. Rev. B* **38** 925, (©) American Physical Society, 1988.)

missing in the so-called *negative-U* Hubbard model.

We have recently quantified the EPI strength, the phonon-induced electron-electron attraction, and the carrier mass renormalization in layered superconductors at different doping using a continuum approximation for the renormalized carrier energy spectrum and the RPA dielectric response function [5]. If, for instance we start with a parent insulator as La_2CuO_4, the magnitude of the Fröhlich EPI is unambiguously estimated using the static, ϵ_s and high-frequency, ϵ_∞ dielectric constants. To assess its strength, one can apply an expression for the polaron binding energy (polaronic level shift) E_p, which depends only on the measured ϵ_s and ϵ_∞, $E_p = (e^2/2\epsilon_0\kappa)\int_{BZ} d^3q/(2\pi)^3 q^2$. Here, the integration goes over the Brillouin zone (BZ), $\epsilon_0 \approx 8.85 \times 10^{-12}$ F/m is the vacuum permittivity, and $\kappa = \epsilon_s\epsilon_\infty/(\epsilon_s - \epsilon_\infty)$. In the parent insulator, the Fröhlich interaction alone provides the binding energy of two holes, $2E_p$, *an order of magnitude* larger than any magnetic interaction ($E_p = 0.647$ eV in La_2CuO_4).

Recent observations of the quantum magnetic oscillations in some cuprate superconductors [6] are opening up a possibility for a quantitative assessment of EPI in these and related doped ionic lattices with the quasi two-dimensional (2D) carrier energy spectrum. The oscillations revealed almost cylindrical Fermi surfaces, enhanced effective masses of carriers (ranging from $2m_e$ to $6m_e$) and the astonishingly low Fermi energy, E_F, which is well below 40 meV in Y-Ba-Cu-O. Such low Fermi energies make the Migdal-Eliashberg (ME) adiabatic approach to EPI inapplicable in these compounds since the characteristic oxygen vibration energy (about $\hbar\omega_0 = 80$ meV) turns out larger than the carrier kinetic energy. Since carriers in cuprates are in the non-adiabatic (underdoped), $E_F < \hbar\omega_0$, or near-adiabatic (overdoped) regimes, their energy spectrum renormalized by EPI and the polaron-polaron interactions can be found with the familiar small-polaron canonical transformation at *any coupling* λ [7].

With doping the attraction and the polaron mass drop [5]. Nevertheless, on-site and inter-site attractions induced by EPI remain several times larger than the superexchange (magnetic) interaction J (about 100 meV) at any doping since the non-adiabatic carriers cannot fully screen high-frequency electric fields. The polaron mass [5] agrees quite well with the experimental masses measured in magnetic quantum oscillations experiments. Hence the Fröhlich EPI with high-frequency optical phonons turns out to be the key pairing interaction in underdoped cuprates and remains the essential player at overdoping. What is more surprising is that EPI is clearly beyond the BCS-ME approximation since its magnitude is larger than or comparable with the Fermi energy and the carriers are in the non-adiabatic or near-adiabatic regimes. Since EPI is not local in the nonadibatic electron system with poor screening it can provide the d-wave symmetry of the pairing state [8].

There are other independent pieces of evidence in favor of (bi)polarons and 3D BEC in cuprate superconductors. Most compelling evidence for (bi)polaronic carries in cuprate superconductors is the substantial isotope effect on the carrier mass [9]. High resolution ARPES [10] provides another piece of evidence for a strong electron-phonon interaction (EPI) in cuprates and related oxides apparently with c-axis-polarised optical phonons. These as well as recent pump-probe experiments [11] unambiguously show that the Fröhlich EPI is important in those highly polarizable ionic lattices.

Magnetotransport data strongly support preformed bosons in cuprates. In particular, many high-magnetic-field studies revealed a non-BCS upward curvature of the upper critical field $H_{c2}(T)$, predicted for the Bose-Einstein condensation of charged bosons in the magnetic field [12]. Nonlinear normal-state diamagnetism of quite a few hole-doped cuprates has a profile characteristic of normal state real-space composed bosons (i.e. bipolarons) [13], rather than "preformed" Cooper pairs, vortex liquid and the Kosterlitz-Thouless phase transition hypothesized by some authors.

Single polarons, *localised* within an impurity band-tail, coexist with bipolarons in charge-transfer doped Mott-Hubbard insulators. They account for sharp "quasi-particle" peaks near $(\pi/2, \pi/2)$ of the Brillouin zone and high-energy "waterfall" effects observed with ARPES in cuprate superconductors [14]. This "band-tail" model also accounts for two energy scales (superconducting and pseudo-gaps) in ARPES [14] and in the extrinsic and intrinsic tunnelling [15].

All these and many other observations point to a crossover from the bipolaronic to polaronic superconductivity [1] in high-temperature superconductors with doping.

References

[1] Alexandrov A S 1983 *Zh. Fiz. Khim.* **57** 273 [1983 *Russ. J. Phys. Chem.* **57** 167]
[2] Alexandrov A S and Devreese J T 2009 Advances in Polaron Physics (Heidelberg-Dordrecht-London-New York: Springer)
[3] Hirsh J E 2009 *Phys. Scr.* **80** 035702
[4] Alexandrov A S and Ranninger J 1981 *Phys. Rev. B* **23** 1796; 1981 *Phys. Rev. B* **24** 1164
[5] Alexandrov A S and Bratkovsky A M 2010 *Phys. Rev. Lett.* **105** 226408
[6] Doiron-Leyraud N *et al* 2007 *Nature (London)* **447**565
[7] Alexandrov A S and Kornilovitch P E 1999 *Phys. Rev. Lett.* **82** 807
[8] Alexandrov A S 2008 *Phys. Rev. B* **77** 094502
[9] Alexandrov A S and Zhao G M 2009 *Phys. Rev. B* **80** 136501
[10] Meevasana W *et al* 2010 *New J. Phys.* **12** 023004
[11] Gadermaier C *et al* 2010 *Phys. Rev. Lett.* **105** 257001
[12] Alexandrov A S 1993 *Phys. Rev. B* **48** 10571
[13] Alexandrov A S 2006 *Phys. Rev. Lett.* **96** 147003; 2010 *J. Phys.: Condens. Matter* **22** 426004
[14] Alexandrov A S and Reynolds K *Phys. Rev. B* **76** 132506
[15] Alexandrov A S and Beanland J 2010 *Phys. Rev. Lett.* **104** 026401

Quantum Detection Applications of NanoSQUIDs fabricated by Focussed Ion Beam

L Hao

National Physical Laboratory, Teddington, TW11 0LW, UK

E-mail: ling.hao@npl.co.uk

Abstract. The increasing demands of quantum metrology and nanoscience are driving the requirements for single particle detection capability across a wide variety of physics applications, including QIP, single photon detection, NEMS, nanomagnetism and spintronics. NanoSQUIDs represent a new manifestation of an old but exciting superconducting technology which addresses some of these requirements. We will describe a straightforward approach to fabricating Nb microbridge weak-links which may be used in nanoscale SQUIDs. These devices have demonstrated very low noise, even at operating temperatures above 4.2 K. We will also describe the performance of prototype devices made using this technique for energy resolving single photon detection, nanoparticle spin detection and NEMS resonator readout.

1. Introduction

SQUIDs (Superconducting QUantum Interference Devices) are macroscopic quantum objects which are capable of detecting and measuring a wide range of physical parameters with unequalled sensitivity. This extends from magnetic flux (the natural most straightforward quantity that a SQUID responds to directly) through spatial displacement to photon detection and mass detection, for example [1]. Interestingly the SQUID devices which were first developed and that demonstrated an approach to the standard quantum limit were of relatively macroscopic size (typically a millimetre or so in linear dimension). It is really only rather recently that it has been appreciated that SQUID size may be radically reduced towards the nanoscale and that such devices will not only retain exceptional sensitivity but will, through their size, find applications in a whole new range of detection and measurement areas [2,3,4,5,6].

In this paper we describe a particularly simple fabrication method for SQUIDs which we have applied to producing nanoscale SQUIDs, based on focussed ion beam fabrication. We further illustrate that the exceptional measurement sensitivity and low noise performance which has long been known from macroscopic devices extends down, unabated, to the nanoscale. We then go on to describe some prototype applications of these simple microbridge based nanoSQUIDs in the areas of single photon spectroscopy, nanomagnetic particle detection and nanoelectromechanical systems (NEMS) resonator readout. We close with some speculation on further applications and some limits to performance.

2. Fabrication

The fabrication process combines conventional optical lithography with an ion etch step followed by a subsequent focussed ion beam (FIB) milling step. The process both simplifies the fabrication of nanoSQUIDs and allows the loop diameters to be reduced down to 200 nm, with the incorporation of microbridge junctions of the order of 60 nm width in a single lithography step on a bilayer Nb/W film. A Nb thin film (~200 nm thick) is first deposited by sputtering and patterned into a series of lines around 1–2 μm in width by conventional photolithography. An important part of our process involves the *in situ* deposition of a tungsten layer (~150 nm thick) by e-beam over the entire region to be patterned, immediately before the FIB milling. This use of a normal-metal layer affects the subsequent properties of the SQUID in three crucial aspects, as demonstrated by Lam and Tilbrook [7]: it initially provides a protective barrier to minimize ion beam damage to the Nb bridge, thus preserving its superconducting property; in the operation of the SQUID, the tungsten acts as both a thermal shunt, minimizing hot spot generation and the resulting thermal hysteresis, and as a resistive shunt, which limits the McCumber parameter:

$$\beta_c = \frac{2\pi I_c R^2 C}{\Phi_0} \tag{1}$$

thus removing the effect of electrical hysteresis. Tungsten (W) was used as the overcoating material rather than, for example, Au [8] because of its resistivity, which is both high and temperature independent. These SQUIDs, which we estimate to have an inductance of ~0.6 pH, exhibit non-hysteretic current- voltage characteristics with critical currents I_c of order of 50–300 μA in the temperature range of 5–9 K, with regular periodic I_c-B curves in magnetic fields up to 10 mT. More details of the fabrication process are given by Hao *et al.* [9, 10].

3. Noise Performance of Microbridge nanoSQUIDs

Figure 1 shows the I-V characteristic of one such FIB fabricated nanoSQUID, an SEM image of which is shown in the inset. Note that the characteristic accurately follows the predictions of the resistively shunted junction model (see for example [11]), with parameters I_c = 120 μA, R = 5.5 Ω.

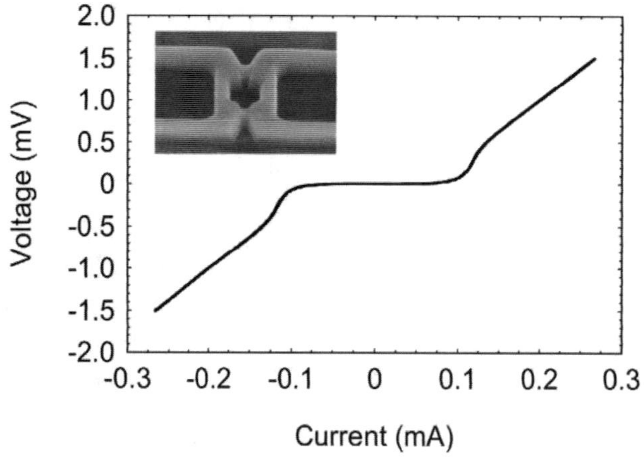

Figure 1. The I-V characteristic of a FIB-fabricated nanoSQUID at 4.65K, an SEM image of nanoSQUID is shown in the inset and the internal loop diameter is approximately 300 nm.

The thermal rounding parameter Γ is given by

$$\Gamma = \frac{2\pi k_B T_{eff}}{\Phi_0 I_c} \tag{2}$$

and, treating the effective temperature T_{eff} as a fitting parameter the result is $T_{eff} \sim$ 55K, well above the operating temperature, suggesting that higher temperature black-body radiation or other artificial electromagnetic interference is incident on the SQUID. Improved filtering of the bias leads and shielding may be expected to reduce this effective temperature. The absence of observed hysteresis predicts that $\beta_c < 1$, implying that the junction capacitance $C < 0.1$pF. The effective flux noise in this SQUID (that is the noise observed at the output, converted to an effective flux noise at the input) is shown in figure 2. This was measured at PTB (Berlin) using a series SQUID array (SSA) [12] as a cryogenic preamplifier. This work has been reported in more detail elsewhere [4]. Note that in the white noise regime around 1kHz the flux noise $\delta\Phi \sim 2\text{x}10^{-7}\Phi_0/\text{Hz}^{1/2}$. A classical analysis of thermal noise arising in the shunt resistors of the Josephson junctions implies that the limiting flux noise Φ_n is set by the expression

$$\Phi_n^2 \,/\, 2L = \varepsilon = 16k_BT(\,LC\,)^{1/2} \qquad (3)$$

where ε is the minimum detectable energy in unit bandwidth [11]. The thermal noise predicted by equation (3) for this device is about $6.6\ \text{x}10^{-9}\Phi_0/\text{Hz}^{1/2}$ at 6.8K, which is about 30 times lower than the observed white noise value.

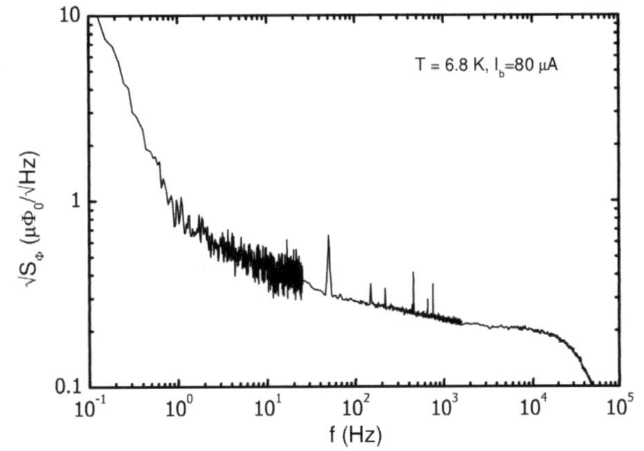

Figure 2. Flux noise spectral density in a nanoSQUID, as a function of frequency from 10mHz to 100kHz.

4. Inductive Superconducting Transition Edge Detection (ISTED)

SQUIDs using the type of microbridge junctions described above have been developed as a novel form of sensitive bolometer for energy resolving single photon measurements. The basic principle of the ISTED has some common features with superconducting transition edge detectors (TEDs). The device consists of a photon absorber which is inductively coupled to a SQUID (the absorber generally consists of a thin film of superconductor with a transition temperature T_c somewhat less than that of the SQUID) (see figure 3). When this thin film absorbs an incident photon an appreciable temperature rise is produced, due to the low heat capacity of the absorber. As a result pairs dissociate, reducing the pair density n_s and thus resulting in an increase in the absorber superconducting penetration depth $\lambda(T)$. This in turn results in an increased inductance of the SQUID and a resulting change in the voltage read out from the SQUID. Unlike a TED detector the SQUID and absorber both remain below their T_c and consequently there is no added Johnson noise associated with the detection process.

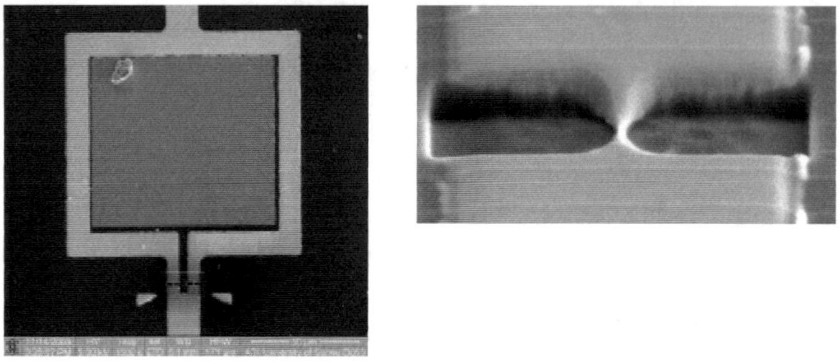

(a) (b)

Figure 3. (a) SEM image of ISTED sensor with 50x50 μm^2. The FIB-bed junctions are shown close to the arrow marks near the bottom of the image. (b) A single junction is shown enlarged and the width of junction is about 80nm.

The ISTED process has been analysed in some detail elsewhere [13]. Here we present new data demonstrating single photon energy resolution using an ISTED. The pulse height output from a SQUID surrounding a 50 μm absorber, operated at 8.15 K was analysed for 3 different situations. First it was irradiated with a weak source from a 633 nm laser, coupled from room temperature via an optical fibre. Next the noise was measured with the laser off but the fibre still coupled to the ISTED. Finally the warm end of the fibre was cooled to 77 K, with the laser still not connected. The results for the pulse height histograms in the three situations are shown in figure 4. The energy resolution is estimated from this data to be ~0.2 eV, even at this high operating temperature of 8.15 K. The noise equivalent power has been estimated as $1x10^{-17}$ $W/Hz^{1/2}$. It has not proved possible to accurately measure the quantum efficiency of the device at present but it has been estimated to be only around 0.3%. This low figure is partly due to the relatively high optical reflectance and transmission of the Nb thin film absorber, which is only ~ 60-80 nm thick.

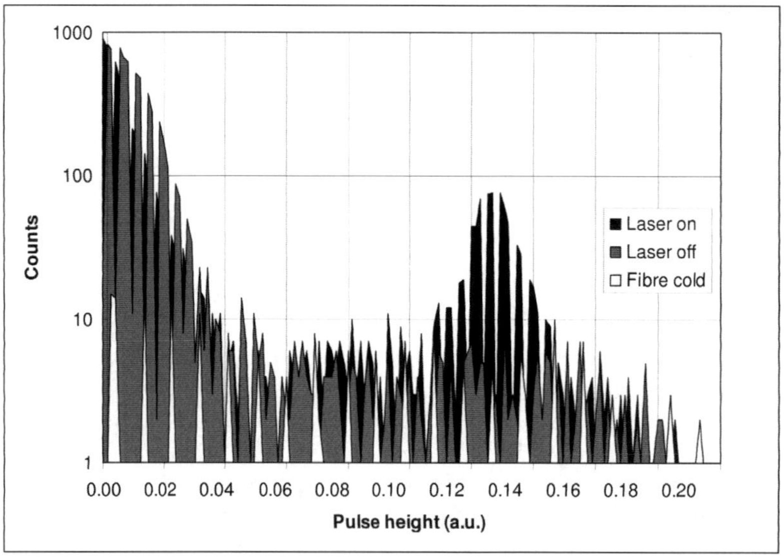

Figure 4. Histograms of pulse height distributions for three different situations: a) weak 633nm laser irradiating the ISTED through a room temperature fibre (black bars), b) the same fibre with the laser turned off (grey bars) and c) with the laser turned off and the open end of the fibre cooled to 77K (white bars).

Further work is planned to try to optimise absorber performance to improve the quantum efficiency of this type of device.

5. NanoSQUIDs as Nanomagnetic Particle Detectors

With the increasing importance of the spin component of electron transport for future devices as manifested in the coming spintronics revolution the optimization of SQUID capability for spin or magnetization detection becomes highly desirable [14-18]. Ultimately, a nanoSQUID may enable single electronic spin flips to be detected, allowing electronic spin resonance (ESR) and other spin manipulations to be done on a single spin. Recently single spin detection has been achieved using a mechanical magnetic resonance detection device [19], but this technique is rather specialized and of limited applicability. As well as setting the lower limiting size of a single magnetic domain storage element, single spins are also regarded as one of the most promising possible realizations of the *qubits* on which quantum computation will depend [20], if it becomes feasible. In such a single-spin detection situation the input coil and flux transformer are dispensed with and a 'bare' SQUID is used.

Reducing the loop area of the SQUID, which reduces its inductance, can increase the energy sensitivity of the device to near quantum limit operation. Furthermore, a device incorporating a SQUID of small loop area has a reduced sensitivity to external magnetic fields, making it an ideal probe of samples situated within the SQUID loop. When it comes to measuring magnetization of a small particle the SQUID design requirements are in some sense the opposite to those for magnetic field sensing. It is preferable to use the SQUID loop to directly sense the magnetisation rather than employing a separate flux transformer. In addition the SQUID loop area should be minimized, in order to approach optimal coupling [21-23]. There are two major issues which must be taken into account. First, achieving optimal coupling, given the physical size, geometric shape and expected magnetisation of the particles to be detected and second, a method for bringing the magnetic nanoparticle and the SQUID into close proximity. Since the SQUID only detects relative, not absolute, flux changes it is preferable to be able to manipulate either the position or the magnetization direction of the magnetic particle.

The electronic spin sensitivity is given by

$$S_n = d\frac{(S_\Phi)^{1/2}}{2\pi\mu_B\mu_0} \tag{4}$$

in units of spins (of moment μ_B) per √Hz, where S_Φ is the power spectral density of flux noise, related to the energy resolution by $S_\Phi = 2\varepsilon L$, and d is the dimension of the SQUID loop, scaled by a geometry-dependent factor between 1 and 10. In measurements made on a set of Nb dc SQUID devices ranging from d = 100 μm down to 3 μm [16] it has been demonstrated that the flux noise sensitivity exhibits the expected linear scaling with d. The smallest device has a spin sensitivity of about 40 spins in a 1 Hz bandwidth limited by noise in the first-stage room temperature preamplifier. Extension of this scaling to SQUIDs with $d \approx 200$ nm, as demonstrated in the section on noise performance above, should enable single electronic spin flips to be detected.

We are considering a number of ways of reading out the spin flip of a nanomagnetic particle placed in close proximity to a nanoscale SQUID loop. All of these configurations require manipulation of single nanoparticles close to the SQUID loop. This may be achieved most simply by mounting the particle on a scanned probe system. Modelling supports the common sense realisation that the particle should be asymmetrically placed with respect to the axis through the centre of the SQUID loop. Recently we have demonstrated that a micromanipulator mounted in a scanning electron microscope (SEM) can be used to attach (see inset to figure. 6) and then remove a nanomagnetic particle from a nanoSQUID, without affecting the SQUID performance or physical parameters [24].

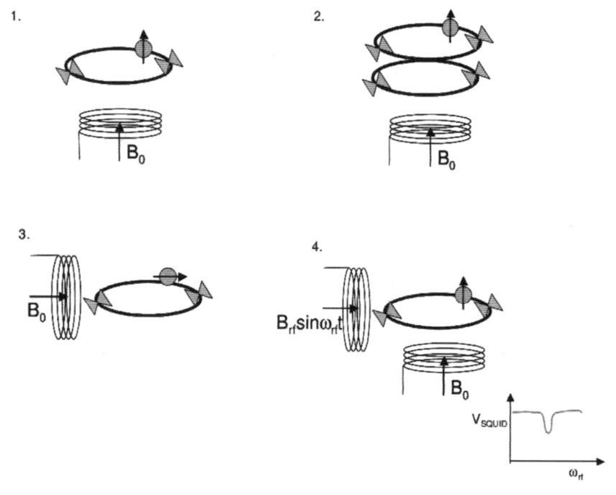

(a)

(b)

Figure 6. (a) Schematic layout of various methods for nanomagnetic spin flip detection using a nanoSQUID, directly coupled to the magnetisation. (b) SEM image of a nano-SQUID with a single magnetic particle bead attached. The nano-SQUID loop size is ~ 350 nm, junction size ~50 nm x 80nm. The particle diameter is ~ 120 nm.

6. Nano-Electromechanical Systems (NEMS) Resonator Readout

As electromechanical system devices become smaller, approaching the nanoscale, the oscillation displacement amplitude scales down in proportion to size. Thus ultra-sensitive transducer techniques and low dissipation excitation schemes are needed to operate NEMS sensors. Developing suitable excitation and readout methods is crucial and may have spin-offs into a number of other nanotechnology applications. We propose here a novel form of NEMS displacement readout provided by our on-going development of nanoSQUIDs [4,13,25]. As in the ISTED this readout is based on the modulation of the SQUID loop inductance by the movement of the NEMS resonator in its vicinity.

The resonator support structure and its surroundings are important in that they will determine the energy transfer between the resonator and the environment, both wanted and unwanted. Characterisation of the resonator's properties can be carried out using a range of conventional techniques such as electrical and thermal transport methods. A disruptive technology such as nanomechanical resonators (NMRs) brings new requirements for enabling technologies. It is necessary to excite and detect the resonant modes of the NMRs. The sub-wavelength size of the resonators relative to optical wavelengths means that the simple optical readout which is conventionally used in, for example, scanned probe microscopes will not be appropriate. But a generic technique, expected to be even more sensitive than any other, involves the use of the novel nanoSQUID readout method under development at NPL. By positioning an electrically conducting NMR close to the sensing loop of a nanoSQUID the mutual inductance between NMR and SQUID loop will be modulated by their relative movement. This in turn will modulate the loop inductance of the SQUID, leading to a change in output voltage, arising from either (or both) an external flux applied to the SQUID or the internally generated flux arising from an asymmetric bias current distribution. The SQUID readout will then carry sideband information concerning the time dependent movement of the NMR. We estimate *sub picometre* sensitivity will be readily achievable by this method.

Alternatively a SQUID may be used as a sensitive displacement sensor in a number of other ways [26]. For example if a ferromagnetic particle is attached to the vibrating element of the NMR cantilever the adjacent SQUID can detect the enclosed magnetic flux change as the particle moves.

There is another possible SQUID detection method for the case of a conducting NMR cantilever clamped at both ends. A direct current flowing through the cantilever will produce a magnetic flux at the SQUID loop which will be modulated by cantilever movement. This would produce an alternating output signal from the SQUID.

Close inductive coupling between SQUID and NMR will be necessary, regardless of the specific readout mechanism implemented. This implies that the SQUID loop should be comparable in size to the linear dimension of the NMR, i.e of nanoscale dimensions. We have already demonstrated how to make SQUIDs with a loop size as small as 200 nm [13] and more recently have shown that these SQUIDs have exceptional low noise properties at relatively high operating temperatures [4]. It is also important for high sensitivity that the cantilever may be brought sufficiently close to the SQUID loop to maximise the change in inductance or coupled magnetic flux change. With this in mind it appears clear that the cantilever and SQUID would best be deposited, mounted or grown on the same substrate. The cantilever displacement could be read out by any of the methods outlined above which do not need connections to a conducting cantilever.

A second possibility incorporates the cantilever (superconductor coated) directly into the SQUID loop, allowing rf current excitation to be driven through the cantilever without significantly affecting the SQUID bias. Multi-layer deposition is also quite straightforward so much more complex systems are possible in principle.

In order to test these ideas a prototype SQUID structure has been fabricated, using FIB milling, on the same chip as a paddle-shaped Si NEMS resonator (see figure 7). The vibration of the paddle will modulate the inductance of the nanoSQUID loop, allowing sub picometre amplitude sensitivity to be achieved, according to a simple model of expected coupling between the two components. We will present elsewhere detailed calculations of the performance of these readout and excitation methods and also preliminary device fabrication details [27].

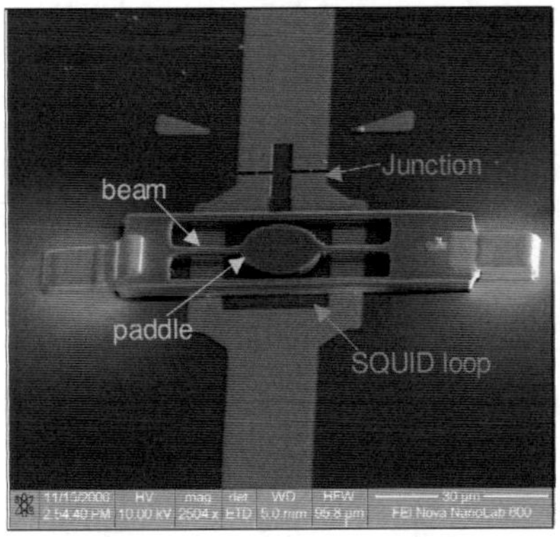

Figure 7. SEM micrograph of Si paddle NEMS resonator coupled microbridge d.c. SQUID.

It is straightforward to give an estimate of the potential sensitivity of the inductively coupled nanoSQUID readout system. Using finite element modelling techniques we may readily estimate the self-inductance of a SQUID loop and how it is affected by the presence of a conducting cantilever. Movement of this cantilever element corresponding to a chosen expected vibrational mode can be factored in to this model to characterise the inductance variation as a result of the oscillation at a given amplitude. For an angle ϕ between the plane of the SQUID loop and the paddle and an angle dependent self-inductance $L(\phi)$ of the SQUID loop (see figure 9a) it is straightforward to estimate that the minimum detectable angular displacement $\delta\phi$ is related to the minimum detectable magnetic flux change of the SQUID $\delta\Phi$ by the relationship

$$\delta\phi = \delta\Phi L(\phi)\left[\frac{dL(\phi)}{d\phi}\right]^{-1} \qquad (5)$$

For a paddle shaped resonator we assume a filling factor of around 80% of the SQUID loop and a nanoSQUID with a flux sensitivity $\delta\Phi$ of around 1 $\mu\Phi_0/\mathrm{Hz}^{1/2}$ (a realistic value for a nanoSQUID). Then the spatial movement detection sensitivity to a torsional oscillation mode is expected to be better than $10^{-13}\mathrm{m/Hz}^{1/2}$. The variation of displacement sensitivity of the paddle edge as a function of mean angle ϕ is shown in figure 9b).

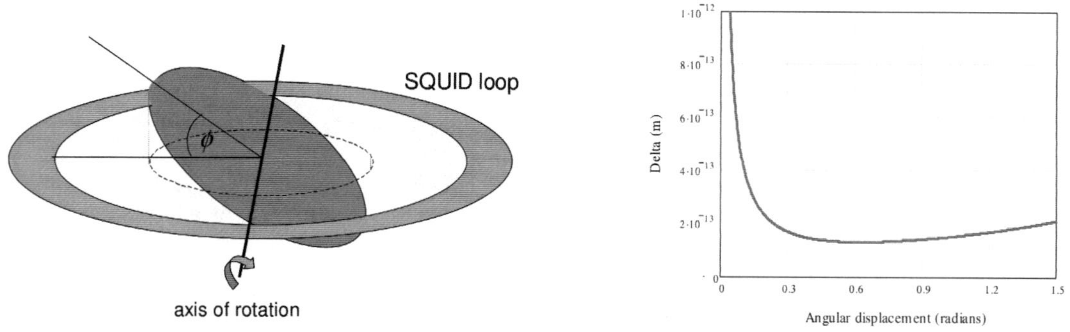

(a) (b)

Figure 8. (a) Schematic of paddle SQUID in torsional motion b) dependence of inductive SQUID signal on angle of rotation ϕ. (b) Variation of minimum displacement sensitivity as a function of ϕ calculated from equations 4 and 5.

7. Future Prospects

SQUIDs have found widespread applications in metrology over the past thirty years. Other applications in medicine and instrument development have also emerged. Much more recently the growing importance of nanoscience and technology has led to a realisation that SQUIDs reduced to nanoscale size have an entire new range of applications. In this paper we have summarised some preliminary developments at NPL in producing nanoscale SQUIDs with weak-link microbridge junctions fabricated by focussed ion beam milling. We have demonstrated the low-noise properties of these detectors and shown that they may be used for exceptional bolometric performance. We also present some initial work on using these nanoSQUIDs for magnetic nanoparticle detection with the long term aim of being able to detect single electronic spin flips, a goal which would be of potential importance in QIP. SQUID displacement sensors have been known for many years (see for example Harry et al. [28]) but the idea of using nanoSQUIDs to sense NEMS readout is proposed and modelled here. We expect to provide experimental evidence for this sensitivity in the near future.

Acknowledgements

The author would like to thank many colleagues at NPL, especially John Gallop, John Macfarlane and David Cox, who were involved in many of the experiments reported here. Also she has benefited greatly from collaborations and discussions with co-workers at CSIRO (Sydney), PTB (Berlin) and University College London. The work reported here was funded under the UK NMO *Pathfinder* Programme.

References

[1] Clarke J and Braginski A I 2004 *The SQUID Handbook Vol. 1* (Weinheim: Wiley–VCH, GmbH and Co. KGaA)
[2] Foley C P and Hilgenkamp H 2009 *Supercond. Sci. Technol.* **22** 064001
[3] Wernsdorfer W 2009 *Supercond. Sci. Technol.* **22** 064013
[4] Hao L, Macfarlane J C, Gallop J C, Cox D, Beyer J, Drung D and Schurig T 2008 Appl. Phys. Lett. 92 192507
[5] Troeman A G P, Derking H, Borger B, Pleikies J, Veldhuis D, and Hilgenkamp H 2007 *Nano Letters.***7 7 2152**
[6] Josephs-Franks P, Hao L, Tzalenchuk A, Davies J, Kazakova O, Gallop J C, Brown L and Macfarlane J C 2003 *Supercond. Sci. Technol.* **16** 1570
[7] Lam S K H and Tilbrook D L 2003 *Appl. Phys. Lett.* **82** 1078
[8] Lam S K H 2006 *Supercond. Sci. Technol.* **19** 963
[9] Hao L, Macfarlane J C, Gallop J C, Romans E, Cox D, Hutson D and Chen J 2007 *IEEE Trans. Appl. Supercond.* **17** 742
[10] Hao L, Macfarlane J C, Gallop J C, Cox D, Hutson D, Josephs-Franks P, Chen J and Lam S K H 2007 *IEEE Trans. Instrum. & Meas.* **56 2** 392
[11] Tesche C D and Clarke J 1977 *J. Low Temp. Phys.* **29** 301
[12] Drung D, Aßmann C, Beyer J, Kirste A, Peters M, Ruede F, and Schurig Th 2007 *IEEE Trans. Appl. Supercond.* **17**, 699
[13] Hao L, Macfarlane J C, Lam S K H, Foley C P, Josephs-Franks P and Gallop J C 2005 *IEEE Trans. Appl. Superconductivity* **15** 514
[14] WernsdorferW, Orozco E B, Hasselbach K, Benoit A, Barbara B, Demoncy N, Loiseau A, Pascard H and Mailly D 1997 *Phys. Rev. Lett.* **78** 1791
[15] Jamet M, Wernsdorfer W, Thirion C, Mailly D, Dupuis V, Mélinon P, and Pérez A 2001 *Phys. Rev. Lett.* **86** 4676
[16] Gallop J C, Josephs-Franks P W, Davies J, Hao L and Macfarlane J C 2002 *Physica C* **368** 109
[17] Lam S K H, Yang W, Wiogo H T R and Foley C P 2008 *Nanotechnology* **19** 285303
[18] Granata G, Esposito E, Vettoliere A, Petti L and Russo M 2008 *Nanotechnology* **19** 275501
[19] Rugar D, Budakian R, Mamin H J and Chui B W 2004 *Nature* **430** 329
[20] Tejada J, Chudnovsky E M, del Barco E, Hernandez J M and Spiller T P 2001 *Nanotechnology* **12** 181
[21] Gallop J C 2003 *Supercond. Sci. Technol.* **16** 1575
[22] Ketchen M B, Awschalom D D, Gallagher W J, Kleinsasser A W, Sandstrom R L, Bozen J R and Bumble B 1989 *IEEE Trans. Magn.* **25** 1212
[23] Narasimhan L R, Takigawa M and Ketchen M B 1994 *Appl. Phys. Lett.* **65** 1305
[24] Hao L, Cox D, See P, Gallop J and Kazakova O, 2010, *J. Phys. D: Appl. Phys.* **43** 474004
[25] Hao L, Cox D, Gallop J 2009 *Supercond. Sci. Technol.***22** 064011
[26] Etaki S, Poot M, Mahboob I, Onomitsu K, Yamaguchi K and van der Zant H S J 2008 *Nature Physics.* **4** 785
[27] Hao L, Gallop J C, Cox D, Romans E J, Macfarlane J C and Chen J 2009 *IEEE Trans. Appl. Supercond.* **19** 639
[28] Harry G M, Jin I, Paik H J, Stevenson T R and Wellstood F C 2000 *Appl. Phys. Lett.* **76** 1446

Superconductivity for hydrogen economy

P. Mikheenko[1]
School of Metallurgy and Materials, University of Birmingham, Edgbaston, Birmingham B15 2TT, UK

E-mail: p.mikheenko@bham.ac.uk

Abstract. The emerging hydrogen economy is expected to deal with a large amount of liquid hydrogen produced from the renewable energy resources. The main advantage of *liquid* hydrogen in comparison with other forms of its storage and transportation is in allowing wide use of superconductivity, which would optimise energy efficiency of the economy. The basic element of the infrastructure for hydrogen economy is a network of superconducting pipelines carrying simultaneously liquid hydrogen and loss-free electricity. The most likely material for such infrastructure is MgB_2, the only superconductor efficiently working at boiling temperature of liquid hydrogen and not showing strong critical current reduction on grain boundaries. The cheap techniques for the preparation of MgB_2 are hot isostatic pressing, resistive sintering and paint coating. These and other advanced techniques are able to provide MgB_2 with suitable for the infrastructure structural and superconducting properties. The preparation of a large-area superconducting joint between two pieces of MgB_2 as a technique enabling this infrastructure is reported. A potential of synergy between liquid hydrogen and superconductivity is revealed in a range of possible new energy applications.

1. Introduction

In a few tens of years, the age of fossil fuels with their wide use in transport and energy generation will mainly be over. Still, the demand for energy is expected to grow, and only a short time is available to reshape the energy economy making it fossil fuels-free. Moreover, a significant reduction in the use of fossil fuels is needed to decrease CO_2 emissions leading to global warming. The only reasonable response to growing energy demand and necessity to remove fossil fuels from transport and energy generation is development of hydrogen economy.

Hydrogen is a convenient clean fuel giving no CO_2 emissions. It could be produced in abundance from renewable energy resources such as solar, wind, marine, hydro, geothermal, biomass and also from controllable nuclear and thermonuclear reactions. The abundance of liquid hydrogen would allow wide use of superconductivity, making energy use more efficient and eliminating its unnecessary losses. This work describes materials basis for the development of hydrogen economy starting with its infrastructure in the form of a network of superconducting pipelines carrying simultaneously liquid hydrogen and loss-free electricity. The MgB_2 pipes can be produced by hot-pressing, hydro-extrusion or by coating MgB_2-compatible materials with a superconducting paint. The superconducting properties of hot-pressed bulks, paint-coated surfaces and superconducting MgB_2

[1] To whom any correspondence should be addressed

joint are reported. A potential of the synergy between liquid hydrogen and superconductivity is described in a range of applications.

2. Experimental

Three experimental techniques for the preparation of MgB_2 are used in this work; they are hot isostatic pressing (HIP), resistive sintering (RS) and paint coating (PAC). HIP is a standard technique that was very successful in application to MgB_2 [1-6]. RS [6,7] is an original technique. Its closest analogue is spark plasma sintering (SPS) [8,9]. PAC is an original innovation of the author.

HIP and RS techniques are schematically shown in figures 1 and 2. In HIP (figure 1), a pre-compacted powder is enclosed in a steel container, evacuated and pressed in a high-pressure (typically at 1 kbar) argon chamber at temperature of 940–1000 °C for 4–12 hours. A medium size chamber allows to process thousands of cubic centimetres of MgB_2. Much bigger volume could be processed in industrial chambers.

The RS (figure 2) is performed in a vacuum chamber. The MgB_2 powder is uniaxially pressed in a graphite die with tungsten rods simultaneously acting as electrodes carrying high electrical current up to 1000 A. The primary role of the electrical current is to heat the powder when, on the initial stage, MgB_2 particles are in weak contact and the heat mainly comes from the current flowing in the carbon die. At a temperature of about 700 °C the powder becomes dense and current flows mainly through MgB_2. Moreover, high current ignites plasma between the isolated MgB_2 particles. The term 'spark plasma sintering', which is frequently used to describe the process [8,9], comes from this phenomenon. The plasma welds particles into dense sintered pellet with strong superconducting properties. RS is an express-method allowing to produce MgB_2 samples quickly. It is a method of choice for testing nanoparticle additions to MgB_2. Typically porosity in RS samples is about 15%.

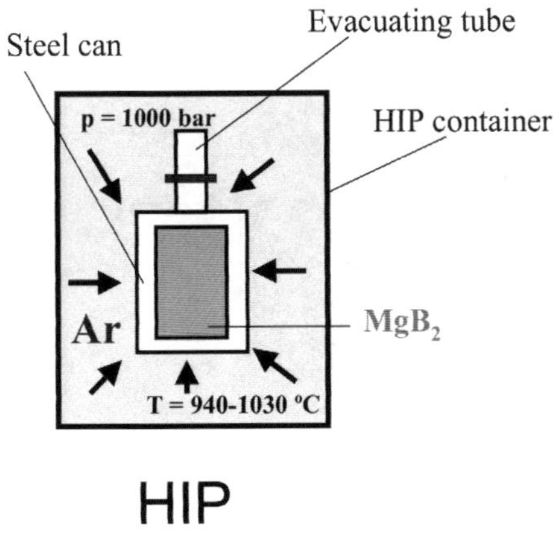

Figure 1. Schematics of hot isostatic pressing. The MgB_2 in the centre of the evacuated stainless steel can is subjected to high pressure and high temperatures. The arrows show the direction of isostatic pressure applied through the pressurised argon.

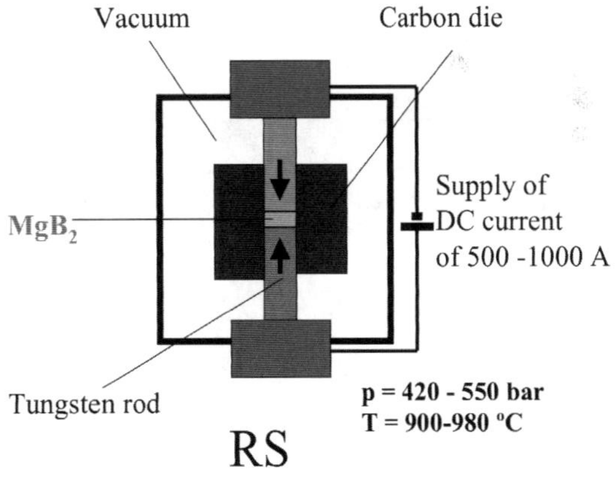

Figure 2. Schematics of resistive sintering. The MgB_2 is in the centre of the carbon die between the tungsten rods. It is subjected to the high pressure, high temperatures and high current. The arrows show the direction of uniaxial pressure applied by the tungsten rods that simultaneously carry high electrical current.

HIP allows producing dense samples with lower porosity (5-7%), bigger volume and better superconducting properties than in RS samples. However, it is a slow process involving long preparation procedures.

MgB_2 samples prepared by HIP or RS are cut on a wire spark erosion unit into small, about 0.2 × 0.2 × 3-4 mm^3, bars and their magnetic moment (m) is measured on a Quantum Design Magnetic Properties Measurement System MPMS-XL in DC field $\mu_0 H$ up to 7 T. The critical current density is derived from m using equation:

$$J_c = \frac{4m}{a^2 bc(1 - \frac{a}{3b})}, \qquad (1)$$

where a, b and c are dimensions of the bar ($c>b>a$) and magnetic field is applied along the longest c-axis. The small size of the samples is necessary because a very high J_c in MgB_2 generates large magnetic moment exceeding the limit of m measurable in MPMS.

The equation (1) represents a critical state model, which assumes that the current is constant and equal to J_c in every point of the cross-section of the sample. The J_c is considered to be field (H) dependent and the gradients of H are ignored. This equation can be obtained by integration from the definition of m. In specific case of the MgB_2 superconducting joint, a more general formula has been derived treating the case of a weak link between the two pieces of a superconductor.

The superconducting joint was produced using the RS set-up. First, two 5-mm diameter and 3-mm thick sintered pellets of MgB_2 were produced by RS. The jointing surfaces were polished and the pellets were placed back into the carbon die with a thin layer of MgB_2 powder between them. After that, routine resistive sintering has been performed on the stack. The jointed pellets were sliced by the spark wire cutting and the bars were cut perpendicularly to the joint surface as shown in figure 3.

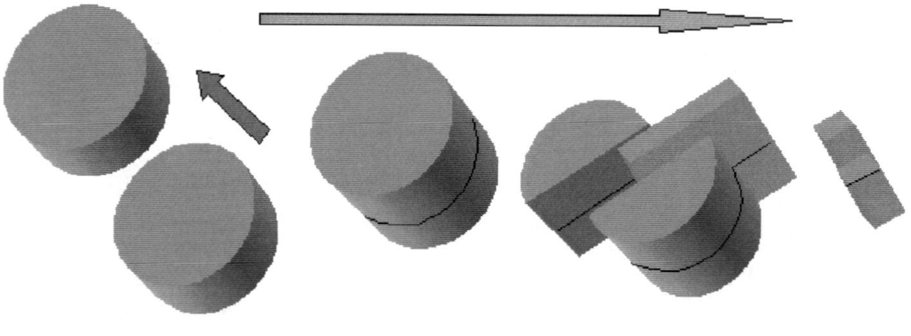

Figure 3. Procedure of jointing two pellets of MgB_2. The position of the joint is shown by thin black line. The red arrow shows the direction of pressing and the green arrow shows the order of the jointing and cutting the sample.

The paint in the PAC method was prepared by mixing fine MgB_2 powder with a liquid hydrocarbon. The paint is of a brown colour, has good adhesion to the surfaces and can be spread evenly using a brush. When treated at a high temperature in oxygen-free atmosphere, it produces a thick superconducting MgB_2 cover.

The magnetisation measurements have been performed on all samples and a variety of visualisation techniques was applied to investigate their microstructure, namely polarised optical microscopy, scanning electron microscopy, transmission electron microscopy (TEM), electron backscatter diffraction (EBSD), atomic force microscopy (AFM) and scanning conductivity probe microscopy (SCPM).

The described methods, together with other techniques developed for MgB_2 and high temperature superconductors (HTS), form a critical mass for wide penetration of superconductors into developing hydrogen economy. The examples of other successful MgB_2 techniques include magnesium infiltration [10], reactive evaporation [11] and shock consolidation [12]. Together with high demand for energy, scarcity of fossil fuels and dangers of CO_2 emissions, superconductivity is posed to be a major driving factor of hydrogen economy.

3. Results and discussion

The main parameters in applications of superconductivity are critical temperature (T_c) and critical current density J_c, or, more precisely, total critical current I_c and total critical current per centimetre of width: $I_{c\text{-}w} = J_c d$, when a flat superconductor of the thickness d is considered. The value of J_c in high H is of primary importance due to the dominant use of superconductors for high magnetic field generation. Because J_c rapidly decreases when temperature approaches T_c, liquid hydrogen applications require critical temperature well above boiling temperature of liquid hydrogen. Both MgB_2 and HTS satisfy this requirement. However, in HTS J_c exponentially decreases with the increase of misorientation angle between the grains. This restricts the use of HTS to their low-$I_{c\text{-}w}$ applications with samples prepared on textured substrates or high-I_c applications with small-volume textured bulks. In addition to MgB_2 and HTS there are several other superconducting materials, including recently discovered iron-based superconductors [13], but all of them are sensitive to grain misorientations. The only exception is MgB_2 and this makes it most suitable superconducting material for hydrogen applications. The following sections describe structural and superconducting properties of MgB_2 obtained by techniques described in section 2 in comparison with other materials and techniques.

3.1. Structural properties of MgB_2 samples

Figure 4 shows a dense MgB_2 sample produced by hot isostatic pressing. The material is of gold-like appearance in the center of bakelite disk surrounded by the white stainless steel ring. Several bars are cut from the HIP MgB_2 and placed on the disk. MgB_2 is a hard light material three times lighter than stainless steel. The density of the samples is close to that of single crystalline MgB_2 with the pores accounting for about 7% of the volume.

Figure 5 shows an AFM image of the surface of HIP MgB_2 on a sub-micron scale. Few pores and inclusions are seen. Nano-inclusions of secondary or external phases play an important role in grain boundaries generation. One of such nano-inclusions is shown in the center of the image marked by the red arrow. The nano-inclusion forms two straight grain boundaries with an angle of about 50° between them. Such grain boundaries contain large number of dislocations important for increasing J_c in the sample. The AFM imaging allows to visualize few grain boundaries etched by the water in the process of sample polishing, while an ultimate method to image grain boundaries that does not require etching is EBSD. The AFM and EBSD images of the same area of MgB_2 are shown in figures 6 and 7. The EBSD allows to determine accurately the orientation of all grains on the surface of the sample. The application of EBSD to MgB_2 was very successful and clarified the role of grain boundaries in pinning providing large J_c in high magnetic fields [6,14].

More detailed information about grain boundaries could be obtained by the application of scanning conductivity probe microscopy (SCPM) that measures local conductivity of the surface with the nanometer resolution. The comparison of SCPM and AFM is given in figures 8 and 9. In SCPM, contrast represents local conductivity of the surface. This method allows to identify grain boundaries that are not seen in AFM, like the boundary in figure 9, which crosses the center of SCPM image at the angle of about 15°. The combination of AFM, CSPM and EBSD gives complete information that allows determining relation between the local resistance of the grain boundaries and the angle of their misorientation, as shown in figure 10.

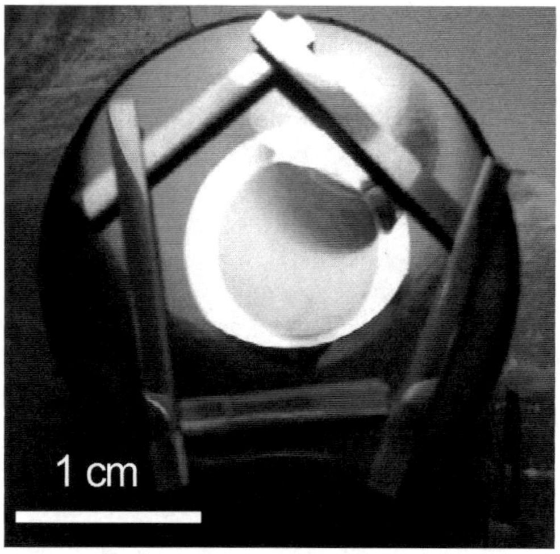

Figure 4. MgB_2 sample produced by hot isostatic pressing. It is in the centre of bakelite disk surrounded by the stainless steel ring and several bars cut from pressed HIP MgB_2. These are placed around the MgB_2 disk in a form of pentagon.

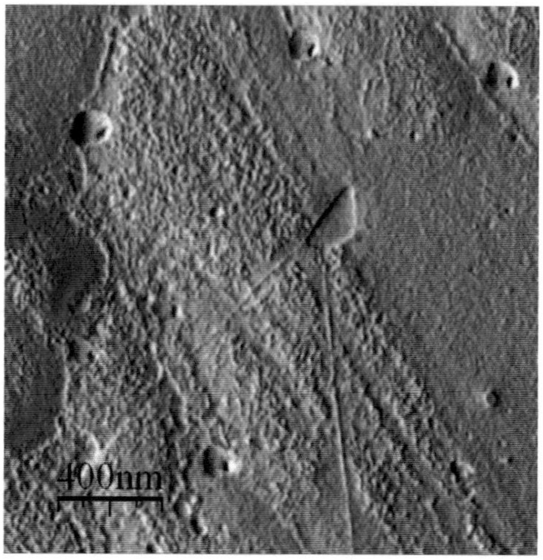

Figure 5. A sub-micron scale AFM image of the surface of HIP MgB_2. The red arrow shows a nano-inclusion that forms two straight grain boundaries of the angle of about 50°. Such grain boundaries are important for increasing J_c in the material.

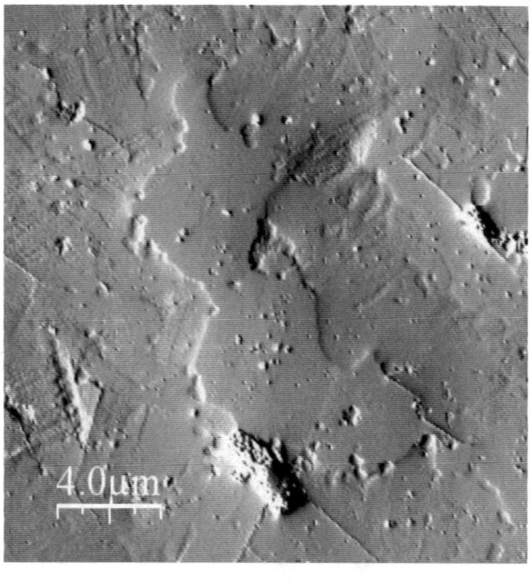

Figure 6. AFM image of a polished surface of HIP MgB_2. The areas of different orientation have different rate of polishing, those with *c*-axis perpendicular to the surface having the highest rate.

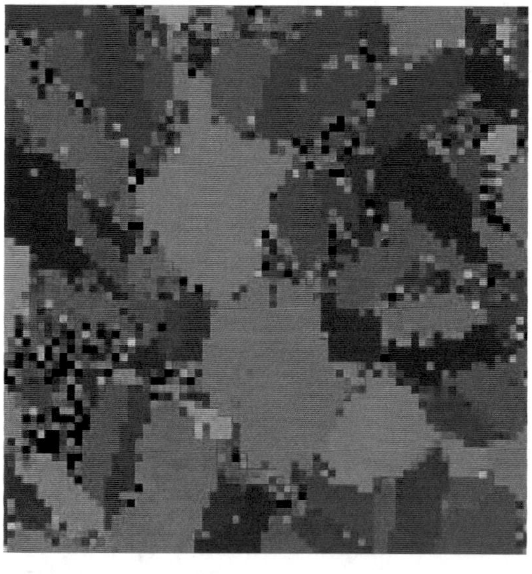

Figure 7. EBSD image of the surface of HIP MgB_2 shown in figure 6. The areas of different orientation are clearly distinguished and shown in the different colours.

Journal of Physics: Conference Series **286** (2011) 012014 doi:10.1088/1742-6596/286/1/012014

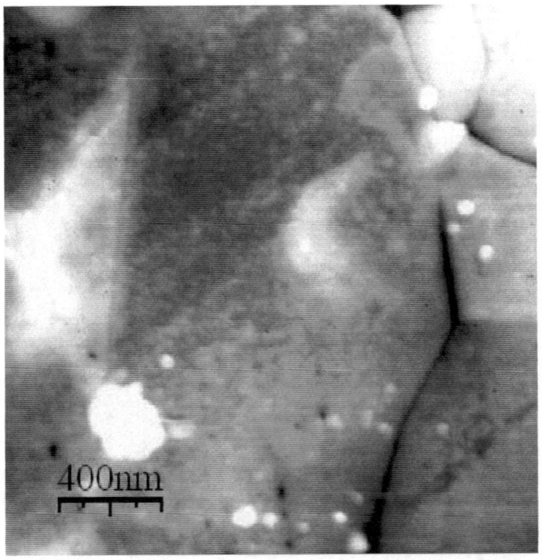

Figure 8. AFM image of a surface of HIP MgB_2. Several grain boundaries are exposed due to water etching during polishing.

Figure 9. SPCM image of the surface of HIP MgB_2 shown in figure 8. The contrast of the lines reflects local conductivity of the boundaries.

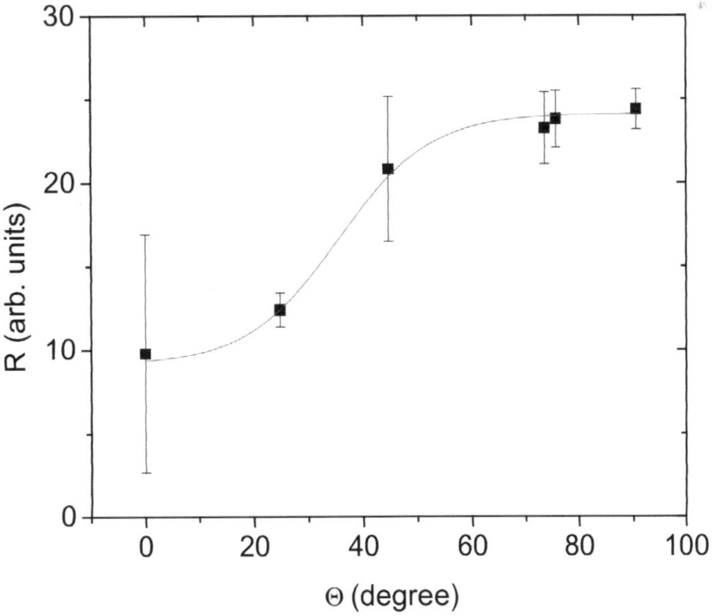

Figure 10. Surface resistance as a function of misorientation angle for grain boundaries shown in figures 8 and 9.

Figure 10 shows that a strong change in local resistance of grain boundaries takes place at an angle of about 40°. It is likely that at this angle the density of dislocations in the grain boundaries is high enough to overlap their cores, which, in its turn, decreases J_c. This identifies a subset of highly misoriented grain boundaries that have negative effect on J_c. Engineering samples without these boundaries would improve effectiveness of MgB_2 in practical applications.

3.2. Superconducting properties of MgB_2 and $YBa_2Cu_3O_x$ samples

Figure 11 compares J_c at temperature of 20 K and different magnetic fields for a range of superconducting samples that are important for the development of hydrogen economy. The black filled squares show $J_c(H)$ of conventionally sintered (CS) MgB_2 sample. It has the lowest critical current density in a wide range of magnetic fields. The HIP procedure (black open squares) significantly increases J_c. The addition of SiC [3] (blue open squares) makes J_c even higher in high magnetic fields. The RS sample (filled green squares) has J_c comparable with that of HIP samples in low magnetic fields, but shows a lower J_c in high fields.

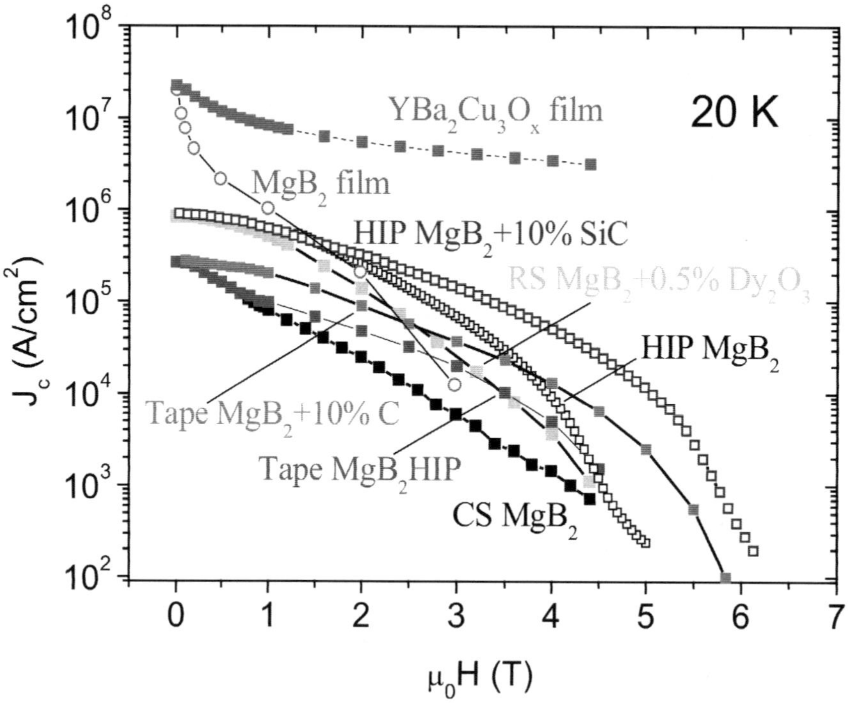

Figure 11. Magnetic field dependence of critical current density in MgB_2 and $YBa_2Cu_3O_x$ samples at temperature of 20 K.

The MgB_2 powder-in-tube tape with addition of nano-carbon [15] (filled magenta squares) has lower J_c than in bulk HIP in low field, but high J_c in high fields. When HIP procedure is applied to the power-in-tube tapes [1], it increases J_c in high field (filled olive squares) above that in CS MgB_2. The red open circles show $J_c(H)$ for an epitaxial MgB_2 film [16]. Such film does not have high-angle grain boundaries. It is likely that this is the reason why J_c of the film is much higher than in all other MgB_2 samples in low magnetic fields. However, due to small amount of dislocations residing on grain boundaries, J_c in high fields is low. The specially designed HTS films, for example $YBa_2Cu_3O_x$ film composed of nano-columns [17] (filled red squares) have much higher J_c than in any other sample in high fields. Although the thickness of HTS films is limited by few microns, they can be used in many liquid hydrogen applications. Overall, critical current density in the samples shown in figure 11 is sufficient for most liquid hydrogen applications and the relevant techniques only need a scale-up to the industrial level.

3.3. Superconducting MgB_2 joint

The superconducting properties of the samples shown in figure 11 were measured on small pieces of material, typically few millimetres in length. The infrastructure for hydrogen economy would, however, require much longer superconductors. The pipelines, for example, may need to be thousands kilometres long. Although it is easy to maintain high J_c on long distances in continuous pipes, they are likely to be constructed from sections, and the joints between these sections may strongly affect J_c. HTS are unlikely candidates for the pipelines because of difficulties in making HTS superconducting joints. Fortunately, low MgB_2 sensitivity to grain misorientations makes its jointing much easier than in HTS. As it is described in section 2, the procedure could be based on the combination of moderately high pressure of few hundred bars, high temperature of few hundred centigrade and a buffer layer of MgB_2 powder. This is now a general approach to MgB_2 jointing, see for example [18].

Figure 12 shows the joint area of the sample produced by the RS technique shown in figure 2. The buffer layer, shown by arrows in the lower part of the plot, is about 200 μm thick. It is hardly distinguishable from the rest of the sample.

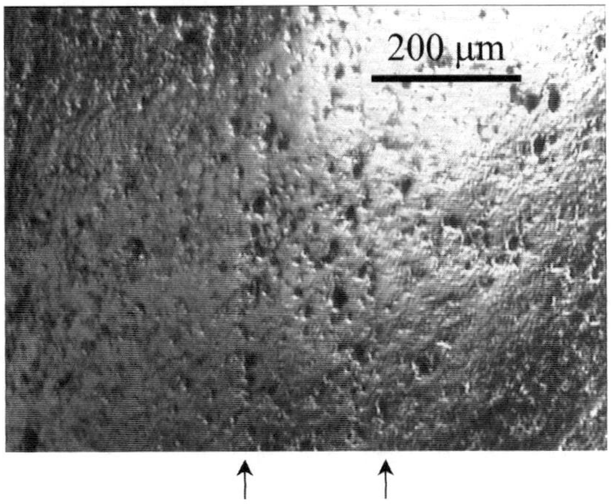

Figure 12. Optical image of the MgB_2 joint. The buffer layer is about 200 μm thick. It is marked by the arrows in the lower part of the plot.

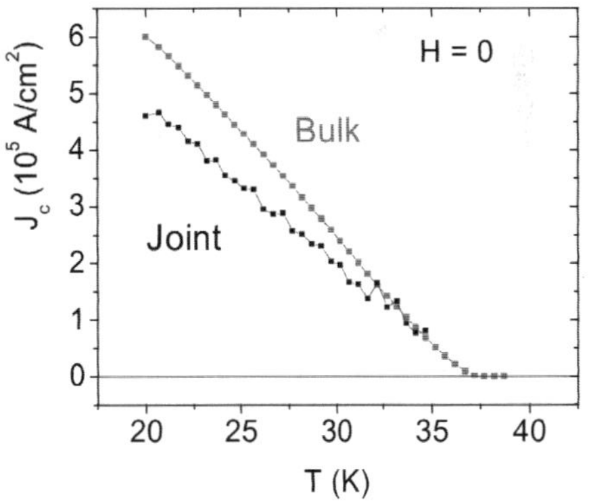

Figure 13. Temperature dependence of the critical current density of the joint shown in figure 12 (black squares) and bulk MgB_2 adjacent to the joint (red squares).

The joint in figure 12 is prepared from the material containing nano-inclusions of Dy_2O_3 [19]. In spite of high porosity partially coming from the non-evenly distributed Dy_2O_3, the critical current density in this sample is high, as shown by the temperature dependence of J_c in figure 13. Most importantly, J_c of the current through the joint (black squares) is close to J_c in the bulk (red squares).

3.4. MgB_2 paint coating

The critical current density of HIP MgB_2 is so large that a pipeline of inner diameter of 18 centimetres and outer diameter of 20 centimetres would be able to carry a power of up to 10 GW without losses. This is sufficient to connect a large city and a big power station. However, only a small amount of hydrogen could be pumped through such a pipeline. A long distance pipeline supplying large amount of liquid hydrogen would require MgB_2 pipes of a bigger diameter. Such pipes would be difficult to produce by HIP, even if this method is extended to the continuous process of hot hydro-extrusion. The ideal solution for such a pipeline would be paint technology. A simple painting of a stainless-steel pipe by MgB_2-based paint followed by thermal treatment may produce a thick superconducting layer suitable for the transport of supercurrent. Such a technology would be intrinsically free from the superconducting joint problem.

An optimal solution would be a pipe of inner diameter of 1.3 meters delivering 10 GW of electrical energy and 20 GW of energy in the form of liquid hydrogen. Such a pipeline will be suitable for transportation of electricity and hydrogen on distances of several thousand kilometres. Similarly to the small-diameter pipeline described above, this pipeline would feature double layer of MgB_2 for the two-way transport of current necessary to close delivery loop and to cancel strong magnetic field induced by the current. An important feature of the pipeline would be its magnetic suspension based on the property of superconductor to levitate above the magnet. Such a suspension would reduce stresses and vibrations that may affect its integrity.

To clarify the viability of large-diameter pipes, the paint technology was tested with MgB_2 paint prepared as described in the section 2. Figure 14 shows temperature dependence of the real (lower part of the plot) and imaginary (upper part of the plot) AC magnetic moment of a thick paint cover on Ta (black squares) and stainless steel (blue and red squares). Although T_c of the cover is lower than that of pure MgB_2, the diamagnetic response at 20 K is strong. The first layer of MgB_2 on Ta is superconducting (black squares), whereas first layer on stainless steel shows only traces of superconductivity (blue symbols) due to chemical interaction of the paint with iron. However, second layer above the first one shows strong superconductivity (red symbols). There is no restriction on the number of layers or the thickness of the paint cover.

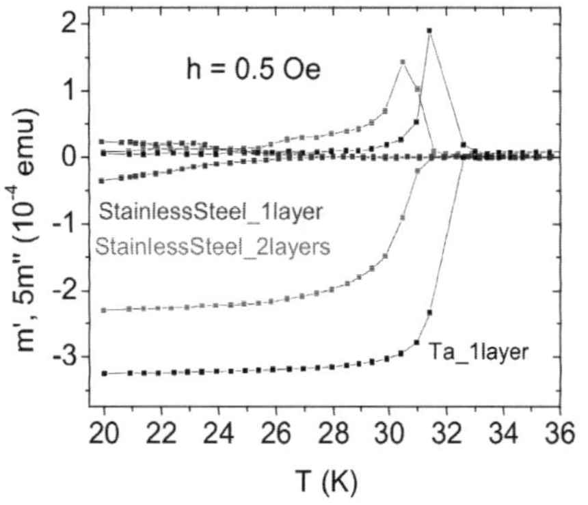

Figure 14. Temperature dependence of real (lower part of the plot) and imaginary (upper part of the plot) AC magnetic moment of thick paint cover on Ta (black squares) and stainless steel (blue and red squares).

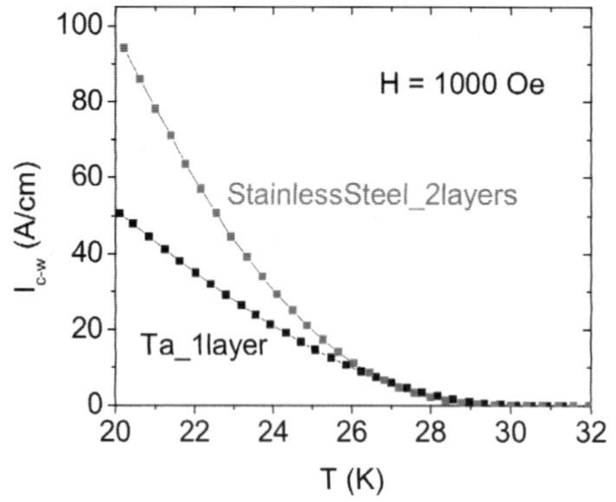

Figure 15. Total critical current per centimetre of width for the layers shown in figure 14 with the matching colours in magnetic field of 0.1 T. The I_{c-w} of the first MgB_2 layer on stainless steel is not shown.

Figure 15 shows total critical current per centimetre of width in 0.1 T magnetic field for the layers shown in figure 14. It is the first test of the paint and it is believed that its optimisation together with proper treatment of the substrate would significantly increase $I_{c\text{-}w}$.

The hydrogen pipeline infrastructure is designed to serve carbon-free economy. Unexpectedly, production of MgB_2 offers another possibility to tackle CO_2 emissions. A disadvantage of pure MgB_2 in comparison with HTS is in its relatively low critical magnetic field. This could restrict its use in high-field applications. A way to increase critical field is incorporation of a small amount of carbon into the crystal lattice of MgB_2 [20]. In coated paint technique, it is achieved by mixing hydrocarbons with MgB_2 powder. During the thermal treatment hydrocarbons decompose, carbon enters crystal lattice of MgB_2 and is trapped there permanently (no CO_2 emission, no necessity for CO_2 storage) while hydrogen is released and can be collected and used as a fuel. This process could be one of the most effective ways of decomposing hydrocarbons. With well-developed MgB_2 powder metallurgy producing thousands (possibly millions) tons of MgB_2, this process would significantly reduce emissions of CO_2 from otherwise burned hydrocarbons.

4. Synergy of hydrogen economy and superconductivity

The pipeline infrastructure is one of the applications of superconductivity for hydrogen economy. When supply of liquid hydrogen from renewable resources is secured, superconductivity would find wider applications. It could be implemented in car manufacture. There are already several models of cars using liquid hydrogen. The tank of a car could contain up to 100 litres of liquid hydrogen and, utilising its low temperature, car could be fitted with superconducting devices.

The car can feature compact and efficient superconducting motor; superconducting generator to cover electrical needs of the car; superconducting energy storage unit; loss-free superconducting wiring; superconducting current limiter, protecting electrical system of the car; superconducting (possibly quantum) computer etc. Some of the superconducting components, for example, flexible flat MgB_2 superconducting cable [21], are already developed. It would also be possible to design superconducting train, plane, ship, submarine, spaceship, and built superconducting home energy unit. The home unit would contain some 200-300 litres of liquid hydrogen and have superconducting devices as in the superconducting car adapted for the home use. The house would be self-sufficient in energy for some time and immune to blackouts. This would be of special importance for hospitals and defence installations.

Superconductors could also find wide use in the screening of magnetic field. In particular, MgB_2 is attractive for Space applications, since it is a light superconductor and due to low temperatures in Space, it would work there with limited amount or even without liquid hydrogen. The MgB_2 paint could be a method of choice for these applications, especially on the Moon's surface or Earth's orbit, where it can be treated in oxygen-free environment. Small MgB_2 cylinders showing complete screening of magnetic field have already been demonstrated [22].

When liquid hydrogen is produced on a large-scale, magnetic screening could be extended to larger objects. From materials point of view, it would even be possible to magnetically protect the whole planet by a single superconducting pipeline. For Earth, for example, necessary amount of boron for MgB_2 pipeline could be mined just in 5 years. Other global projects like generating energy from Solar wind or extracting it from below the crust in magnetically active volcanic regions would also be possible.

5. Conclusions

The disappearance of fossil fuels, global warming and transition to renewable energy resources lead to synergy of superconductivity and hydrogen economy. The superconductivity offers compactness, efficiency, energy savings and wide range of applications that are not possible without using this macroscopic-scale quantum phenomenon. It gives hope to cope with threatening energy crisis and may addresses global problems on the planetary scale.

References

1 Serquis A, Civale L, Hammon D L, Liao X Z, Coulter J Y, Zhu Y T, Jaime M, Peterson D E, Mueller F M, Nesterenko V F and Gu Y 2003 *Appl. Phys. Lett.* **82** 2847
2 Shields T C, Kawano K, Holdom D and Abell J S 2002 *Supercond. Sci. Technol.* **15** 202
3 Wang X L, Soltanian S, James M, Qin M J, Horvat J, Yao Q W, Liu H K, and Dou S X 2004 *Physica C* **408-410** 63
4 Murai K, Hori J, Fujii Y, Shaver J, Kozlowski G, 2005 *Cryogenics* **45** 415
5 Mikheenko P, Bevan A I, and Abell J S, 2006 *Journal of Physics: Conference Series* **43** 535
6 Mikheenko P, Martínez E, Bevan A, Abell J S and MacManus-Driscoll J L 2007 *Supercond. Sci. Technol.* **20** S264
7 Dancer C E J, Mikheenko P, Bevan A, Abell J S, Todd R I, Grovenor C R M 2009 *J. Europ. Ceramic Soc.* **29** 1817
8 Dancer C E J, Prabhakaran D, Basoglu M, Yanmaz E, Yan H, Reece M, Todd R I and Grovenor C R M 2009 *Supercond. Sci. Technol.* **22** 095003
9 Miu L, Aldica G, Badica P, Ivan I, Miu D and Jakob G 2010 *Supercond. Sci. Technol.* **23** 095002
10 Giunchi G, Ripamonti G, Cavallin T, Bassani E 2006 *Cryogenics* **46** 237
11 Moeckly B H and Ruby W S 2006 *Supercond. Sci. Technol.* **19** L21
12 Ohashi W, Takenaka K, Kondo T, Tamaki H, Matsuzawa H, Kai S, Kakimoto E, Takano Y, Minehara E 2006 *Physica C* **444** 5
13 Aswathy P M, Anooja J B, Sarun P M and Syamaprasad U 2009 *Supercond. Sci. Technol.* **23** 073001
14 Martínez E, Mikheenko P, Martínez-López M, Bevan A, Abell J S, Millán A 2007 *Phys. Rev. B* **75** 134515
15 Yeoh W K, Kim J H, Horvat J, Xu X, Qin M J, S. X. Dou, Jiang C H, Nakane T, Kumakura H and Munroe P 2006 *Supercond. Sci. Technol.* **19** 596
16 Xu S Y, Li Q, Wertz E, Hu Y F, Pogrebnyakov A V, Zeng X H, Xi X X and Redwing J M 2003 *Phys. Rev. B* **68** 224501
17 Mikheenko P, Dang V-S, Tse Y Y, Awang Kechik M M, Paturi P, Huhtinen H, Wang Y, Sarkar A, Abell J S and Crisan A 2010 *Supercond. Sci. Technol.* **23** 125007
18 Li X H, Ye L Y, Jin M J, Du X J, Gao Z S, Zhang Z C, Kong L Q, Yang X L, Xiao L Y and Ma Y W 2008 *Supercond. Sci. Technol.* **21** 025017
19 Mikheenko P, Chen S K, MacManus-Driscoll J L 2007 *Appl. Phys. Lett.* **91** 202508
20 Dou S X, Shcherbakova O, Yeoh W K, Kim J H, Soltanian S, Wang X L, Senatore C, Flukiger R, Dhalle M, Husnjak O and Babic E 2007 *Phys. Rev. Lett.* **98** 097002
21 Yung C S and Moeckly B H 2010 *IEEE/CSC & ESAS European Superconductivity News Forum (ESNF),* **14** October 2010 1
22 Rabbers J J, Poomen M, Bassani E, Ripamonti G and Giunchi G 2010 *Supercond. Sci. Technol.* **23** 125003

Acknowledgement

Author thanks Dr R Oliver for help in running SCPM.

Integrated pinning centers in $YBa_2Cu_3O_x$ thick films on single-crystalline and textured metal substrates

P. Mikheenko[1,2], V-S. Dang[1], A. Sarkar[1], J. S. Abell[1] and A. Crisan[1,3]
[1] School of Metallurgy and Materials, University of Birmingham, Edgbaston, Birmingham B15 2TT, UK
[3] National Institute of Materials Physics, Bucharest 077125, Romania

E-mail: p.mikheenko@bham.ac.uk

Abstract. The second-generation coated tapes of high temperature superconductors (HTS) deposited on textured metal substrates strongly rely on the introduction of extended nano-defects in order to enhance their critical current. A proven effective way to introduce such defects is growing $BaZrO_3$ (BZO) nano-rods and an alternative way is to generate HTS columnar growth using nanoparticles of noble metals. Here we report the combination of these methods. It allows achieving controlled pinning and high critical current in $YBa_2Cu_3O_x$ films deposited on single crystal substrates and significantly improves critical current in coated conductors deposited on rolling-assisted biaxially-textured metal substrates (RABiTS). The superconducting properties of thick (up to 5 micron) conductors are analysed using DC-magnetisation, AC-susceptibility and angle-dependent transport measurements. TEM imaging is used to confirm the presence of extended defects in the tapes.

1. Introduction

The tapes of high temperature superconductors (HTS) deposited on textured metal substrates known as second-generation HTS wires promise to revolutionise applications of superconductivity. Strong research efforts to develop cheap and effective way of the introduction of extended defects to enhance pinning and critical current in these wires is under way. A proven effective way to introduce nano-defects in films deposited on single-crystal and technical metal substrates is growing $BaZrO_3$ (BZO) nano-rods [1,2]. An alternative way already applied to films on single crystal substrates is to use nanoparticles of noble metals to generate columnar growth of $YBa_2Cu_3O_x$ (YBCO) films [3]. This method allows depositing thick films without strong decrease in critical current density with thickness.

Recently a new method has been introduced for YBCO films deposited on single crystalline $SrTiO_3$ substrates that combined the two methods described above [4]. It allows to achieve controlled pinning and high critical current in low magnetic fields along the *c*-axis, which is important for cable applications, and high magnetic fields in the *ab*-plane of YBCO, crucial for magnet application of superconductivity. The transmission electron microscopy (TEM) revealed entangled nano-threads of BZO and YBCO in these films. In a 5 μm-thick film, large critical current per centimetre of width $I_{c\text{-}w}$ (the actual important parameter for practical applications) at 77.3 K in high fields has been obtained. The achieved value of $I_{c\text{-}w}$ was 170 A/cm-w in 1 T, 90 A/cm-w in 1.5 T, and 60 A/cm-w in field of 2 T [4]. A combination of high $I_{c\text{-}w}$ with suitable behaviour in magnetic field of different orientation

[2] To whom any correspondence should be addressed

promises to make it a valuable method for films deposited on metal substrates. Here we clarify its potential by growing integrated films on rolling-assisted biaxially textured substrates (RABiTS).

The texture in the metal substrates is necessary to eliminate high-angle grain boundaries in HTS films since they strongly affect critical current [5]. The critical current density (J_c) decreases exponentially with misorientation angle, and an average misorientation of about 7° is sufficient to suppress J_c by an order of magnitude [5,6]. The texturing of long-length metal tapes is not a trivial issue [7] and the process significantly contributes to the price of coated conductors. Especially difficult is to promote grain misorientations with the angle below 7° that would allow to achieve the highest possible critical current in the tapes. Preliminary to the growth of HTS films, tapes require deposition of a buffer layer that would prevent diffusion of metal ions into HTS. The buffer layer preserves orientation of the metal tape and the local HTS misorientations are typically close to those in the buffer layer. There is currently no reliable mechanism that would allow the reduction of misorientations in HTS film below those in the RABiTS tape. The integrated method developed in [4] has a specific mechanism of film growth in which nanoparticles of noble metals coherently move perpendicularly to the film surface. Such a motion is not linked to the local misorientations and could reduce them in the film, thereby increasing J_c in the coated conductor. In order to clarify whether it is possible and whether it could work for thick HTS films, we have grown a wide range of YBCO films on RABiTS substrates with and without nanoparticles of noble metals and additions of BZO.

2. Experimental

A range of experimental techniques was used in this investigation. The crystallography of the RABiTS substrates, supplied by 3C-s Ltd., was mapped by electron backscatter diffraction (EBSD) on a scanning electron microscope (SEM) JEOL7000. The buffer layer in the form of a trilayer: CeO_2 (0.1 μm), yttria-stabilised zirconia (YSZ, 0.6 μm) and CeO_2 (0.1 μm) was deposited on the substrate by pulsed laser deposition (PLD). In the integrated method, the buffer layer was densely covered by Ag nanoparticles (also using PLD) and after that a YBCO/BZO layer was laser-deposited from a YBCO target with 2% of nano-dispersed BZO. All depositions were done in-situ by an excimer KrF 248 nm laser with pulse duration of 30 ns. The substrate temperature was 700 °C for the buffer layer, 780 °C for Ag and 800 °C for YBCO/BZO. The repetition rate of the laser was 8 Hz for the buffer layer and Ag and 5 Hz for YBCO/BZO. For the latter, the repetition rate is lower and the substrate temperature is higher than in preparations of pure YBCO films (6-10 Hz and 780 °C). Such conditions were chosen to promote growth of BZO nanorods [2].

A variety of Ag/YBCO/BZO films of different thickness have been grown in configurations as described above or in the form of quasi-superlattices, in which deposition of the quasi-layer of Ag nanodots was periodically repeated after the growth of YBCO/BZO layer of a certain thickness. The total thickness of the films and quasi-superlattices was between 0.75 and 5 μm.

The critical current density in magnetic field along the c-axis of YBCO was determined from DC magnetization loops of rectangular samples measured on a Quantum Design Magnetic Properties Measurement System MPMS-XL in DC field $\mu_0 H$ up to 4.4 T. For a film of planar dimensions *a*, *b* ($a<b$), critical current density J_c was calculated from the averaged magnetic moment (m) using equation:

$$J_c = \frac{4m}{a^2bd(1-\frac{a}{3b})}. \quad (1)$$

This equation represents a critical state model, in which current is constant and equal to J_c in every point of the cross-section of the sample. The J_c is considered to be field-dependent and the gradients of magnetic field are not taken into account when considering $J_c(H)$. Equation (1) can easily be obtained by simple integration from the definition of *m*. However, gradients of magnetic field can be ignored at high fields only, and this equation is not applicable below a certain characteristic field [8]. In low fields, J_c can be calculated by an extrapolation from the higher magnetic fields. For non-magnetic substrates the characteristic field usually does not exceed 0.1 T, but the magnetism of RABiTS substrate shifts it to a higher value. This restricts the ability to correctly determine J_c from DC measurements in zero magnetic field. In this field we calculated J_c from temperature (T) dependence of AC magnetic moment, which is measured at frequency of 100 Hz and excitation fields (h) from 5 10^{-6} T to 4 10^{-4} T.

In this technique, the imaginary part of AC magnetic moment $m''(T)$ shows a maximum at certain temperature T_{max} and J_c at this temperature is connected with the excitation field h by the equation:

$$J_c(T_{max}) = \frac{h}{\alpha d} , \quad (2)$$

where d is the film thickness and α is a constant between 0.8 and 0.9 that slightly depends on the geometry of the film [9]. Due to the low value of h available in MPMS and the high value of d in our samples, only small values of J_c are accessible at temperatures close to critical temperature (T_c).

These magnetisation techniques are only suitable for magnetic field along the c-axis of YBCO or perpendicular to the surface of the film. In magnetic fields of arbitrary orientation, J_c was derived from the angle-dependent transport measurements on 550 μm long and 18-30 μm wide micro-bridges produced by conventional photolithography and chemical etching. The measurements have been done on a Quantum Design Physical Properties Measurement System (PPMS).

The micro and nanostructure of the samples was analysed on scanning (SEM, JEOL 7000) and transmission (TEM, TECHNAI F20 operated at 200 kV) electron microscopes. The samples for TEM were prepared by focused ion beam milling (FIB) on a FEI Quanta 3D FEG.

3. Results and discussion

Figure 1 shows EBSD grain maps of a typical section of Ni5%W RABiTS substrate tape. The grains are shown by different colour if they are misoriented with neighbour grains by an angle above 1°, 3°, 5°, and 7° for plots a, b, c and d, respectively. If a superconducting film copies the grain map of RABiTS tape, the superconducting current flowing in this section would need to cross several grain boundaries with misorientations at least up to 7°. The misorientation of the last boundary between the blue and green areas in figure 1c is 7.4°. Each grain boundary restricts supercurrent and the final J_c is expected to be [6] of about one order of magnitude below J_c of the film without grain boundaries. The supercurrent may flow around highest-angle grain boundaries, which would keep an intermediate value of J_c.

The J_c can be derived from DC magnetisation measurements. In addition to DC signal from superconductor, the RABiTS substrate gives a strong magnetic signal. Figure 2 shows magnetisation loops of the RABiTS substrate at two temperatures: 77.3 K (black squares) and 20 K (red circles). These temperatures are important for applications in liquid nitrogen and hydrogen, respectively. The Ni5%W is a soft magnetic material with weak amplitude dependence of magnetic moment below 77.3 K. The amplitude of magnetic moment from RABiTS tape is large comparable with the magnetic moment of most superconducting films, and to obtain magnetisation loop of the superconductor, it is necessary to accurately subtract magnetic moment of the substrate. Figure 3 shows total m as function of field for a 3.75 μm-thick integrated Ag/YBZO film deposited on RABiTS at 77.3 K (black squares) and 20 K (red circles).

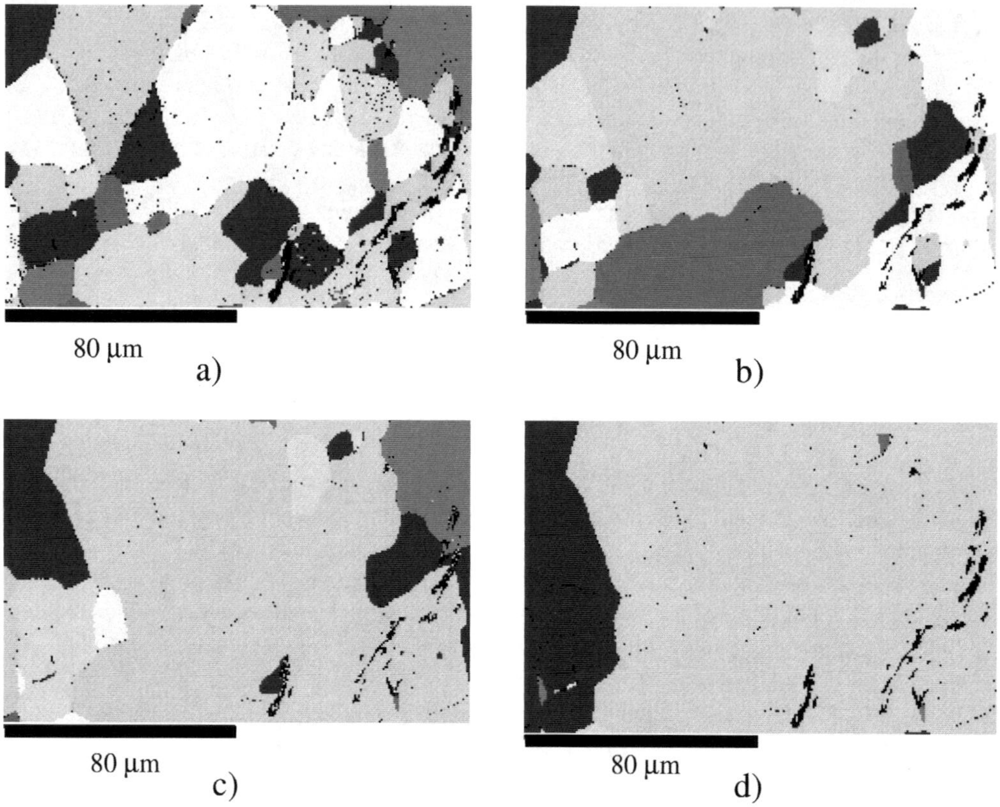

Figure 1. EBSD grain maps of a section of RABiTS substrate in which grains are shown by different colour if they are misoriented with neighbour grains by an angle above a) 1°, b) 3°, c) 5° and d) 7°

Because the loops in figures 2 and 3 belong to the samples of slightly different size, an adjustment is necessary to subtract the magnetic moment of the substrate. It was done by recording the temperature dependence of the magnetic moment of the substrate (filled black squares in figure 4) and film on substrate (filled red squares in figure 4) at an intermediate field of 1 T. These dependencies were plotted together and the former was multiplied by a coefficient of 0.765 to match values of magnetic moment above T_c as shown by open black squares in figure 4. This coefficient was further used for this film to reduce magnetisation loops in figure 1 and subtract them from the loops in figure 2. The result of subtraction is shown in figure 5. The unusual drop in magnetic moment in the loop at 77.3 K and tilted to the right loop at 20 K reflect the influence of the field gradients enhanced by the magnetic substrate. Using equation (1), the loops in figure 5 were used to calculate J_c and total critical current density per centimetre of width $I_{c\text{-}w} = J_c d$. The $I_{c\text{-}w}(H)$ for the film at two temperatures are shown in figure 6.

The difference between the lines represented by filled red squares and open black symbols in figure 4 gives the net magnetic moment of the superconducting film and can be used for calculation of the temperature dependence of J_c or $I_{c\text{-}w}$ at field of 1 T. We performed these calculations using equation (1) for several key samples deposited on RABiTS substrates.

The example given in figure 6 shows that $I_{c\text{-}w}$ in the film of moderate thickness can reach a high value above 10^3 A/cm at 20 K, which is the operating temperature in liquid hydrogen applications.

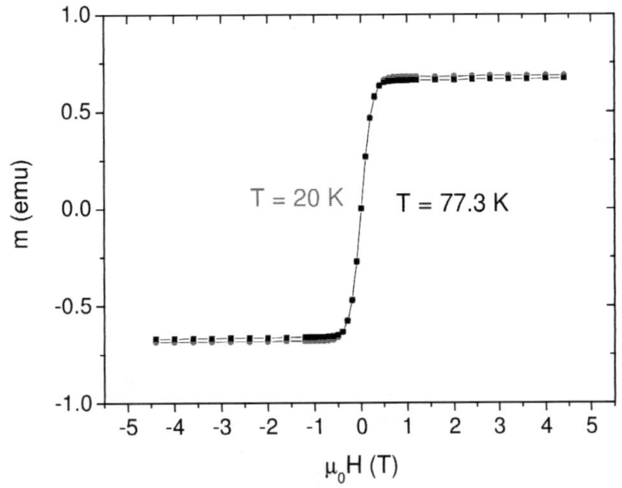

Figure 2. Magnetisation loops of the Ni5%W RABiTS substrate at two temperatures: 77.3 K (black squares) and 20 K (red circles).

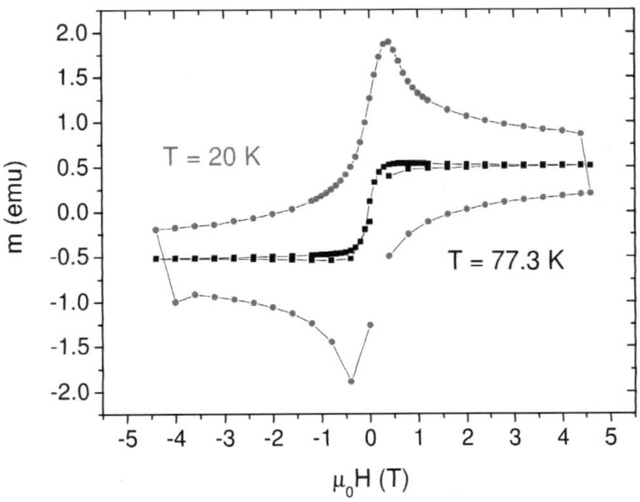

Figure 3. Magnetisation loops of a 3.75 μm thick integrated Ag/YBZO film deposited on RABiTS substrate at 77.3 K (black squares) and 20 K (red circles).

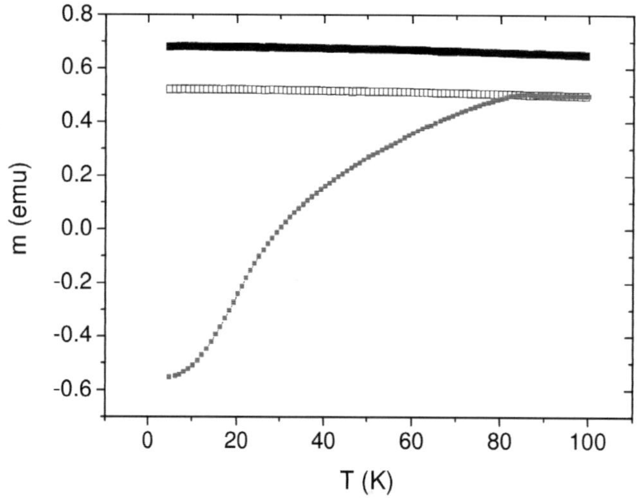

Figure 4. Temperature dependence of the DC magnetic moment of the substrate (filled black squares) and YBCO integrated film on the substrate (filled red squares) at magnetic field of 1 T. The open black squares show the reduced *m* of the substrate adjusted to *m* of integrated film at *T* above T_c.

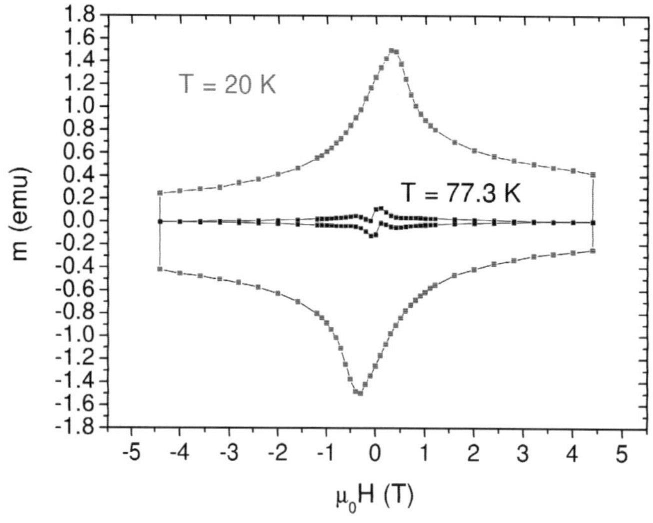

Figure 5. Subtracted magnetisation loops for a 3.75 μm thick integrated Ag/YBZO film at 77.3 K (black squares) and 20 K (red squares).

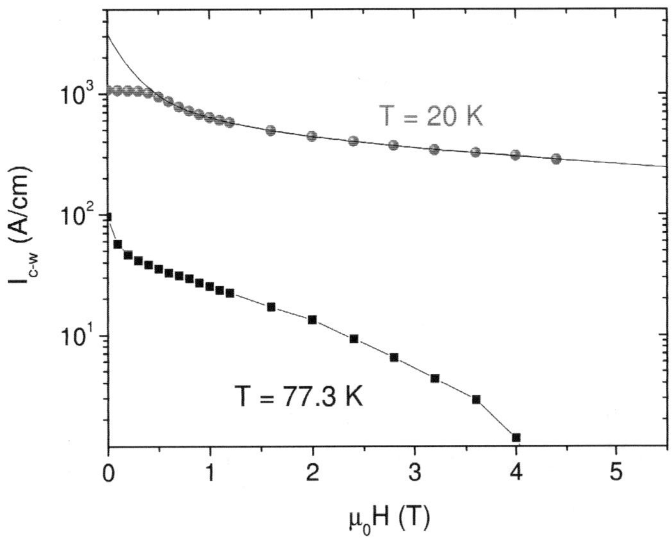

Figure 6. Magnetic field dependence of total critical current per centimetre of width for Ag/YBZO film at 77.3 K (black) and 20 K (red) derived from the magnetisation loops in figure 5. Thin solid line shows an extrapolation for high-field $I_{c\text{-}w}(H)$.

The field dependence of $I_{c\text{-}w}$ for 20 K is weak and the $I_{c\text{-}w}$ at this temperature can only be determined from equation (1) at fields above 0.5 T as shown by the extrapolation line from higher magnetic fields. The underestimation of $I_{c\text{-}w}$ at low fields from (1) is significant. The $I_{c\text{-}w}$ at 77.3 K is more than an order of magnitude below $I_{c\text{-}w}$ at 20 K and it has a strong dependence on magnetic filed.

Figures 7 and 8 show the temperature dependence of J_c and $I_{c\text{-}w}$ for a range of the films at a magnetic field of 1 T, important for applications. It is also the field at which (1) can reliably be used to determine J_c and $I_{c\text{-}w}$ from DC measurements. The black filled squares show J_c and $I_{c\text{-}w}$ for pure YBCO film deposited on RABiTS substrate. As expected, for the misorientations of substrate shown in figure 1, J_c and $I_{c\text{-}w}$ are well below that of pure YBCO film deposited on single crystal $SrTiO_3$ substrate: two examples are shown by open black triangles and squares. The ratio of $I_{c\text{-}w}$ in these films is about one order of magnitude. There is little improvement in J_c and $I_{c\text{-}w}$ by the deposition on RABiTS substrate of YBCO with added 2% of BZO (red filled squares), and the highest values of J_c and $I_{c\text{-}w}$ are achieved in Ag/YBCO+2%BZO single-layer (olive and dark cyan spheres) and quasi-multilayer (blue) films.

As seen in figure 7, J_c in integrated films reaches that in the YBCO films deposited on single-crystal substrates. It is likely to be due to the correlated growth of BZO and YBCO nano-columns, which could even reduce misorientations in the superconducting film comparable with misorientations in the substrate. As seen from figures 7 and 8, the BZO nanorods alone are not able to improve the crystallinity of the film. The slightly different relative arrangement of the lines in figures 7 and 8 is due to different thickness of the films.

In zero field, J_c was measured by the AC method. Figure 9 shows the temperature dependence of imaginary AC magnetic moment m'' for one of the best pure YBCO samples (open squares) deposited on single-crystal $SrTiO_3$ (STO) substrate and integrated Ag/YBCO+2%BZO sample (filled squares) deposited on RABiTS substrate. $m''(T)$ was measured at four different values of h: 0.05 Oe (black), 0.1 Oe (red), 0.5 Oe (green) and 1 Oe (blue) and in the plot it is reduced by the value of h.

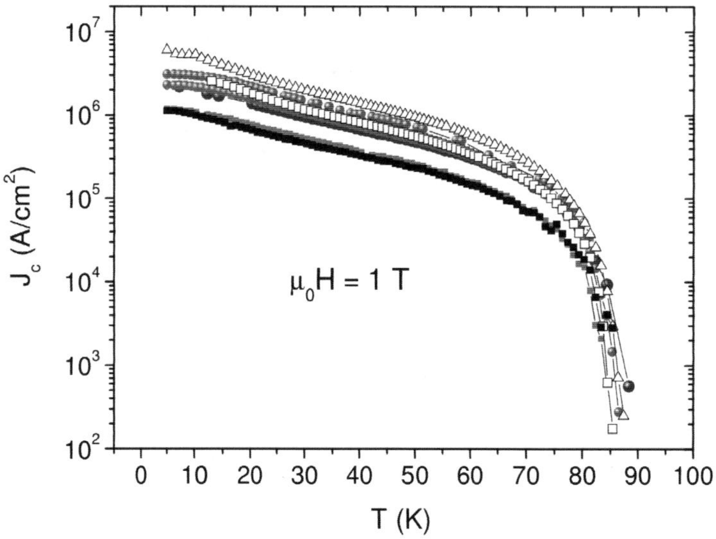

Figure 7. Temperature dependence of J_c for a range of films at magnetic field of 1 T. Black filled squares show J_c for Ag/YBCO film deposited on RABiTS substrate. Open black triangles and squares are for pure YBCO films deposited on $SrTiO_3$ substrate. Red filled squares show J_c for YBCO+2% BZO film on RABiTS substrate and blue, olive and dark cyan symbols are for three different integrated Ag/YBCO+2%BZO films on RABiTS substrate.

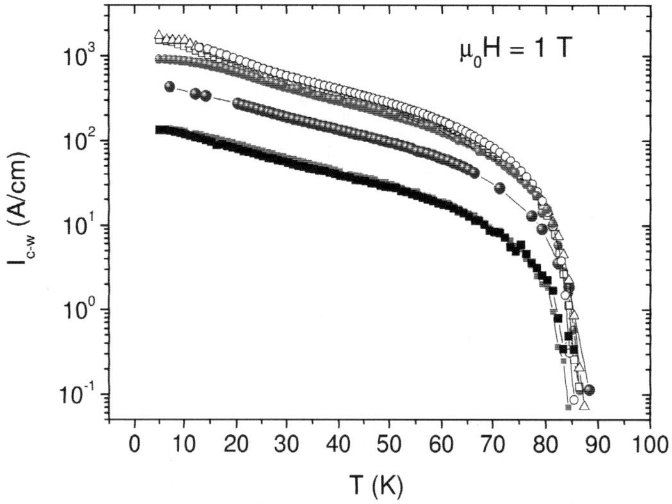

Figure 8. Temperature dependence of $I_{c\text{-}w}$ for a range of the films at magnetic field of 1 T. The symbols are the same as in figure 7.

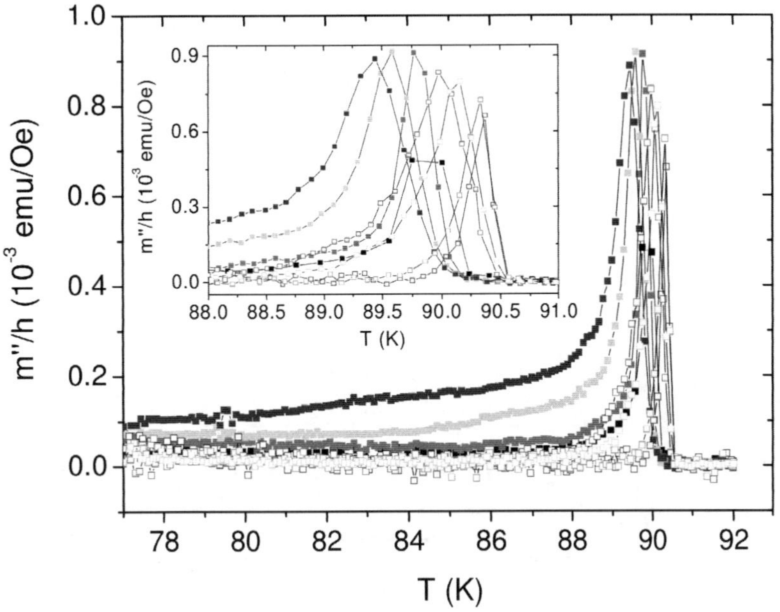

Figure 9. Temperature dependence of reduced by excitation field imaginary AC magnetic moment m'' for one of the best pure YBCO samples (open squares) deposited on single-crystal $SrTiO_3$ substrate and integrated Ag/YBCO+2% BZO sample (filled squares) deposited on RABiTS substrate. The inset shows $m''/h(T)$ of the main plot in the vicinity of T_c. $m''(T)$ has been measured at four different values of h: 0.05 Oe (black), 0.1 Oe (red), 0.5 Oe (green) and 1 Oe (blue).

The position of the maximum on each curve gives the temperature at which J_c is equal to a certain value defined by h according to equation (2). It is the motion of the maximum to lower T with the increase of h that characterises the temperature dependence of J_c. Figure 9 shows that in both films the maximum remains close to T_c in spite of significant increase in h. Although the film deposited on STO has a higher T_c than in the Ag/YBCO+2%BZO film on RABiTS substrate, the rate of the peak motion is comparable in both films, as can be seen in the inset. The wider peaks in Ag/YBCO+2%BZO film reflect the less homogeneous distribution of current and/or the influence of misorientations between grains enhanced by the deformations developed at cutting samples for magnetisation measurements. The measurements similar to those shown in figure 9 have been performed for several samples grown on RABiTS and STO substrates and their $J_c(T)$ are shown in figure 10.

In this figure, the open triangles show $J_c(T)$ for two representative YBCO films deposited on single-crystal STO substrates. Filled blue squares are for a pure YBCO film deposited on RABiTS substrate. It shows somewhat lower J_c than in films on STO substrate at low temperatures. The YBCO+2%BZO film deposited on the tape (black diamonds) shows even lower J_c. The Ag/YBCO+2%BZO film deposited on Ni5%W RABiTS (red stars) shows the highest J_c among other samples except when comparing with higher-T_c YBCO tape shown by open down triangles. The Ag/YBCO film on the metal tape with YBCO nanocolumns [4] (orange circles) is an exceptional sample with very high T_c. However, the rate of decrease in J_c with decrease of T is higher in this sample comparable with others, and it cannot compete with Ag/YBCO+2%BZO at low temperatures.

Figure 10 confirms that the superconducting properties of the samples deposited on RABiTS substrates, the best being the integrated Ag/YBCO+2%BZO, are comparable with properties of the samples on STO substrates.

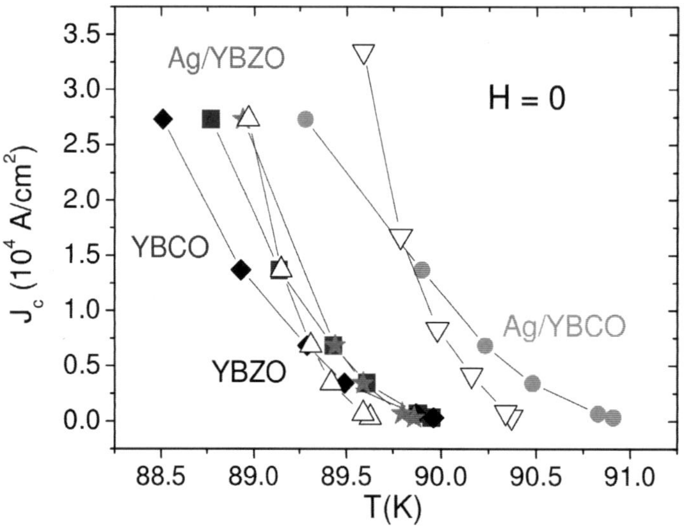

Figure 10. Temperature dependence of J_c for several samples deposited on single-crystalline STO substrates (open symbols) and RABiTS flexible metal tape (filled symbols). The blue squares are for pure YBCO film deposited on RABiTS substrate. The YBCO+2%BZO film deposited on the tape is shown by filled black diamonds and the Ag/YBCO+2%BZO film on Ni5%W is shown by red stars. The Ag/YBCO film on metal tape with YBCO nanocolumns is shown by orange circles.

The DC measurements discussed above were performed in a magnetic field perpendicular to the substrate, which is the direction of field with the strongest suppression of J_c. The extended defects like BZO nanorods or YBCO columns reduce the effect of magnetic field and the best way to clarify how it is done is to measure the angle (Θ) dependent transport J_c or $I_{c\text{-w}}$. The $I_{c\text{-w}}$ (Θ) for integrated Ag/YBCO+2%BZO sample on RABiTS tape at temperature of 77.3 K and magnetic field of 2 T is shown in figure 11. The angle dependence clearly demonstrates the increase of $I_{c\text{-w}}$ in the direction perpendicular to the substrate or along the *c*-axis of YBCO (maxima at angles of 0°, 180° and 360°). However, the pinning along the c-axis is not dominant in this sample as it is in the integrated samples deposited on STO substrates [4]. This indicates that there is still room for improvement in the properties of superconducting films deposited on RABiTS substrates.

Finally, the presence of extended defects responsible for the increase in $I_{c\text{-w}}$ along the *c*-axis of YBCO is confirmed by TEM imaging of the cross-section of an Ag/YBCO+2%BZO sample deposited on Ni5%W RABiTS tape as shown in figures 12 and 13. Figure 12 is focused on the buffer layer composed of two layers of CeO_2 and one layer of YSZ in-between, while figure 13 is a TEM image of the YBCO+2%BZO (YBZO) layer. The extended vertical nano-defects perpendicular to the substrate are clearly seen in this layer.

The described method of deposition features double additions to YBCO: Ag and BZO nanoparticles. Similar methods, for example, with nano-additions of BZO and Y_2O_3 are already described in the literature [10,11]. The difference between our method and methods described in

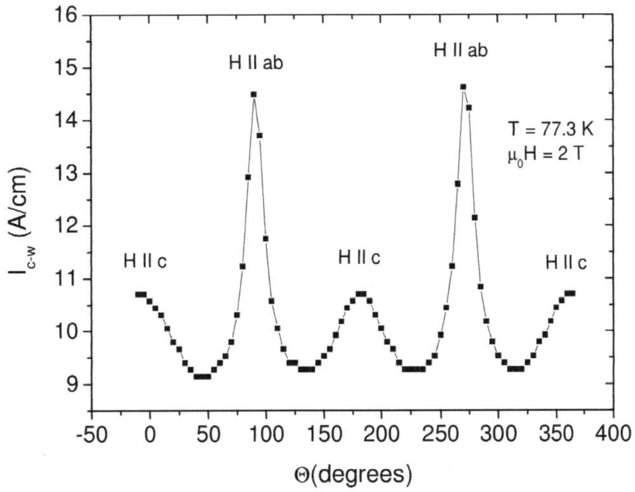

Figure 11. Angle dependence of transport $I_{c\text{-}w}$ for integrated Ag/YBCO+2%BZO sample deposited on Ni5%W RABiTS tape.

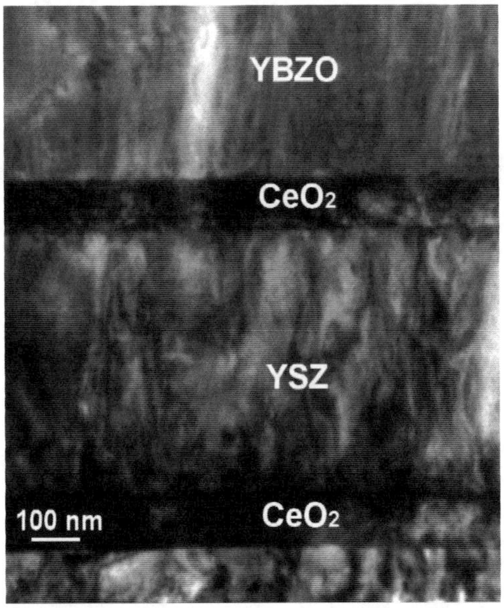

Figure 12. TEM image of the cross-section of an Ag/YBCO+2%BZO sample deposited on Ni5%W RABiTS tape in the vicinity of the buffer layer.

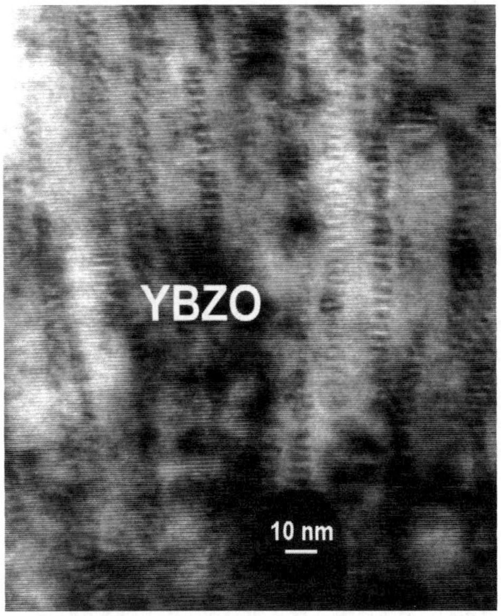

Figure 13. TEM image of YBCO+2%BZO layer of the sample shown in figure 12. The extended nano-defects perpendicular to the substrate are clearly seen in the layer.

[10,11] is that Ag nanoparticles do not pin directly the magnetic flux. Instead, they promote the growth of YBCO nanocolumns. The motion of Ag nanoparticles perpendicularly to the substrate may decrease the misorientations in YBCO films and reduce costs of texturing metal substrates.

4. Conclusions

The integrated method for the deposition of superconducting films that unites the growth of $BaZrO_3$ nano-rods and generation of HTS columns by nanoparticles of noble metals is applied to rolling-assisted biaxially-textured metal substrates with relatively high misorientations between grains. The $YBa_2Cu_3O_x$ superconducting coatings of the thickness up to 5 μm show strong increase in total critical current per centimetre of width in comparison with pure $YBa_2Cu_3O_x$ films. The method can be considered as a way to overcome the negative effect of strong grain misorientations in the flexible metal substrates and may lead to a cheap technology for the deposition of coated conductors.

5. References

[1] Macmanus-Driscoll J L, Foltyn S R, Jia Q X, Wang H, Serquis A, Civale L, Maiorov B, Hawley M E, Maley M P, and Peterson D E 2004 *Nat. Mater.* **3** 439
[2] Maiorov B, Baily S A, Zhou H, Ugurlu O, Kennison J A, Dowden P C, Holesinger T G, Foltyn S R and Civale L 2009 *Nat. Mater.* **8** 398
[3] Mikheenko P, Tanner J L, Bowen J, Sarkar A, Dang V-S, Abell J S and Crisan A 2010 *Physica C* **470** S234
[4] Mikheenko P, Dang V-S, Tse Y Y, Awang Kechik M M, Paturi P, Huhtinen H, Wang Y, Sarkar A, Abell J S and Crisan A 2010 *Supercond. Sci. Technol.* **23** 125007
[5] Dimos D, Chaudhari P and Mannhart J 1990 *Phys. Rev. B* **41** 4038
[6] Durrell J H and Rutter N A 2009 *Supercond. Sci. Technol.* **22** 013001
[7] Eickemeyer J, Hühne R, Güth A, Rodig C, Gaitzsch U, Freudenberger J, Schultz L and Holzapfel B 2010 *Supercond. Sci. Technol.* **23** 085012
[8] Shantsev D V, Galperin Y M and Johansen T H 2000 *Phys. Rev. B* **61** 9699
[9] Brandt E H 1994 *Phys. Rev. B* **49** 9024
[10] Mele P, Matsumoto K, Horide T, Ichinose A, Mukaida M, Yoshida Y, Horii S and Kita R 2008 *Physica C* **468** 1631
[11] Feldmann D M, Holesinger T G, Maiorov B, Zhou H, Foltyn S R, Coulter J Y and Apodoca I 2010 *Supercond. Sci. Technol.* **23** 115016

Acknowledgements

Financial support of the European Commission through the Marie Curie Excellence Grant MEXT-CT-2006-041111 "NanoTechPinningHTS", ESF-NES Network and Romanian Ministry of Education and Research are gratefully acknowledged. Authors thanks Dr. Eamonn Maher and 3C-s Ltd. for supplying RABiTS tape and Prof. P. Paturi for supplying target for the deposition of YBCO+2%BZO.

The principles and evolution of magnetic resonance imaging

M. Lakrimi[1], A.M. Thomas, G. Hutton, M. Kruip, R. Slade, P. Davis, A.J. Johnstone, M.J. Longfield, H. Blakes, S. Calvert, M. Smith, C.A. Marshall
Siemens Magnet Technology, Wharf Road, Eynsham, Witney, Oxon OX29 4BP, UK

E-mail: mhamed.lakrimi@siemens.com

Abstract. Magnetic Resonance Imaging (MRI) offers the best textbook example for the exploitation of both Heike Kamerlingh Onnes's major discoveries, namely the liquefaction of liquid helium and superconductivity. This paper will briefly describe how MRI images are acquired, give a historical account about the early days, and then focus on the magnet. It provides a window into the world of physics and engineering of magnet technology and it will chart the magnet design evolution since inception. For example, over the past 20 years, magnet designers have taken 1.5T MRI magnets from a weight of 13tonnes and a length of 2.40m in 1989 to a weight of 3.2tonnes and a length of 1.37m in 2009. The 3T products have also undergone similar developments. Coils experience mechanical forces of order 380tonnes. The cryogenics for MRI magnets have also undergone major improvements. Today's magnets do not use liquid nitrogen and most importantly have zero helium loss, thus needing no refill. Combined with the weight, this makes the body scanner accessible to developing countries and easy to site in any room. Although the future may eliminate the need for liquid helium completely, MRI will always continue to depend on superconductivity.

1. Introduction and superconductivity work by Onnes

The achievement of Heike Kamerlingh Onnes to liquefy helium in 1908, which won him the Nobel Prize in 1913 [1], and his discovery of superconductivity in 1911 opened up a plethora of new physics and engineering. The units Ohm [2] and Volt [3] have been redefined in terms of physical observables at low temperatures as is much of the research on quantum encryption and QUBITs [4]. Several Nobel Prizes have also been awarded for discoveries made at low temperatures or in superconductivity.

The world waited 50 years before the first commercial research magnet was built, thanks to the breakthrough in manufacturing superconducting wire and the efforts of Sir Martin Wood and his wife Audrey who founded Oxford Instruments in 1959 with the blessing of Sir Nicholas Kurti [5]. Magnetic Resonance Imaging (MRI) uses the same underlying physics as Nuclear Magnetic Resonance (NMR) and exploits the spin relaxation of a small percentage of protons in response to a specially sequenced external excitation. Since its introduction in the early 1980's, MRI quickly became accepted as one of the leading diagnostic imaging modality in healthcare. Being a safe and non ionising technique, almost all the tissues in the body could be non invasively and routinely scanned, enabling the diagnosis of tumours, tendonitis, stroke, multiple sclerosis, infections of the brain, spine and joints, visualisation of torn ligaments, or shoulder injuries, brain trauma (bleeding or swelling), and evaluation of bone tumours, cysts, bulging/herniated discs, structure of the heart or aorta or

[1] To whom any correspondence should be addressed.

masses in soft tissues. Preventative care and the problem of an ageing population will continue to drive new applications for MRI.

Superconductivity, an electrical state discovered by Onnes at Leiden, manifests itself by zero (electrical) resistance when the temperature drops below a certain critical temperature. The cooling of the sample is achieved by means of liquid helium, which was first liquefied by Onnes.

After liquefying helium, Onnes dreamed that [1] "*in the future, I see all over the Leyden laboratory measurements being made in cryostats, to which liquid helium is transported just as the other liquid gases now are, and in which this gas also, one might say, will be as freely available as water*". Nowadays, helium is an expensive commodity and the advent of cryogen-free system and higher temperature superconductors are gnawing into this niche application.

Shortly after discovering superconductivity, Onnes thought of magnets. He told that [1] "*mercury at 4.2K has entered a new state, which can be called the state of superconductivity... Now that we are able to use these metals, all types of electrical experiments with resistance-free apparatus have become possible... A self-contained coil, cooled in the magnetic field, should, when the field is removed, be able to simulate for some time an Ampère molecular current... The absence of Joule heat makes feasible the production of strong magnetic fields using coils without iron, for a current of very great density can be sent through very fine, closely wound wire spirals. Thus we were successful in sending a current of 0.8 amperes, i.e. of 56A/mm^2, through a coil, which contained 1,000 turns... All that I have said so far about superconductors applies only to currents below a certain threshold value, which lies all the higher the lower one drops in temperature*". The far sighted Onnes was already using notions which nowadays are taken for granted, namely a current density, a critical current, a current-temperature relation. In April-June 1914, he achieved persistence for hours which he destroyed using a heater [1]. However, a practical superconducting electromagnet had to await the discovery of type-II superconductors that could withstand high magnetic fields. The first successful superconducting magnet was built by Yntema [6] in 1954 using niobium wire and achieved a field of 0.71T at 4.2K. On return from a conference in the USA, Martin Wood and his wife Audrey of Oxford Instruments built their first NbZr 4T in March 1962 [5]. Nb_3Sn, a higher current material and its tolerance of high background fields accelerated interest in building magnets [7]. Human MRI scanners became widely available in the late 1980s, producing images of the inside of the body. Current MRI scanners can produce highly 3-dimensional images and video of the human anatomy showing moving organs. A 3T scanner has an image resolution of better than 0.5mm. We shall first start with a brief review of the physics, the historical events, and then deal with the design and evolution of magnets over the past few decades and the role they now play in MRI.

2. The physics of MRI

2.1. Introduction

MRI has its roots in NMR. The word *nuclear* was in reference to the nucleus and has nothing to do with radioactivity. It is for this reason that the medical community dropped the word (nuclear) and it simply became MRI and not NMRI. Magnetism is a property of matter that is a result of orbiting electrons in atoms. Working independently, Block [8] and Purcell [9] made the first successful NMR experiments in 1946. The first patent in the field of MRI was granted in 1974 to Damadian. He performed a form of MRI prototype exam on a human being on July 3, 1977, and it took almost five hours to produce one image! Today a whole body can be scanned in about 20 minutes, thanks to faster computers and improved hardware for producing the necessary switched field gradients and improved radio frequency transmit/receive antennas.

The nucleus of a hydrogen atom is a single proton. The human body consists of 60% water which represents the biggest source of hydrogen protons in the body, followed by fat. The orientation of these tiny proton magnets is random in the absence of a magnetic field. In analogy to a pixel in 2D images, MRI uses a volumetric pixel called voxel. Since one mole of water fills a volume of 18ml and that there are two hydrogen atoms per mole, a voxel of $1\times1\times1$mm^3=.001ml would contain $2\times6.02\text{x}10^{23}$

(molecules per mole) $\times 0.001/18 = 6.7\text{x}10^{19}$ total protons. The population between the two spin up and down levels obey Maxwell-Boltzman statistics. At 3T and 37°C (the body temperature), the total number of excess protons = $6.7\times10^{19}\times\Delta E/(2k_BT) = 6.02\times10^{15}$, where k_BT is the thermal energy, and $\Delta E=\gamma B$ is the magnetic energy. B is the magnetic field flux density and γ is the gyromagnetic ratio associated with a Larmor precession frequency taking a value of 42.576MHz/T for the hydrogen nucleus. One can see that the frequency is in the low Radio Frequency (RF) range and this is why MRI uses RF excitation. Figure 1 gives the population distribution for every 2 million protons from 0T to 11.75T. This means that for every two million protons available in the 1mm^3 volume, there are, at 3T, 18 more protons aligned **with** the field than **against** it. The vast majority of spins cancel each other and hence it is this small spin excess which is probed in MRI to yield the images. It is incredible that such a small number of spins contribute to the MRI images that we know of today, which is a testament to the remarkable sensitivity of the detection electronics developed for MRI.

2.2. RF pulse, T1, T2, and T2 relaxations*

When a patient is loaded in the bore of an MRI magnet, the magnetic field, aligned along the z or longitudinal direction, settles the spin population between the low and higher energy states as given in figure 1. Under this steady state, not much happens except that there is a net magnetisation along the z-direction. The magnetisation in the transverse xy-plane is zero because of the randomness of the spins phases. When an external RF pulse excitation, matching or resonant with the Larmor frequency and oriented orthogonal to the static field is applied, a spin from the lower energy (up or ↑) state can be stimulated to jump into the higher energy (down or ↓) state. This is similar to intraband transitions in quantum wells. The net magnetisation experiences a torque which causes it to spiral away from the z-axis alignment; its vector will move a distance proportional to the duration of the RF field pulse, as shown in figure 2. The net magnetisation vector rotates an angle, α, called the flip or tip angle and maximum signal amplitude is achieved when this angle is set to 90°. After the RF pulse the spins precess around the static field like a spinning top.

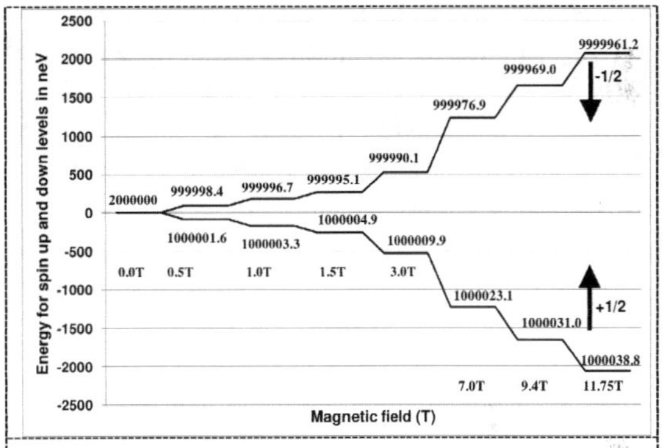

Figure 1. Distribution of 2million protons for spin up and spin down levels for a number of field intensities.

When the RF pulse is switched off, the excited protons gradually return to their original lower energy state and MRI exploits this relaxation or decay back to the equilibrium state. The spins return to their equilibrium in two ways. They can relax to their lower energy state by loosing energy to the surrounding molecular lattice and hence the name spin-lattice relaxation, as known in solids. This relaxation is characterised by a time constant T1 and a recovery in the longitudinal magnetisation, $M_z(t) \propto M_z(t=0) \times e^{-t/T1}$, where $M_z(t=0)$ the magnetisation at equilibrium, just before the pulse is applied. The recovery rate in $M_z(t)$ towards equilibrium follows an exponential decay. T1 is defined as the time when the magnetisation has returned to 63% of its original value. The relaxation is considered complete after 5T1 duration.

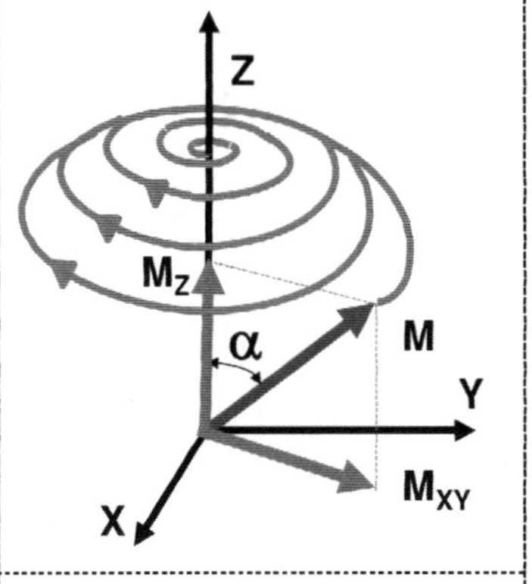

Figure 2. The magnetisation spirals away from B_z-alignment and decomposes into M_z and M_{xy}.

Following the RF pulse, the number of spins in the upper state is increased and thus there is increased interaction between the spins spinning in the xy-plane, leading them to lose their phase coherence over time and resulting in a reduction in the net transverse magnetisation, $M_{xy}(t)=M_{xy}(t=0)\times\sin\alpha(1-e^{-t/T2})$. The precessing spins generate an alternating magnetic field which can be detected using a suitable pick up coil. The signal emitted during this relaxation is called a Free Induction Decay (FID). This timescale is defined by the T2 spin-spin or transverse relaxation time constant. In this dephasing process, there is no loss of energy to the surroundings. T2 is also characterised by the time taken to lose 63% of the transverse magnetisation and is always smaller or equal to T1. T1 and T2 are defined by properties of the materials under study and hence may be used to distinguish between types of tissue by differing image contrast. Table 1 gives some values for T1 and T2. In terms of the black and white MRI images, T1 weighted images depict fat and normal anatomic planes and fat suppressed T2 weighted images are used for abnormalities.

The magnet's imaging field can never be perfectly homogeneous and is also influenced by differences in magnetic susceptibility between tissues and cavities, for example in the sinus region, or if a patient has metallic inclusions. This residual inhomogeneity causes a further dephasing or loss of transverse magnetisation with a time constant shorter than T2, called T2*. At first this appears to be a significant problem, but fortunately a sequence call Hahn-echo or spin echo can be used to rephase those spins which have undergone a T2* decay. To illustrate the spin-echo imagine protons precessing at different frequencies; let's select for the sake of clarity just three spins precessing at different rates due to the small variations in local magnetic field. These spins are analogous to a tortoise, a hare, and a snail racing. At t=0sec, all three are standing on the starting line and, focused on the race, they are said to be in coherence. If left to race in a straight line, the hare would cover a much longer distance than the tortoise and the snail would finish last. As the gaps between them are increasing, they are said to lose coherence. However, imagine that after t=TE/2 they are summoned to return to the starting line. It follows that the hare will have a longer distance to travel back and the snail the shortest. In the end, all three would cross the starting line at the same time despite having covered different distances. As all three stand on the starting line again, they are said to have regained phase coherence causing the formation of an echo signal in the RF pick up coil. The event which causes the direction of the race to be reversed is the application of a further RF pulse which flips the magnetisation vectors around the third axis by 180°, rotating their precession.

Table 1. T1 and T2 values at 1T.

Material	**T1(s)**	**T2(s)**
Cerebro-Spinal Fluid	2.50	1.40
Fat	0.24	0.09
Spleen	0.46	0.08
White matter	0.68	0.09
Gray matter	0.81	0.10
Water	2.50	2.50
Liver	0.49	0.04
Blood	0.80	0.18
Muscle	0.73	0.04

A great variety of MRI pulse sequences have been developed, some basic ones being Spin Echo, Inversion Recovery, Gradient Echo, Turbo Spin Echo [10], etc. Important pulse sequence parameters are TR (Repetition Time between applications of the initial tipping pulse), TE (Time to Echo), TI (Time for magnetisation Inversion) and Flip Angle. The values of these parameters are chosen depending on the tissues and MRI examination. Finally, scientists also use contrast agents such as paramagnetic materials at the right dose to engineer a preferential T1 enhancement. For example gadolinium contrast agents cause preferential T1 relaxation enhancement, causing increase in signal on T1-weighted images and thus achieve better resolution. Contrast agents are used for bowel, liver, and lymph node imaging.

In section 3, we shall give a short description as to how the MR signals are used to give us the Cartesian two dimensional images. But how did the science of MRI start?

2.3. The history and race to MRI

As far back as 1845, Faraday started looking into the magnetic properties of dry blood and even noted in his book that he must try recent fresh fluid. However, Damadian was the first to show in 1971 that the nuclear magnetic relaxation times of tissues and tumours differed, thus motivating scientists to consider magnetic resonance for the detection of disease. In 1973, Lauterbur proposed the use of magnetic field gradient coils to distinguish between NMR signals from different locations combining this with a form of reconstruction from projections as used in computed tomography [11]. He went on to demonstrate MRI for the first time on small test tube samples. In 1974, Mansfield's group developed selective spin-echo excitation [12]. The University of Aberdeen followed Lauterbur's article and built a small scale MR imager which produced an image of a mouse in 1974 [13]. In 1975, Ernst's group proposed MRI using phase and frequency encoding and the Fourier Transform [14]. In July 1977, Damadian demonstrated a form of MRI called field-focusing nuclear magnetic resonance [15]. Aberdeen University (UK) conducted the first clinical whole-body MRI scan in the world on 28 August 1980 [13]. Edelstein et al developed the spin warp sequence [16] and Mansfield and co-workers the Echo-Planar Imaging (EPI) technique [17]. Nowadays images are produced at video rates of 30ms per image. Figure 3 shows a photograph of a modern MRI body scanner. The MRI images can be in any plane: axial (xy-plane), coronal (zx-plane), or sagittal (zy-plane) or any arbitrary angled plane as illustrated in the inset to figure 3. MRI creates cross-sectional images or slices of a body part.

In this next section, we will look at the history of the MRI superconducting magnet.

2.4. A short history of the MRI magnet

From the famous date of 1911, several superconducting materials have been discovered, the list and chronology of which is beyond the scope of this paper. Currently, NbTi, Nb_3Sn, and MgB_2 have emerged as the materials of choice for building superconducting magnets with NbTi being the true workhorse; its current cost is less than one fifth of that for Nb_3Sn and MgB_2.

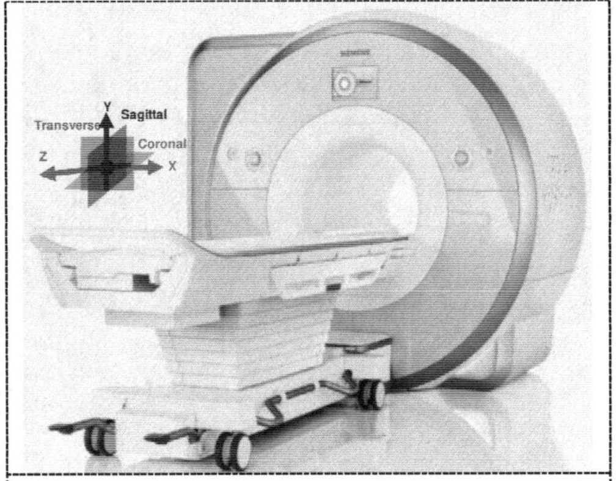

Figure 3. A modern MRI scanner. Sagittal, coronal, and transverse planes are shown.

The use of superconductors benefited from the fact that copper was expensive and consumed much greater electricity and the resulting heating was very challenging to address. Surprisingly, it was the high energy physicists in need of high magnetic fields who championed and propelled the use of superconductors. Rutherford Appleton Laboratory (UK) built the first superconducting accelerators using copper clad NbTi in 1960 and American scientists followed in 1967. Commercial NbTi was only developed in 1962 and in the same year General Electric (GE) built the world's first 10T magnet in 1962. Although this was looked on as an achievement, the commercial aspect was a disaster as Bell Labs only paid GE the fixed price contract of $75k for a magnet that cost $200k to realise [18]. Intermagnetics General Corporation (IGC, a spin off from GE in 1971) and Oxford Instruments (OI, the second spin off from The University of Oxford) emerged as champions of the commercialisation of superconductors. In 1967, OI demonstrated a 10T magnet and a 30mK dilution refrigerator. However, it is MRI and to some extent NMR that expanded the use of superconductors.

OI built the first superconductor magnet with a bore diameter of 1m. It delivered, in 1980, the first MRI whole body magnets (0.15T) to Hammersmith hospital (UK) and the University of California in San Francisco and the third to GE. Orders for MRI magnets increased from £1million in 1981 to £25million in 1982 and OI commanded 90% of the market [18]. Following this achievement the company set up in 1982 Oxford Magnet Technology (OMT) to exploit its technology for MRI body scanners; the magnet volume was 50 systems a year. In 1984, a purpose built factory was established with a production capacity of 300 magnet systems per year. In May 1989, OMT became a joint venture between Siemens AG (51%) and OI (49%). In November 2003, Siemens acquired whole ownership of OMT and its name changed to Siemens Magnet Technology (SMT) in June 2004. By the end of the joint venture, OMT had supplied more than 7,000 MRI magnets which amounted to at least a third of all magnets installed in world hospitals. SMT continues to supply Siemens and other third party customers and following two further factory expansions, the manufacturing capacity is stretched to 1500 systems per year. In 2006, Philips Medical Systems acquired IGC. The top five MRI magnet suppliers are Siemens, GE, Philips, Toshiba, and Hitachi.

NMR and MRI remain the most significant profitable application of superconductivity to date. More than 75% of MRI magnets are made using superconducting wires. More than 2,500 superconducting MRI systems are now sold annually and more than 75,000 people are imaged daily. In 2008, the installed base of superconducting MRI systems was about 26,500 units compared to 14,600 in 2002 [19]. We shall now review the underlying physics and engineering behind the magnet design.

3. The design of an MRI magnet

Although, the design of the MRI magnet has undergone a major revolution, magnets only came of age recently. In 1980, a standard superconducting MRI magnet achieved only 0.5Tesla compared with 1.5Tesla today. Furthermore, in 1980, a 3Tesla MRI magnet was almost unthinkable compared with today when it is gradually becoming the standard and systems with field strengths above 7T have been successfully built and 11.75T is in progress [20].

3.1. The magnet design

The magnet could be designed as a set of discrete coils or solenoids. However, the former geometry is preferred to make the magnet commercially viable, occupy the least volume and have a better control over the field homogeneity. Although a customer may only be interested in the specifications of the magnetic field intensity, homogeneity, drift, and temperature of operation, the magnet designer has to adhere to several other requirements for critical current, strain/stress, stray field, and other factors to balance the technical and economical risks. Designers continually refine their design limits through observing the performance of magnets and the measurements of certain parameters such as strain, temperatures and voltages during quench. Within one field system, there may be several products and the customer chooses the magnet to suit their purposes.

Magnets are designed by a team of people. Upon being issued with a specification for an imaging volume and the field strength, first the design and location of the coils to an accuracy of less than 0.1mm and use wire of a particular geometry and insulation is carried out by the magnet designer. The mechanical engineer designs how the coils will be supported and how the magnet will be suspended within the cryostat. The cryogenic engineer will carry out the necessary calculation with regards to minimising conduction, radiation, and convection heating loads which affect the helium boil-off.

Magnets are designed using wire capable of carrying a few 1000A/mm^2 at a certain field; domestic wiring carry typically 10A/mm^2. At low temperatures, say 4.2K, the wire specific heat drops by a factor of 4,000 compared to that at room temperature and this is why the minimum heating energy required to destroy completely superconductivity in the magnet can be less than 20μJoules [21].

3.2. The magnet stresses and quench

Superconducting magnets operate at low temperatures and thus experience two types of stresses. Firstly, during thermal cooldown, the coils contract axially and radially, resulting in large negative strains and stresses. Secondly, during energisation, the coils experience intense Lorentz magnetic forces whereby the coils expand radially thus inducing a positive strain and simultaneously contract axially resulting in a negative strain. In addition, because of their proximity from one another the coils exert a force of a few hundred tonnes on each other and their supporting structure. It is also well known that if a local temperature exceeds that of superconducting wire critical temperature over a critical length known as the minimum propagation zone, the wire undergoes a transition to normal state and current will be dissipated through the resistive copper matrix, leading to excessive heating of the coil. This is termed a quench and the magnet suddenly releases its stored energy, typically tens of megaJoules, and a complex quench scenario ensues whereby a particular coil could be magnetically pumped up by its adjacent coils, leading it to experience higher stresses than intended [21]. To prevent any wire movement the coils are typically impregnated in wax or epoxy resin. Without adding other coil refinements, it suffices to say that each individual coil is actually a complex composite that may not necessarily be easy to model *a priori*.

Onnes himself also pondered over the origin of quench as to whether [1] "*this first impulse proceeds from bad sections of the wire or is also produced in pure, tension-free, evenly crystallised metal. If the potential phenomena are to be attributed to bad spots, we shall learn to eliminate them and then it is perhaps possible that the conductivity could be increased still higher than to the value reached at present, which is thousand million times that at normal temperature*". Fortunately, engineers rely on known rules to design magnets to avoid coils suffering mechanical damage and elasto-plastic deformation. If necessary, a quench protection circuit is designed to manage the stresses, the temperatures, and voltages arising during a quench [21].

3.3. The magnet stray field and homogeneity

The variation in homogeneity of the field within the bore of the magnet may be described by a series of Legendre polynomials. The designer adjusts the coils in order to minimise all impurities to an acceptable level while managing the other critical parameters. The very large inner diameter of 1m comes at a heavy price. The horizontal and vertical distances over which the stray magnetic field intensity for a 1.5T drops to less than 5G, deemed safe for patients with a heart pacemaker and electronics, can be very large indeed and need to be tightly controlled. In early magnets this was of the order of 10m axially, however this could be reduced by shielding consisting of a hundred or so tonnes of iron, making the requisite for reinforced floors and spacious rooms necessary. In 1986, OMT revolutionised MRI by introducing the Active Shield (AS) technology, allowing the stray field to drop to less than 5G within less than 4m axially and less than 2.5m radially, and thus enabling MRI systems to be installed at sites previously thought impossible. Active Shielding is achieved by engineering two outer coils in reverse polarity to the inner coils which generate the central magnetic field. This OMT innovation has become the industry standard. A typical MRI suite is sketched in figure 4. OMT/SMT have continued to focus on technological and manufacturing excellence. The business has won several Awards in recognition of its achievements in these areas and also for export success since 95% of the magnets produced at SMT in Eynsham are exported. The company has made significant contributions to the development of MRI technology and some of the notable world's firsts to emerge from OMT/SMT are given in table 2.

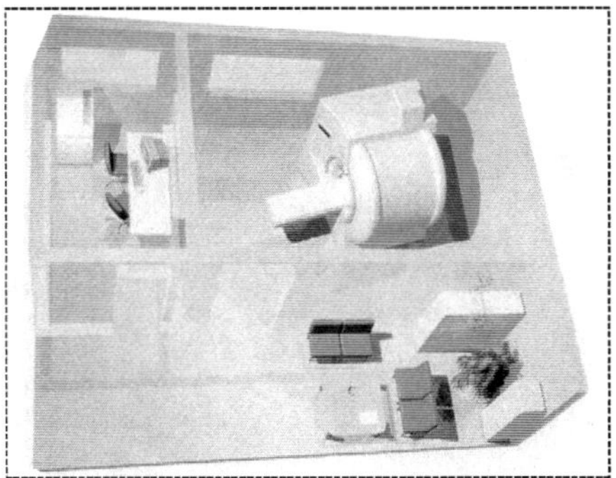

Figure 4. Sketch of an MRI suite using an Actively Shielded magnet.

For a high quality image, the magnetic field must not vary over the imaging volume by more than a few parts per million as shown in figure 5. This high homogeneity field is required to minimise the dephasing of the spins given by T2*.

Table 2. World's firsts from OMT/SMT.

Year	**Innovative Technology**
1980	First MRI superconducting magnet
1986	Actively shielded magnet
1989	1.5T AS magnet
1994	Open magnet
1997	New generation compact magnets
1997	3T AS magnet
1999	1T Open magnet
2000	4T AS magnet
2000	Shortest 1.5T AS magnet
2004	1.5T AS open bore magnet

3.4. The magnet stability

The coils are then linked in series and the wires jointed together and there is also a stability specification that the field should not decay by ≥-0.1ppm/hr. This means that the resistance of the joints must be much better than $10^{-12}\Omega$ at the operating current (typically several hundred amperes) and in a background field. The MRI in a hospital is run in persistent mode, i.e there is no power supply attached. In order to achieve this, a superconducting switch, which in normal state has a resistance of several ohms, is assembled in parallel to the magnet. When the MRI magnet is commissioned, the installation engineer first quenches the switch using a heater, thus causing its resistance to rise to a few Ω, and the current from the magnet power supply goes straight in the magnet which is in a superconducting state (0Ω). The rate at which the magnet is energised depends on the compliance voltage of the power supply and the inductance of the magnet. After reaching the desired operating current, the heater of the switch is turned off. After allowing sufficient time for the switch to revert to a superconducting state (0Ω), the current is ramped down to zero. Again given the finite internal resistance of the power supply, the current prefers to flow through the switch and it is only the current in the leads that drops to zero. The dissipation of the current with time is dictated by the quality of the resistance of the joints. A decay measurement is carried out to make sure that the drift stability is met.

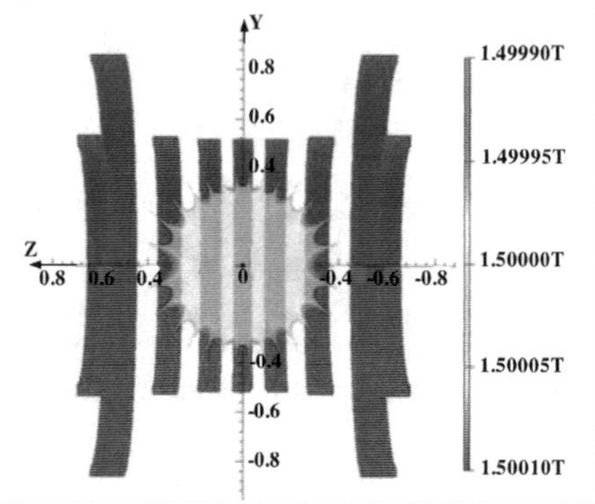

Figure 5. Homogeneity of 1.5T AS system. The bar shows ±67ppm. Distances are in metres.

The magnet engineer also includes in the design an external interference screening (EIS) circuit to protect the magnet against external disturbances such as a lifts, vehicles driving by, electrical cables in walls, etc. because Faraday law of induction states that, in a magnetic field, a current or voltage will be induced in a fast moving conducting material. Every MRI magnet is equipped with an Emergency Run Down Unit (ERDU) to deliberately quench it in case of emergency.

3.5. The magnet cryogenics and mechanical stability

The magnet is then put in a cryostat as shown in figure 6. The cryogenic engineer is concerned with limiting the three types of heating: conduction, convection, and radiation. These are minimised by the use of suitable materials. Magnets have a vacuum chamber that is pumped to ~10^{-5}mbar at room temperature which then cryopumps to ≤10^{-7}mbar on cooldown. The cryostat is checked for leaks. Multi-layer insulation is used to reflect back any radiation heating that finds its way through the cryostat. The mechanical engineer has to deal with the mechanical stability and suspension issues. The MRI magnet could be transported by plane, ship, and road to its final site. Note that there are also mobile systems which are shared between different sites. These spend their lifetime on the road with the magnet always at field.

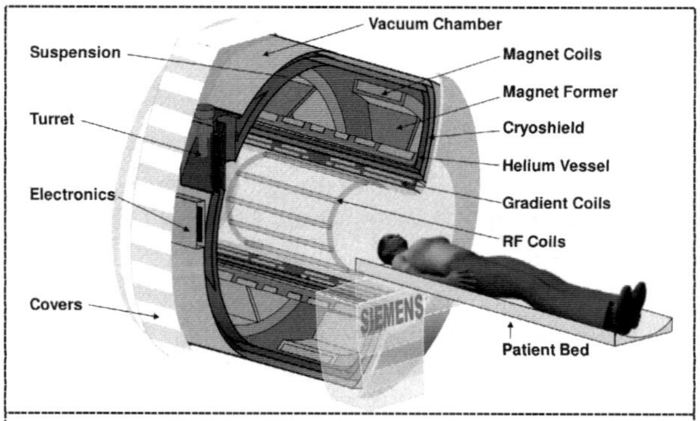

Figure 6. The MRI magnet and its components

Early magnets made use of liquid nitrogen and needed a refill of nitrogen on a weekly basis and helium on a monthly basis. On modern magnets known as zero boil-off magnets, smart cryogenic engineering has eliminated the use of liquid nitrogen. These have a cryocooler inserted through the turret (shown in figure 6) which acts as a mini-liquefier whereby any helium gas boiling off is recondensed back into the magnet and so the systems only need topping up with liquid helium every few years! The cryostat holds approximately a thousand litres of helium

3.6. The design evolution

Table 3 lists the design evolution for 1.5T over a period of 20 years. It can be seen that the length has been reduced to 1.37m, easing a patient's claustrophobic feelings. The weight has been cut from 13 to 3.23tonnes, thus making the magnet easy to site on any floor of a building. Figure 7 shows a picture of an old (left) and modern (right) MRI systems. In the short length magnet the patient's head and feet can be seen and thus combined with the large 70cm diameter clear bore magnet reduces the claustrophobia felt by patients.

Table 3. Design evolution of 1.5T AS magnets over 20 years

Year	Product name	Length (m)	Weight (kg)
1989	A	2.40	13,000
1994	B	2.20	8,650
1997	C	1.60	4,070
2003	D	1.50	4,050
2004	E	1.25	4,350
2009	F	1.37	3,230

4. The gradient coils and cause of audible noise

So far we have only referred to the RF pulse and the magnet. As mentioned before, it was Lauterbur who invented the concept of imaging using a gradient coil [11]. The gradient coils, as the name implies, are used to produce magnetic field gradients in specific directions. The gradient coils are in pairs with equal and opposite polarity about the centre. One coil decreases the field whilst the other increases it, like an incline on a seesaw. These field gradients are $\partial B/\partial z$, $\partial B/\partial x$, and $\partial B/\partial y$, where z is along the axis of the magnet bore, y is the vertical, and x is the horizontal. They are used to provide spatial localisation of the MR signal by using complex sequences of gradient coil and RF pulses. The spins in the higher field region will precess at a faster frequency than those located in a low field region. In figure 8, a transverse slice of spins is first selected to be in the xy-plane by applying the z-gradient simultaneously with an RF tipping pulse. The flip angle of spins within the slice is dependent on the resonant frequency and thus the thickness of the slice can be changed by using an RF pulse with a bandwidth and fixing the field gradient or by using a fixed frequency RF pulse which is held constant for a period of time whilst the field gradient is varied. Similarly applying the x-gradient will lead to a sagittal slice and the y-gradient will result in a coronal slice. The image resolution is dependent on several factors including the strength of the magnetic field. Increased image resolution usually comes at a penalty of increased acquisition time but this is mitigated by higher magnetic field strengths and modern innovations such as multiple parallel receivers. The image is typically a matrix of 256×256 pixels, i.e. made from 65,536 voxels.

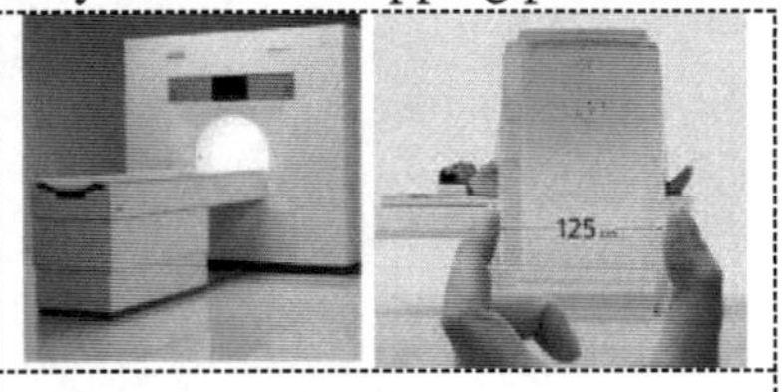

Figure 7. Long bore tube versus a modern MRI. The patient is not fully confined in the magnet.

The signal from different voxels within the slice can be split into 256 strips of say along the x-direction and one pixel height in the y-direction by application of suitable x and y gradient pulses during image acquisition. If a gradient is applied in the x-direction, the spins will precess at different rates dependent on their x-position; this is known as frequency encoding gradient. To differentiate the spins' y-location, the y-gradient may be switched on for a very brief period of time to *dial in* a phase shift between spins in the y-direction, as shown in the top part of figure 8. In this example, the y-gradient is known as the phase encoding gradient. Table 4 gives the phase encoding gradient for any imaging plane. The signal acquired by the RF pickup or receiver coil is an integration of signals from all excited spins in the slice. Earlier descriptions referred to frequency, phase coherence, and relaxation times, it is better to switch from spatial locations in the (x,y,z) Cartesian to a reciprocal *k*-space where these are easily described. In *k*-space, the axes within a slice are frequency and phase. The computer generates an intensity histogram from the signal acquired by the receiver RF coil for all 256 voxels forming a strip. The location of spins in a particular row is determined by their different frequencies and in a

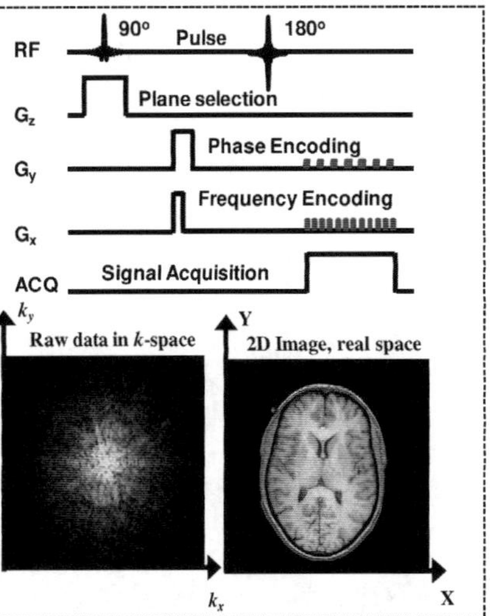

Figure 8. Typical MRI sequence and images in *k* and real spaces.

particular column by their relative phases. Two-dimensional Fast Fourier Transform (2D-FFT) is used to convert the k-space signal to a Cartesian image, as shown in Figure 8.

Table 4. Image plane and encoding gradient

Image	Gradient		
Plane	Select	Phase	Frequency
XY	X	X or Y	Y or X
XZ	Z	X or Z	Z or X
YZ	Y	Y or Z	Z or Y

People who have experienced examination or simply entered the imaging room during an examination procedure often remark on the noise. This is caused by the gradient coils being switched on for the sequences described above. The gradient coils sit inside the background field of the magnet and so when current is changed in the (gradient) coils to generate the field gradients they experience significant Lorentz forces which cause the coils to expand radially and contract axially as explained before in the context of the magnet design. As the switching is typically in the audible frequency range, the resulting vibration produces loud noises (clicking or beeping during scanning). The sound intensity is about 110dB but can reach 120dB (equivalent to the noise from a jet engine at take-off). Patients wear appropriate ear protection during the examination.

5. The other MRI sub-systems

Figure 6 shows a cross-section of an MRI magnet and a view of all its components; we have already discussed the magnet, the gradient coils, and the RF pulse. The installation engineer only uses a magnet power supply during commissioning of a magnet or servicing. There is electronics on the system to monitor the general state of the system. The gradient coils have their own power supplies which stay on site.

The bore contains the RF body coil which is used to generate the intense and uniform pulses of transverse RF magnetic field required to tip spins in the imaging volume. In figure 6, the RF pulse coil has the shape of a bird cage coil.

The RF receive coils are used to pickup and transmit the RF signal emitted from the hydrogen protons during relaxation back to equilibrium. The RF-receive coils are tuned to the magnet's Larmor frequency and fashioned as local antennas which are body-part specific, as shown in figure 9. They exploit Faraday's law of induction which states that a current is induced in a coil made from conducting material placed in a changing magnetic field. The signal is increased by close proximity to the region being imaged. Modern MRI systems use arrays of coils operating in parallel to reduce image acquisition or increase resolution.

The shim coils and iron are required to improve the uniformity of the magnetic field because realistic manufacturing tolerances make it impossible to achieve the required homogeneity without fine adjustments to the magnetic field, called shimming. Passive shimming hardware is simple and uses an array of small iron shim plates loaded in cartridges, typically located inside slots within the gradient coil. However, it has the drawback that the magnet must be ramped down to load the iron for obvious safety reasons and then re-energised again. The passive shims are not changed after set-up. There may also be active shimming utilising either superconducting or resistive coils.

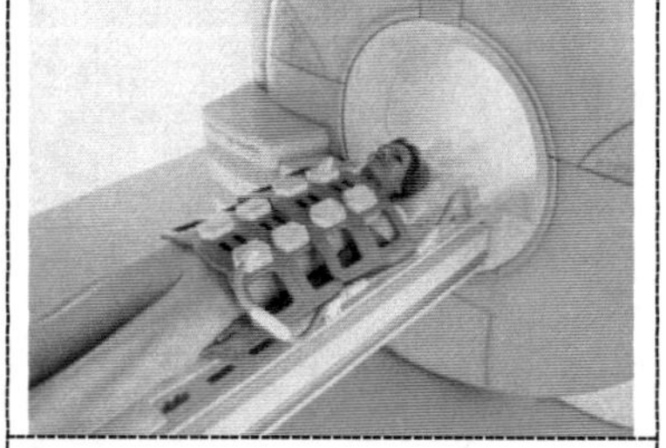
Figure 9. RF receiver coils.

The patient table or bed is used to load the patient into the central bore of the magnet. It can be fixed or demountable and often contains receiver coils (e.g. spine array).

The computer is for programming the gradient coils to select different imaging planes. The computer also processes the 2D-FFT images and creates the MRI images.

Finally the MR signal is extremely small as the earlier calculation showed and this is why it is important to minimise external RF interference. Thus, the MRI magnet is located in a Faraday cage which prevents external RF signals from reaching the patient under examination.

6. Conclusions

This paper reviewed a number of topics. It is in celebration of Onnes's centenary for the discovery of superconductivity to illustrate the commercial and profitable business that is MRI. It provides the basics behind MRI and describes how images are acquired. It offers an insight into how commercial MR magnet production started and the history of SMT. It gives a description of how magnets are designed and how the design has evolved over a period of 20years and finally an explanation of the various pieces of hardware that are used on a hospital MR whole body scanner.

7. Acknowledgements

We would like to thank Siemens Healthcare for the MRI images of the systems.

8. References

[1] http://nobelprize.org/nobel_prizes/physics/laureates/1913/onnes-lecture.pdf

[2] Klitzing Kvon, Dorda G and Pepper M 1980 New Method for High-Accuracy Determination of the Fine-Structure Constant Based on Quantized Hall Resistance *Phys. Rev. Lett.* **45** 494–497 and see also http://www.bourbaphy.fr/klitzing.pdf and references therein

[3] Burroughs CJ and Hamilton CA 1990 Voltage calibration systems using Josephson junction arrays, *IEEE Trans. on Inst. and Meas.* **39** 972–5 and references therein

[4] Morton JJL and Lovett BW 2011 Hybrid Solid-State Qubits: The Powerful Role of Electron Spins, *Annual Review of Condensed Matter Physics* **2** (publication date March 2011)

[5] Wood A 2001 *Magnet Venture* Oxford University Press, Oxford, UK

[6] Yntema GB 1955 *Physics Review* **98** 1197

[7] Kunzler JE, Buehler E, Hsu FSU and Wernick JH 1961 *Phys. Rev. Let.* **6** 890

[8] Bloch F, Hansen WW and Packard M 1946 Nuclear Induction *Phys. Rev.* **69** 127

[9] Purcell EM, Torrey HC and Pound RV 1946 Resonance Absorption by Nuclear Magnetic Moments in a Solid *Phys. Rev.* **69** 37-38

[10] Bernstein MA, King KF, Zhou XJ 2004 *Handbook of MRI pulse sequences* Elsevier Academic Press. See also McRobbie DW, Moore EA, Graves MJ and Prince MR 2007 *MRI: from picture to proton* 2nd Edition Cambridge University Press

[11] Lauterbur PC 1973 Image Formation by Induced Local Interactions: Examples Employing Nuclear Magnetic Resonance *Nature* **242** 190-191

[12] Mansfield P 1977 Multi-planar image formation using NMR spin-echos *J. Phys. C: Solid State Physics* **10**, L55-L5

[13] Mallard JR 2006 The Contribution of Medical Physicists and Doctors in Aberdeen to the Evolution of Modern Medical Imaging - SPECT, PET and MRI, 1965-1992, *Scot. Med. J.* **51** 44-48

[14] Ernst RR and Anderson WA 1966 Application of Fourier Transform Spectroscopy to Magnetic Resonance *Rev. Sci. Instr.* **37** 93-102

[15] Damadian R, Goldsmith M and Minkoff L 1977 NMR in cancer: XVI. Fonar image of the live human body *Physiological Chem. and Phys.* **9** 97-100

[16] Edelstein WA, Hutchison JMS, Johnson G and Redpath T 1980 Spin warp NMR Imaging and Applications to Human Whole-Body Imaging. *Phys. Med. Biol.* **25**,751-756

[17] Mansfield P 1988 Imaging by Nuclear Magnetic Resonance *J. Phys. E: Sci. Instrum.* **21**18-30

[18] Abetti PA and Haldar P 2009 One hundred years of superconductivity: science, technology, products, profits and industry structure *Int. J. Technol. Management* **48** 423-47

[19] Cosmus CT and Parizh M 2011 Advances in whole body MRI magnets Applied Superconductivity, *IEEE Trans. Appl. Supercond.* In preparation

[20] Sinanna A, Aguiar PM, d'Amico F, Bermond S, Bredy P, Donati A, Hugon C, Lannou H, Sakellariou D, Schild T, Tixador P, 2010 *IEEE Trans. Appl. Supercond.* **20** 790-793

[21] Wilson MN 1997 Superconducting Magnets, Oxford University Press

Nonequilibrium density operator approach to domain wall resistivity

Jun-ichiro Kishine[1], A S Ovchinnikov[2], and I V Proskurin[2],

1 Department of Basic Sciences, Kyushu Institute of Technology, Kitakyushu 804-8550, Japan
2 Department of Physics, Ural State University, Ekaterinburg 620000, Russia

Abstract. We derive the static resistivity due to the Bloch domain wall using the nonequilibrium density operator technique by Zubarev in the ballistic limit of electronic transport.

An electronic transport in magnets with an inhomogeneous magnetic order is an active research topic due to promising applications in spintronics. A presence of a domain wall (DW) in a ferromagnet enables to control a DW motion by a spin polarized current of conducting electrons flowing through the domain wall [1]. A feedback effect is a change of transport properties that has been observed in a number of experiments. The excess resistivity due to DW (DWR) has been a subject of a number of studies. The most relevant theoretical treatments of DWR are based on the following concepts. (i) The mixing of spin states due to the magnetization twist in the DW [2]. The phenomenological spin dependent *impurity* scattering mixes the two-spin channels. The extra resistivity arises since the channels exhibit unequal conductivities in a ferromagnet in the absence of the wall. (ii) A redistribution of the charge carriers between spin-majority and spin minority channels due to the *domain wall* scattering [4, 5]. The works perform a diagrammatic evaluation of the Kubo's formula over the DW scattering potential. The *impurity* scattering is introduced in the unperturbed Green's functions through two phenomenological parameters, the momentum scattering times for spin-up and spin-down channels. (iii) A destruction of the weak localization corrections to conductivity by the domain wall [3]. It was shown that a DW contributes to the decoherence of electrons, leading to a decrease of the resistance. (iv) Spin accumulation around the DW which leads to an additional potential drop [6, 7, 8].

In the present paper we propose a microscopic calculation of the Bloch DW resistivity using the nonequilibrium density operator technique in the ballistic regime of electronic transport, when the electron mean free path is longer than the system size. The approach takes into account a domain-wall scattering [type-(ii) theory], but it is valid in the clean limit (the electronic relaxation time $\tau \to \infty$), where the resistivity is dominated by a spin structure. The case has been previously addressed by using the Mori formula [9].

We use the following model Hamiltonian to describe the system of conduction electrons

$$\hat{\mathcal{H}} = -\sum_{\langle i,j \rangle,\sigma} t_{ij} c^{\dagger}_{i\sigma} c_{j\sigma} - \mu \sum_{i\sigma} c^{\dagger}_{i\sigma} c_{i\sigma} - J_{sd} \sum_{i} \boldsymbol{S}_i \cdot \boldsymbol{s}_i, \tag{1}$$

where t_{ij} is a hopping integral between the nearest neighbour sites, μ is the chemical potential, $c^{\dagger}_{i\sigma}$ ($c_{i\sigma}$) is an electron creation (annihilation) operator at the site i with spin $\sigma = \uparrow, \downarrow$. The conduction electrons interact with the inhomogeneous magnetic texture via the sd exchange interaction with the coupling strength $J_{sd} > 0$, $\boldsymbol{S}_i = S(\cos\varphi_i \sin\theta_i, \sin\varphi_i \sin\theta_i, \cos\theta_i)$ denotes the localized spin of the texture parametrized by the polar angles θ and ϕ. The local spin of conduction electrons $\boldsymbol{s}_i = (1/2)\sum_{\sigma\sigma'} c^\dagger_{i\sigma}\boldsymbol{\sigma}_{\sigma\sigma'}c_{i\sigma'}$ is treated as a quantum operator, where $\boldsymbol{\sigma}$ stands for Pauli matrices.

Since the length scale of the DW is very large compared with the length scale of the electron (~ 0.1 nm at the Fermi level), the back reflection of free carriers by the DW can be ignored [10] and one can use the "adiabatic frame" approximation to study electron transport in such a slow varying in space magnetic background [9, 11].

The first step of analysis is to perform a local SU(2) unitary transformation $c_{i\sigma} = \sum_{\sigma'}(\hat{U}_i)_{\sigma\sigma'}b_{i\sigma}$, which removes the inhomogeneity of $\boldsymbol{S}$

$$\hat{U}_i^{-1}(\boldsymbol{S}_i\cdot\boldsymbol{\sigma})\hat{U}_i = S\sigma_z. \tag{2}$$

The transformation matrix is given by $\hat{U}_i = \boldsymbol{m}_i\cdot\boldsymbol{\sigma}$ with the unit vector

$$\boldsymbol{m}_i = (\cos\varphi_i \sin\frac{\theta_i}{2}, \sin\varphi_i \sin\frac{\theta_i}{2}, \cos\frac{\theta_i}{2}). \tag{3}$$

Applying the local unitary rotation to the kinetic part of the Hamiltonian (1) and using the identity

$$\hat{U}_i^{-1}\hat{U}_{i+a_\nu} = 1 + ia_\nu\left(\boldsymbol{A}_\nu(\mathbf{r}_i)\cdot\boldsymbol{\sigma}\right) + o(A^2), \tag{4}$$

where a_ν is the lattice spacing along the ν axis, and

$$\boldsymbol{A}_\nu(\mathbf{r}_i) = [\boldsymbol{m}(\mathbf{r}_i)\times\partial_\nu\boldsymbol{m}(\mathbf{r}_i)] \tag{5}$$

is the non-Abelian gauge field created by the non-uniform magnetic texture ($\nu = x, y, z$), we obtain the Hamiltonian in the rotated frame as $\hat{\mathcal{H}} = \hat{\mathcal{H}}_0 + \hat{\mathcal{H}}_{int}$.

After the Fourier transformations

$$b_{i\sigma} = \frac{1}{\sqrt{V}}\sum_{\mathbf{k}} e^{i\mathbf{k}\mathbf{r}_i}b_{\mathbf{k}\sigma}, \quad \boldsymbol{A}_\nu(\mathbf{q}) = \frac{1}{V}\sum_{\mathbf{r}_i} e^{i\mathbf{q}\mathbf{r}_i}\boldsymbol{A}_\nu(\mathbf{r}_i), \tag{6}$$

the first part is given by

$$\hat{\mathcal{H}}_0 = \sum_{\mathbf{k}\sigma}\varepsilon_{\mathbf{k}\sigma}b^\dagger_{\mathbf{k}\sigma}b_{\mathbf{k}\sigma} \tag{7}$$

with the quasiparticle energies

$$\varepsilon_{\mathbf{k}\sigma} = -2\sum_{\nu=1}^{3} t_\nu\cos\left(k_\nu a_\nu\right) - \sigma M - \mu, \tag{8}$$

where $M = J_{sd}S/2$ is half the exchange splitting, and t_ν is the hopping integral along the ν axis in the real space.

The interaction part has the form

$$\hat{\mathcal{H}}_{int} = \hbar\sum_{\nu,\alpha}\sum_{\sigma\sigma'}\sum_{\mathbf{k},\mathbf{k}'} A^\alpha_\nu(\mathbf{k}'-\mathbf{k})\,\sigma^\alpha_{\sigma\sigma'}\,e^{\frac{i}{2}(k'_\nu-k_\nu)a_\nu}v_\nu\left(\frac{\mathbf{k}'+\mathbf{k}}{2}\right)b^\dagger_{\mathbf{k}\sigma}b_{\mathbf{k}'\sigma'}, \tag{9}$$

where the group velocity is

$$v_\nu(\mathbf{k}) = \frac{1}{\hbar}\frac{\partial\varepsilon_{\mathbf{k}\sigma}}{\partial k_\nu} = \frac{2}{\hbar}t_\nu a_\nu\sin\left(k_\nu a_\nu\right). \tag{10}$$

After the rotation the Hamiltonian (1) is transformed to that of electrons uniformly polarized and interacting with the SU(2) gauge field **A** localized around the inhomogeneity [9]. The non-adiabaticity causes a scattering between the spin-up and spin-down channels and plays a role of spin-dependent scattering potential.

The current operator is

$$\mathbf{J} = e \sum_{\mathbf{k}\sigma} \mathbf{v}(\mathbf{k}) b^{\dagger}_{\mathbf{k}\sigma} b_{\mathbf{k}\sigma}, \tag{11}$$

where $e < 0$ is the electron charge. Its time derivative is

$$\dot{\mathbf{J}} = \frac{i}{\hbar}[\hat{\mathcal{H}}, \mathbf{J}] = ie \sum_{\alpha,\nu} \sum_{\mathbf{k}\sigma} \sum_{\mathbf{k}'\sigma'} A^{\alpha}_{\nu}(\mathbf{k}' - \mathbf{k}) \sigma^{\alpha}_{\sigma\sigma'} e^{\frac{i}{2}(k'_{\nu}-k_{\nu})a_{\nu}} v_{\nu}\left(\frac{\mathbf{k}' + \mathbf{k}}{2}\right) (\mathbf{v}(\mathbf{k}') - \mathbf{v}(\mathbf{k}))\, b^{\dagger}_{\mathbf{k}\sigma} b_{\mathbf{k}'\sigma'}. \tag{12}$$

The dynamical resistivity per volume in isotropic materials obtained within the nonequilibrium density operator method [12] is given by

$$\rho(\omega) = \frac{3k_B T}{(\mathbf{J}, \mathbf{J})} \left[-i\omega + \frac{\langle \dot{\mathbf{J}}; \dot{\mathbf{J}} \rangle_{\omega+i\varepsilon}}{(\mathbf{J}, \mathbf{J}) + \langle \dot{\mathbf{J}}; \mathbf{J} \rangle_{\omega+i\varepsilon}} \right], \tag{13}$$

where T is the temperature and k_B is he Boltzmann constant. In contrast to Kubo's formula for a linear response of isolated systems, Eq.(13) describes a response of open systems that in contact with a heat bath. What concerns external mechanical perturbations (for instance, electric or magnetic fields) the both approaches are equivalent. However, the method of the nonequilibrium density operator is naturally extended to thermal perturbations (f.e., local fluctuations of temperature or chemical potential) [12].

Provided the *adiabatic condition*, $M/t_{\nu} \gg a_{\nu}/\lambda_0$, where λ_0 is a scale of magnetic inhomogeneity, that means a smoothness of a spin texture, the time derivative (12) becomes a small parameter of the theory. This makes (13) is more convenient to evaluate the resistivity in comparison with the linear response theory, where an interaction with the gauge field is treated by using the standard diagrammatic perturbation theory [3].

The resistivity includes the static correlation function of the current operators

$$(\mathbf{J}, \mathbf{J}) = \sum_{\alpha=x,y,z} \int_0^1 dx\, \mathrm{Tr}\left\{ \mathbf{J}_{\alpha} \mathbf{J}_{\alpha}(i\hbar\beta x)\, \rho_0 \right\}, \tag{14}$$

where $\beta = 1/(k_B T)$, $\rho_0 = \exp\left(-\beta\hat{\mathcal{H}}\right) / \mathrm{Tr}\exp\left(-\beta\hat{\mathcal{H}}\right)$ is the equilibrium density matrix and $\mathbf{J}_{\alpha}(i\hbar\beta x) = e^{-\beta x \hat{\mathcal{H}}} \mathbf{J}_{\alpha} e^{\beta x \hat{\mathcal{H}}}$.

The symbol $\langle \ldots \rangle$ represents the Laplace transform of the time correlation function

$$\langle A; B \rangle_{\omega+i\varepsilon} = \int_0^{\infty} dt\, e^{i(\omega+i\varepsilon)t} \left(A(t), B \right), \tag{15}$$

where A, B mean either $\mathbf{J}$ or $\dot{\mathbf{J}}$ operators.

In the following we consider the static resistivity at $\omega = 0$ and focus on spin textures varying only in one spacial direction, say along the z axis, and uniformly magnetized within the perpendicular xy plane. Then, $\mathbf{A}_{\nu}(\mathbf{q}) \to A_z(q_z)$. Likewise we assume that electrons move only along the selected direction with the energy $\varepsilon_{k_z\sigma} = -2t_z\cos(k_z a_z) - \sigma M - \mu$, i.e. the hopping terms t_x, t_y are neglected and the factor 3 in Eq.(13) should be omitted.

Plugging the expression in Eqs.(17,15) we get

$$\langle \dot{J}_z; \dot{J}_z \rangle_{i\varepsilon} = \pi\hbar e^2 \sum_{\alpha_1\alpha_2} \sum_{k_1\sigma_1} \sum_{k_1'\sigma_1'} A_z^{\alpha_1}(k_1' - k_1) A_z^{\alpha_2}(k_1 - k_1') \sigma^{\alpha_1}_{\sigma_1\sigma_1'} \sigma^{\alpha_2}_{\sigma_1'\sigma_1}$$

$$\times v_z^2 \left(\frac{k_1' + k_1}{2} \right) \left(v_z(k_1') - v_z(k_1) \right)^2 f_{k_1\sigma_1}(1 - f_{k_1'\sigma_1'}) \delta\left(\varepsilon_{k_1\sigma_1} - \varepsilon_{k_1'\sigma_1'} \right), \tag{16}$$

and after some calculations the expression is simplified to be

$$\langle \dot{J}_z; \dot{J}_z \rangle_{i\varepsilon} = 2\pi\hbar e^2 \sum_{kk'} A_z^{(-)}(k'-k) A_z^{(+)}(k-k') v_z^2 \left(\frac{k'+k}{2} \right) \left(v_z(k') - v_z(k) \right)^2 f_{k\uparrow}(1-f_{k\uparrow}) \delta(\varepsilon_{k\uparrow} - \varepsilon_{k'\downarrow}), \tag{17}$$

where $A_z^{(\pm)} = A_z^x \pm i A_z^y$. As is obvious, the spin flip scattering of conduction electrons is provided by the transverse components of the gauge field.

The static correlation function is obtained as

$$(J_z, J_z) = e^2 \sum_{k\sigma} v_z^2(k) f_{k\sigma} (1 - f_{k\sigma}), \tag{18}$$

where $f_{k\sigma} = (\exp \beta\varepsilon_{k\sigma} + 1)^{-1}$ is the Fermi distribution function. By the same manner we get the trivial result $\langle \dot{J}_z; J_z \rangle_{i\varepsilon} = 0$.

Replacing in Eqs.(17,18) the sums over k by the appropriate integrals $\sum_k \to a_z/(2\pi) \int_{BZ} dk$ and carrying out straightforward calculations we get the eventual result for the dc resistivity per volume

$$\rho = \frac{\hbar k_B T M^2}{2 e^2 a_z^2 t_z^3} \frac{\mathcal{K}_1}{\mathcal{K}_2^2}. \tag{19}$$

Here we introduce the dimensionless quantities determined as

$$\mathcal{K}_1 = \frac{a_z}{2\pi} \sum_\sigma \int_{k_{\mathrm{min}}}^{\pi/a_z} dk \, \frac{\mathcal{A}(k - \sigma k_0) \cos^2\left(\{ka_z + \sigma k_0 a_z\}/2 \right)}{\left[1 - \{M/t_z + \cos(ka_z)\}^2 \right]^{1/2}} f_{k\uparrow}(1 - f_{k\uparrow}), \tag{20}$$

$$\mathcal{K}_2 = \frac{a_z}{2\pi} \sum_\sigma \int_{-\pi/a_z}^{\pi/a_z} dk \sin^2(ka_z) f_{k\sigma}(1 - f_{k\sigma}), \tag{21}$$

where $k_0 a_z = \cos^{-1}[M/t_z + \cos(ka_z)]$, $k_{\mathrm{min}} a_z = \cos^{-1}[1 - M/t_z]$, and

$$\mathcal{A}(q) = a_z^2 \left[A_z^{(+)}(q) A_z^{(-)}(-q) + A_z^{(+)}(-q) A_z^{(-)}(q) \right]. \tag{22}$$

The explicit form of $\mathcal{A}(q)$ depends on a spin configuration. In the case of the Bloch domain wall, $\varphi(z) = 2\tan^{-1} e^{z/\lambda}$ and $\theta = \pi/2$, where λ is the thickness of the DW, it is given by

$$\mathcal{A}(q) = 2(\pi q \lambda)^2 \frac{\cosh(\pi q \lambda)}{\sinh^2(\pi q \lambda)}. \tag{23}$$

The last result is derived by using the vector gauge field created by the Bloch DW

$$\mathbf{A}_z(z) = \frac{1}{2\lambda} \left(\frac{\sinh(z/\lambda)}{\cosh^2(z/\lambda)}, -\frac{1}{\cosh^2(z/\lambda)}, \frac{1}{\cosh(z/\lambda)} \right). \tag{24}$$

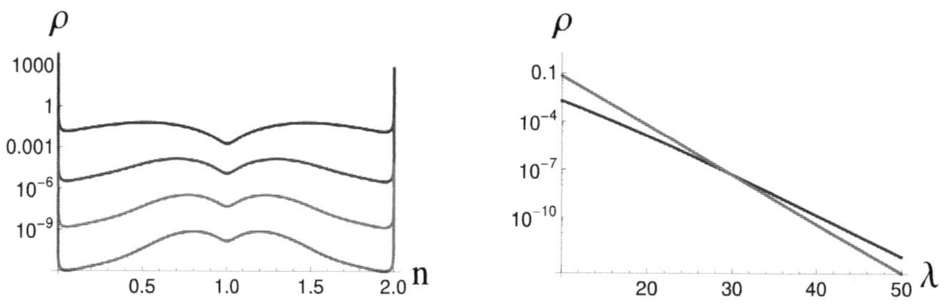

Figure 1. The static DW resistivity (in arb. units) at different domain wall thicknesses (λ varies from 10 till 40 with the step 10) as a function of the electronic concentration n (left panel). Dependence of ρ on the DW thickness for the concentrations $n = 0.5$ (or 1.5) and $n = 1$ (right panel). The parameters are chosen as $M/t_z = 0.2$, $k_BT/t_z = 0.1$ at $t_z = 1.0$.

The answer for the resistivity becomes transparent at zero temperature

$$\rho = \frac{\hbar\pi M^2}{e^2 a_z^2 t_z^2} \frac{\sum_\sigma \mathcal{A}(k_{F\uparrow} - \sigma k_{F\downarrow}) \cos^2\left([k_{F\uparrow} + \sigma k_{F\downarrow}]a_z/2\right)}{|\sin(k_{F\downarrow}a_z)||\sin(k_{F\uparrow}a_z)|\left(\sum_\sigma |\sin(k_{F\sigma}a_z)|\right)^2}, \tag{25}$$

where $k_{F\sigma}$ is the Fermi momentum of the polarized electrons, i.e. the DW gauge field supports the spin-flip scattering of electrons at the Fermi level. We pay attention that spin flip processes *without momentum transfer* are responsible for an emergence of the spin-trasfer torque from spins of itinerant electrons to local moments [11] whereas spin flip processes *with momentum transfer* results in the resistivity due to the DW.

The finite temperature resistivity is plotted in Fig.1 as a function of the electronic concentration per site n and the DW thickness λ. To obtain the dependencies we calculate the chemical potential from the condition $\sum_{k\sigma} f_{k\sigma} = n$, where $0 \leq n \leq 2$.

As a summary, by using the nonequilibrium density operator technique we have calculated the static domain-wall resistivity of a Bloch wall situated in a wire of quasi-one-dimensional geometry when the current flows perpendicularly to the wall plane.

Acknowledgments

J. K. acknowledges Grant-in-Aid for Scientific Research (C) (No. 19540371) from the Ministry of Education, Culture, Sports, Science and Technology, Japan. A. S. O. and I. V. P. acknowledge Grant RFBR No. 10-02-00098-a.

References

[1] Slonczewski J.C. 1996 *J. Magn. Magn. Mater.* **159** L1; Berger L., 1996 *Phys. Rev. B* **54** 9353.
[2] Levy P.M. and Zhang S. 1997 *Phys. Rev. Lett* **79** 5110.
[3] Tatara G, Fukuyama H. 1997 *Phys. Rev. Lett.* **78**, 3773.
[4] van Gorkom R.P., Brataas A., and Bauer G.E.W. 1999 *Phys. Rev. Lett.* **83**, 4401.
[5] Brataas A., Tatara G., and Bauer G.E.W. 1999 *Phys. Rev. B* **60** 3406.
[6] Simanek E., 2001 *Phys. Rev. B* **63**, 224412.
[7] Rebei A., Simanek E., 2005 *Phys. Rev. B* **71**, 172405.
[8] Wickles C. and Belzig W., 2009 *Phys. Rev. B* **80**, 104435.
[9] Tatara G., Kohno H., Shibata J. 2008 *Physics Reports* **468**, 213.
[10] Cabrera G.G., Falicov L.M. 1974 *Physica Status Solidi* (b) **62**, 217
[11] Kishine J., Ovchinnikov A.S., Proskurin I.V. 2010 *Phys. Rev. B* **82** 064407.
[12] Zubarev D.N., Morozov V., Röpke G., *Statistical Mechanics of Nonequilibrium Processes*, Akademie-Verlag, Berlin 1996.

A structural phase transition in the intermetallic compound $Tm_3Cu_4Sn_4$

S Muñoz-Perez, R Cobas, R A Susilo and J M Cadogan

Department of Physics and Astronomy,
University of Manitoba, Winnipeg, Manitoba, R3T 2N2, Canada

E-mail: r_cobas@yahoo.com

Abstract. The ternary intermetallic system $R_3T_4X_4$ (where R = rare earth; T= Mn, Cu, Ag, Au, Pd; X=Si, Ge, Sn) comprises an extensive series of isostructural compounds that form in the orthorhombic $Gd_3Cu_4Ge_4$-type structure (space group *Immm*, #71). The exceptions to this generic structure are $Tm_3Cu_4Sn_4$ and $Lu_3Cu_4Sn_4$ which have a monoclinic structure (space group *C*12/*m*1, #12) at room temperature. Here, we report a structural phase transition in $Tm_3Cu_4Sn_4$ from monoclinic *C*12/*m*1 to orthorhombic *Immm* upon heating above room temperature. We have carried out X-ray powder diffraction and differential scanning calorimetry on $Tm_3Cu_4Sn_4$ and we observe the structural transition at 458(2) K.

1. Introduction

The orthorhombic $R_3Cu_4Sn_4$ (R = rare earth; T= Mn, Cu, Ag, Au, Pd; X=Si, Ge, Sn) compounds exhibit a rich variety of magnetic ordering behaviour of the R sublattices [1]. The space group of this generic orthorhombic structure is *Immm* (#71) and the R atoms occupy two crystallographically inequivalent sites [2, 3]. The magnetic moments observed at the two R sites usually exhibit quite different magnetic ordering temperatures and ordering modes.

The exceptions to this generic orthorhombic crystallography are $Tm_3Cu_4Sn_4$ and $Lu_3Cu_4Sn_4$ which form in a monoclinic *C*12/*m*1 (#12) cell at room temperature (RT). With regard to $Tm_3Cu_4Sn_4$ it was shown that a single-crystal sample formed with the monoclinic *C*12/*m*1 cell [4] whereas a polycrystalline sample annealed at 670 K formed in the orthorhombic *Immm* cell [5]. In this paper we show that the monoclinic structure of $Tm_3Cu_4Sn_4$ transforms to the generic orthorhombic structure at 458(2) K. To the best of our knowledge, this is the first clear determination of such a crystallographic transformation in this intermetallic family.

2. Experimental Methods

The $Tm_3Cu_4Sn_4$ sample was prepared in an argon-arc furnace. Stoichiometric amounts of the pure elements [Tm (99.9%), Cu (99.999%), Sn (99.998%)] were melted several times under pure (less than 1-ppm impurity) argon to ensure homogeneity. The resulting ingot was annealed *in vacuo* at 800° C for two weeks and water-quenched. X-ray powder diffraction measurements were made at room temperature using Cu-Kα radiation on a PANalytical X'Pert Pro diffractometer equipped with a high-temperature stage. Diffraction patterns were obtained over the temperature range 295 K – 495 K and all patterns were refined using the Rietveld method and the FullProf/WinPlotr suite [6, 7]. Differential Scanning Calorimetry (DSC) was carried out on a Perkin-Elmer DSC-4 system at a constant heating rate of 2 K/minute.

3. Results

In figure 1 we show the refined XRD pattern of $Tm_3Cu_4Sn_4$ obtained at 295 K and in table 1 we give the structural data determined from this refinement. The space group at 295 K is monoclinic C12/*m*1. The conventional R-factors for this refinement are $R_F = 2.9$, $R_{struct.} = 3.9$ and GoF = 2.6.

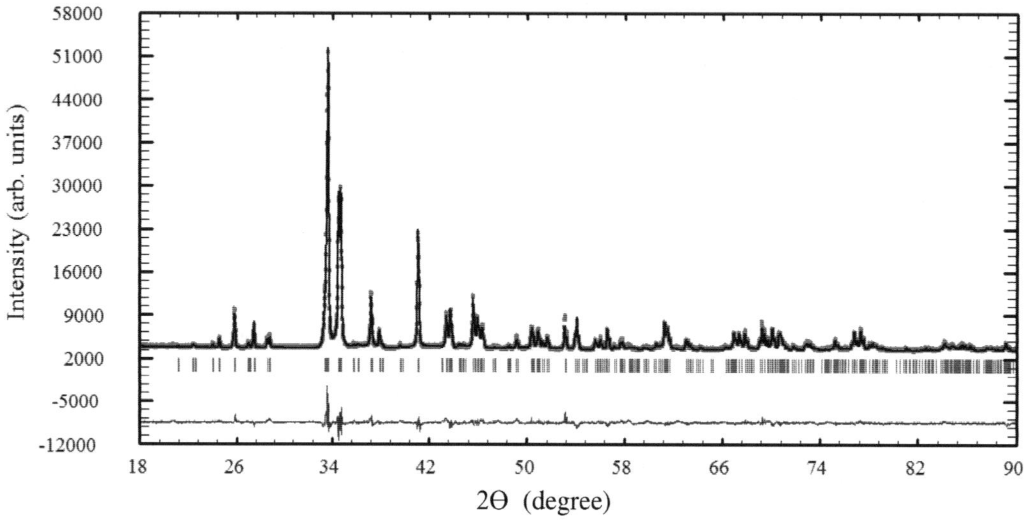

Figure 1. Cu-Kα X-ray powder diffraction pattern of $Tm_3Cu_4Sn_4$ (at 295 K) and refined in the *C*12/*m*1 monoclinic space group.

Table 1. Crystallographic data (at 295 K) for $Tm_3Cu_4Sn_4$ in the monoclinic phase

		C 12/*m*1 cell *a* = 15.9887(3) Å, *b* = 4.3898(1) Å, *c* = 6.8931(1) Å, β = 115.0(1)°			*I* 12/*m*1 conventional cell *a* = 14.4911(3) Å, *b* = 4.3898(1) Å, *c* = 6.8931(1) Å, β = 90.54(1)°		
Atom	**Site**	x	y	z	x	y	z
Tm (1)	2*a*	0	0	0	0	0	0
Tm (2)	4*i*	0.1298(1)	0	0.6373(4)	0.8702(1)	0	0.5075(4)
Cu (1)	4*i*	0.3358(4)	0	0.0233(8)	0.6642(3)	0	0.6875(7)
Cu (2)	4*i*	0.3208(4)	0	0.6297(8)	0.6791(4)	0	0.3088(7)
Sn (1)	4*i*	0.2157(2)	0	0.2056(5)	0.7843(2)	0	0.9899(5)
Sn (2)	4*i*	0.5055(2)	0	0.2997(4)	0.4944(2)	0	0.7941(4)

This monoclinic *C*-centred cell is not a *conventional* cell and we identified the body-centred *conventional* cell using the SETSTRU program on the Bilbao Crystallographic Server [8]. The space group of the *conventional* cell is *I*12/*m*1 (#12). The lattice parameters and atomic positions in this *conventional* cell are given in table 1. We refer the reader to the work of Mighell [9] for a detailed discussion of the *conventional* and *reduced* cells in monoclinic systems.

In figure 2 we show the temperature dependence of a section of the XRD pattern of $Tm_3Cu_4Sn_4$. The structural transformation from the RT monoclinic phase to the higher-temperature orthorhombic phase is clearly seen as the monoclinic peaks merge due to the formation of the higher-symmetry orthorhombic structure.

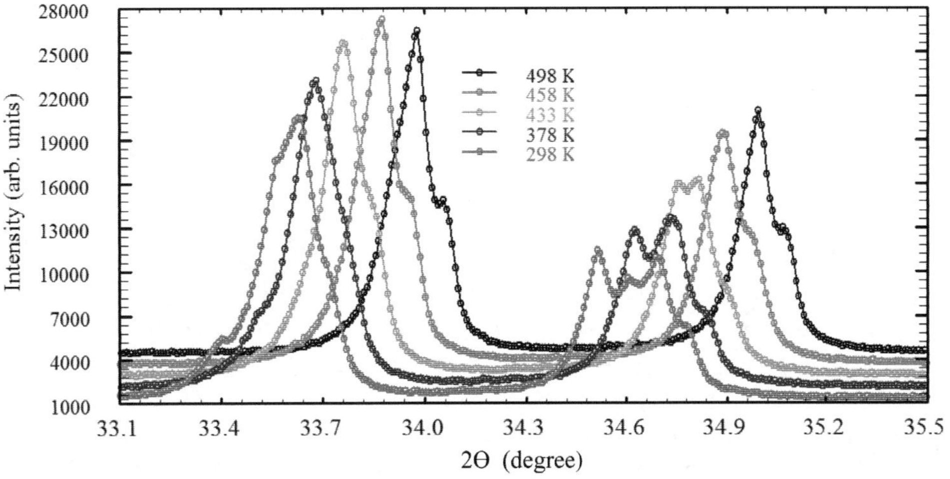

Figure 2. Thermal evolution of a section of the X-ray powder diffraction pattern of $Tm_3Cu_4Sn_4$

In figure 3 we show the angular positions of the $(\bar{5}11)$ and (311) monoclinic peaks as they merge to form the (114) orthorhombic peak, as a function of temperature. These data indicate the completion of the transformation to the orthorhombic structure occurring at 458(2) K. We can also follow the crystallographic transformation by monitoring the peak width upon heating and in figure 4 we show the temperature dependence of the full-width at half-maximum intensity for the peaks monitored in figure 3. Once again, the formation of the orthorhombic structure is observed at 458(2) K.

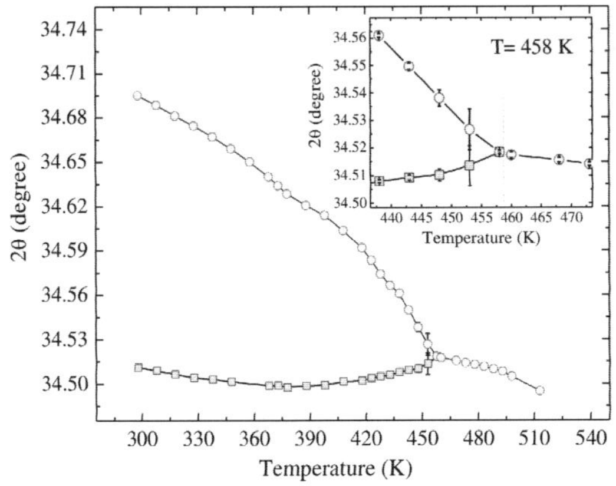

Figure 3. Temperature dependence of the $(\bar{5}11)$ and (311) peak positions of $Tm_3Cu_4Sn_4$.

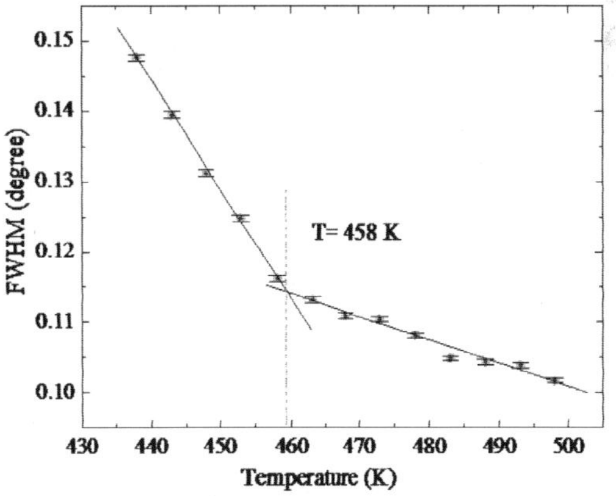

Figure 4. Temperature dependence of the width of the $(\bar{5}11)$ and (311) peaks of $Tm_3Cu_4Sn_4$.

An independent determination of the structural transition temperature can be found in the DSC data shown in figure 5 where the onset of the transformation is clearly observed at 457(3) K.

In table 2 we give the structural data determined from the refined XRD pattern of $Tm_3Cu_4Sn_4$ in its orthorhombic *Immm* phase, obtained at 473 K. The conventional R-factors for this refinement are $R_F = 4.3$, $R_{struct} = 5.9$ and GoF = 2.2.

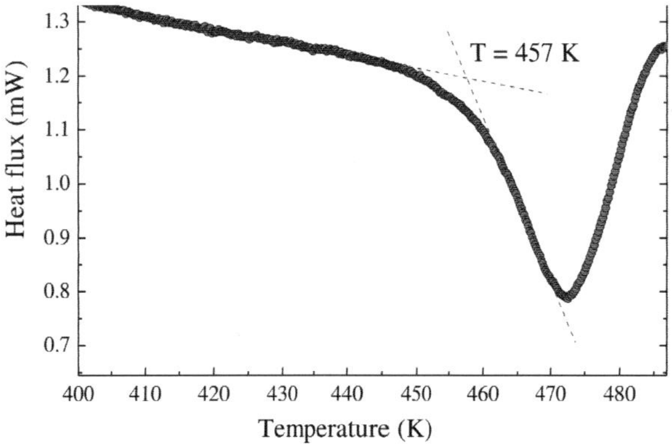

Figure 5. Differential scanning calorimetry trace of $Tm_3Cu_4Sn_4$ obtained at a heating rate of 2 K/minute.

Table 2. Crystallographic data (at 473 K) for $Tm_3Cu_4Sn_4$ in the orthorhombic *Immm* cell, a = 4.4028(1) Å, b = 6.9139(2) Å, c = 14.5261(4) Å.

Atom	Site	x	y	z
Tm (1)	2*a*	0	0	0
Tm (2)	4*j*	0.5	0	0.3718(2)
Cu	8*l*	0	0.3149(5)	0.3249(2)
Sn (1)	4*i*	0	0	0.2175(2)
Sn (2)	4*h*	0	0.2018(3)	0.5

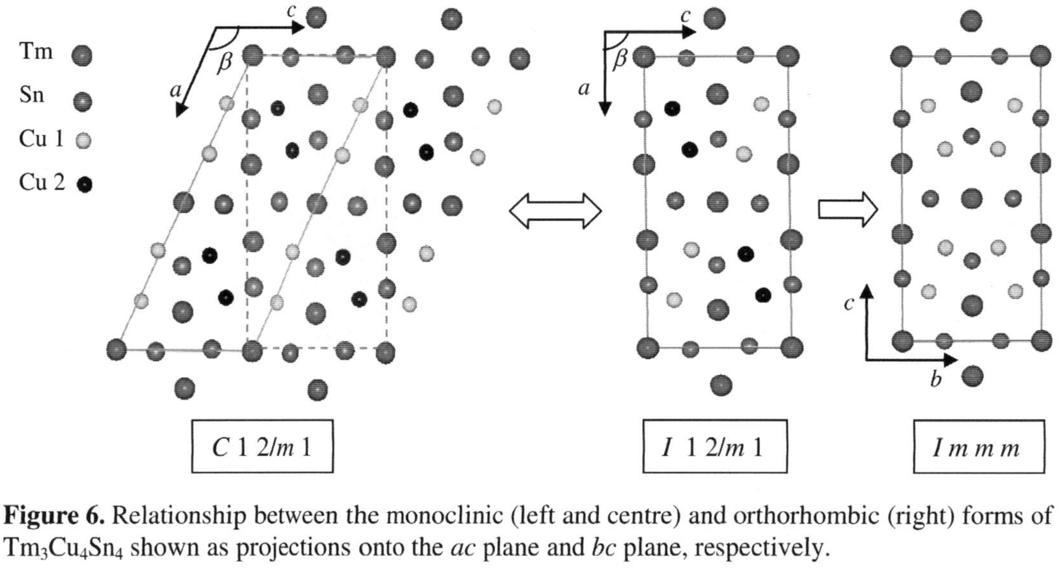

Figure 6. Relationship between the monoclinic (left and centre) and orthorhombic (right) forms of $Tm_3Cu_4Sn_4$ shown as projections onto the *ac* plane and *bc* plane, respectively.

The relationship between the monoclinic (*C* 12/*m*1 & *I* 12/*m*1) and orthorhombic (*Immm*) forms of $Tm_3Cu_4Sn_4$ is shown in figure 6. The most significant difference is in the atomic positions of the Cu(2) sites whose relative displacement is nearly three times those of the other sites. In the orthorhombic *Immm* structure, the Cu atoms occupy a single crystallographic site (8*l*) with the point group *m*. In the monoclinic cells, the Cu atoms split into two 4*i* sites. Finally, the fact that only the Tm and Lu

$R_3Cu_4Sn_4$ compounds form the monoclinic phase suggests that the size of the R ion might play a major role in determining the crystallography in the $R_3T_4X_4$ intermetallic series.

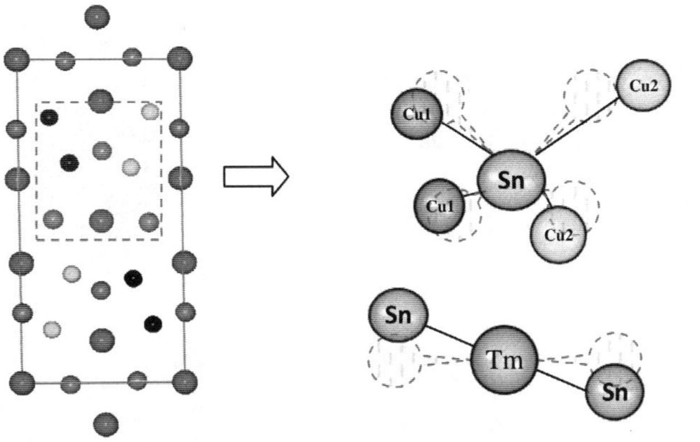

Figure 7. Schematic representation of the distortion of the Cu and Sn atomic environments in the monoclinic cell (solid figure) from the symmetric arrangement in the orthorhombic phase (dotted line).

4. Conclusions

We have used X-ray powder diffraction, complemented by differential scanning calorimetry, to show that the room temperature monoclinic crystal structure of $Tm_3Cu_4Sn_4$ transforms to the generic orthorhombic structure of the $R_3T_4X_4$ intermetallic family at 458(2) K.

Acknowledgments

We are grateful for financial support from the Natural Sciences and Engineering Research Council of Canada and the Canada Foundation for Innovation. We also thank Dr M. Bieringer (Chemistry, U. Manitoba) for giving us access to his diffractometer. Finally, JMC acknowledges support from the Canada Research Chairs programme.

5. References

[1] Ryan D H, Cadogan J M, Voyer C, Napoletano M, Riani P and Cranswick L M D 2010 *Mod. Phys. Lett. B.* **24** 1–28.
[2] Rieger W 1970 *Monatsch. Chem.* **101** 449–62.
[3] Hanel G and Nowotny H 1970 *Monatsch. Chem.* **101** 463–8.
[4] Thirion F, Steinmetz J and Malaman B 1983 *Mat. Res. Bull.* **18** 1537–42.
[5] Romaka L, Romaka V V and Davydov V 2008 *Chem. Met. Alloys* **1** 192–7.
[6] Rodríguez-Carvajal J 1993 *Physica* B **192** 55–69.
[7] Roisnel T and Rodríguez-Carvajal J 2001 *Mater. Sci. Forum* **378–81** 118–23.
[8] Bilbao Crystallographic Server, http://www.cryst.ehu.es.
[9] Mighell A D 1976 *J. Appl. Cryst.* **9** 491–8; 2002 *J. Res. Natl. Inst. Stand. Technol.* **107** 373–7; 2003 *Acta Cryst.* **B59** 300–2.

Vector Network Analysis of Dielectric and Magnetic Materials in the Millimetre Wave Band

Bin Yang*, Hansheng Su and Robert S. Donnan*
School of Electronic Engineering and Computer Science, Queen Mary University of London, Mile End Road, London, E1 4NS, UK.

E-mail*: bin.yang@elec.qmul.ac.uk; robert.donnan@elec.qmul.ac.uk

Abstract. A well-designed, quasi-optical spectrometer (transmissometer and reflectometer) driven by a HP N5244A Vector Network Analyzer is introduced for characterization of solid condensed materials such as silicon, Perspex and hexaferrite in the millimetre wave band. A brief note is made on quasi-optical transmissometry being adapted for study of soft condensed phase (biological) systems.

1. Introduction

High quality measurement systems in millimetre and sub-millimetre wave bands have critical design requirements in a number of major research fields. One of these fields is spectrometric characterization of materials, metamaterials and biomaterials. In this paper, a well-designed, quasi-optical spectrometer (transmissometer and reflectometer) driven by a HP N5244A Vector Network Analyzer (QO+VNA) is introduced for characterization of the solid condensed materials such as silicon, Perspex and hexaferrite. Sample vessels prototyped for handling soft condensed material such as liquids and proteins have also been characterized for the calibration (or reference) condition of empty-vessel.

2. Measurement System

Vector Network Analysis has been extensively used in microwave measurement systems. It is a heterodyne receiver that is capable of measuring the amplitude and phase-difference between a test and reference path to very high accuracy and phase-stability. Ongoing advances in microwave electronics has extended operation of the VNA beyond low millimetre wave bands upto 1 THz. Vector Network Analysis has distinct advantages over Time Domain Spectroscopy (TDS) and Fourier transform Infrared spectroscopy (FTIR). These are superior spectral resolution (down to 1MHz), a huge dynamic range (100 to 140 dB at low sub-THz band to 50-80 dB near 1THz) and, a rapid sweeping efficiency (spot rate less than picoseconds) [1]. A well-designed, quasi-optical spectrometer (transmissometer and reflectometer) [2] delivers a very pure collimated plane-wave, and in this case is established at the geometric centre of the measurement bench. It is here that sample-location is set. A large sample-plate (width 100 mm) is used to ensure negligible diffraction at the edges of the plate, even though the horns deliver a beam with high taper. It is quite straight forward to extend the spectrometer to THz energies by using the same quasi-optical circuit, by replacing only the corrugated feed-horns with a higher frequency band pair.

3. Theory

The transmittance and reflectance of the plate, $t(\omega,d)$ and $r(\omega,d)$ respectively [3] are given by

$$t(\omega,d)=\frac{(1-r_1^2(\omega,d))\cdot\exp i\phi(\omega,d)}{1-r_1^2(\omega,d)\cdot\exp i2\phi(\omega,d)}, \quad (1)$$

$$r(\omega,d)=\frac{r_1(\omega,d)\cdot(1-\exp i2\phi(\omega,d))}{1-r_1^2(\omega,d)\cdot\exp i2\phi(\omega,d)}. \quad (2)$$

d is the sample's thickness and c is speed of light in vacuum; $\phi_\pm(\omega,d)$ is the complex angle $\omega.n(\omega).d/c$ and n is the refractive index; $r_1(\omega,d)$ is the single-surface reflectance which is related to the reduced wave-impedance, $r_1=(z-1)/(z+1)$.

For a layered structure such as biological liquid sandwiched by two dielectric layers, ABCD matrix methods can be used to set up the transmittance and reflection equations. In this paper, the regression analysis fitting is used to characterize the materials' constants.

4. Results and Discussion

The measured transmittances and reflections of silicon and Perspex are presented in Fig.1. The best fitted permittivity of silicon is 11.65 ± 0.02, and of Perspex is 2.59 ± 0.01. These are very close to those reported from high-precision measurements on high-purity silicon [4] and Perspex [5].

The inherent optical properties of hexaferrite materials support circularly-polarized propagating waves, i.e. $t_\pm(\omega,d)$ and $r_\pm(\omega,d)$, rather than linearly-polarized waves, where the '+' indicates clockwise rotation of the electric vector in the direction of propagation and '-' anti-clockwise rotation. It is, however, very time consuming to directly measure material properties using circularly-polarized waves [6]. Instead, in this spectrometer, the plate's *linear-polarisation* transmittance and reflectance, both co- and cross-polar, respectively, are first measured and the results then transformed into circularly-polarised transmittance and reflectance. The results for each sense are illustrated in Fig.2. The detailed experimental and analysis methods are published elsewhere [2].

Vector network analysis of biological liquid samples is another interesting application of the measurement system. Our work has therefore developed a new bespoke cuvette system enabling analysis of bio-liquids. The first cuvette consisted of two 1.1 mm thick boron glass plates (a material generally transparent over millimeter and sub-millimeter wavelengths) having a 0.28 mm separation (established by four metallic corner-located pads). Inlet and outlet valves are placed at opposite ends of the plates to allow for flushing without the need to dismantle the cuvette. Its performance is presented in Fig. 3(left) and shows that the loss is very low when the wave propagates through the empty cuvette. The theoretical 3-layer ABCD model yield a high quality fit to the experimental results. The second design of cuvette incorporates a vacuum layer made of a Perspex vessel. The sample can be located in the middle of the container. The vacuum establishes an adiabatic environment to ensure internal thermal stability for stable and controllable test conditions. This is particularly important for biological samples. In Fig.3 (right), a 5.74 mm thick Perspex plate is put in the sample-position for the test. Theoretical calculations are very close to the measured transmittance of the 5-layer structure.

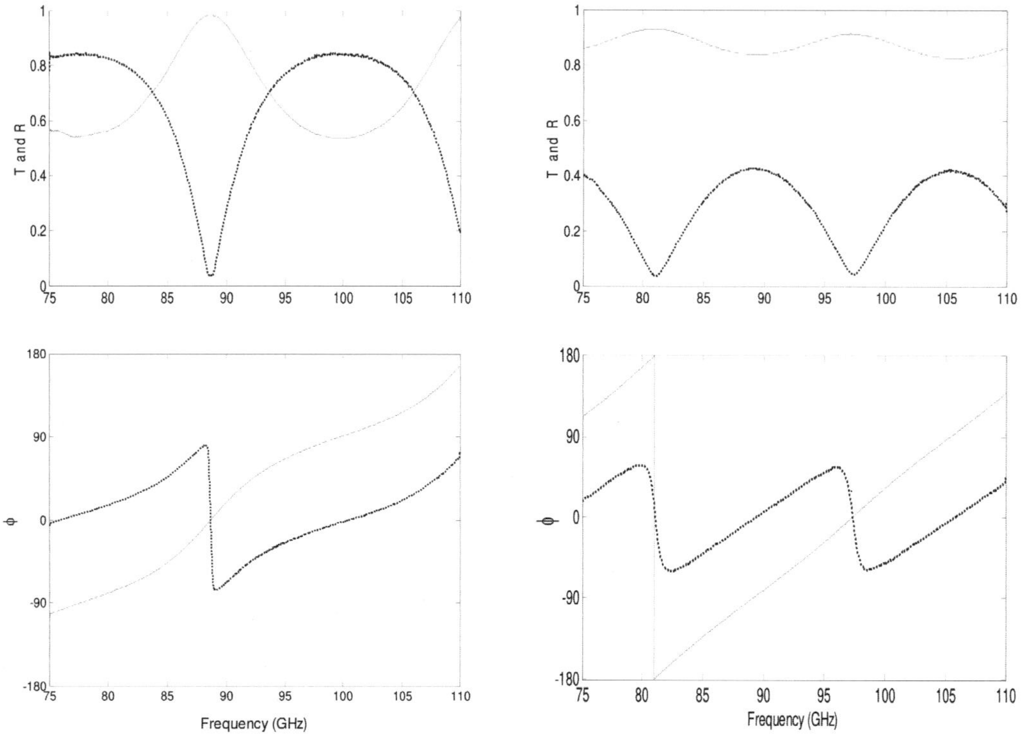

Fig.1: (left) measured complex transmittances (line) and reflections (dashed line) of 1.98 mm thick silicon; (right) measured complex transmittance (line) and reflection (dashed line) of 5.74 mm thick plastic. Upper plots, left and right, are of amplitude and the lower plots are of phase.

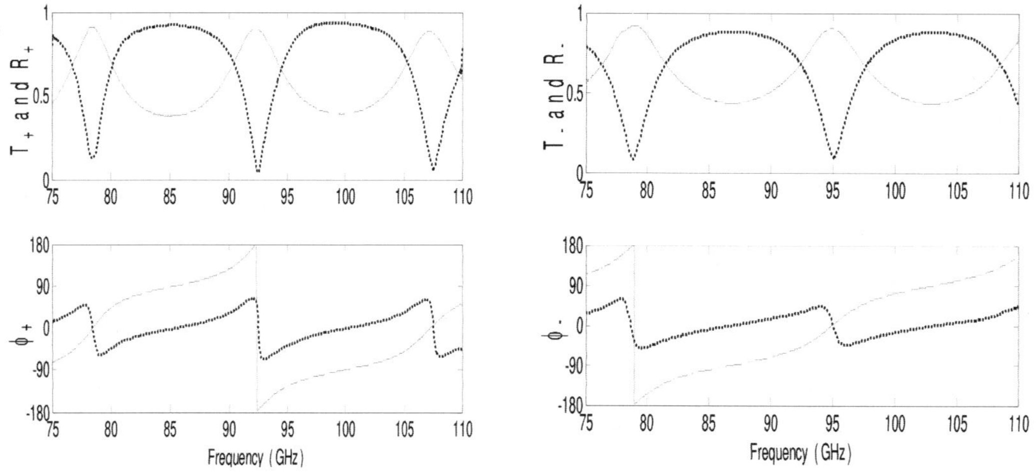

Fig.2: Complex circularly-polarised transmittances (lines) and reflections (dashed lines) (upper for the amplitudes, lower is for the phases) of hexaferrite sample; the plots on the left are for positive circular-polarisation and the plots on the right for negative.

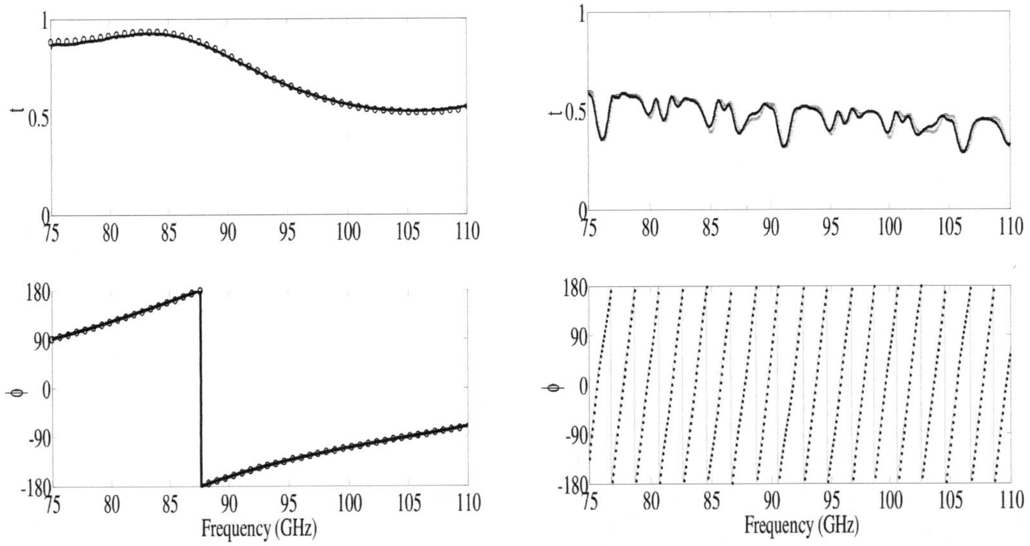

Fig.3: Comparison of theoretical transmittance (circles) with measured data (lines); upper plots are for the amplitudes, lower for phase): (left-plots) boron glass | air | boron glass 3 layer structure. Constituent thicknesses (mm) are 1.1 |0.28 |1.1; (right-plots) Perspex | air | Perspex | air | Perspex. Constituent thicknesses (mm) are 19.05 | 40.10 | 5.74 | 40.10 | 18.89.

5. Conclusions

Complex transmission and reflection spectra from a well-designed, quasi-optical spectrometer (transmissometer and reflectometer), driven by a HP N5244A Vector Network Analyzer (QO+VNA), have been presented on calibration materials (high-purity silicon and Perspex) and to further materials supporting biological fluid measurements. The sample holders developed for coming biological sample studies display the required low-loss and environmental stability.

References

[1]Manual of HP N5244A Vector Network Analyzer.
[2]B.Yang, R.J.Wylde, D.H.Martin, P.Goy, R.S.Donnan and S.Caroopen,IEEE Trans Microw. Theory.Tech, DOI: 10.1109/TMTT.2010.2086290.
[3]D.M.Pozar, Microwave Engineering, 3rd edition, Wiley.
[4]M.N.Afsar and H.Chi, Int J.Infr.Millimeter waves, vol.15, no.7,1994.
[5]J.W.Lamb, Int J.Infr.Millimeter waves, vol.17, no.12,1996.
[6]G.M.Smith, C.P.Unsworth,M.R.Webb and J.C.G.Lesurf, Microwave symposium digest, 1994, IEEE MTT-S International vol.1.pp.293-296.

Long-range interactions and information transfer in spin chains

Rebecca Ronke[1*], **Tim Spiller**[2†], **Irene D'Amico**[1‡]

[1] Department of Physics, University of York, York, YO10 5DD, UK

[2] School of Physics and Astronomy, E. C. Stoner Building, University of Leeds, Leeds, LS2 9JT, UK

E-mail: *rr538@york.ac.uk, †t.p.spiller@leeds.ac.uk, ‡irene.damico@york.ac.uk

Abstract. One of the main proposed tools to transfer information in a quantum computational context are spin chains. While spin chains have shown to be convenient and reliable, it has to be expected that, as with any implementation of a physical system, they will be subject to various errors and perturbative factors. In this work we consider the transfer of entangled as well as unentangled states to investigate the effects of various errors, paying particular attention to unwanted long-range interactions.

1. Introduction

The ability to reliably transfer quantum information in a solid state system is one of the key ingredients to quantum information processing, particularly when considering the construction of networked and distributed systems on a larger scale. Spin chains are able to respond to this challenge, while the mathematical framework that they are based on can be applied to a variety of physical systems. Examples of successful applications include for instance encoding into soliton-like packets of excitations [1], electrons or excitons trapped in nanostructures [2, 3, 4], strings of fullerenes [5] or nanometer scale magnetic particles [6].

Previous work on linear spin chains has shown that in order to ensure perfect state transfer (PST), the coupling strength $J_{i,i+1}$ between two neighboring sites i and $i+1$ on a chain of length N should be pre-engineered according to [7]

$$J_{i,i+1} = J_0\sqrt{i(N-i)}, \tag{1}$$

with J_0 the characteristic coupling constant. The time-independent perfect-transfer Hamiltonian, where we assume that any single excitation energies are site-independent, is thus:

$$\mathcal{H} = \sum_{i=1}^{N-1} J_{i,i+1}[|1\rangle\langle 0|_i \otimes |0\rangle\langle 1|_{i+1} + |0\rangle\langle 1|_i \otimes |1\rangle\langle 0|_{i+1}]. \tag{2}$$

This Hamiltonian also preserves the number of excitations. In order to assess the quality of the state transfer of entangled states, we use the entanglement of formation (EoF) [8]. For unentangled states, we define a fidelity F corresponding to mapping the initial state $|\psi_{in}\rangle$ over a time t into the desired state $|\psi_{fin}\rangle$ via the chain natural dynamics

$$F = |\langle\psi_{fin}|e^{-i\mathcal{H}t/\hbar}|\psi_{ini}\rangle|^2, \tag{3}$$

so that PST is achieved when $F = 1$. This system dynamics leads in particular also to the so-called mirroring rule [9], which guarantees perfect transfer of any one state to its mirrored image with respect to the centre of the chain. We define the time it takes for a state to then return to its original form as the system or revival time, t_S.

2. Effect of perturbations

When considering perturbations, we consider fabrication defects in the spin chains themselves on the one hand, and on the other also the effect of non-synchronous or imperfect input operations [10]. For all perturbations, we consider three types of initial input states: (i) unentangled states (such as $|\psi_{in}\rangle = |110000\rangle$), (ii) initially entangled states (such as $|\psi_{in}\rangle = 1/\sqrt{2}(|100000\rangle + |010000\rangle)$) and (iii) states where the entanglement is created by the system dynamics (such as $|\psi_{in}\rangle = 1/2(|000000\rangle + |100000\rangle + |000001\rangle + |100001\rangle)$). After reviewing the effect of various perturbative factors, we will focus on the central topic of this contribution, the effect of long-range interactions.

2.1. Imperfect input operations

In the vast majority of quantum information protocols, it is assumed that the system under investigation is governed by a universal, perfect clock. In an experimental context however, this might be overly optimistic, especially when the preparation of an input state involves accessing two or more separate sites of the chain simultaneously. For a type (i) chain, this is relevant when there are multiple excitations involved as otherwise the entire system dynamics is merely shifted in time. The actual effect of non-synchronous input depends greatly on the input mechanism used. We consider the two main possibilities for injection into spin chains, which are injection via SWAP operation (an additional particle is injected into the chain, e.g. via a waveguide) and excitation via a Rabi-flopping pulse (e.g. flipping the spin of an electron confined in the chain) [10]. Both these methods allow for some correction via measurement of the system but we find that with increasing delay times between injections, systems with Rabi flopping perform noticeably worse than those where the injection was done via SWAP operation. An exception are input states of type (iii), where the access sites are far apart and there is therefore virtually no difference between the effects of the two injection mechanisms.

2.2. Random noise

It has to be expected that any fabrication process of spin chains may lead to random, but time-independent errors in the coupling values of the system. We model these by adding to all non-zero entries of the Hamiltonian a random energy $\eta d_{l,m} J_0$ for $1 \leq l,m \leq number\, of\, basis\, states$. The scale is fixed by η which we set to 0.1 and for each $l \leq m$ the different random number $d_{l,m}$ is generated with a flat distribution between zero and unity. In order to preserve hermiticity, we set $d_{l,m} = d_{m,l}$. The specific weight of the noise would have to be determined depending on the experiment under consideration, but we found that for a perturbation of 10% of J_0, near-perfect transfer is still achieved during the first few periods of the system for all considered types of initial input.

2.3. Single excitation energies

We previously assumed that single excitation energies are site-independent, and thus did not need to be explicitly included into the Hamiltonian (Eq. (2)). However, local magnetic fields or additional single site fabrication defects may result in a loss of this independence, and so we would have to add to (2) the term

$$H_1 = \sum_{i=1}^{N} \epsilon_i |1\rangle\langle 1|_i, \tag{4}$$

where ϵ_i is not independent of the site i any more. We find that this perturbation has little effect on all input types for small values of ϵ_i, resulting in a loss of less than 5% for $\epsilon_i \leq 0.1J_{max}$, J_{max} the maximum value of the coupling between neighbouring sites, at the first revival. However for larger values the effect becomes very detrimental, in particular for type (iii) input.

2.4. Excitation-excitation interactions

Spin chains containing multiple excitations may also be prone to interactions between excitations in neighbouring sites. To represent this, we consider the perturbative term

$$H_2 = \sum_{i=1}^{N-1} \gamma J_0 |1\rangle\langle 1|_i \otimes |1\rangle\langle 1|_{i+1}. \tag{5}$$

An example of this would be biexcitonic interaction in quantum dot-based chains [11, 12]. While this perturbation is not relevant for input type (ii) (due to there only being a single excitation), we find that for the other two types, the transfer fidelity is again very well preserved for values of γ up to $0.1J_{max}$, resulting in less than a 5% loss of the original input state at revival time. It can also be shown that the loss in fidelity is exponential in N with Gaussian dependence on the characteristic noise parameters [10].

2.5. Next-nearest neighbor interactions

In our original Hamiltonian (2), we assumed that interactions would be restricted to the nearest neighbour of any one spin, which is a fair assumption if the transfer between sites is based on tunnelling. If however instead we consider dipole-dipole interactions, we can model the resulting perturbation as

$$H_3 = \sum_{i=1}^{N-2} J_{i,i+2}[|1\rangle\langle 0|_i \otimes |0\rangle\langle 1|_{i+2} + |0\rangle\langle 1|_i \otimes |1\rangle\langle 0|_{i+2}], \tag{6}$$

with $J_{i,i+2} = \Delta(J_{i,i+1} + J_{i+1,i+2})/2$. We have confirmed in [10] that this is a realistic model for example for semiconductor quantum dots with excitonic qubits. Of all the perturbations considered so far, this is the most influential. Again, the loss in fidelity is exponential in N with Gaussian dependence on the characteristic noise parameters. Not only do values of Δ as small as 0.05 lead to a considerable loss in transfer fidelity (for a 10 spin chain, 40% loss with input (i), 20% loss with input (ii), 5% loss with input (iii)), but even for the most stable input (iii) the periodicity of the system is entirely lost after a single period (for any chain length). As some protocols rely on continual periodicity, this is a potentially very serious issue (see also [13]).

2.6. Longer range interactions

The main reason for next-nearest neighbours being the most detrimental perturbation is that it is the only perturbation we have considered so far that effectively opens up "new channels" for the system dynamics. We can consider this phenomenon in a more general context by formulating a perturbation factor that opens up all available channels of the system. Assuming that these unwanted longer range interactions are random, we add to all zero entries in the Hamiltonian a coupling $\chi d_{l,m} J_{max}$ for $1 \leq l,m \leq$ *number of basis states*. For each $l \leq m$ the different random number $d_{l,m}$ is generated with a flat distribution between zero and unity and, in order

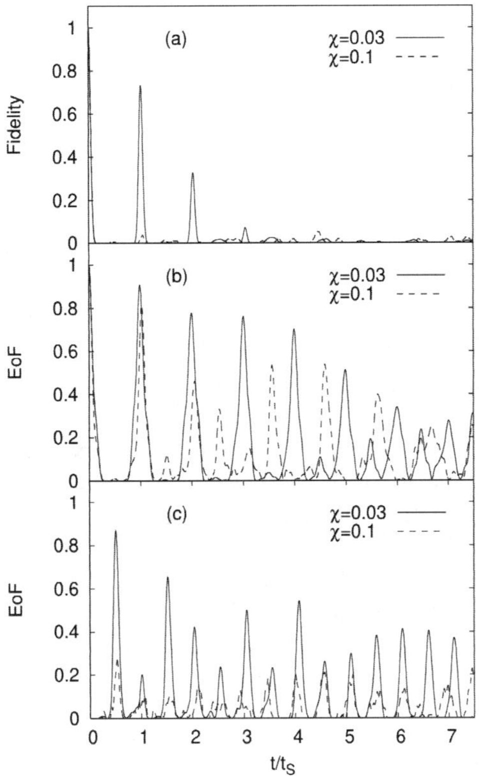

Figure 1. Plots of fidelity and EoF for a 10-spin chain for two values of χ vs. rescaled time t/t_S: (a) with input type (i), (b) with input type (ii), (c) with input type (iii).

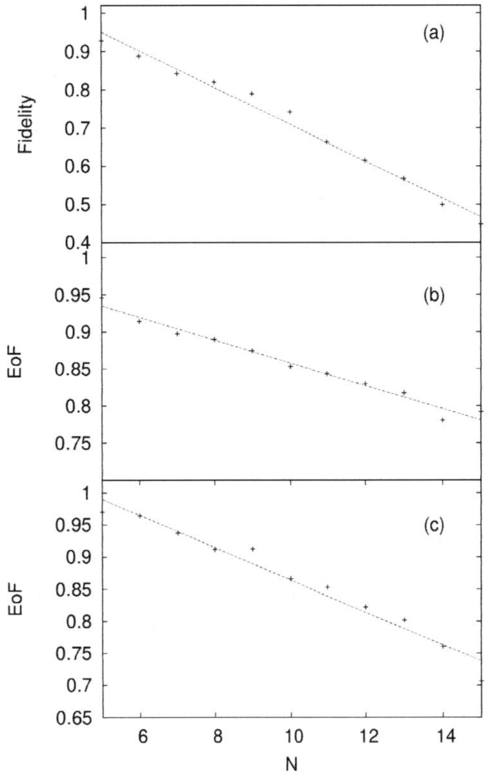

Figure 2. Plots of fidelity and EoF at the first expected revival time for $\chi = 0.03$ vs N: (a) with input type (i), (b) with input type (ii), (c) with input type (iii). Lines are best fit to numerical data.

to preserve hermiticity, we set $d_{l,m} = d_{m,l}$. In figure 2, every point on the graph corresponds to an average obtained from 100 realisations. The result of these additional channels has a different effect on the three types of initial input that we consider. In figure 1 we consider the dynamics of a typical realisation. We see in figure 1 (a) that the unentangled input type (i) suffers by far the most: even for small χ ($\chi = 0.03$) the initial state fidelity is not recovered well after a single period, and subsequently entirely lost. For $\chi = 0.1$, the initial state is not even recovered at all. By contrast, input type (ii) in figure 1 (b) conserves an acceptable amount of EoF for $\chi = 0.03$ for a few periods before becoming erratic, whereas the plot for $\chi = 0.1$ shows that the EoF peaks are shifted from their expected times, making the system entirely unpredictable after the first revival. A similar phenomenon can be observed in figure 1 (c) for $\chi = 0.03$ where the EoF peaks of input type (iii) should occur at $0.5t_S, 1.5t_S, \cdots$ but are actually starting to be shifted after just the first revival, with additional peaks forming at integer multiples of t_S. However for $\chi = 0.1$, the EoF never reaches decent values. Overall it is worth noting that this is in-keeping with the detrimental effects of next-nearest neighbour interaction we have observed, where the most detrimental effect was also seen for input type (i). However the opening up of new channels even just on a small scale has almost immediate and serious consequences for all systems considered.

The qualitative effect of opening up new, undesired channels in a system is also dependent on the number of spins in the chain, N, as a higher dimensional Hamiltonian also allows for more perturbative elements. We see this confirmed in figure 2, where all three systems show a clear decreasing trend in transfer quality with increasing N. Again, this data was sampled and averaged over 100 random runs, keeping a constant $\chi = 0.03$. We see in figure 2 (a) that again, it is the unentangled input type (i) that suffers most from this perturbation, showing that, for $N = 15$, the first revival is barely over 40% of the initial input state. Input types (ii) and (iii) perform better, with type (ii) in figure 2 (b) losing less than 25% EoF for $N = 15$ and similarly with type (iii) in figure 2 (c) losing less than 30% EoF for $N = 15$. We see therefore that unentangled states are much more prone to the detrimental effects of new open channels in a system, but also that shorter chains always perform much better than longer ones, regardless of the initial input.

3. Conclusions

We have considered the performances of the transfer of both unentangled as well as entangled states under various physically relevant perturbation factors. Overall, we found spin chains to be a very robust system, provided that the errors we considered remain of the order of a few percent of the characteristic coupling constant and that imperfect input will be corrected via subsequent measurement of the system. In particular, we were concerned with the opening up of new channels for the system dynamics, beyond next-nearest neighbour interaction, as this had proved to have the most detrimental effect on PST of any type of input. We have shown that longer chains suffer significantly more from this type of perturbation, with unentangled initial states being particularly affected. Large errors lead to a complete loss of the periodicity of the system, while small errors are tolerable up to a minimum of the first revival for all input types considered. Our studies demonstrate the impact of various potential difficulties with the implementation of spin chains, while giving encouraging results for modest errors. This demonstrates that spin chains are an encouraging candidate for future experiments on information transfer in solid state quantum information systems.

RR was supported by EPSRC-GB and HP. IDA acknowledges partial support by HP. IDA and RR acknowledge the kind hospitality of the HP Research Labs Bristol.

References

[1] Osborne T J and Linden N 2004 *Phys. Rev.* A **69** 052315
[2] D'Amico I 2007 *Relatively short spin chains as building blocks for an all-quantum dot quantum computer architecture* (Nova Science Publishers) chap 9, pp 281–300
[3] D'Amico I 2006 *Microelectron. J.* **37** 1440–1441
[4] Nikolopoulos GM Petrosyan D and Lambropoulos P 2004 *J. Phys. Condens. Matter* **16** 4991–5002
[5] Twamley J 2003 *Phys. Rev.* A **67** 052318
[6] Tejada J, Chudnovsky E, del Barco E, Hernandez J and Spiller T 2001 *Nanotechnology* **12** 181–186
[7] Christandl M, Datta N, Dorlas T, Ekert A, Kay A and Landahl A 2005 *Phys. Rev.* A **71** 032312
[8] Wootters W 1998 *Phys. Rev. Lett.* **80** 2245
[9] Albanese C, Christandl M, Datta N and Ekert A 2004 *Phys. Rev. Lett.* **93**
[10] Ronke R, Spiller T and D'Amico I 2010 (*Preprint* quant-ph/1007.4746)
[11] D'Amico I and Rossi F 2001 *Appl. Phys. Lett.* **79** 1676
[12] De Rinaldis S, D'Amico I and Rossi F 2002 *Appl. Phys. Lett.* **81** 4236
[13] Avellino M, Fisher A and Bose S 2006 *Phys. Rev.* A **74** 012321

Magnetic templating by superconducting films, disks and rings

A Amthong and S Crampin

Department of Physics, University of Bath, Bath BA2 7AY, UK

E-mail: aa281@bath.ac.uk

Abstract. Dilute magnetic semiconductors (DMSs), exhibiting both semiconducting and magnetic properties, are interesting materials for novel spintronic devices. Because of the giant Zeeman effect in the paramagnetic state, non-uniform magnetic fields can manipulate the spin texture of carriers in the DMS. A tight-binding model is developed to describe the energy bands formed due to the magnetic field associated with a type-II superconducting film above a DMS layer. Magnetic flux concentrated at the edges of superconducting disks and rings above DMS layers is shown to result in edge-confined states.

1. Introduction

Following the discovery of the giant Zeeman effect in dilute magnetic semiconductors (DMSs) [1], there has been much interest in DMSs in nonuniform magnetic fields [2, 3] because the giant Zeeman splitting can play an important role not only in confining charge carriers, but also in manipulating their spin degrees of freedom, with applications in spintronics. It has been first predicted by Berciu and Jankó [4] that the inhomogeneous magnetic field created by nanoscale permalloy disks can trap spin polarised carriers in DMSs, leading to Zeeman bound states. Another possibility for obtaining non-uniform fields on the nanoscale is to use fields associated with superconductors (SCs). A sufficiently strong uniform external magnetic field can penetrate into type-II SC films in the form of periodic lattices of Abrikosov vortices. A hybrid system consisting of a SC film above a DMS layer illustrated in Fig. 1 has been explored by Rappoport *et al* [5], who found the nearly spin polarised localised states induced by the magnetic field due to a single vortex can interact to form a band structure. Because each vortex carries a fixed unit of magnetic flux, $\phi_0/2 = h/2e$, the distance between vortices depends on the magnitude of the external field. Hence the applied field can control the band structure by moving vortices, illustrating one benefit of such a hybrid device.

As part of a joint theoretical and experimental research programme in Bath we have revisited this heterostructure, as well as beginning an exploration of magnetic templating due to other SC nanostructures. Below we present a tight-binding theory of the energy spectrum induced by the magnetic fields associated with a periodic lattice of SC vortices, developed to validate a "brute-force" diagonalisation of the Hamiltonian using a Landau-level basis, and which will form the basis of future modelling of low-dimensional vortex structures and spin-charge pumps. We also present preliminary results illustrating localised states associated with SC disks and rings.

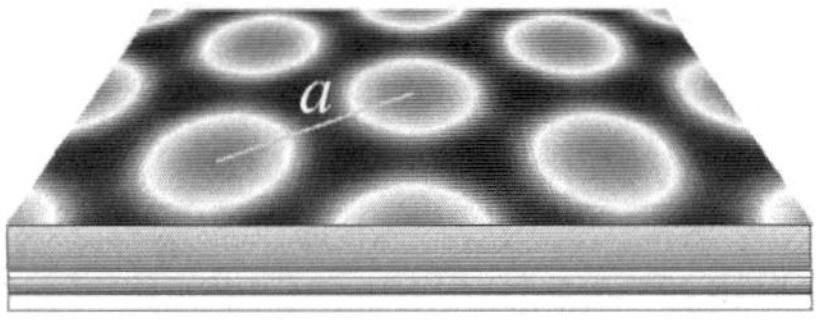

Figure 1. Schematic illustration of the periodic lattice of vortices, spacing a, in a type-II superconducting film above a DMS (blue layer).

2. Energy spectrum due to a vortex lattice using a Landau level basis

An ideal type-II SC film will allow a magnetic field B_0 between the lower and upper critical field strengths to penetrate through the formation of the triangular Abrikosov vortex phase. With each vortex carrying a magnetic flux quantum, the distance between vortices a is fixed by $B_0 = \phi_0/\sqrt{3}a^2$. Assuming that the magnetic field $\vec{B}_L$ within the thin DMS a distance z beneath the SC film is the summation of the fields for periodically distributed single vortices [6], $\vec{B}_v$, one can write $\vec{B}_L(\vec{r}; z) = \sum_{\vec{R}} \vec{B}_v(\vec{r} - \vec{R}; z) = B_0\hat{z} + \sum_{\vec{G}\neq 0} e^{i\vec{G}\cdot\vec{r}}\vec{B}_{\vec{G}}(z)$, where $\vec{R}$ and $\vec{G}$ are direct and reciprocal lattice vectors, with an analogous Fourier decomposition of the associated vector potential $\vec{A}_L(\vec{r}; z) = \vec{A}_0(\vec{r}) + \sum_{\vec{G}\neq 0} e^{i\vec{G}\cdot\vec{r}}\vec{A}_{\vec{G}}(z)$, where $\vec{A}_0(\vec{r}) = (0, B_0 x, 0)$ in the Landau gauge. Electrons (charge $-e$) within the DMS may then be described by the Hamiltonian

$$\hat{H} = \frac{1}{2m^*}[\hat{p} + e\vec{A}_L(\vec{r}; z)]^2 - \frac{1}{2}g_{\text{eff}}\mu_B\vec{\sigma}\cdot\vec{B}_L(\vec{r}; z), \tag{1}$$

and because g_{eff} is typically extremely large in DMSs, $\vec{A}_L$ can be replaced by $\vec{A}_0$ without significant loss of accuracy. Eigenstates of (1) may be found by numerical diagonalisation [5], using a basis of spin-polarised Landau level (LL) eigenstates corresponding to the solution of (1) for a homogeneous field $\vec{B}_L = B_0\hat{z}$. A magnetic unit cell is adopted containing two vortices, corresponding to the lattice $\vec{R} = n\vec{a} + \ell\vec{b}$ with $\vec{a} = (a, 0)$ and $\vec{b} = (0, b)$, $b = \sqrt{3}a$. This ensures that $\hat{H}$ commutes with magnetic translation operators [7, 8, 9] $\hat{T}_M(\vec{R}) = e^{ieB_0R_xy/\hbar}\hat{T}(\vec{R})$ that satisfy $\hat{T}_M(\vec{R})\hat{T}_M(\vec{R}') = \hat{T}_M(\vec{R} + \vec{R}')$, and so its eigenstates may be labelled by a wavevector $\vec{k}$ within the associated magnetic Brillouin zone (inset, Fig 2a). Basis functions [10]

$$\Psi_{n,\vec{k},\sigma}(\vec{r}) = \sum_\lambda e^{-ik_x l^2(k_y + \lambda\frac{2\pi}{b})}\phi_{n,k_y+\lambda\frac{2\pi}{b},\sigma}(\vec{r}), \tag{2}$$

where $l = \sqrt{ab/2\pi}$ is the magnetic length and $\phi_{n,k_y,\sigma}(\vec{r})$ are LL eigenstates in the Landau gauge with energies $E_{n,\sigma} = (n+1/2)\hbar\omega_C - (1/2)g_{\text{eff}}\mu_B B_0\sigma$, represent suitable eigenstates of the magnetic translation operators. Using these to expand the eigenfunctions $\psi_{\vec{k}}(\vec{r})$ of (1) containing the non-uniform field as

$$\psi_{\vec{k}}(\vec{r}) = \sum_{n,\sigma} d_{n\sigma}(\vec{k})\Psi_{n,\vec{k},\sigma}(\vec{r}), \tag{3}$$

substituting into $\hat{H}\psi_{\vec{k}}(\vec{r}) = E_{\vec{k}}\psi_{\vec{k}}(\vec{r})$, multiplying by $\Psi^\dagger_{n,\vec{k},\sigma}$ and using orthonormality gives the matrix equation

$$-\frac{1}{2}g_{\text{eff}}\mu_B\sum_{n'\sigma'}\vec{\sigma}_{\sigma\sigma'}\cdot\vec{b}_{n,n'}d_{n'\sigma'}(\vec{k}) = (E_{\vec{k}} - E_{n,\sigma})d_{n\sigma}(\vec{k}), \tag{4}$$

where $\vec{\sigma}_{\sigma\sigma'} = \chi^\dagger_\sigma\vec{\sigma}\chi_{\sigma'}$, and

$$\vec{b}_{n,n'} = \delta_{\vec{k},\vec{k}'}\sqrt{\frac{T!}{S!}}\sum_{\vec{G}\neq 0}\vec{B}_{\vec{G}}(z)(i\sqrt{g})^{S-T}L_T^{S-T}(g)e^{-g/2}e^{i\pi pq}e^{i(k_xG_y - k_yG_x)l^2}\left(\frac{G_x - iG_y}{G}\right)^{n-n'} \tag{5}$$

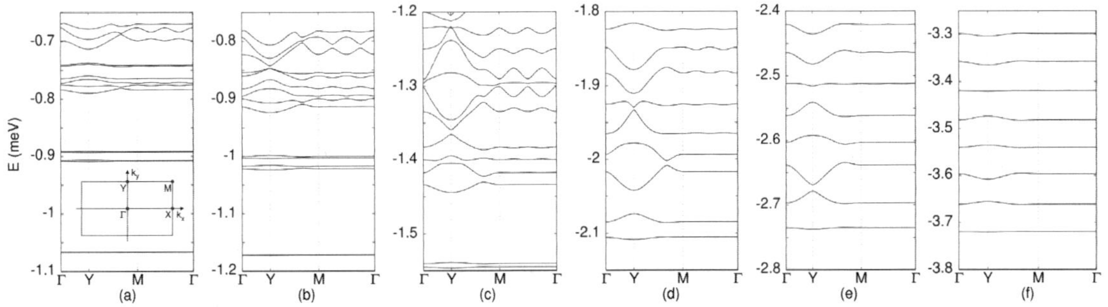

Figure 2. Band structure of electrons in the DMS for applied fields B_0 of (a) 0.06 T, (b) 0.07 T, (c) 0.10 T, (d) 0.15 T, (e) 0.19 T, and (f) 0.26 T, corresponding to vortex spacings a/λ of (a) 5.0, (b) 4.6, (c) 3.8, (d) 3.2, (e) 2.8, and (f) 2.4. The inset in (a) shows the Brillouin zone.

with $\vec{G} = (p\frac{2\pi}{a}, q\frac{2\pi}{b})$, $g = l^2G^2/2$, S (T) the greatest (smallest) of n and n', and $L_n^m(x)$ an associated Laguerre polynomial.

Fig. 2 shows results obtained with $\lambda = 40$ nm, $\xi = 35$ nm (corresponding to Nb), $z = 4$ nm, and typical DMS parameters $m^* = 0.5m_e$, $g_{\text{eff}} = 500$. At high fields the spectrum approaches Landau levels as the corrugation in $\vec{B}_L$ becomes insignificant. At low fields, flat bands correspond to the energy levels of Zeeman bound states associated with isolated vortices, because the separation a is large. At intermediate fields, the behaviour is more complex. Band energies shown in the figure are converged to 6 d.p. using basis set dimensions ranging from 170 for the lowest fields, down to 50 for the highest.

Although these results appear sensible, they differ from those in Ref. [5]. Expressions for the basis functions and our matrix elements $\vec{b}_{n,n'}$ are also different. Partly to confirm our analysis, and to develop an alternative framework for the weakly interacting vortex limit, we have considered a tight-binding description presented in the following section.

3. Energy bands using a tight-binding model

When the distance between vortices is large, the band structure tends to the energy levels of states trapped by isolated vortices. We therefore expect that a tight-binding (TB) model can be developed to describe the spectrum in the low field limit. To do so we express the wavefunction in the presence of the vortex lattice as a superposition of isolated vortex states [5] which we labelled by an angular momentum index m, unit cell $\vec{R} = (na, \ell b)$ which is a magnetic translation vector, and location within the unit cell through $\tau \in \{0, 1\}$, with $\vec{r}_\tau = \tau(a/2, b/2)$:

$$|\psi\rangle = \sum_{\vec{R},\tau,m} a_{m\tau\vec{R}} |m, \tau, \vec{R}\rangle. \tag{6}$$

The influence of the vector potential on the Hamiltonian matrix elements is included as [11]

$$\langle \vec{R}'|\hat{H}^A(\vec{r}, \hat{p} + e\vec{A})|\vec{R}\rangle = \exp\left[\frac{-ie}{\hbar}\int_{\vec{R}}^{\vec{R}'} \vec{A}(\vec{r})\cdot \mathrm{d}\vec{r}\right] \langle \vec{R}'|\hat{H}^0(\vec{r}, \hat{p})|\vec{R}\rangle, \tag{7}$$

where $\hat{H}^A$ and $\hat{H}^0$ are the Hamiltonian with and without the vector potential respectively. Using $\vec{A}_0(\vec{r}) = B_0 x\hat{j}$ for $\vec{A}$ gives

$$\hat{H}^A = \sum_{\vec{R}',\tau',m'} \sum_{\vec{R},\tau,m} |m, \tau, \vec{R}\rangle e^{\frac{-ie}{2\hbar}(\vec{R}_\tau - \vec{R}'_{\tau'})\cdot(\vec{A}_0(\vec{R}_\tau) + \vec{A}_0(\vec{R}'_{\tau'}))} \langle m, \tau, \vec{R}|\hat{H}^0|m', \tau', \vec{R}'\rangle\langle m', \tau', \vec{R}'| \tag{8}$$

where $\vec{R}_\tau = \vec{R} + \vec{r}_\tau$ and $\hat{H}^0 = \frac{1}{2m^*}\hat{p}^2 - \frac{1}{2}g_{\text{eff}}\mu_B\vec{\sigma}\cdot\vec{B}_L$. A magnetic translation operator for magnetic translation vector $\vec{L}$ may be written as

$$\hat{T}_M(\vec{L}) = \sum_{\vec{R},\tau,m} |m,\tau,\vec{R}\rangle e^{\frac{-ie}{\hbar}\vec{R}_\tau\cdot\vec{A}(\vec{L})}\langle m,\tau,\vec{R}-\vec{L}|, \tag{9}$$

which commutes with (8) and satisfies $\hat{T}_M(\vec{L})\hat{T}_M(\vec{L}') = \hat{T}_M(\vec{L}+\vec{L}')$, so its eigenvalues can be implied to be $e^{i\vec{k}\cdot\vec{L}}$. Hence we propose wavefunctions of electrons in the presence of the field due to a vortex lattice of the form,

$$|\psi_{\vec{k}}\rangle = \sum_{\tau,m} a_{m,\tau,\vec{k}}|m,\tau,\vec{k}\rangle, \tag{10}$$

where $|m,\tau,\vec{k}\rangle = \frac{1}{\sqrt{N}}\sum_{\vec{R}} e^{-i\vec{k}\cdot\vec{R}_\tau}e^{-\frac{ie}{\hbar}\vec{r}_\tau\cdot\vec{A}_0(\vec{R})}|m,\tau,\vec{R}\rangle$ is an orthonormal eigenstate of $\hat{T}_M(\vec{L})$ in the nearest neighbour TB approximation, $\langle m',\tau',\vec{R}'|m,\tau,\vec{R}\rangle = \delta_{mm'}\delta_{\vec{R}_\tau,\vec{R}'_{\tau'}}$, and N is the number of unit cells. The Schrödinger equation can be derived straightforwardly,

$$\sum_{\tau',m'} \langle m,\tau,\vec{k}|\hat{H}^A|m',\tau',\vec{k}'\rangle a_{m'\tau',\vec{k}'} = E_{\vec{k}}a_{m\tau,\vec{k}}\delta_{\vec{k}\vec{k}'}, \tag{11}$$

where analysis gives

$$\begin{aligned}\langle m,\tau,\vec{k}|\hat{H}^A|m',\tau',\vec{k}'\rangle &= \delta_{\vec{k}\vec{k}'}\sum_{\vec{R}} e^{i(\vec{k}\cdot\vec{R}_\tau-\vec{k}\cdot\vec{r}_{\tau'}+\frac{e}{\hbar}\vec{r}_\tau\cdot\vec{A}_0(\vec{R}))}e^{-\frac{ie}{2\hbar}(\vec{R}_\tau-\vec{r}_{\tau'})\cdot[\vec{A}_0(\vec{R}_\tau)+\vec{A}_0(\vec{r}_{\tau'})]}\\ &\quad\times\langle m,\tau-\tau',\vec{R}|\hat{H}^0|m',0,0\rangle.\end{aligned} \tag{12}$$

In the spirit of the nearest neighbour TB approximation, we assume that the hopping term in (12) vanishes except when $\vec{R}_{\tau-\tau'}$ is zero ("on-site", $\langle m,0,0|\hat{H}^0|m',0,0\rangle = E_m\delta_{mm'}$) or points to a vortex nearest neighbour to the origin, when we have $\langle m,\tau-\tau',\vec{R}|\hat{H}^0|m',0,0\rangle = e^{i\phi(m'-m)}t_{mm'}$ where $\vec{R}_{\tau-\tau'} = a(\cos\phi,\sin\phi)$.

If we only include $m=0$ states, then $\hat{H}^A$ can be represented by a 2-by-2 matrix which reduces at $\vec{k}=0$ to

$$\hat{H}^A = \begin{pmatrix} E_0+2t_{00} & 0 \\ 0 & E_0-2t_{00}\end{pmatrix}. \tag{13}$$

from which the TB parameters E_0, t_{00} may be determined by fitting to the eigenvalues at Γ found with the LL basis. Fig. 3 compares the resulting TB bands along high symmetry directions with those found using the LL basis, for various vortex lattice spacings, and the variation in the associated TB parameters. Similarly modelling the $m=\pm1$-derived bands ignoring couplings to other isolated-vortex states gives a 4-by-4 Hamiltonian which involves 5 unique parameters, E_{-1}, E_1, $t_{-1,-1}$, $t_{-1,+1} = -t_{+1,-1}$, and $t_{+1,+1}$, which we fix by requiring that the 4 eigenvalues at $\vec{k}=\Gamma$ and the lowest eigenvalue at $\vec{k}=X$ agree with those obtained using the LL basis expansion. Fig. 4 shows the subsequent agreement between the TB bands and the LL-basis bands along high symmetry directions, and the TB parameters obtained.

As a increases the magnitudes of the hopping parameters $t_{mm'}$ decreases, as it becomes increasingly difficult for electrons to jump between Zeeman potential wells. The on-site energies also saturate with increasing vortex spacing at the energy level of the associated bound state for an isolated vortex, the deviation for smaller a being due to the influence of stray fields from neighbouring vortices. Both Fig. 3 and Fig. 4 show that the TB description becomes

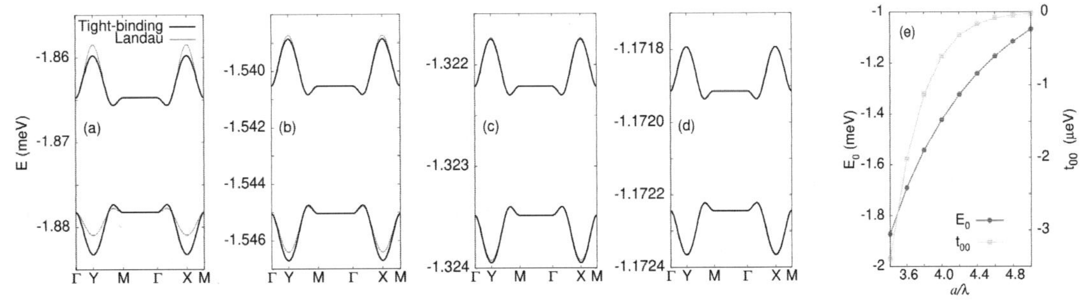

Figure 3. Comparison between the lowest energy bands obtained using tight-binding with isolated-vortex $m = 0$ states only and those found using the Landau-level basis for $a/\lambda =$ (a) 3.4, (b) 3.8, (c) 4.2, and (d) 4.6. Also shown in (e) is the dependence of the tight-binding parameters on the vortex spacing.

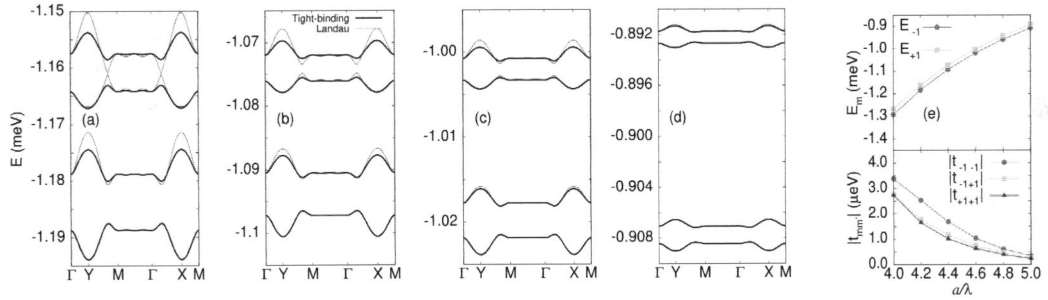

Figure 4. Comparison between the bands derived from $m = \pm 1$ states using the tight-binding method, and those found with the Landau-level basis for $a/\lambda =$ (a) 4.2, (b) 4.4, (c) 4.6, and (d) 5.0. In (e) we show the variation of the tight-binding parameters on the vortex spacing.

increasingly good as the vortex spacing increases, with the bands derived from the more tightly bound $m = 0$ states being described generally better than the $|m| = 1$-derived bands. Even better agreement can be achieved by including hopping between states with different $|m|$, at the expense of additional parameters. However, the results shown here are sufficient to demonstrate the success of the TB description, and the agreement between the two sets of energy band calculations using the LL basis and the TB method confirm that our results with the former are correct.

4. Confined states due to superconducting nanostructures

Screening currents induced in SC disks and rings above a DMS when placed in a uniform applied field can also result in inhomogeneous total magnetic fields that can influence electrons in the DMS. As illustrated in Fig. 5a, flux focussing results in typical $\sim |r - r_0|^{-1/2}$ variations in magnetic field components near an edge at r_0. Due to the giant Zeeman interaction in the DMS, these field profiles induce a series of bound states, in which spin-$\uparrow$ and spin-$\downarrow$ components exhibit different spatial distributions, Fig. 5b,c. In the absence of an applied field, flux trapped within the ring can also cause bound states, with energies that depend upon the number of flux quanta that are trapped. Investigations into these and related systems are currently underway.

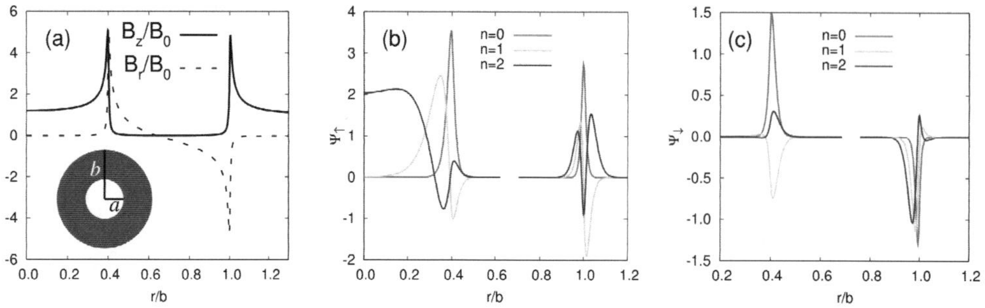

Figure 5. (a) Radial and normal component magnetic field profiles in a DMS beneath a superconducting ring with ratio of inner to outer radius a/b=0.4. (b) Radial dependence of the spin-$\uparrow$ component of the lowest 3 states confined at each of the inner and outer edges. (c) As (b), but spin-$\downarrow$.

5. Summary

We have developed and validated a tight-binding model to describe the electronic structure of states in a DAMS beneath a type-II superconducting film. This will form the basis of studies of other low-dimensional systems. We have also identified states localised by the concentration in magnetic flux that occurs at the edges of superconducting disks and rings.

References

[1] Furdyna J K 1988 *J. Appl. Phys.* **64** R29
[2] Berciu M, Rappoport T G and Jankó B 2005 *Nature* **435** 71
[3] Redliński P, Wojtowicz T, Rappoport T G, Libál A, Furdyna J K and Jankó B 2005 *Phys. Rev.* B **72** 085209
[4] Berciu M and Jankó B 2003 *Phys. Rev. Lett.* **90** 246804
[5] Rappoport T G, Berciu M, and Jankó B 2006 *Phys. Rev.* B **74** 094502
[6] Carneiro G and Brandt E H 2000 *Phys. Rev.* B **61** 6370
[7] Brown E 1964 *Phys. Rev.* **133** A1038
[8] Opechowski W and Tam W G 1969 *Physica* **42** 529
[9] Wal A 2008 *Physica* B **404** 1040
[10] Pfannkuche D and Gerhardts R R 1992 *Phys. Rev.* B **46** 12606
[11] Graf M and Vogl P 1994 *Phys. Rev.* B **51** 4940

Entangled-Photon Pair Emission from a Light-Emitting Diode

Cameron L. Salter[1,2], R. Mark Stevenson[1], Ian Farrer[2], Christine A. Nicoll[2], David A. Ritchie[2] and Andrew J. Shields[1]

[1]Toshiba Research Europe Limited, 208 Cambridge Science Park, Cambridge, CB4 0GZ, UK.

[2]Cavendish Laboratory, University of Cambridge, JJ Thomson Avenue, Cambridge, CB3 0HE, UK

E-mail: cls68@cam.ac.uk

Abstract. Electrically-driven entangled-photon generation is demonstrated for the first time using a single InAs/GaAs quantum dot embedded in a light emitting diode structure. Under alternating-current injection, we found that the entanglement fidelity was of sufficient quality for quantum information applications such as quantum key distribution.

1. Introduction

Quantum entanglement plays a central role in quantum information technology as sources of entangled-photon pairs are required for entanglement-based quantum key distribution [1] and efficient linear-optical quantum computers [2]. Existing sources of entangled-photon pairs all require a laser for optical excitation, imposing a practical limit on their potential for large-scale quantum information applications. For Poissonian sources, including the most widely used parametric down-conversion, zero or multiple-photon pairs can be emitted per laser excitation pulse due to the probabilistic nature of the non-linear process. This presents an additional fundamental limitation on large-scale systems, due to the limited success probability of quantum operations and errors. For successful future applications, these complications can be overcome with an electrically-driven on-demand source of entangled-photon pairs [3], which currently does not exist.

Here we report the realisation of the first electrically-driven source of entangled-photon pairs, which consists of a quantum dot embedded in a semiconductor light emitting diode (LED) structure [4]. We show that the device emits entangled-photon pairs under direct- and alternating-current (dc and ac) injection; the latter indicates its potential function as an on-demand source and without the need for a complicated laser driving system.

2. Experiments

2.1. Entangled-LED design and operation

Our device contains a single quantum dot, which in recent years have been manipulated to emit single pairs of entangled photons via the radiative decay of the biexciton state [5] (Figure 1 (a)). The biexciton state is formed by capture of two electrons and two holes via electrical injection. The biexciton state radiatively decays to the ground state via one of two bright-exciton states, which determines the polarisations of the resulting pair of photons. If the fine-structure splitting between the two exciton states is close to zero then the decay path can only be determined by measuring the polarisation of the photons [6]. Consequently, the photons are entangled in polarisation and the emission is characterised by the two-photon Bell-state, $|\psi^+\rangle = (|RL\rangle + |LR\rangle)/\sqrt{2}$, where $|RL\rangle$ corresponds to right- and left-hand circularly polarised photons being emitted from the biexciton and exciton states, respectively.

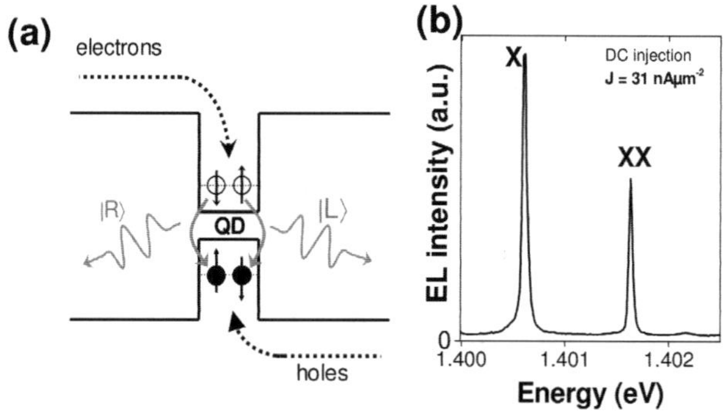

Figure 1. (a) Schematic of the active region of the entangled-LED showing the emission of a polarisation entangled-photon pair via the biexciton cascade. Here the diode is in its 'off' state, just after non-resonant electrical injection of two electrons (open circles) and two holes (closed circles) into the quantum dot. (b) CCD-recorded electroluminescence (EL) spectrum of the quantum dot investigated in this report using dc electrical injection with current density (J) = 31 nAμm^{-2}.

The design of the LED is based on a single layer of quantum dots embedded in a *p-i-n* doped planar microcavity. Details of device growth, by molecular beam epitaxy, and fabrication can be found in reference [4]. Two features of the design are crucial to successful operation as an entangled-LED. The first is an unusually thick cavity (2λ) and intrinsic region (~ 400nm). This is to suppress electrons tunneling into the dot from the n-doped region during biexciton decay, thereby minimising destruction of entanglement after emission of the first photon caused by charging of the intermediate exciton state and maximising the amount of light emitted from the neutral excitonic states. Secondly, careful control of the growth conditions created InAs quantum dots emitting ~ 1.4eV, which are suitable for entangled-light generation due to their very small fine-structure splitting [5][7].

Carrier injection into the quantum dots is achieved by biasing the diode beyond its turn-on voltage. We studied a quantum dot in our device which had almost zero fine-structure splitting, with magnitude 0.4±0.1μeV [8]. Its 5K electroluminescence (EL) spectrum is shown in Figure 1 (b), with emission from the exciton (X) and biexciton (XX) states labelled. The emission lines were identified by power dependence, time-resolved EL, and correlation measurements [9].

2.2. *dc experiment*

Photons from the X and XX transitions were resolved in polarisation and in time in order to measure the co- and cross-polarised second order pair-correlation functions $g^{(2)}_{XX,X}(\tau)$ and $g^{(2)}_{XX,\bar{X}}(\tau)$, where τ is the time delay between the XX and X photons [10].

Figure 2 shows the measurement of $g^{(2)}_{XX,X}(\tau)$ and $g^{(2)}_{XX,\bar{X}}(\tau)$ in the rectilinear (a), diagonal (b) and circular (c) polarisation bases when a direct current with density 31 nAμm^{-2} was injected into the device. The co-polarised (VV) correlations in Figure 2 (a) show the characteristic shape of a radiative cascade. We expect the emission of an exciton photon to follow the emission of a biexciton photon. As a result, $g^{(2)}_{XX,X}(\tau)$ is increased for small positive delays and suppressed for small negative delays. Away from $\tau = 0$ $g^{(2)}_{XX,X}(\tau)$ and $g^{(2)}_{XX,\bar{X}}(\tau)$ both tend to 1 due to uncorrelated emission events. A dip is seen for cross-polarised pairs (HV) in Figure 2 (a) due to the selection rule that the cascade must produce pairs with the same linear polarisation. Similar correlation behaviour is seen for photon pairs polarised in the diagonal basis in Figure 2 (b). Significantly we see the opposite correlation behaviour for circularly polarised photon pairs (Figure 2 (c)) than in the linear and diagonal cases. This is expected for entangled-photon pairs in the Bell state $|\psi^+\rangle$ [5][10], as the two-photon wavefuntion can be expressed as the superposition of co-linearly, co-diagonally or cross-circularly polarised photon pairs.

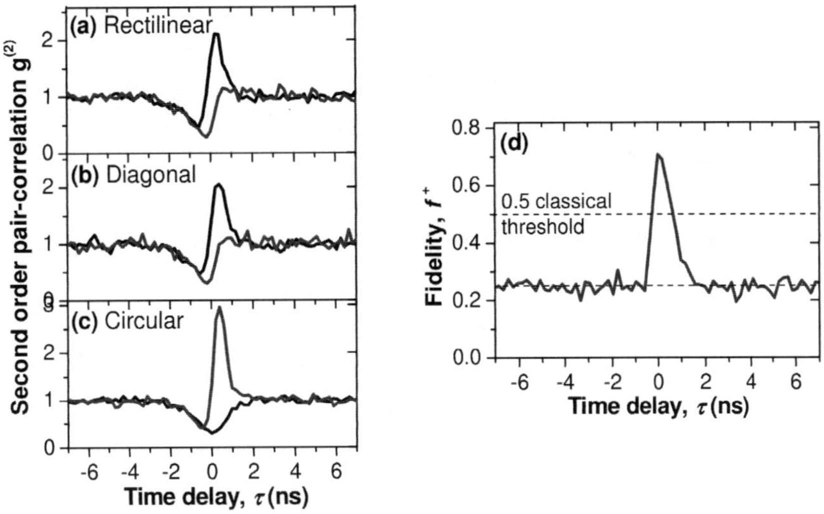

Figure 2. Polarised pair-correlation results from dc electrical injection into the LED. $g^{(2)}_{XX,X}(\tau)$ and $g^{(2)}_{XX,\bar{X}}(\tau)$ measured in the (a) rectilinear, (b) diagonal and (c) circular bases. Correlations measured for photons of the same (orthogonal) polarisation are shown in black (blue). (d) Fidelity, f^+, as a function of time delay, τ.

The entanglement fidelity (f^+) of the emitted light, projected onto the maximally entangled state $|\psi^+\rangle$, is plotted in Figure 2 (d). The fidelity was determined directly by combining correlations measured in the rectilinear, diagonal and circular polarisation bases [6]. The peak at $\tau = 0$ gives a maximum dc fidelity $f^+ = 0.707 \pm 0.023$. This exceeds the 0.5 threshold for a source emitting a classically polarisation-correlated state by 9 standard deviations, proving that entangled photons have been electrically generated for the first time. The measured fidelity is limited by several factors [6], including the timing jitter on the photon detectors and re-excitation of the quantum dot part way through the cascade.

2.3. *ac experiment*

Figure 3 shows the results of polarised photon-pair correlation experiments conducted when the device was pulsed with an alternating current at a repetition rate of 80MHz. As in the dc case above, for entangled-photon pairs in the Bell state $|\psi^+\rangle$ we expect to measure co-polarised photon pairs in the rectilinear and diagonal bases and cross-polarised photon pairs in the circular basis. Figure 3 (a) shows the measurement of $g^{(2)}_{XX,X}(\tau)$ and $g^{(2)}_{XX,\bar{X}}(\tau)$ in the rectilinear (a), diagonal (b) and circular (c) polarisation bases. The dominant peak at zero time delay in each of these traces has a higher number of coincidences than the others, corresponding to emission of an X photon shortly after a XX photon. In both Figure 3 (a) and (b) this large peak belongs to the co-polarised trace, indicating measurement of co-polarised photon pairs in the rectilinear and diagonal bases. In contrast the cross-polarised peak dominates in Figure 3 (c), indicating measurement of cross-polarised photons pairs in the circular basis. This data gives a peak ac fidelity of $f^+ = 0.785 \pm 0.022$. This ac fidelity is larger than the dc fidelity (0.707) presented above in Figure 2 (d). This direct comparison is possible as the timing resolution is the same in both experiments at 0.2 ns. It is likely that the ac fidelity is larger due to the lower level of re-excitation of the quantum dot when driven by a short voltage pulse.

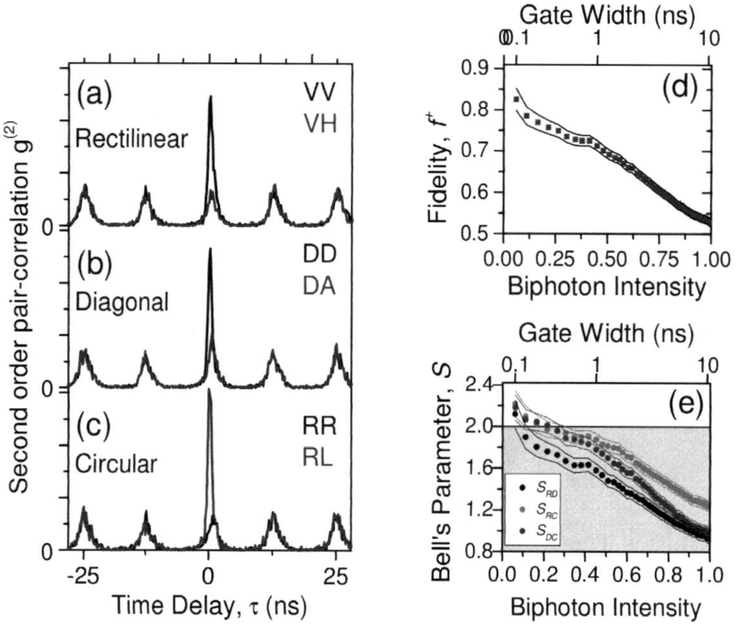

Figure 3. Polarised pair-correlation results from ac electrical injection into the diode. $g^{(2)}_{XX,X}(\tau)$ and $g^{(2)}_{XX,\bar{X}}(\tau)$ measured in the (a) rectilinear, (b) diagonal and (c) circular polarisation. (d) Fidelity and (e) Bell's parameters at $\tau = 0$ as a function of the proportion of the total biphoton intensity that is analysed. The biphoton intensity is varied by changing the gate width (top axis).

The entanglement fidelity available to an application may be enhanced by limiting the time the detectors remain active. We explore this possibility in Figure 3 (d) by changing the gate width used in the measurement of f^+. For the zero period, gate width is defined as the maximum permitted time delay between the two photons to register a detection event. This changes the proportion of the detected biphotons to be accepted in the measurement of f^+. Without rejecting *any* detected photon pairs (i.e. biphoton intensity = 1.0), $f^+ = 0.530 \pm 0.010$, which is above the classical threshold of 0.5, proving entangled-light emission. The fidelity rises quickly as the proportion of biphoton intensity accepted is reduced. This leads to a maximum ac fidelity of $f^{+} = 0.826 \pm 0.027$, when 6% of biphoton coincidences closest to $\tau = 0$ are accepted (0.1ns gate width).

Using these results it is also possible to determine Bell's parameters within a quantum mechanics framework [8]. The three different Bell parameters measured here, S_{RD}, S_{DC} and S_{RC} defined in reference [8], are non-equivalent and correspond to three different planes of the Poincaré sphere. A value greater than 2 for any of these parameters represents a violation of Bell's inequality on the corresponding plane. Figure 3 (e) displays these three Bell parameters measured as a function of the biphoton intensity (bottom axis) and gate width (top axis). Bell's parameters increase as we decrease the gate width due to an improving degree of correlation in all three polarisation bases. S_{RD} is less than S_{RC} and S_{DC}, as the circular basis has the highest degree of correlation. At the smallest gate width of 0.1ns we find; $S_{RD} = 2.12 \pm 0.13$, $S_{DC} = 2.18 \pm 0.12$ and $S_{RC} = 2.22 \pm 0.12$. The latter two violate Bell's inequality by more than one standard deviation. Overcoming this threshold shows that the entangled-LED is of high enough quality for applications such as a quantum relay and entanglement-based quantum key distribution [1].

3. Conclusions

We have eliminated the need for large and complicated laser driving systems in future quantum information applications by demonstrating high fidelity entangled-light emission triggered from an entangled-LED. Entangled-photon pairs with such high fidelity are sufficient for teleportation [12] and entanglement swapping [13], which are important components in a quantum computer. Improvements to the LED such as reducing the background light emission [10][14] and increasing the speed of the device to minimise re-excitation during pulsing could be made to push the fidelity even higher to eventually realise electrically-operated fault-tolerant quantum computing.

4. Acknowledgements

We would like to acknowledge funding for this work from EPSRC, QIPIRC, QAP and NanoSci-ERA.

References

[1] Ekert A 1991 *Phys. Rev. Lett.* **67** 661
[2] Knill E, Laflamme R and Milburn G J 2001 *Nature* **409** 46
[3] Benson O, Santori C, Pelton M, and Yamamoto Y 2000 *Phys. Rev. Lett.* **84** 2513
[4] Salter C L, Stevenson R M, Farrer I, Nicoll C A, Ritchie D A and Shields A J 2010 *Nature* **465** 594-597
[5] Stevenson R M, Young R J, Atkinson P, Cooper K, Ritchie D A and Shields A J 2006 *Nature* **439** 179
[6] Hudson A J, Stevenson R M, Bennett A J, Young R J, Nicoll C A, Atkinson P, Cooper K, Ritchie D A and Shields A J 2007 *Phys. Rev. Lett.* **99** 266802
[7] Young R J, Stevenson R M, Shields A J, Atkinson P, Cooper K, Ritchie D A, Groom K M, Tartakovskii A I and Skolnick M S 2005 *Phys. Rev. B* **72** 113305
[8] Young R J, Stevenson R M, Hudson A J, Nicoll C A, Ritchie D A and Shields A J 2009 *Phys. Rev. Lett.* **102** 030406
[9] ed Michler P 2003 *Single Quantum Dots* (Berlin Heidelberg: Springer-Verlag)
[10] Young R J, Stevenson R M, Atkinson P, Cooper K, Ritchie D A and Shields A J 2006 *New J. Phys.* **8** 29
[11] Moreau E, Robert I, Manin L, Thierry-Mieg V, Gérard J M, and Abram I 2001 *Phys. Rev. Lett.* **87** 183601
[12] Bouwmeester D, Pan J -W, Mattle K, Eibl M, Weinfurter H and Zeilinger A 1997 *Nature* **390** 575–579
[13] Pan J -W, Bouwmeester D, Weinfurter H, and Zeilinger A 1998 *Phys. Rev. Lett.* **80** 3891
[14] Nicoll C A, Salter C L, Stevenson R M, Hudson A J, Atkinson P, Cooper K, Shields A J and Ritchie D A, 2008 *J. Crystal Growth* **311** 1811

Low-temperature transport in ultra-thin tungsten films grown by focused-ion-beam deposition

O Chiatti and P A Warburton

London Centre for Nanotechnology, University College London, 17-19 Gordon Street, London, WC1H 0AH, United Kingdom

E-mail: `o.chiatti@ucl.ac.uk`

Abstract. We have fabricated tungsten-containing films by focused-ion-beam (FIB)-induced chemical vapour deposition. By using ion-beam doses below 50 pC/μm^2 on a substrate of amorphous silicon, we have grown continuous films with thickness below 20 nm. The low-temperature electron transport properties were investigated by measuring current-voltage characteristics for temperatures down to 400 mK and in magnetic fields up to 3 T. FIB-deposited tungsten films are known to have an enhanced transition temperature compared to bulk tungsten, and films with thickness down to 50 nm have been investigated previously. The films in this work are closer to the limit of two-dimensional system and are superconducting below 1 K. Magnetoresistance measurements yield upper critical fields of the order of 1 T, and the resulting coherence length is smaller than the film thickness.

The use of a focused ion- or electron-beam (FIB or FEB) to induce chemical vapour deposition (CVD) of composite materials from metal-organic precursors has emerged in recent years as an important nanofabrication technique. In particular, using tungsten hexacarbonyl ($W(CO)_6$) as a precursor, it is possible to deposit tungsten-containing materials, which are amorphous and undergo a superconducting transition at $T_c \approx 5$ K [1, 2]. Tungsten-composite films fabricated by FIB-induced CVD ("FIB-W"), have been found to be superconducting at low temperatures for thickness down to 25 nm [3]. This opens the possibility of investigating a thickness-dependent superconductor-insulator transition and superconductivity in the two-dimensional limit. However, up to now an investigation of the superconductivity of ultra-thin FIB-W films has been lacking, because the traditional methods cannot grow continuous ultra-thin films [4].

By using ion-beam doses below 50 pC/μm^2 on a substrate of amorphous silicon, we have grown continuous ultra-thin FIB-W films with thickness below 20 nm. The resulting films are closer to the two-dimensional limit than previous films. We have investigated the low-temperature transport properties by measuring both current-voltage characteristics and differential resistance for temperatures down to 400 mK and in magnetic fields up 3 T. The temperature-dependent resistance measurements yield superconducting transition temperatures between 1 and 4 K, lower than thick FIB-W films. The magnetoresistance measurements yield upper critical fields of the order of 1 T, and the resulting coherence length is between 15 and 20 nm, larger than the film thickness.

Acknowledgments
This work is supported by EPSRC.

Journal of Physics: Conference Series **286** (2011) 012023 doi:10.1088/1742-6596/286/1/012023

References

[1] Sadki E S, Ooi S and Hirata K 2004 *Appl. Phys. Lett.* **85** 620
[2] Luxmoore I J, Ross I M, Gullis A G, Fry P W, Orr J, Buckle P D and Jefferson J H 2007 *Thin Solid Films* **515** 6791
[3] Li W, Fenton J C, Wang Y, McComb D W and Warburton P A 2008 *J. Appl. Phys.* **104** 093913
[4] Li W, Fenton J C and Warburton P A 2009 *IEEE Trans. Appl. Supercond.* **19** 2819

Materials for superconducting nanowires for quantum phase-slip devices

J C Fenton[1,2], C H Webster[2] and P A Warburton[1]

[1] London Centre for Nanotechnology, UCL, 17–19 Gordon Street, London WC1H 0AH, UK
[2] National Physical Laboratory, Hampton Road, Teddington TW11 0LH, UK

E-mail: `j.fenton@ucl.ac.uk`

Abstract. Quantum phase-slip processes in superconducting nanowires of suitably small cross-section have been proposed as the basis for a new current standard, based on physics dual to that for the Josephson voltage standard. The practical realisation of such devices presents several challenges. We consider the requirements which need to be met in constructing a nanowire quantum-phase-slip device and in particular the need to maximise R_ξ, the normal-state resistance of a length of nanowire equal to the superconducting coherence length. Titanium and niobium–silicon are promising materials for the nanowires.

1. Introduction

In 2006 Mooij and Nazarov demonstrated that circuits containing quantum phase slip (QPS) junctions are exactly dual to circuits containing Josephson junctions [1]. For example, a voltage-biased circuit containing a QPS element, an inductance and a resistance in series is dual to a current-biased circuit containing a Josephson element in parallel with a capacitance and a resistance; this example has particular technological relevance since it raises the possibility of a quantum current standard based on the dual of the Shapiro effect in Josephson junctions on which the international voltage standard is now based.

There have been experimental efforts by a number of groups to realise nanowires in which QPS can be observed, on a number of different materials. Several groups have reported a larger resistance for $T < T_c$ than would be expected in a homogeneous material from thermally activated phase slips. Similar observations might, however, also arise due to sample inhomogeneities. Observation of a critical voltage below which no current flows would be a less ambiguous demonstration of QPS but to date, there has been only one isolated such report [3]. It is important, therefore, to carefully consider the experimental criteria needed to observe QPS. Here we focus particularly on choice of nanowire material.

2. Candidate materials

A narrow wire along whose length occurs a sufficiently high rate Γ_S of quantum phase slips is suitable as a QPS element. T_c should be high enough that measurements are possible at temperatures significantly less than T_c. Phase slips occur over a lengthscale ξ and nanowires for devices should therefore have transverse dimensions less than ξ. QPSs are expected to be only weakly interacting [2] and therefore observable when $Z_c > R_q \equiv h/4e^2 = 6.45$ kΩ, where $Z_c = \sqrt{L'/C'}$ is the characteristic impedance of the nanowire, L' is the kinetic inductance of the nanowire per unit length and C' is its capacitance per unit length. For a QPS device in which Shapiro steps might be observed, the series inductance L should be much larger than $R_q/2\Gamma_S$ to make the charge a well-defined variable and the series resistance R large enough that the damping parameter $\beta_L \equiv 2\pi^2(2L\Gamma_S/R_q)(R_q/R)^2$ is $\lesssim 1$. It has been suggested [1] that $\Gamma_S \sim 10^{11}\ \mathrm{s}^{-1}$ with achievable values of R and L would be suitable for a QPS device.

Experimental studies on Al (BCS coherence length $\xi_0 = 1600$ nm, $T_c = 1.19$ K), Ti ($\xi_0 = 6200$ nm, $T_c = 0.4$ K) and Nb–Si (typical $\xi_0 = 175$ nm, $T_c = 1.5$ K) are among those previously reported [2, 3] — see the review by Arutyunov [2] for a detailed survey.

Theoretical analysis of the problem has generated a number of formulas for the QPS rate, depending on the details of the situation and the underlying assumptions. One example [4] is

$$\Gamma_S = b\frac{\Delta_0}{\hbar}\frac{l}{\xi}\frac{R_q}{R_\xi}\exp\left(-a\frac{R_q}{R_\xi}\right), \tag{1}$$

where Δ_0 is the superconducting energy gap, l is the length of the wire, $a \sim 1$, $b \sim 1$ and R_ξ, the normal-state resistance of a length ξ of the superconducting nanowire. Note that, for a BCS superconductor with $T_c = 1K$, $\Delta_0/\hbar = 2.3.10^{11}\ \mathrm{s}^{-1}$. A common feature of these formulas for the QPS rate is the exponential factor dependent on R_ξ. Therefore, for making a QPS nanowire, it is a crucial requirement that $R_\xi/R_q \gtrsim 1$ otherwise the QPS rate will be exponentially suppressed.

3. Maximising R_ξ

To maximise $R_\xi \equiv \rho\xi/A$, we should clearly maximise the resistivity ρ and minimise the cross-sectional area A. For a given material, ρ increases as the mean free path λ decreases, i.e. as the material becomes more dirty. For a free electron metal, $\rho\lambda$ is set by band-structure properties of the material. Noting that the Ginzburg–Landau coherence length ξ_{GL} in the dirty-limit is $\sim \sqrt{\xi_0\lambda}$, it follows that $R_\xi(\lambda) \sim 1/\sqrt{\lambda}$ in the dirty limit.[1] Samples with small λ are therefore the most suitable for QPS nanowires.

We can estimate the maximum obtainable resistivity of some candidate materials for QPS nanowires. We use experimentally reported values for ξ and $\rho\lambda$ [2, 3, 5, 6], assuming that this product would remain constant even for short λ and we assume that the shortest achievable mean free path is 0.2 nm, of the order of the interatomic spacing. Assuming a cross-sectional area $A = 100\ \mathrm{nm}^2$, the maximum value of R_ξ/R_q at low temperature is ~ 0.1 in Al, ~ 1 in Ti and ~ 1 in NbSi. This suggests that, unless A can be further reduced, even the most resistive Al nanowires will have R_ξ too small for the QPS rate to be significant. Ti and NbSi nanowires with very short λ are more promising. Since T_c may be below its value for a pure, bulk material in such dirty materials, the doping and composition need to be carefully controlled to obtain the highest ρ while maintaining an acceptable T_c and good sample homogeneity.

Acknowledgments
The work is funded by EPSRC and the UK Department for Business, Innovation and Skills.

References
[1] Mooij J E and Y V Nazarov 2006 *Nat. Phys.* **2** 169
[2] Arutyunov K Y 2008 *Phys. Rep.* **474** 1
[3] Van der Sar T 2007 *Master's thesis* (Delft)
[4] Golubev D S and Zaikin A D 2001 *Phys. Rev. B* **64** 014504
[5] Mayadas A F 1968 *J. Appl. Phys.* **39** 4241
[6] Caballero J (1984) *Thin Solid Films* **117** 1

[1] Additional scattering from the boundaries has been shown to cause increases in the resistivity of thin-film samples when the thickness of the sample is $\sim \lambda$. This effect should be small for samples in which the mean free path is similar to the atomic spacing. Other changes away from the bulk 3–D properties due to the small sample dimensions are possible, but are neglected here.

Condensed Matter and Materials Physics Conference (CMMP10) IOP Publishing
Journal of Physics: Conference Series **286** (2011) 012025 doi:10.1088/1742-6596/286/1/012025

Study of the localization-delocalization transition for phonons via transfer matrix method techniques

S D Pinski and R A Roemer

Department of Physics and Centre for Scientific Computing, University of Warwick, Coventry, CV4 7AL, UK

E-mail: `s.d.pinski@warwick.ac.uk`

Abstract. We use a transfer-matrix method to study the localization properties of vibrations in a 'mass and spring' model with simple cubic lattice structure. Disorder is applied as a box-distribution to the force-constants k of the springs. We obtain the reduced localization lengths Λ_M from calculated Lyapunov exponents for different system widths to roughly locate the squared critical transition frequency ω_c^2. The data is finite-size scaled to acquire the squared critical transition frequency of $\omega_c^2 = 12.54 \pm 0.03$ and a critical exponent of $\nu = 1.55 \pm 0.002$.

The disorder-induced metal-insulator transition (MIT) and the concept of Anderson localization [1] for electrons has been studied extensively for many years. During this time highly accurate results for the critical disorder and critical exponent have been obtained through numerous numerical methods, such as multifractal analysis [2, 3], energy level statistics [4, 5] and the transfer matrix method (TMM) [6, 7]. Although phonon localization has been a topic of research for a similar duration, where the first study of a disordered phonon is credited to Lifshitz et al. in 1954 [8], comparatively, less research has been conducted in the field. Analysis of the dynamical matrix through participation ratios [9, 10], level-spacing statistics [11, 12] and multifractal analysis [13] requires highly accurate numerical results on large system sizes which is not trivial and remains challenging even with todays computing resources.

There are many features of interest in phononic research, such as the origins of the boson peak [11, 14], the name given to the peak in the vibrational density of states (VDOS) $g(\omega)$ that is an excess contribution compared to the usual Debye behaviour $g(\omega) \propto \omega^2$. The nature of the modes within the peak are of importance as they are strongly related to the mechanisms of thermal transport in the $\approx 10K$ temperature range [15]. Schirmacher et al. [11, 15] successfully demonstrated that the boson peak consists of extended phonon modes. The locations of the boson peak, Ioffe-Regel transition between weak and strong elastic-scattering regimes and the localization-delocalization transition (LDT) are also of great interest in particular how these are related with each other. Scorpigno et al. [16] state that the link between the locations of the boson-peak and Ioffe-Regel transition is still debatable, whereas through level statistics Kantelhardt et al. [14] estimate the critical phonon frequency to be $\omega_c \approx 3\omega_{peak}$. Taraskin et al. [17] declare that the links between vibrational localization, the boson peak, the Ioffe-Regel crossover and the zero-energy spectral singularity are yet to be fully established.

Here, we take a simple cubic lattice structure made up of point masses connected by springs. We assume that the masses are displaced by only small amplitudes and apply the harmonic approximation. This gives rise to the classical model of vibration, where the displacements of the masses from their equilibrium positions are governed by the displacements of their nearest neighbours. Due to the symmetry of the lattice we can de-couple the spatial dimensions of the system into three identical problems and massively reduce the time of calculations. This

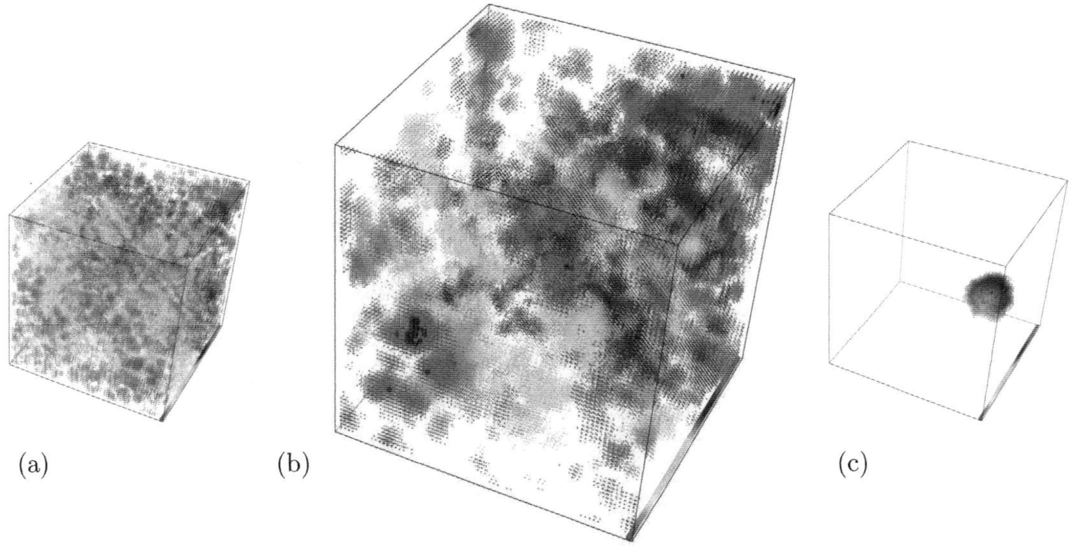

(a) (b) (c)

Figure 1. Amplitudes of vibration in a box of length $N = 70$ with periodic boundary conditions and evenly distributed force constant disorder with mean $\langle k \rangle = 1$ and width $\Delta k = 1$. (a) Extended state $\omega^2 = 12$, (b) critical state near the LDT transition $\omega^2 = 12.5$ and (c) localized state $\omega^2 = 13.03$. Colours are an indication of the depth within the box that a particular amplitude is situated, the scale for which is included in the lower right hand side of the boxes. Sizes of internal boxes are determined from vibrational amplitudes at lattice site. Amplitudes less than the mean are not displayed.

is achieved by assuming that all springs have identical central and non-central force constants reducing the 3×3 force constant matrices to scalars. Therefore, solving for just one spatial dimension achieves all three components of the vibrational amplitude for all lattice sites.

We apply the above approximations and hence have to solve the standard equation of motion for a single lattice site with a plane wave solution to obtain

$$\begin{aligned} -\omega^2 m_{l,m,n} u_{l,m,n} &= k_{l,m,n+1}(u_{l,m,n+1} - u_{l,m,n}) + k_{l,m,n-1}(u_{l,m,n-1} - u_{l,m,n}) + \\ & k_{l,m+1,n}(u_{l,m+1,n} - u_{l,m,n}) + k_{l,m-1,n}(u_{l,m-1,n} - u_{l,m,n}) + \\ & k_{l+1,m,n}(u_{l+1,m,n} - u_{l,m,n}) + k_{l-1,m,n}(u_{l-1,m,n} - u_{l,m,n}), \end{aligned} \quad (1)$$

where ω is the frequency of vibration, $m_{l,m,n}$ and $u_{l,m,n}$ are the mass and amplitude of vibration of the lattice site denoted by indices l, m, n. k is the force constant of the spring connecting two adjacent lattice sites, where the indices of the current lattice site l, m, n are omitted to simplify notation. The system of equations for all lattice sites is converted to matrix form by substitution into the generalised equation of motion

$$-\omega^2 \mathbf{u} = [\mathbf{m}]^{-1}[\mathbf{k}]\mathbf{u} \quad (2)$$

The above matrices $[\mathbf{m}]$ and $[\mathbf{k}]$ contain all the masses and force constants of the eigensystem and $\mathbf{u}$ is a vector of all vibrational amplitudes of the lattice sites. $[\mathbf{m}]^{-1}[\mathbf{k}]$ is the dynamical matrix and is diagonalized to obtain the normal modes of vibration $\mathbf{u}$ with eigenvalues $-\omega^2$. Although the vibrational amplitudes $\mathbf{u}$ are for only a single spatial dimension, they are also valid as the total magnitude of the amplitude. Recombining the individual amplitudes to obtain the total requires multiplication of single components by $\sqrt{3}$, renormalization of the eigenvector removes these factors.

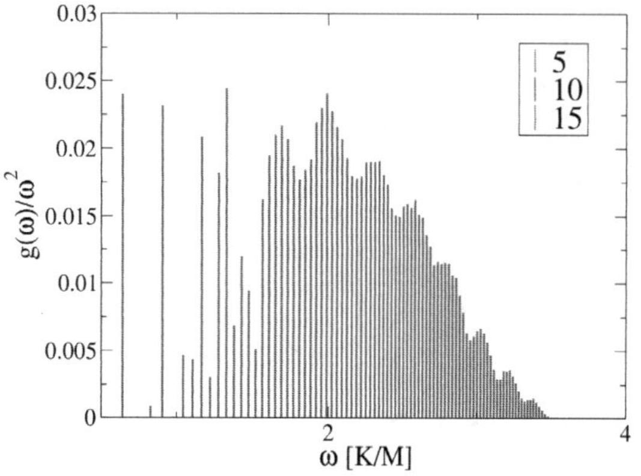

Figure 2. Reduced VDOS for evenly distributed spring disorder with mean $\langle k \rangle = 1$ and width $\Delta k = 1$. 150 bins are used and the data is normalized by total number of states.

Several methods can be used to analyse both the eigenvectors and eigenvalues obtained from diagonalization of the dynamical matrix to distinguish between extended and localized states. Just looking at the amplitude eigenvectors is a viable method of studying the localization lengths of the phonons, although it is not trivial due to the localization lengths diverging as the critical frequency is approached from the localized regime. Therefore in finite sized systems the localized states can appear extended as the localization length can be greater than the length of the system. We diagonalize the dynamical matrix for a box of size 70^3 with the inclusion of box distribution disorder with width $\Delta k = 1$ and mean $\langle k \rangle = 1$ applied to the force constants of the springs. In figure 1(a)-(c) we visualise three amplitude eigenvectors for squared frequencies of $\omega^2 = 12$, $\omega^2 = 12.5$ and $\omega^2 = 13.03$ respectively. Figure 1(a) is that of an extended state as the amplitudes exhibit short range periodicity within the box and figure 1(c) is that of a localized state where the amplitudes of vibration are confined to a small section of the box. Figure 1(b) exhibits properties of both extended and localized states and is close to the transition.

We calculate the reduced VDOS $g(\omega)/\omega^2$ for cubes with size 5^3, 10^3 and 15^3 with 1360, 170 and 50 random seeds respectively, resulting in roughly 170,000 states for each system size. The VDOS plot in figure 2 demonstrates the latter stages of the broadening of the Van Hove singularities due to disorder and the emergence of the characteristic low-frequency boson peak at $\omega_{\text{peak}} \approx 2$. The frequencies $\omega^2 = 12$, 12.5 and 13.03 of the states plotted in figure 1 are located close to the upper band edge of the VDOS, therefore we can safely assume that the LDT will be in this region.

We introduce the TMM which calculates the localization length λ of a quasi-one dimensional bar. For simplicity the bar geometry has an identical height and width M and in all cases the length of the bar $L \gg M$. L is not preset as the transfer matrix calculations continue along the length of L until the desired accuracy of the Lyapunov exponent is met, this convergence criterion is chosen before starting the calculation.

To calculate the Lyapunov exponents, the system of equations (1) is converted to a form suitable for the TMM. This requires the equations to be re-arranged so that the amplitudes of a particular layer of lattice sites can be calculated solely from the amplitudes of the previous layer and the force constants of the surrounding springs. Computationally this is far less memory consuming than storing the whole system of masses and springs, as the TMM calculation only requires parameters for 2 layers of the system to be stored in memory at any one time. Equation (1) for a single lattice site is rearranged into a form where the amplitude of vibration of the nearest neighbour in layer $l+1$ is calculated solely from parameters of sites in layers l and $l-1$.

$$u_{l+1,m,n} = -\frac{1}{k_{l+1,m,n}}\left[(\omega^2 m_{l,m,n} + k_{\text{all}})u_{l,m,n} - H_{\text{l}}\right] - \frac{k_{l-1,m,n}}{k_{l+1,m,n}}u_{l-1,m,n} \quad (3)$$

We introduce $H_{\text{l}} = k_{l,m,n+1}u_{l,m,n+1} + k_{l,m,n-1}u_{l,m,n-1} + k_{l,m+1,n}u_{l,m+1,n} + k_{l,m-1,n}u_{l,m-1,n}$ as a collection of the in-plane contributions to the final amplitude and k_{all} is the sum of all surrounding force constants for site l, m, n. We define U_l, U_{l+1} and U_{l-1} as vectors containing

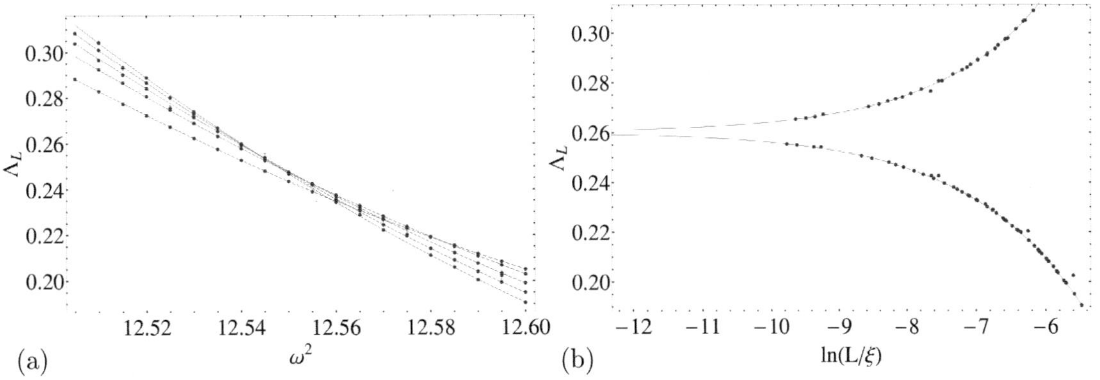

Figure 3. Spring disorder $\Delta k = 1.0$: (a) reduced localization lengths for a range of frequencies and system sizes (12-20 in steps of 2). (b) Scaling function (solid line) and scaled data points for the expansion $n_\mathrm{r} = 4$, $m_\mathrm{r} = 1$, $n_\mathrm{i} = 2$, $m_\mathrm{i} = 0$.

the amplitudes of the constituent sites of layers l, $l+1$ and $l-1$ respectively. Therefore, $U_l = (u_{l,1,1}, u_{l,1,2}, u_{l,2,1}, \ldots, u_{l,M,M})$, where M is the width and height of the layer. Equations (3) is now expressed in the following matrix form, where $\mathbf{H}_\mathrm{l}$ is a matrix containing all in-layer contributions, $\mathbf{0}$ and $\mathbf{1}$ are the zero and unit matrices, ω^2 is single valued and all other parameters are vectors.

$$\begin{bmatrix} U_{l+1} \\ U_l \end{bmatrix} = \underbrace{\begin{bmatrix} -\frac{[(\omega^2 m_l + k_{all})\mathbf{1} - \mathbf{H}_l]}{k_{l+1}} & -\frac{k_{l-1}}{k_{l+1}}\mathbf{1} \\ \mathbf{1} & \mathbf{0} \end{bmatrix}}_{\mathbb{T}_n} \begin{bmatrix} U_l \\ U_{l-1} \end{bmatrix} \tag{4}$$

The transfer matrix used to progress from one slice to the next is given as $\mathbb{T}_n$ and repeated multiplication of this gives: $\mathbb{T}_N = \prod_{n=1}^{N} \mathbb{T}_n$ which according to Oseledec [18] guarantees existence of the matrix $\Gamma \equiv \lim_{N\to\infty} \left(\mathbb{T}_N \mathbb{T}_N^\dagger\right)^{\frac{1}{2}N}$ with eigenvalues e^γ. Where γ's are Lyapunov exponents and an estimation of localization length is given by the inverse of the minimum exponent $\lambda(M, \omega^2) = 1/\gamma_{\mathrm{min}}$. The reduced (dimensionless) localization length may then be calculated as $\Lambda_M(\omega^2) = \lambda(M, \omega^2)/M$. We numerically calculate the reduced localization lengths for a range of squared frequencies and even system widths M between 12 and 20 to a convergence criterion of 0.1 percent of the variance. The reduced localization lengths Λ_M as function of the squared frequency ω^2 are plotted in figure 3(a).

The reduced localization lengths from the TMM are finite size scaled by applying the procedure outlined in reference [19]. The finite size scaling (FSS) procedure obtains the correlation length ξ for an infinite system from the reduced localization length $\Lambda_M(\omega^2)$ of finite sized systems by using the one-parameter scaling law $\Lambda_M = f(M/\xi)$ [20]. We assume that the LDT is characterised by a divergent correlation length, so that at a fixed disorder Δk, $\xi(\omega^2) \propto |\, \omega^2 - \omega_\mathrm{c}^2 \,|^{-\nu}$, where ν is the critical exponent and ω_c^2 is the critical frequency squared. Highly accurate numerical studies of the Anderson model for electron localization have found the critical exponent $\nu \equiv 1.5 \pm 0.1$ [21, 22, 19] for box distribution disorder. Whereas in the phononic case Monthus and Garel [23] have demonstrated that their participation ratio data for high disorder at an LDT collapsed fairly well onto their scaling function using a critical exponent $\nu = 1.57$ taken from reference [24]. Akita and Ohtsuki [25] find a critical exponent of $\nu \approx 1.2 \pm 0.2$ by finite size scaling TMM data at an LDT for box distribution spring disorder $\Delta k = 1.8$ with a convergence criterion of 2 percent. Table 1 contains the critical parameters obtained from FSS for particular orders of Taylor expansions of the relevant and irrelevant scaling variables n_r, n_i,

Table 1. Table of critical parameters obtained from finite size scaling for $\Delta k = 1$ on even numbered system widths between 12 and 20 and a range of squared frequencies $\omega_c^2 = 12.5 - 12.6$. NDF is the number of degrees of freedom. We note that preliminary errors for ν are too small as they have not been fully checked with respect to stability and robustness.

n_r	n_i	m_r	m_i	ω_c^2	ν	χ^2	NDF	Γ_q
3	1	1	0	12.538 ± 0.05	1.5023 ± 0.002	51.12	80	0.995
3	1	2	0	12.540 ± 0.07	1.5507 ± 0.002	48.13	79	0.997
3	1	3	0	12.540 ± 0.08	1.5782 ± 0.002	46.71	78	0.998
4	1	2	0	12.540 ± 0.07	1.5573 ± 0.002	47.62	77	0.997

m_r and m_i listed in columns 1-4. Figure 3(b) is the scaling function obtained from the highest order stable fit given in table 1 ($n_r = 4$, $n_i = 1$, $m_r = 2$, m_i=0). The mean critical exponent $\langle \nu \rangle = 1.55 \pm 0.002$ which is remarkably close to the MIT critical exponent.

From FSS we achieve a weighted mean critical squared transition frequency of $\omega_c^2 = 12.54 \pm 0.03$ which as we can see from figure 2 occurs in the tail of the VDOS and close to the example state of squared frequency $\omega_c^2 = 12.5$ in Figure 1(b) that exhibits properties of both extended and localized states. It is generally assumed that the LDT shares the same exponent as the MIT, where the current accepted value is $\nu \approx 1.5 \pm 0.1$. Although Akita and Ohtsuki [25] have previously obtained a much lower value $\nu \approx 1.2 \pm 0.2$, thought to be a consequence of the transition occurring close to the upper band edge [26], we have achieved a striking resemblance between critical exponents between the MIT and LDT. Therefore we assume that the phononic model will be of the same orthogonal universality class as the Anderson model.

We gratefully acknowledge EPSRC for financial support (EP-F040784-1).

References

[1] Anderson P W 1958 *Phys. Rev.* **109** 1492–1505
[2] Grussbach H and Schreiber M 1995 *Phys. Rev. B* **51** 663–666
[3] Milde F, Römer R A and Schreiber M 1997 *Phys. Rev. B* **55** 9463–9469
[4] Zharekeshev I K and Kramer B 1997 *Phys. Rev. Lett.* **79** 717–720 ArXiv: cond-mat/9706255
[5] Milde F, Römer R A and Schreiber M 2000 *Phys. Rev. B* **61** 6028–6035 ArXiv: cond-mat/9909210
[6] Kramer B and MacKinnon A 1993 *Rep. Prog. Phys.* **56** 1469–1564
[7] Pichard J L and Sarma G 1981 *J. Phys. C* **14** L127–L132
[8] Lifshitz I and Kosevich A 1954 *Journal of Physics: USSR* **8** 217–254
[9] Canisius J and van Hemmen J 1985 *J. Phys. C* **18** 4873–4884
[10] Ludlam J, Stadelmann T, Taraskin S and Elliott S 2001 *Journal of Non-Crystalline Solids* **293** 676–681
[11] Schirmacher W, Diezemann G and Ganter C 1998 *Phys. Rev. Lett.* **81** 136–139
[12] Shima H, Nishino S and Nakayama T 2007 *Journal of Physics: Conference Series* **92** 012156
[13] Ludlam J, Taraskin S and Elliott S 2003 *Phys. Rev. B* **67** 132203
[14] Kantelhardt J W, Russ S and Bunde A 2001 *Phys. Rev. B* **63**
[15] Schirmacher W and Diezemann G 1999 *Ann. Phys. (Leipzig)* **8** 727–732
[16] Scopigno T, Suck J, Angelini R, Albergamo F and Ruocco G 2006 *Phys. Rev. Lett.* **96** 135501
[17] Taraskin S, Ludlam J, Natarajan G and Elliott S 2002 *Philosophical Magazine B* **82** 197–208
[18] Oseledec V I 1968 *Trans. Moscow Math. Soc.* **19** 197–231
[19] Slevin K and Ohtsuki T 1999 *Phys. Rev. Lett.* **82** 382–385 ArXiv: cond-mat/9812065
[20] Thouless D J 1974 *Phys. Rep.* **13** 93–142
[21] MacKinnon A and Kramer B 1981 *Phys. Rev. Lett.* **47** 1546–1549
[22] MacKinnon A 1994 *J. Phys.: Condens. Matter* **6** 2511–2518
[23] Monthus C and Garel T 2010 *Phys. Rev. B* **81** 224208
[24] Evers F and Mirlin A D 2008 *Rev. Mod. Phys.* **80** 1355–1417
[25] Akita Y and Ohtsuki T 1998 *J. Phys. Soc. Jap.* **67** 2954–2955
[26] Kawarabayashi T, Kramer B and Ohtsuki T 1999 *Ann. Phys. (Leipzig)* **8** 487–496 ArXiv: cond-mat/9907319

Journal of Physics: Conference Series **286** (2011) 012026 doi:10.1088/1742-6596/286/1/012026

Observation of anticrossings in the exciton state of single quantum dots via electrical tuning of the fine-structure splitting

M A Pooley[1,2], A J Bennett[1], R M Stevenson[1], M B Ward[1], R B Patel[1,2], A Boyer de la Giroday[1,2], N Sköld[1], I Farrer[2], C A Nicoll[2], D A Ritchie[2] and A J Shields[1]

[1] Toshiba Research Europe Limited, Cambridge Research Laboratory, 208 Science Park, Milton Road, Cambridge, CB4 0GZ, UK

[2] Cavendish Laboratory, Cambridge University, J J Thomson Avenue, Cambridge, CB3 0HE, UK

E-mail: `map57@cam.ac.uk`

Abstract. Single quantum dots have many potential applications across the field of quantum computation, ranging from the generation of single photons or entangled photon pairs to the storage and manipulation of qubits. Single InAs quantum dots are optically active and thus can be used as an interface between photonic flying qubits and spin-based stationary qubits. Incorporating single InAs quantum dots into semiconductor devices allows the stationary qubits to be manipulated, making this system a promising candidate for quantum computation. It is well known that the exciton state of quantum dots is split into two polarisation dependent states; the energy difference between the states is the fine-structure splitting, (s). A vertical electric field has been used to tune $|s|$ over a large range of $\sim 100\mu$eV, such that a dot which has $|s|$ over 50μeV at zero field has been tuned to emit polarisation entangled photon pairs[1]. We observe coherent coupling between the two polarisation eigenstates. This coupling results in an anticrossing as the two states are tuned close to each other, and a rotation of the eigenstates through 90° as $|s|$ is swept through its minimum value.

1. Introduction

It is well known that the neutral exciton states of quantum dots (QDs) are split into two polarisation dependent energy eigenstates, separated by a fine-structure splitting (s) [2]. This system has been the subject of much research, particularly in the field of quantum computation, however until recently the two eigenstates have been considered independent [3–7]. We have demonstrated that it is possible to electrically induce coherent coupling of the exciton states using a vertically applied electric field [1]. The electric field tunes s by Stark-shifting the two eigenstates at different rates. When the two eigenstates are brought close together an anticrossing is observed indicating coherent coupling. The magnitude of s is important because it is easier to observe entangled photon generation from the biexciton-exciton-empty cascade when s is small. It is therefore often desirable to work with QDs that have as small s as possible, however it is not always possible to grow QDs that naturally have small s. This has motivated many other techniques for tuning s, including the use of magnetic field [2, 3], lateral electric field [6–8], and strong coherent lasers [4, 5]. However, those that have been able to minimize s have not seen an anticrossing. The application of vertical electric field is arguably the most practical technique for tuning s, as it can be achieved by simply placing the QDs in the center of a p-i-n diode and allows easy dynamic control of the applied field using standard voltage sources.

The magnitude of the anticrossing (s_0) is the minimum value to which s can be reduced using this technique, and so determines how suitable a particular QD is for use as an entangled photon pair source. We have measured an ensemble of QDs and found that s_0 varies from dot to dot, with some dots having sufficiently small s_0 to allow entangled photon pairs to be observed.

2. Sample design and methods

Application of a vertical electric field can lead to quenching of the QD emission by increasing the rate at which carriers tunnel out of the QD; once the tunneling time is comparable with the time for radiative recombination the emission is drastically reduced. To overcome this problem InAs QDs are placed inside a 10nm thick two-dimensional GaAs quantum well which is sandwiched between two AlAs/GaAs short-period superlattices equivalent to $Al_{0.75}Ga_{0.25}As$. This increases the carrier confinement in the region of the QDs and limits the rate at which carriers can tunnel away. Distributed bragg reflectors are used to create a weak planar microcavity, with 14/4 periods on the bottom/top of the device. The structure is doped to create a p-i-n device with a $d = 140$nm thick intrinsic region centered around the dot layer. The electric field applied to the QDs can be calculated as $(V - V_{bt})/d$, where V_{bt} is the built-in potential which for these samples is 2.2V. An opaque aluminum film is deposited on the surface and patterned with arrays of μm diameter apertures through which optical excitation and emission collection occurs. These apertures allow individual QDs to be optically addressed. This sample design enables very large electric fields to be applied to the QDs before emission is quenched and so allows the observation of giant Stark effects [9]. Figures 1 and 2 show the device structure and a typical plot of the photoluminescence from a single QD as a function of applied electric field.

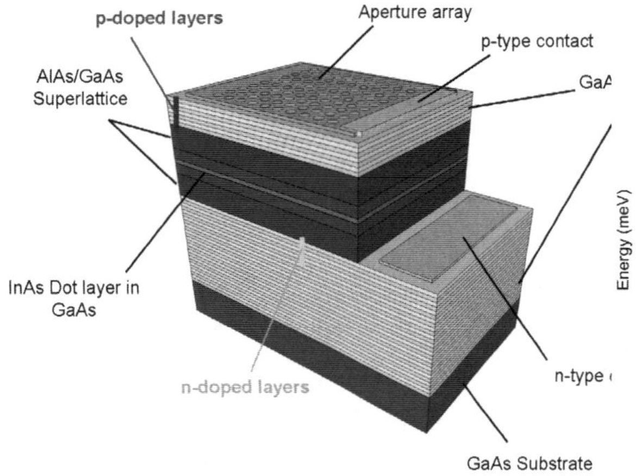

Figure 1. Diagram of device structure. The quantum dots are embedded in a 2D quantum well inside a weak planar microcavity.

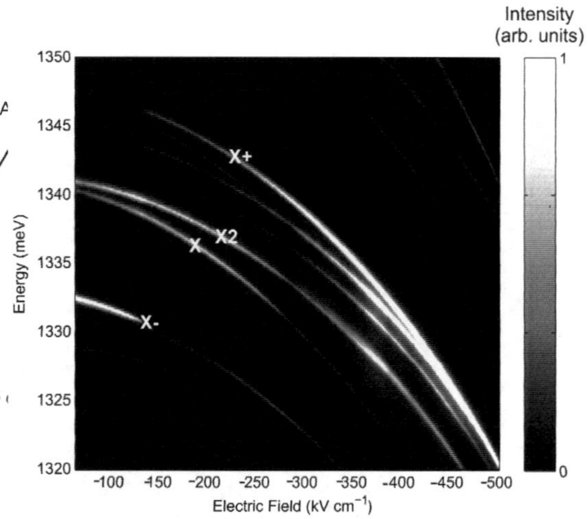

Figure 2. Example of photoluminescence as a function of electric field for a single quantum dot.

The QDs were optically excited with a continuous-wave 850nm diode laser, which creates carriers both in the QDs and the wetting layer. The spectra from single QDs is well understood

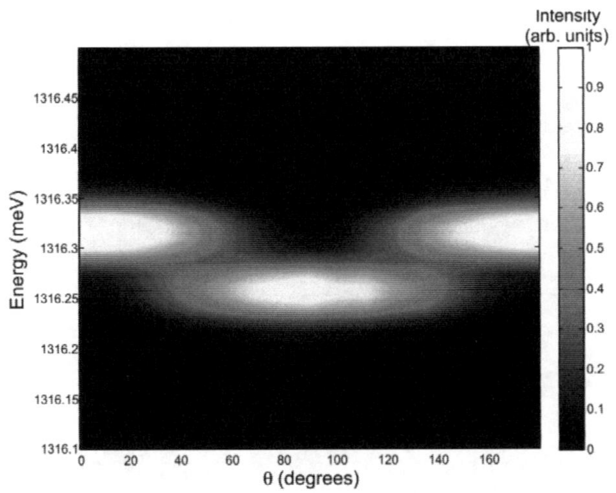

Figure 3. Example of photoluminescence as a function of polarisation angle. There are two clear energy eigenstates separated by 90°.

and the spectral lines can be identified as the biexciton (X_2), exciton (X), and two charged excitons (X^+ and X^-), where this notation refers to the initial state of each transition.

Measurements of $|s|$, and the eigenstate orientation (θ) were obtained by recording polarised spectra for a range of polarisation angles between 0° and 180°. The value of $|s|$ is given by the energy separation of the two orthogonal eigenstates. Figure 3 shows spectra taken from a QD with large $|s|$ where the two different eigenstates can clearly be resolved; light with $\theta = 0°$ has different energy from that with $\theta = 90°$. The linewidth of the spectral peaks is dominated by the resolution of the spectrometer. In cases where $|s|$ cannot be clearly resolved a fitting procedure is used to extract the energy of peak emission at each angle; the peak energies are fit with a sinusoidal function the amplitude of which yields $|s|$. The angles at which the emission from each state is maximum are used to extract their orientation and thus obtain θ. Small energy shifts caused by rotation of the polarisation optics are eliminated by finding the difference between the biexciton and exciton transition energies giving submicroelectronvolt accuracy [3, 10].

3. Results

As can be seen in Figure 2, the application of a vertical electric field (F) induces a Stark shift of the emission lines, this shift takes the usual form,

$$E = E_0 - pF + \beta F^2 \tag{1}$$

where p is the z-component of the permanent dipole moment and β is the polarizability. Far from the anticrossing the eigenstates tend to align with the [110] and [1$\bar{1}$0] crystalline axis and can be labeled H (V) for horizontal (vertical) polarization. Away from their anticrossings s shifts linearly with electric field at a rate of $\gamma = -0.285 \pm 0.019\mu$eV kV^{-1}cm (see figure 4). We attribute this to a slight difference in the confinement energy of the two eigenstates, which results in each eigenstate having a different p; thus γ is given by $\gamma = p_H - p_V$.

As $|s|$ is tuned to a minimum an anticrossing of the two eigenstates is observed, which is a clear indication of coherent coupling. This behavior is well described by a coupling matrix of the same form as that used to describe strong light-matter coupling [11] and anticrossings in the states of molecular systems [12],

$$E\begin{pmatrix}\cos(\theta)\\ \sin(\theta)\end{pmatrix} = \begin{pmatrix}E_0 & s_0/2\\ s_0/2 & E_0 - \gamma(F - F_0)\end{pmatrix}\begin{pmatrix}\cos(\theta)\\ \sin(\theta)\end{pmatrix} \tag{2}$$

where F_0 is the field at which the minimal splitting occurs.

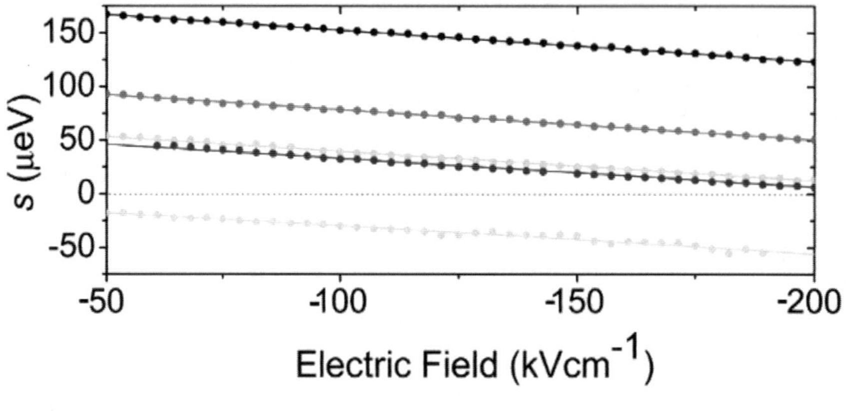

Figure 4. Measurements of fine structure splitting (s) as a function of electric field far from the anticrossing showing five quantum dots in our sample that tune at a rate of $\gamma = -0.285 \pm 0.019\mu$eV kV^{-1}cm. Note that one unusual quantum dot that has the lower energy state orthogonal to all the others in the sample is plotted with negative s. Data points are shown as solid circles, linear fits to the data points are shown as solid lines.

Far from the anticrossing, when $\gamma(F - F_0) >> s_0$, the states are aligned to the crystalline axis with H and V polarized eigenstates. Close to $(F - F_0) = 0$ the eigenstates are a coherent mixture with components $\sin\theta$ and $\cos\theta$ of H and V polarized light. The orientation angle θ is thus the actual angle of the eigenstates relative to the crystalline axis. The eigenvalues, $E_\pm$, and eigenvectors of equation 2 are

$$E_\pm = E_0 - \frac{\gamma(F - F_0)}{2} \pm \frac{1}{2}\sqrt{\gamma^2(F - F_0)^2 + s_0^2} \tag{3}$$

$$\theta_\pm = \pm\tan^{-1}\left(\frac{s_0}{\gamma(F - F_0) \pm (E_- - E_+)}\right) \tag{4}$$

Although there are two degenerate solutions, $\theta\pm$, which describe either clockwise or anticlockwise rotation of the eigenstates, we observe that each QD only ever rotates in one direction. Our measurements are consistent with the ensemble of QDs having no preferred rotation direction.

Figure 5 shows data from three QDs, the data is in excellent agreement with the simple model.

We now discuss how this system could find applications as a solid state quantum memory. Clearly for a memory coherent writing of photon polarization to the spin-state of the exciton is required, and we show this is possible with high fidelity. In addition, we show that the state remains stored coherently on a timescale of several times the radiative lifetime. In order to find practical use this storage time must be increased substantially.

Quasi-resonant excitation of a QD with a linearly polarized laser can be used to write the polarization state of the photons onto the spin state of the carriers created in the QD. If the QD is excited into a superposition of the two polarization eigenstates, time-resolved measurements of the emission from a superposition state yield coherent oscillations. The frequency of the oscillations is given by $|s|/\hbar$, and the visibility is maximal when exciting and measuring maximum superposition states. Figure 6 shows such oscillations for a QD tuned to three different values of $|s|$. The oscillations are visible for longer than the radiative lifetime allowing storage of the qubit for $\sim$ 3ns. In addition, it is possible to dynamically vary s whilst

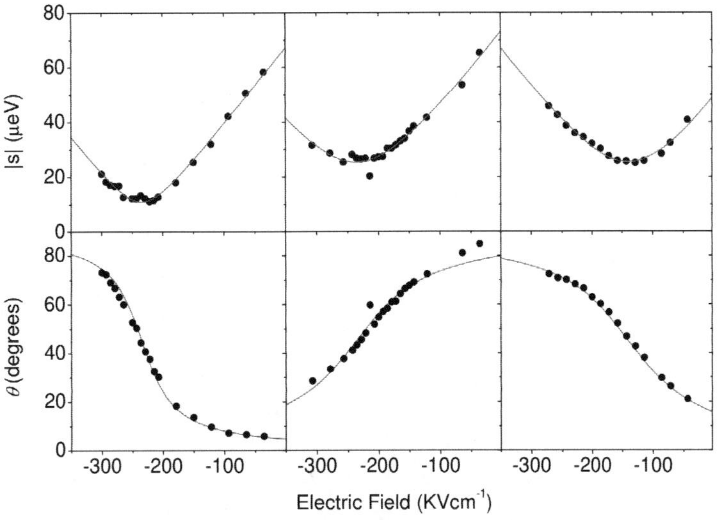

Figure 5. Measurements of fine structure splitting ($|s|$) and eigenstate orientation (θ) as a function of electric field for three quantum dots. The red lines show fits to the data using equations 3 and 4.

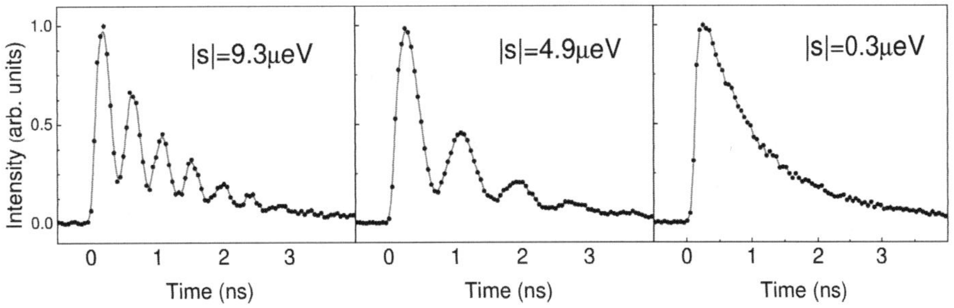

Figure 6. Time-resolved measurements of the exciton emission from a dot tuned to three different values of fine structure splitting showing coherent oscillations of the spin-state. The dot was excited into a maximum superposition state and measurement was along the same orientation as excitation. Data points are shown as solid black circles, the red lines are a quide to the eye.

the qubit is stored in the QD. This allows the orientation of the eigenstates to be varied and the coherent oscillations to be controlled [13].

4. Conclusions

We have used vertical electric fields to tune the fine-structure splitting of single QDs over a very large range of more than 100μeV. Tuning s close to zero has resulted in the observation of an anticrossing of the two polarisation eigenstates, which is attributed to coherent coupling of the states. This has important implications as the magnitude of the anticrossing determines if a particular QD is suitable as a practical entangled photon pair source.

This system also has potential applications in quantum memory and quantum computation. The optically active QDs allow flying photonic qubits to be easily transfered to and from the solid state, where manipulations or logical operations can be performed.

References

[1] Bennett A J, Pooley M A, Stevenson R M, Ward M B, Patel R B, de la Giroday A B, Skold N, Farrer I, Nicoll C A, Ritchie D A and Shields A J 2010 *Nat Phys* **advance online publication** – ISSN 1745-2481 URL `http://dx.doi.org/10.1038/nphys1780`

[2] Bayer M, Kuther A, Forchel A, Gorbunov A, Timofeev V B, Schafer F, Reithmaier J P, Reinecke T L and Walck S N 1999 *Phys. Rev. Lett.* **82** 1748– URL `http://link.aps.org/doi/10.1103/PhysRevLett.82.1748`

[3] Stevenson R M, Young R J, See P, Gevaux D G, Cooper K, Atkinson P, Farrer I, Ritchie D A and Shields A J 2006 *Phys. Rev. B* **73** 033306– URL `http://link.aps.org/doi/10.1103/PhysRevB.73.033306`

[4] Jundt G, Robledo L, Hogele A, Falt S and Imamoglu 2008 *Phys. Rev. Lett.* **100** 177401– URL `http://link.aps.org/doi/10.1103/PhysRevLett.100.177401`

[5] Muller A, Fang W, Lawall J and Solomon G S 2009 *Phys. Rev. Lett.* **103** 217402– URL `http://link.aps.org/doi/10.1103/PhysRevLett.103.217402`

[6] Kowalik K, Krebs O, Lemaitre A, Eble B, Kudelski A, Voisin P, Seidl S and Gaj J A 2007 *Appl. Phys. Lett.* **91** 183104–3 URL `http://link.aip.org/link/?APL/91/183104/1`

[7] Gerardot B D, Seidl S, Dalgarno P A, Warburton R J, Granados D, Garcia J M, Kowalik K, Krebs O, Karrai K, Badolato A and Petroff P M 2007 *Appl. Phys. Lett.* **90** 041101–3 URL `http://link.aip.org/link/?APL/90/041101/1`

[8] Vogel M M, Ulrich S M, Hafenbrak R, Michler P, Wang L, Rastelli A and Schmidt O G 2007 *Applied Physics Letters* **91** 051904 URL `http://dx.doi.org/10.1063/1.2761522`

[9] Bennett A J, Patel R B, Skiba-Szymanska J, Nicoll C A, Farrer I, Ritchie D A and Shields A J 2010 *Appl. Phys. Lett.* **97** 031104–3 URL `http://link.aip.org/link/?APL/97/031104/1`

[10] Young R J, Stevenson R M, Shields A J, Atkinson P, Cooper K, Ritchie D A, Groom K M, Tartakovskii A I and Skolnick M S 2005 *Phys. Rev. B* **72** 113305– URL `http://link.aps.org/doi/10.1103/PhysRevB.72.113305`

[11] Khitrova G, Gibbs H M, Kira M, Koch S W and Scherer A 2006 *Nat Phys* **2** 81–90 ISSN 1745-2473 URL `http://dx.doi.org/10.1038/nphys227`

[12] Stinaff E A, Scheibner M, Bracker A S, Ponomarev I V, Korenev V L, Ware M E, Doty M F, Reinecke T L and Gammon D 2006 *Science* **311** 636–639 URL `http://www.sciencemag.org/cgi/content/abstract/311/5761/636`

[13] Boyer de la Giroday A, Bennett A J, Pooley M A, Stevenson R M, Ward M B, Patel R B, Skold N, Farrer I, Nicoll C A, Ritchie D A and Shields A J *accepted for publication in Phys Rev B Rapid Communications, arXiv:1011.2641v1*

Journal of Physics: Conference Series **286** (2011) 012027 doi:10.1088/1742-6596/286/1/012027

The surface band structure of β-Ga_2O_3

M Mohamed [1], I Unger, C Janowitz [1], R Manzke [1], Z Galazka [2], R Uecker [2] and R Fornari [2]

[1] Humboldt-Universität zu Berlin, Institut für Physik, Newtonstr. 15, 12489 Berlin, Germany
[2] Leibniz-Institut für Kristallzüchtung, Max-Born-Str. 2, 12489 Berlin, Germany

mansour@physik.hu-berlin.de

Abstract. Ga_2O_3 belongs to the group of transparent conducting oxides (TCOs) with a wide band gap and electrical conductivity. It exhibits the largest band gap with $E_g = 4.8$ eV and thus a unique transparency from the visible into the UV region. The information on the electronic structure of β-Ga_2O_3 is very scarce. This is in part due to the challenging problem of growing high purity single crystals. Transparent semiconducting β-Ga_2O_3 single crystals were grown by the Czochralski method from an iridium crucible under a dynamic protective atmosphere to control partial pressures of volatile species of Ga_2O_3. The investigated samples were characterized by different techniques (LEED, Laue, and STM). The experimental valence band structure of the of β-Ga_2O_3 single crystals along Γ-Z and A-M symmetry directions of the (100)-surface of Brillouin zone was determined by high-resolution angle-resolved photoelectron spectroscopy (ARPES) utilizing synchrotron radiation. The experimental band structure is compared and discussed with the theoretical calculations. The effect of changing the temperature from 300K to 20K on the experimental band structure β-Ga_2O_3 was studied.

1. Introduction

β-Ga_2O_3 is a large band gap semiconductor with the largest band gap of about 4.4–4.8 eV [1,2] among transparent conducting oxides (TCOs). Because β-Ga_2O_3 has a unique transparency from the visible into the UV region, it is a very promising candidate for optoelectronic devices operating at short wavelength [3], where the other TCOs have poor transparency. The electrical conductivity of β-Ga_2O_3 is attributed to the existence of oxygen vacancies. The amount of oxygen vacancies induced depends on the growth atmosphere [4,5]. The oxygen vacancies serve as shallow donors with an ionization energy of 0.03- 0.04 eV [6]. At high temperatures the oxygen vacancies concentration depends on the partial pressure of the atmosphere. Therefore, β-Ga_2O_3 is also a promising candidate for high temperature oxygen sensor applications [7]. Ga_2O_3 exhibits polymorphism, and five crystalline modifications have been reported: α, β, γ, δ and ε [8,9]. Of these, β-Ga_2O_3 is the only stable crystalline modification in the range from room temperature up to the melting point of about 1800 °C. All other polymorphs are metastable and transform to the β form at sufficiently high temperature [8]. Additionally, β-Ga_2O_3 has great chemical stability, being unaffected even by concentrated acids such as hydrofluoric acid [1]. β-Ga_2O_3 belongs to the space group C/2m, and its base-centered monoclinic lattice has unit cell parameters a = 12.23, b = 3.04, c = 5.80 and β = 103.7° [10, 11]. The b-axis of the centrosymmetric unit cell is the twofold rotational axis, and the mirror plane (i.e., the (010)) is perpendicular to this axis. There are four Ga_2O_3 units per crystallographic unit cell with two inequivalent Ga sites and three inequivalent O sites. Half the Gallium atoms are in Ga(I) sites which form slightly distorted tetrahedra with four O ions and the other half are in Ga(II) sites which form highly distorted octahedra with six O ions [12].

In view of this rich variety of properties and applications it appears worthwhile to study the surface electronic structure of β-Ga_2O_3. While there is a large body of surface science literature on thin metal oxides films, very few single crystal oxides surfaces have been studied. This is due to a number of obstacles. Among these obstacles, are the difficulties in preparing high purity single crystals and substrates for homoepitaxy, which also limits the attainable progress in device production. Furthermore, the technical difficulty of applying electron-based techniques such as scanning tunneling microscopy (STM) and angle-resolved photoemission spectroscopy (ARPES) to bulk oxides is caused by charging effects arising from the low conductivity. Surface morphology and electronic structure of Ga_2O_3 (100) has been studied with STM, LEED and ARPES on samples prepared by high current annealing, where small terraces were observed and partial correspondence to theoretical band structures has been reported [13]. Thin films obtained by oxidizing CoGa have been studied with STM and low-energy electron diffraction (LEED) [14,15] however, atomic resolution information on single crystal Ga_2O_3 has not been reported. A recent study on β-Ga_2O_3 [16] by ARPES and DFT along Γ-$0.5b^*$ and Γ-$0.5c^*$ directions found a good matching of the experimental band structure with advanced density functional theory calculations. In [16] a slightly smaller indirect band gap of 4.83 eV, with the valence-band maximum (VBM) located slightly away from the M symmetry point has been obtained.

In this paper, we have studied the surface electronic structure of β-Ga_2O_3 along Γ-Z and Γ-M directions by means of ARPES. The electronic band structure presented along the high symmetry lines in the BZ is discussed and compared with density functional theory calculations employing hybrid functionals and projector augmented wave (PAW) potentials [17]. The temperature dependence of the band gap (E_g) of β-Ga_2O_3 determined from ARPES measurements has been studied.

2. Experimental

High resolution photoemission was performed at room temperature on the 5-m-normal incidence monochromator of the beamline BEST with a SES2002 analyzer utilizing the radiation of BESSY II in the photon energy range from 20 eV to 39 eV. An energy resolution of 20 meV and angle resolution of 0.1° was found sufficient to resolve the essential details. The investigated samples were prepared from β-Ga_2O_3 bulk single crystals grown by the Czochralski method using a dynamic growth atmosphere to control partial pressures of volatile species of Ga_2O_3 [2].

3. Sample characterization

β-Ga_2O_3 (100) planes were characterized prior to the measurements by several surface techniques, such as LEED, STM and Laue diffraction. Low energy electron diffraction is used as a standard technique to probe the crystallographic quality of the crystal surface. The sample is cleaved in UHV directly before doing the LEED measurement. The LEED pattern of the β-Ga_2O_3 (100) surface taken at E_{kin} = 188 eV is shown in figure (1a). The LEED picture shows a perfect rectangular lattice and confirms the high quality of the β-Ga_2O_3 crystals. The STM results are shown in figure (1c,1d). Besides the smooth surface observed in STM there are dark spots visible on the surface. These are usually round in shape and are irregularly distributed over the entire surface. They are much smaller than the described terraces in [13]. When the voltage between tip and sample is reversed one sees the same pattern as shown in figure (1c,1d). The origin of the spots is up to now unknown. Figure (1b) shows the Laue diffraction picture. This picture was measured in order to determine the symmetry directions of the (100) surface Brillouin zone. Point contacts of molten tin were fabricated on the backside of the crystals by high voltage yielding good Ohmic contact. The possible charging was checked by applying different photon intensities in the photoemission experiment as shown in figure 2. Absolutely no shift of the spectra was observed, therewith excluding any charging of the sample

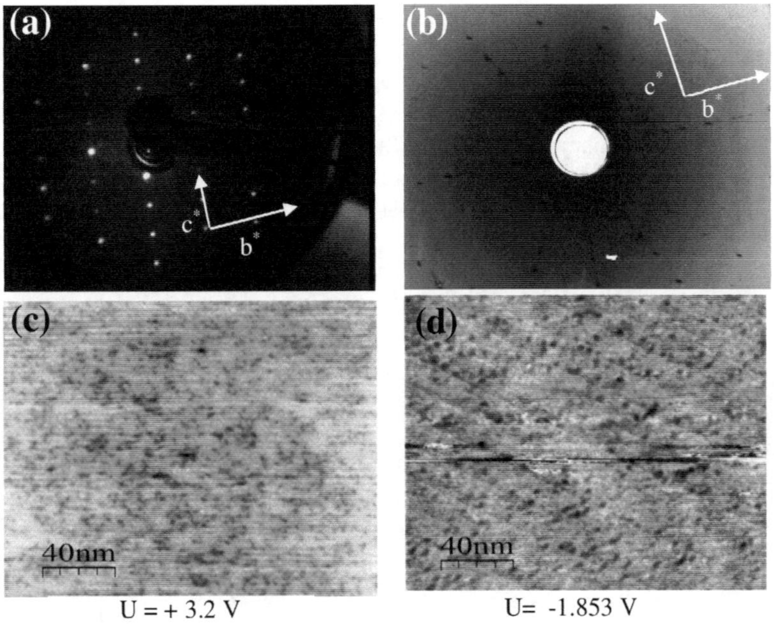

Figure 1. (a), (b) LEED pattern at E_{kin} = 188 eV and Laue picture of the (100)-surface of the β-Ga_2O_3 single crystal. The symmetry directions b* and c* of (100)-surface Brillouin zone are indicated in the LEED and Laue pictures. (c), (d) STM pictures of β-Ga_2O_3 (100)-surface over an area of 200 nm x 200 nm at different potentials between the tip and the sample.

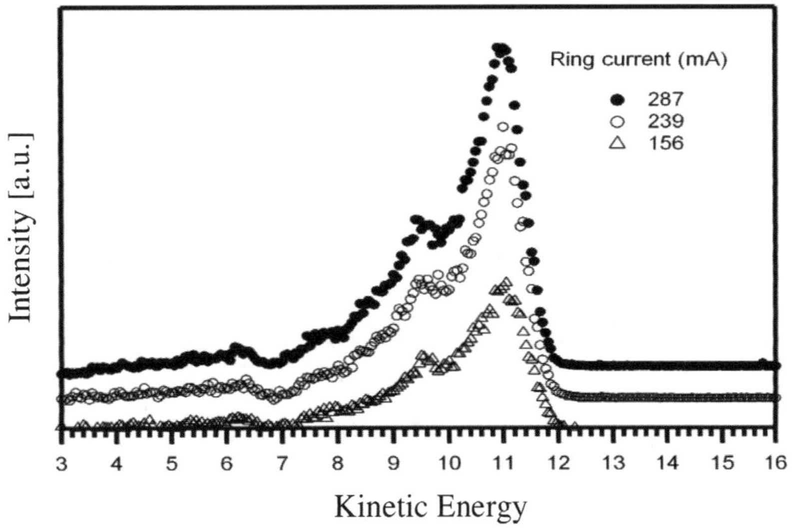

Figure 2. Series of valence band spectra measured at different ring currents at BESSY in order to test the charging effect of the sample.

4. Results and discussion

4.1. The valence band structure along the normal direction (a*):
Measurements performed near normal emission along the a* reciprocal line of the Brillouin zone (the BZ is shown in [16]) of base centered monoclinic β-Ga_2O_3 are shown in figure 3. A series of spectra were taken from 20 eV to 39 eV photon energy in order to determine the exact location of the Γ- point in the BZ. One observes a number of bands with the most prominent feature marked by solid dots. Furthermore, the valence bands at higher binding energies become more pronounced around 30 eV photon energy. Also one can observe clear symmetry of the dispersion around this energy. Thus the Γ-point (the center of the BZ) is at 30 eV.

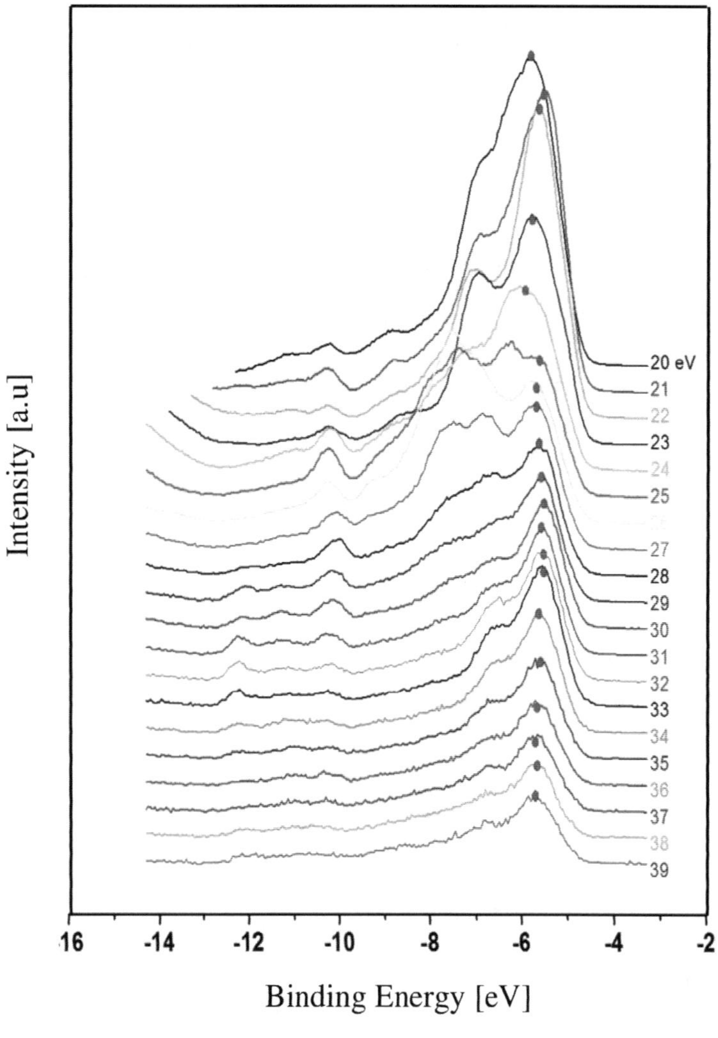

Figure 3. Series of valence band photoemission spectra normal to the cleavage plane of the (100) surface of β-Ga_2O_3 single crystals taken with photon energies from 20 to 39 eV.

Figure (4a,4b) show the photoemission spectrum at the Γ-point and the valence band density of states of β-Ga_2O_3 calculated by density functional theory employing hybrid functionals and projector augmented wave (PAW) potentials [18]. The EDC at the Γ-point in figure (4a) shows many bands from -12.27 to -4.9 eV. This observation confirms the high quality of β-Ga_2O_3 single crystals that have n-type conductivity. One can also observe high intensity bands at lower binding energies than that appearing at higher binding energies. This is in a good agreement with the calculated density of states shown in figure (4b).

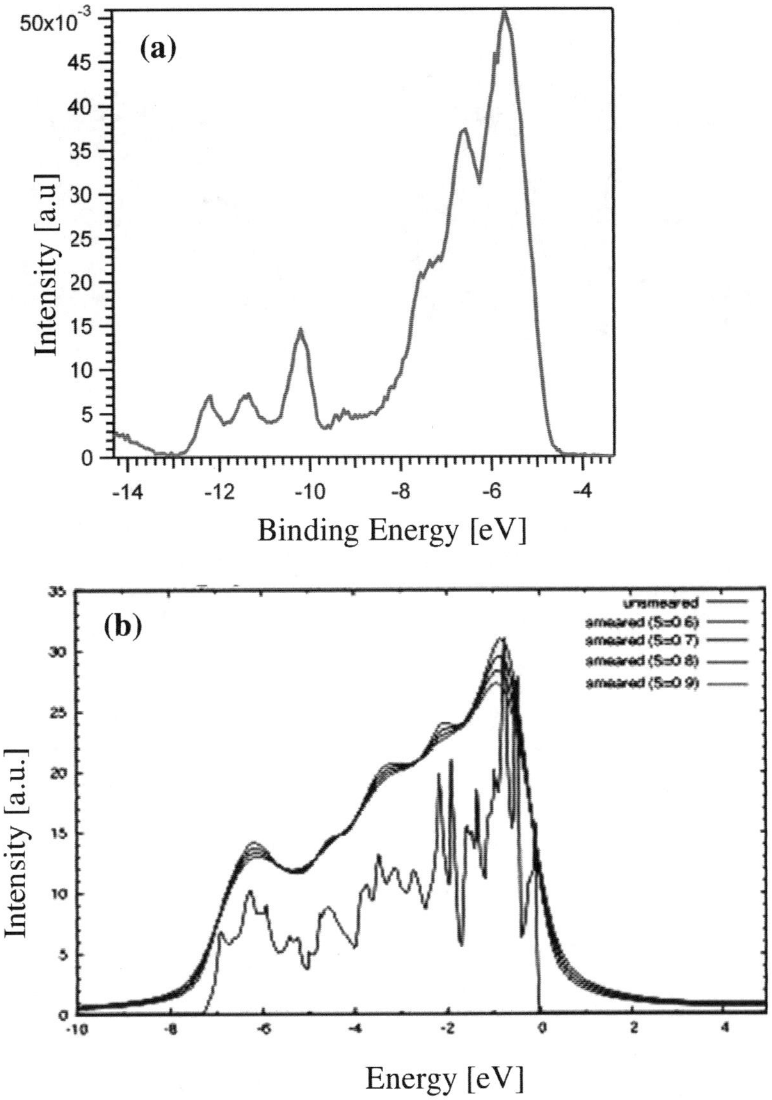

Figure 4. (a) EDC at the Γ-point in the Brillouin zone of β-Ga_2O_3. (b) The valence band density of states of β-Ga_2O_3 calculated by DFT [18].

4.2. The valence band structure along the parallel directions:

From the Γ-point the dispersion relations are studied along the Γ-Z and A-M directions of the Brillouin zone, shown in figure 5. At the Γ-point one finds bands at -12.27, -11.50, and -10.00 eV which show clear k-dispersion. Less intensive bands appear below and above -9 eV and many bands between -4.9 and -8 eV appear as one wide band. The bands near the bottom of the valence band at -12.27 and -10.00 eV come together at 0.5b* in b* direction. The bands at lower binding energy show minor dispersion and high intensity, while the deeper lying bands reveal more dispersion. In the A-M direction the width of the whole valence band is smaller than the width at the Γ-point, where the first band at higher binding energy is at -11.5 eV. Figure 6 compares the experimentally determined bands from Γ to Z and from Γ to M with DFT results [17]. In the Γ- Z and Γ-M directions the band

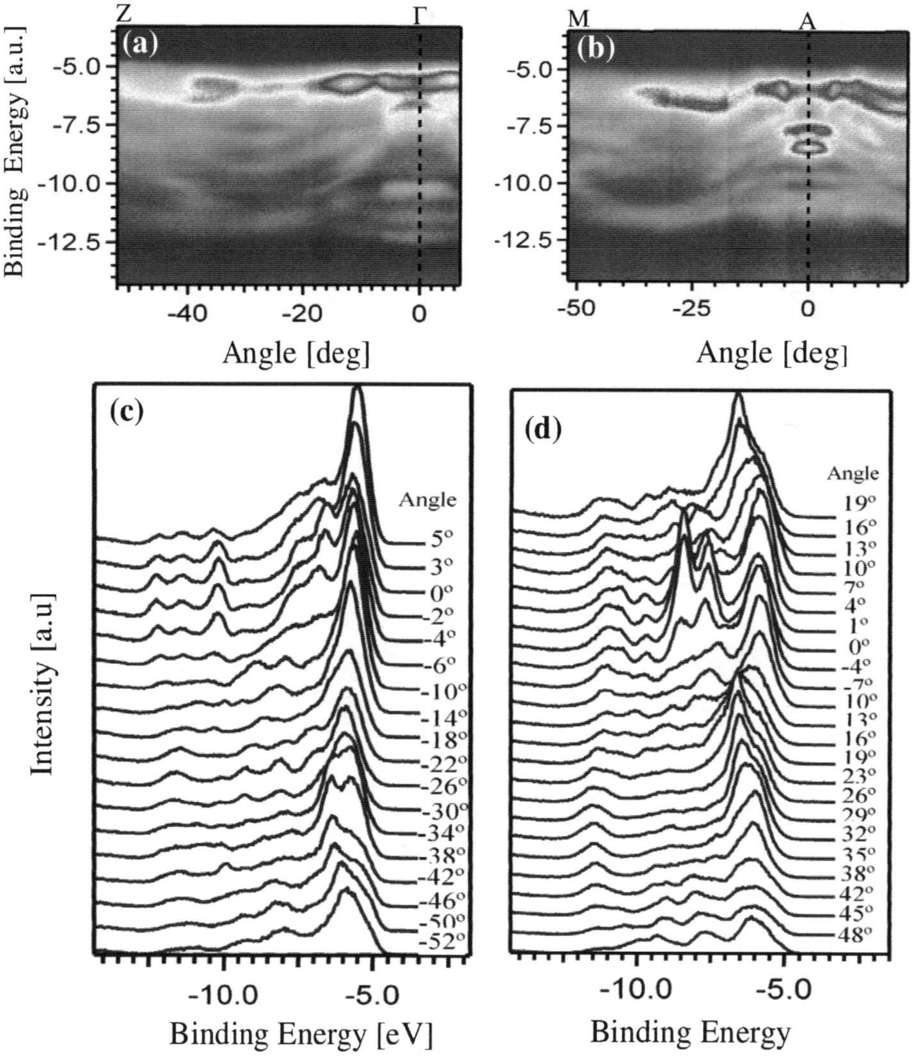

Figure 5. (a), (b) (Intensity map) Experimental band structure along the high symmetry directions Γ-Z and A-M directions of the (100)-surface of β-Ga_2O_3. (c), (d) The corresponding EDC series taken along Γ-Z and A-M for different emission angles.

dispersions show almost excellent agreement with the calculated results. One observes also in the Γ-M symmetry direction a maximum in the valence-band near the M symmetry point. This agrees with the theoretical calculations [16] where an indirect band gap with the valence-band maximum (VBM) located slightly away from the M symmetry point was obtained. In figure (6b) from the Fermi energy down to -4.9 eV binding energy no states appear in the energy gap. This was further checked by scanning this range for a long time without observing any intensity. Since in n-type crystals with the actual electron density of n = 3.5 x 10^{17} cm^{-3} the conduction band is pinned near E_F, the value of 4.9 eV represents within 100 meV the gap value which is in close the value of the optical energy gap of 4.8 eV [19].

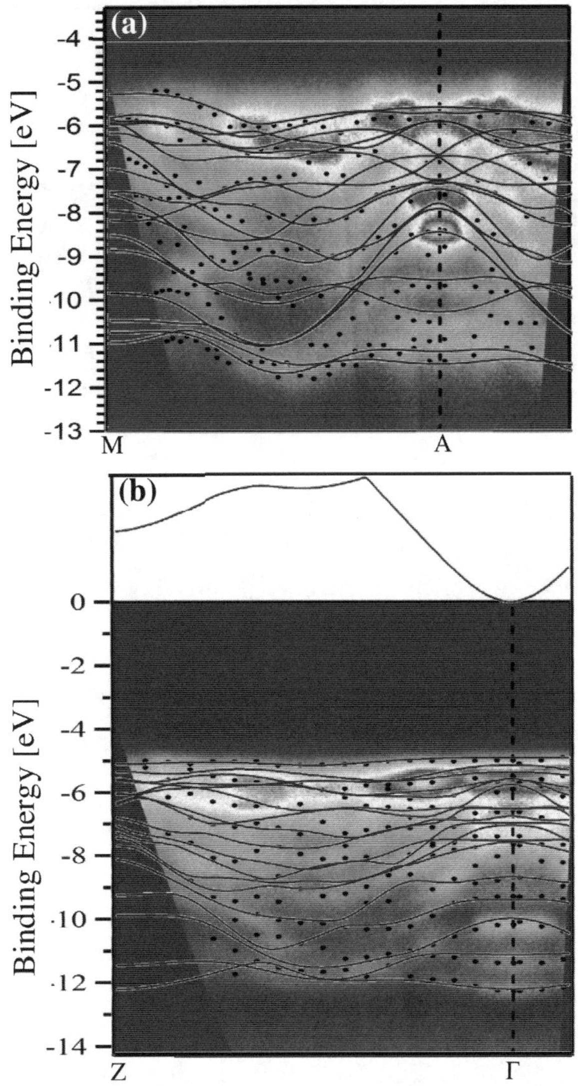

Figure 6. (a), (b) Experimental band structure of β-Ga_2O_3 along the A-M and Γ-Z directions of the Brillouin zone. Black dots show the experimental bands derived from a fit procedure to the spectra. Calculated band structures along Γ-Z and A-M are overlaid on the experimental band structure for comparison.

4.3. The temperature dependence of the band gap (E_g) of β-Ga_2O_3:

Figure (7a) shows a series of photoemission spectra at the Γ-point taken at different temperatures from 300 to 20K. It is observed from these measurements that the photoemission intensity increases with decreasing temperature and the valence bands become much more pronounced and sharper at lower temperatures. The energy band gap determined by linear extrapolations of the valence band maximum shown, in figure (7b), is plotted as a function of the temperature in figure (7c). One can see in figure (7c) that the energy gap decreases with increasing the temperature. This change of the energy gap with temperature is most probably due to electron-phonon interactions [20] that can cause a decreasing band gap of semiconductors with increasing temperature.

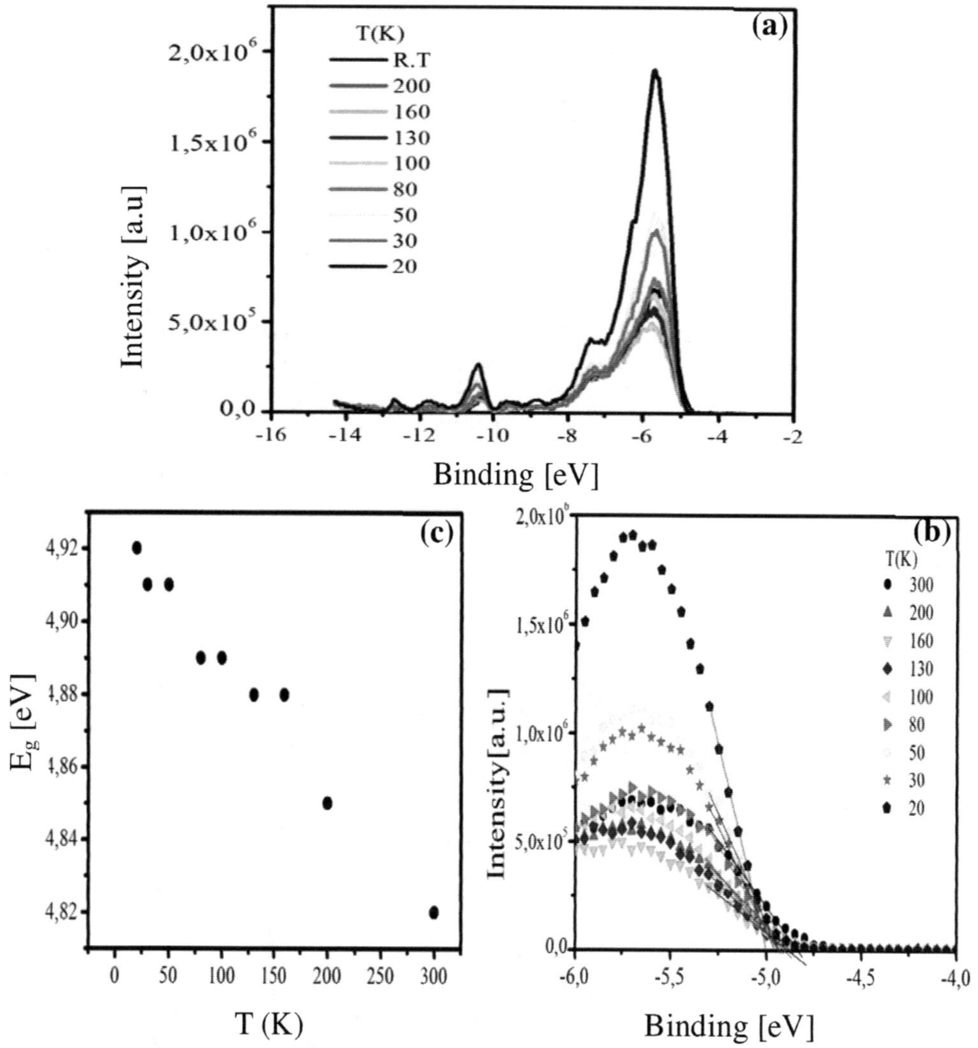

Figure 7. (a) EDCs at the Γ -point taken at different temperatures. (b) The linear extrapolations used to determine the separation between valence band maximum and the Fermi level. (c) Energy band gap (E_g) of β-Ga_2O_3 as a function of the temperature.

5. Conclusion

The (100) cleavage plane of n-type β-Ga_2O_3 grown by the Czochralski method gives a surface of high perfection as became observable in STM, LEED, and ARPES. Comparison of ARPES with the band structure from a DFT calculation shows almost excellent agreement along the Γ-Z and A-M directions of the Brillouin zone. The energy gap of β-Ga_2O_3 decreases with increasing temperature typically due to electron-phonon interactions.

Acknowledgments

This work has been in part conducted at BESSY. We thank the staff for support and in particular Dr. G. Reichardt. Mansour Mohamed thanks Dr. Klaus Irmscher and Mr. Pietsch from IKZ for the valuable help in making Ohmic contact for the samples, the Ministry of Higher Education of Egypt and Assuit University. This work is supported by the DFG (German Research Foundation) under project No. MA2371/8-1.

References

[1] Passlack M, Schubert E F, Hobson W S, Hong M, Moriya N, Chu S N G, Konstadinidis K, Mannaerts J P, Schnoes M L and Zydzik G J 1995 *J. Appl. Phys* **77** 686
[2] Tippins H H 1965 *Phys. Rev.***140** 316
[3] Ueda N, Hosono H, Waseda R and Kawazoe H 1997 *Appl. Phys. Lett.* **70** 3561
[4] Lorentz M R, Woods J F and Gambino R J 1967 *J. Phys. Chem Solids* **28** 403
[5] Harwig T, Wubs G J and Dirksen G J 1976 *Solid State Commun.* **18** 1223
[6] Lorenz M R, Gambino J F W R J 1967 *J. Phys. Chem. Solids* **28** 403
[7] Freisher M and Meixner H *1991 Sens. Actuators B* **4** 437
[8] Roy R, Hill V G and Osborn E F 1952 *J. Am. Chem. Soc.* **74** 719
[9] Roy R, Hill V G and Osborn E F 1953 *Ind. Eng. Chem* **45** 819
[10] Geller S 1960 *J. Chem.Phys.* **33** 676
[11] Geller S 1977 J. *Solid State Chem.* **20** 209
[12] V.M. Bermudez 2006 *Chem. Phys.* **323** 193
[13] Lovejoy T C, Yitamben E N, Shamir N, Morales J, Villora E G, Shimamura K,Zheng S, Ohuchi F S and Olmstead M A 2009 *Appl. Phys. Lett.* **94** 081906
[14] Eumann M, Schmitz G, and Franchy R 1998 *Appl. Phys. Lett.* **72** 3440
[15] Schmitz G, Grassmann P and Franchy R 1998 *J. Appl. Phys.* **83** 2533
[16] Mohamed M, Janowitz C, Unger I, Manzke R, Galazka Z, Uecker R, Fornari R, Weber J R, Varley J B and Van de Walle C G 2010 *Appl. Phys. Lett.* **97** 211903
[17] Varley J B, Weber J R, Janotti A and Van de Walle C G 2010 Appl. Phys. Lett. **97** 142106
[18] Weber J R 2010 *priv. commun.*
[19] Orita M, Ohta H, Hirano M and Hosono H 2000 *Appl. Phys. Lett.* **77** 4166
[20] Manoogian A and Woolley J C 1984 *Can. J. Phys.* **62** 285

Optical response of the Cu(110)/electrolyte interface

E E Barritt[1], C I Smith[1], D S Martin[1], K Gentz[2], K Wandelt[2] and P Weightman[1]*
[1] Department of Physics, University of Liverpool, Liverpool, L69 7ZE, UK, EU
[2] Institut fur Physikalische und Theoretische Chemie der Univeritat Bonn, Wegelerstrasse 12, D-53115 Bonn, Germany, EU

* Corresponding Author

Abstract. It is demonstrated that the angular variation of a positive peak at 2.2 eV in the reflection anisotropy spectrum (RAS) observed from the Cu(110)/electrolyte interface is consistent with its identification as a transition between surface states at the $\overline{Y}$ point of the surface Brillouin zone. The angular variation of a broad negative feature in the RAS spectrum is consistent with the major contribution to its intensity arising from transitions polarised along the principal axes of the Cu(110). The intensity of this feature varies with the potential applied to the Cu(110).

1. Introduction

The electronic structure of metal/electrolyte interfaces is expected to influence the performance of fuel cells and have an effect on important processes such as electrochemical deposition. Although it is difficult to study the electronic structure of such interfaces directly it has recently been shown that the optical technique of reflection anisotropy spectroscopy (RAS) [1] has considerable potential in this field. In particular it has been shown that the RAS of the Au(110)/electrolyte [1,2] and Cu(110)/electrolyte [3] interfaces are similar to the RAS of clean (110) surfaces of these metals prepared in ultra high vacuum (UHV). In particular the RAS of the Cu(110)/electrolyte interface shows a feature at 2.2 eV that has been identified with a transition between surface states at the $\overline{Y}$ point of the surface Brillouin zone [3]. In this work we show that the azimuthal angular variation of the intensity of this feature at 2.2 eV in the RAS spectrum of the Cu(110)/electrolyte interface is consistent with its identification with this transition between surface states. We also report the results of the angular dependence of a broad negative feature observed at ~4.3 eV in the RAS of this interface and its dependence on the potential applied to the Cu(110) electrode.

2. Experimental

The preparation of the Cu(110) crystal and its insertion into the electrochemical cell in an electrolyte of 10mM HCl in ultrapure water (Millipore 18MΩcm) has been described earlier [3]. The electrochemical cell is a three-electrode cell using the copper as the working electrode, a platinum counter electrode and a silver/silver halide electrode as the reference electrode [4]. All potentials quoted are referenced to this reference electrode. The RA spectrometer was of the Aspnes design [5] and the measured RA signal from 1.5 eV to 5.0 eV is given by

$$\mathrm{Re}(\frac{\Delta r}{r}) = 2\frac{(r_{[1\bar{1}0]} - r_{[001]})}{(r_{[1\bar{1}0]} + r_{[001]})} \qquad (1)$$

where $r_{[1\bar{1}0]}$ and $r_{[001]}$ are the reflection coefficients in the $[1\bar{1}0]$ and [001] directions in the (110) surface respectively. The sum of these quantities is given by r and the average by $^{r}/_{2}$. The azimuthal angular variation of the specimen with respect to the direction of the incident light was performed using a rotating platform which ensured the same spot of the crystal was observed over the full 90° rotation.

3. Results

The RAS of the Cu(110) surface in 10mM HCl is shown in figure 1 as a function of the azimuthal angle, θ, in the plane perpendicular to the direction of the incident light. These spectra were obtained with a potential of -0.65 V applied to the Cu(110) electrode. The dependence of the RA spectrum on the applied potential has been reported previously [3] and it is clear that the adsorption of Cl begins at a potential of -0.40 V and that the Cu(110) surface is clean in the range -0.70 V to -0.55 V although it is possible that there are low levels of H adsorption at the more negative potentials. The optical axes of the Cu(110) surface are along $[1\bar{1}0]$ and $[001]$ and the zero of θ used in the experimental results in figure 1 coincides with the $[001]$ direction to ~±10°. The inserts in figure 1 show the variation in the intensity of the peak at 2.2 eV and the broad feature at 4.3 eV as a function of θ. The results follow a cos2θ dependence as shown by the comparisons with theoretical curves. The difference in phase between the origins of the theoretical curves and the experimental results is consistent with an error in determining the origin of the experimental θ of 7°.

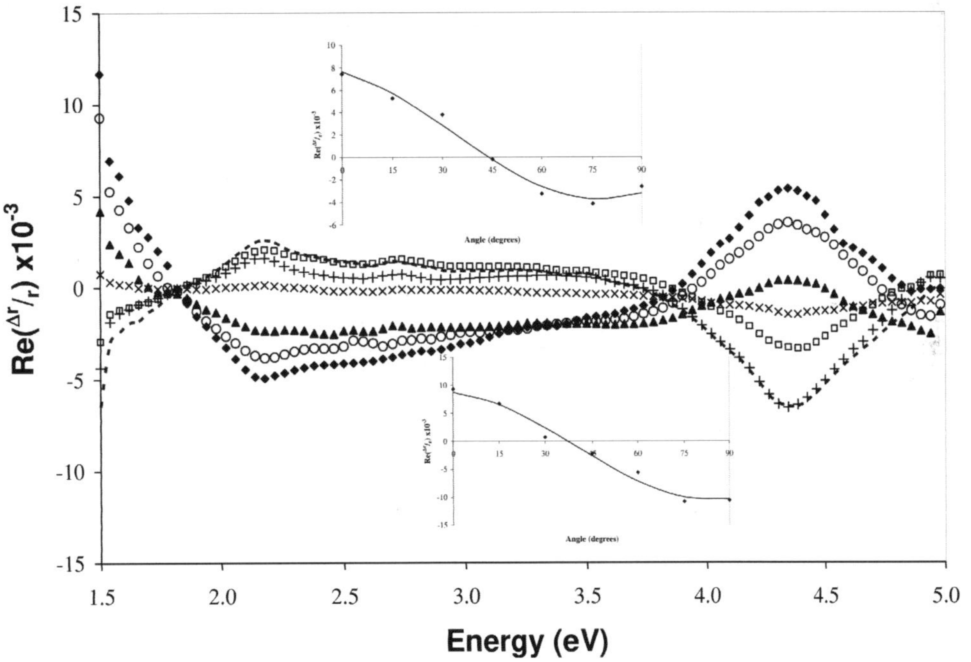

Figure 1. RA spectra of Cu(110) in 10mM HCl at -0.65 V vs Ag/AgCl showing the angular rotation at 90° (◆), 75°(○), 60° (▲), 45° (×), 30° (□), 15° (---) and 0° (+). The inset above shows the data points observed at 2.2 eV compared with a theoretical cos2θ curve and the inset below shows the data points observed at 4.3 eV compared with a theoretical cos2θ curve. In both curves the θ=0° of the theoretical curve has been shifted by 7° to bring it into coincidence with the experimental results.

The dependence of the RAS of the Cu(110)/electrolyte interface on the applied potentials -0.7 V, -0.65 V, -0.60 V and -0.55 V is shown in figure 2. The results for the spectral range 2.0 eV to 2.5 eV encompassing the peak at 2.2 eV have been discussed earlier in detail [3]. The spectral region to low energy of 2.0 eV shows almost no dependence on the applied potential. The strong broad negative peak centred on ~4.3 eV becomes slightly broader and deeper as the potential is made more positive as shown in detail in the inset to figure 2.

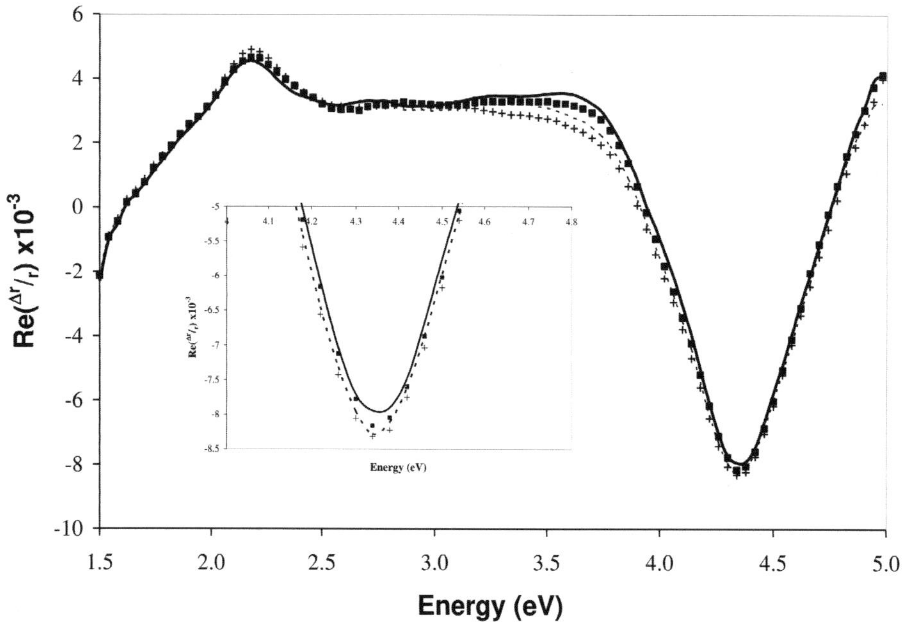

Figure 2. RA Spectra of Cu(110) in 10mM HCl at -0.7 V (solid line), -0.65 V (■), -0.6 V (· · ·) and -0.55 V (+) The inset shows the peak at 4.3 eV in more detail. There is a spurious weak feature ~2.7 eV in the spectra arises from a sharp peak in the output of the Xe lamp at this energy, which due to the less than perfect optical elements, does not cancel completely in the ratio $^{\Delta r}/_{r}$.

4. Discussion

The RAS of the Cu(110)/electrolyte interface has been compared with the RAS of well prepared clean surfaces of Cu(110) in UHV [1,3,6-12]. The positive peak at 2.2 eV has been identified with a stronger feature observed from Cu(110) surfaces in UHV that arises from a transition between surface states at the $\overline{Y}$ point of the surface Brillouin zone [1,6,7]. This identification indicates that the peak is only excited by light incident along the $[001]$ direction [6,7,13]. The azimuthal angular variation of RA spectra can yield considerable insight into the geometry of the optical response of surfaces [14-17]. For the Au(110)/electrolyte interface and for cytosine and adenine adsorbed at this interface [17,18] it has been shown that the RAS response collapses to close to zero across the whole spectrum when the incident plane polarised light is parallel to one of the principal axes in the surface. This implies that the dominant contributions to the RAS spectrum of Au(110) arise from transitions orientated along one of the principal axes [18]. For the absorbed molecules the angular variation of the RAS establishes that both adenine and cytosine are orientated essentially vertical to the Au(110) surface with the long axis of the molecules aligned along one of the principal axes. It has been shown that the RAS of 9-anthracene carboxylic acid [7] and 3-thiophene carboxylic acid [8] are influenced by the azimuthal orientation of the stacked molecules on a Cu(110) surface in UHV. However there has been no previous study of the azimuthal dependence of the RAS of the Cu(110) surface in either UHV or liquid environments. The results of figure 1 show that as found for the Au(110)/electrolyte interface [17,18] the RAS of the Cu(110)/electrolyte interface collapses to close to the zero across the spectral range from 1.5 eV to ~3.5 eV. We will return to the spectral range above 3.5 eV later. An important finding is that the intensity of the peak at ~2.2 eV shows essentially a cos2θ dependence. This is to be expected if this peak arises from the transition between surface states at the $\overline{Y}$ point of the surface Brillouin zone and the result is further confirmation of this assignment. In detail the theoretical cos2θ curve has to be shifted by 7° in order to coincide with the experimental results. This indicates that the experimental values of θ correspond to an origin 7° off the principal axes. This may explain why the RAS in the region of the spectrum above 3.5 eV does not go to zero since it is difficult to find the exact angle at which the spectrum will be a minimum. Alternatively this region of the spectrum could contain weak contributions from transitions that are not orientated along one of the principal directions. The insert to figure 2 shows that the broad negative feature centred on ~4.3 eV becomes broader and more intense as the applied potential is increased from -0.70 V to -0.55 V. A similar increase in intensity with a potential variation across this range was observed for the positive peak at 2.2 eV [3]. There is no clear explanation for either of these changes but they could result from the desorption of H which could lead to a more ordered surface. This would be a similar affect to that observed for the adsorption of CO impurities on the Cu(110) surface in UHV [19-21].

5. Conclusions

The RA spectral profiles of the positive peak observed at 2.2 eV from the electro-polished Cu(110)/electrolyte interface show a cos 2θ dependence which is expected if this transition is associated with a transition between surface states at the $\overline{Y}$ point of the surface Brillouin zone. The broader negative peak at 4.3 eV shows a similar angular variation which indicates that this feature is dominated by contributions that arise from transitions that are polarised along one of the principal axes directions.

Acknowledgements

EEB is supported by a Ph.D. award funded by the UK Engineering and Physical Science Research Council (EPSRC). This work was also support by an award from the UK Biotechnology and Biological Science Research Council (BBSRC).

References

[1] Weightman P, Martin D S, Cole R J and Farrell T 2005 *Rep. Prog. Phys.* **68** 1251
[2] Sheridan B, Martin D S, Power J R, Barrett S D, Smith C I, Lucas C A, Nichols R J and Weightman P 2000 *Phys. Rev. Lett.* **85** 4618
[3] Barritt E E, Smith C I, Martin D S, Gentz K, Wandelt K and Weightman P *In press*
[4] Wilms M, Kruft M, Bermes G and Wandelt K 1999 *Rev. Sci. Inst.* **70** 3641
[5] Aspnes D E, Harbison J P, Studna A A and Florez L T 1988 *J. Vac. Sci. Technol. A.* **6** 1327
[6] Hofmann Ph, Rose K C, Fernandez V, Bradshaw A M and Richter W 1995 *Phys. Rev. Lett.* **75** 2039
[7] Frederick B G, Power J R, Cole R J, Perry C C, Chen Q, Haq S, Bertrams Th, Richardson N V and Weightman P 1998 *Phys. Rev. Lett.* **80** 4490
[8] Frederick B G, Cole R J, Power J R, Perry C C, Chen Q, Richardson N V, Weightman P, Verdozzi C, Jennison D R, Schultz P A and Sears P 1998 *Phys. Rev. B.* **58** 10883
[9] Cole R J, Frederick B G and Weightman P 1998 *J. Vac. Sci. Technol. A.* **18** 3088
[10] Martin D S, Maunder A and Weightman P 2001 *Phys. Rev. B.* **63** 155403

[11] Hansen J –K, Bremer J and Hunderi O 1998 *Surf. Sci.* **418** L58
[12] Stahrenberg K, Hermann Th, Esser N and Richter W 2000 *Phys. Rev. B.* **61** 3043
[13] Martin D S and Weightman P 2001 *Surf. Interface. Anal.* **31** 915
[14] Macdonald B F and Cole R J 2002 *App. Phys. Lett.* **80** 3527
[15] Macdonald B F, Law J S and Cole R J 2003 *J. Appl. Phys.* **93** 3320
[16] Farrell T, Harrison P, Smith C I, Martin D S and Weightman P 2008 *Appl. Phys. Lett.* **93** 191102
[17] Weightman P, Dolan G J, Smith C I, Cuquerella M C, Almond N J, Farrell T, Fernig D J, Edwards C and Martin D S 2006 *Phys. Rev. Lett.* **96** 086102
[18] Mansley C P, Smith C I, Bowfield A, Fernig D G, Edwards C and Weightman P 2010 *J. Chem. Phys.* **132** 214708
[19] Sun L D, Hohage M, Zeppenfeld P, Balderas-Navarro R E and Hingerl K 2003 *Phys. Rev. Lett.* **90** 106104
[20] Weightman P 2008 *Surf. Sci.* **602** 1729
[21] Sun L D, Denk R, Hohage M and Zeppenfeld P 2008 *Surf. Sci.* **602** L1

Journal of Physics: Conference Series **286** (2011) 012029 doi:10.1088/1742-6596/286/1/012029

Two dimensional exclusion process between rough interfaces

J Juntunen and J Merikoski

Department of Physics, University of Jyväskylä, P.O. Box 35, FI-40014 Jyväskylä, Finland

E-mail: `janne.k.juntunen@jyu.fi`

Abstract. Driven flow between rough walls is studied in the framework of the two-dimensional exclusion process by using continuous-time Monte Carlo simulations. In particular we consider the effect of the roughness and other characteristics of the walls on the current and current profiles at different types of locations along the channel.

1. Introduction

Technological advances have made it possible to miniaturize devices down to submicrometric sizes and has enabled manufacturing of numerous devices used for manipulation and transport of matter. Especially the field of microfluidics has obtained a growing amount of interest since it was born about twenty years ago [1]. Microfluidics has been used to answer fundamental questions in physics such as the behavior of single molecules or particles in fluid flows. Transport in some dynamical environment can also be used to probe the properties of the environment. It is a well known fact that the structure, *e.g.* the roughness, of the surface does not cause any significant deviation in the properties of the flow if the system size is big enough. However, surface roughness becomes relevant if typical length scales of the system are comparable to the scale of its variation as is the case of microfluidic setups concerning fluids in rough or structural microchannels [2, 3, 4].

The asymmetric exclusion process is one of the simplest lattice models describing transport far from equilibrium [5, 6]. In one dimension it has been extensively studied during the past decades while its extension to higher and possible more physical (higher) dimensions are almost neglected. Furthermore, the studies on exclusion models in two dimensions are usually studies with semi infinite lattices [7] without any geometrical constraint. When considering transport in realistic environments the geometry is usually restricted. Examples of realistic transport processes vary from fluid flow in channels to vehicular traffic. In addition, from numerous diffusion studies made in past decades we have learned that even the most simple geometrical constraints can cause significant changes in mass transport [8].

Even though the hydrodynamic equations corresponding to the asymmetric exclusion process are not the Navier-Stokes equations, analysis of exclusion process can produce relevant insight in transport in rough channels especially in situations, where the system is far from equilibrium and inertia is not a major factor, but the underlying microscopy can not be forgotten. In this work we consider the two dimensional asymmetric exclusion process (ASEP) in rough channels. The channel walls are provided by interfaces produced by a discrete interface model.

2. Models

2.1. Models for the rough channels

We first consider stationary channel walls h_l and h_u, *i.e.* lower wall and upper wall obeying body-centered solid-on-solid (BCSOS) statistics [9]. For comparison, we also have considered the exclusion process between two BCSOS interfaces driven against each other [10, 11]. The location or the 'height' of a single BCSOS interface [9] is described by a function $h_i(x,t)$ such that, for every site $x = 1, ..., L$, $h_i(x+1,t) - h_i(x,t)_i = \pm 1$, where we without loss of generality assume that the possible values of h_i are integers and $i = l, u$. Only two kinds of processes, adsorption (h_i locally increases) and desorption (h_i locally decreases), are available. The stationary state of a single interface is totally disordered, *i.e.* every configuration allowed by periodic boundary conditions has the same probability to occur.

2.2. Exclusion process

The exclusion process differs from independent random walkers because in it only one particle is allowed to be in a lattice site at any given instant of time. All particles are identical and independently moving in the channel, *i.e.* between walls $h_u(x,t)$ and $h_l(x,t)$, on a lattice (x, y), where the lattice point coordinates in the horizontal direction (x) are the same $x = 1, ..., L$ as for the channel model above, again with periodic boundary conditions and the values of the vertical coordinate (y) coincide with the possible locations of the walls at $h_i(x,t)$.

By denoting the location of the particles by (x_{p_i}, y_{p_i}), the requirement that particles move inside the channel is equivalent to the condition $h_u(x_{p_i}, t) \le y_{p_i} \le h_l(x_{p_i}, t)$. For the moves of the particle, the following exclusion rule is imposed in all cases: Only one particle can be at each site at instant of time so that for a jump $(x_{p_i}, y_{p_i}) \to (x'_{p_i}, y'_{p_i})$ to be possible, the arrival site must be empty.

For moves on the square lattice there are a few natural choices for the possible particle jumps $(x_p, y_p) \to (x'_{p_i}, y'_{p_i})$. Here we concentrate on the case where only nearest-neighbor jumps are allowed such that the particle jumps in the horizontal direction $(x_p, y_p) \to (x_p + 1, y_p)$ with the attempt rate α and in the vertical direction $(x_p, y_p) \to (x_p, y_p \pm 1)$ with rate β. The dynamics used in this study is sequential, *i.e.* one particle jumps at a time.

For tracer diffusion of independent particles between similar walls the diffusion coefficient for different choices of dynamics do not differ significantly [11]. The same is found for dispersion around the flow in the present case. However, for lattice gas models of fluid flow with parallel dynamics including particle collisions, it is well known that in two dimension the number of jump directions must be at least eight in order to retain the conservation laws. Collision phenomena could also be included in exclusion models by allowing longer jumps of the particles. Our model best describes the strongly dissipative case.

3. Results

In figure 1 we show snapshots of the scaled occupation probability between the walls for different average particle densities ρ. The spatial inhomogeneity decreases for increasing β, which is the rate of diffusion in the vertical direction. With flat (in the BCSOS sense) walls, spatial homogeneity is achieved in the simulation for all values of $\beta > 0$, while for increasing roughness the value of β required for approximate homogeneity is larger. Homogeneity gets broken because the particles tends to drift towards the wall at locations with long slopes. Spatial homogeneity, or deviations from it, becomes a particularly important explaining factor for understanding the properties of the current through the system.

In Fig. 2 we show the expected value of the particle current I crossing a lattice site as a function of the wall roughness $W = \langle (h_i - \bar{h}_i)^2 \rangle$, where the average $\bar{h}_i$ is taken over the configuration. The value of the current for a given wall roughness depends strongly also on other properties of the walls. Wall configurations corresponding to the extrema of the current

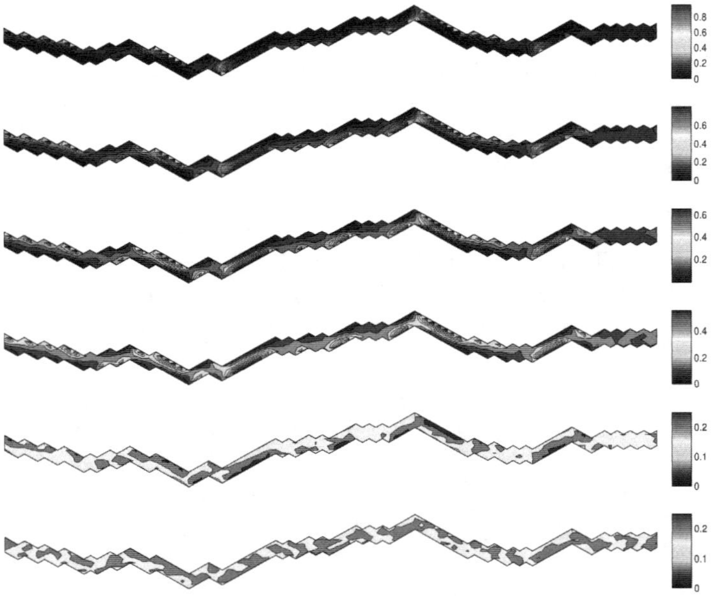

Figure 1. Probability to find particle at location (i, j) with several values of the parameter β values. Here the red color indicates highest density and blue color lowes density. The average density is $\rho = 0.2$ and the drift $\alpha = 1$. The system size is $L = 100$ and the widh of the channel four lattice units. The parameter β controlling diffusion in the vertical direction increases from top to bottom from 0.25 to 20.

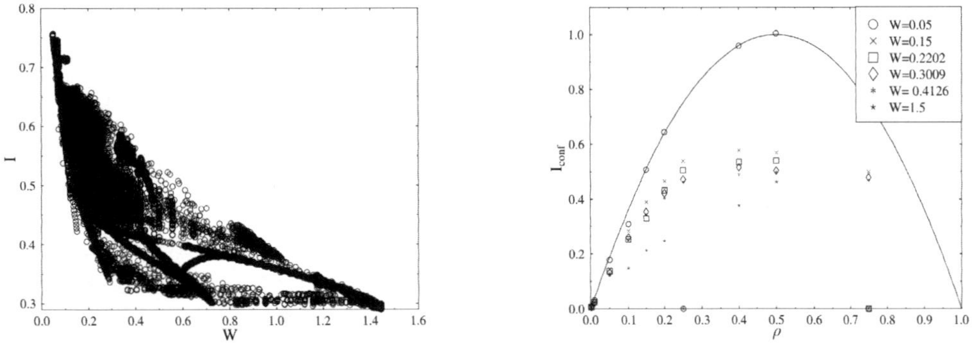

Figure 2. In the figure on the left we show average current through the cross-section measured at $x = L$ as a function of the wall roughness W with $\rho = 0.20$ for individual wall configurations. In the figure on the right we show the averaged current over configurations corresponding to a given wall roughness as a function of density. The system size is $L = 100$ and the jump rates are $\alpha = 1$ and $\beta = 0.5$.

were created by running variations of the usual BCSOS dynamics for a while starting from certain initial configurations, a disordered one, a flat one and a faceted one. For example, the smallest currents were obtained by generating the wall configurations by depositing 'particles' anywhere on the interface [9] such that they slide down to the closest wall height minimum. The maximum current for a given roughness was obtained by producing walls such that the BCSOS dynamics corresponds to one-dimensional ASEP (one-to-one mapping where a particle corresponds to upward step), where in the ASEP interpretation particle a chosen randomly hops to a location as far as possible from the closest particles with conserving the order of particles in the system. This way it was possible to achieve wall configurations with very different topography and efficiently scan the available phase space for this plot. In Fig. 2 we show also the expected current averaged over the different configurations corresponding to the same roughness. For spatially homogeneous density the mean-field prediction $I(\rho) = \alpha\rho(1-\rho)d$ for the averaged current works well, and increasing deviations are observed for increasing the wall roughness. The amount of this deviation decreases for increasing β.

In Fig. 3 we show the ensemble averaged current profiles at special points of the wall

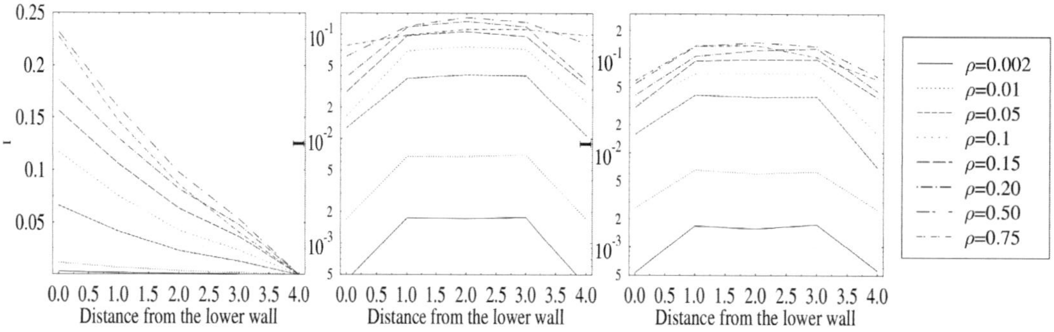

Figure 3. The time-averaged current profile in y direction at locations in x direction corresponding to $\max\{h_i\}$ (left) and $\text{mean}\{h_i\}$ (middle) and at any location (right) for several values of the density ρ. Here $L = 100$, $\alpha = 1$ and $\beta = 0.5$.

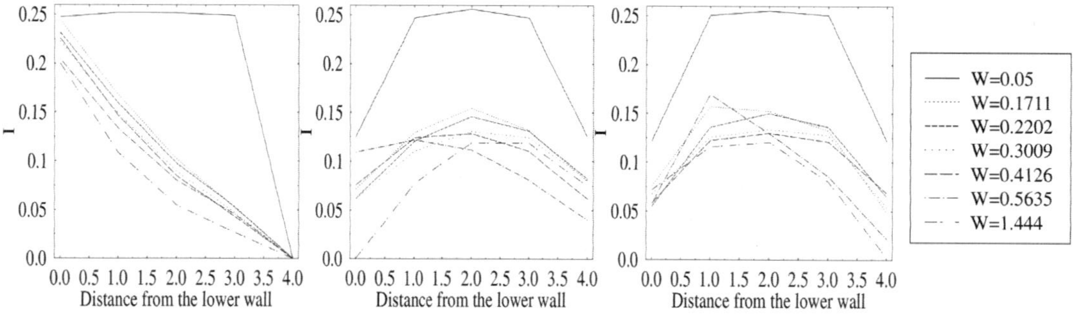

Figure 4. The time-averaged current profile in y direction at locations in x direction corresponding to $\max\{h_i\}$ (left) and $\text{mean}\{h_i\}$ (middle) and at any location (right) for several values of the wall roughness W. Here $L = 100$, $\alpha = 1$ and $\beta = 0.5$.

configurations and also averaged over the system for different values of the particle density ρ. At the locations corresponding to the extrema, *i.e.* $\max(h_1)$ and $\min(h_1)$, the current tends to have a monotonic profile. The general shape of the profiles on those locations is easily understood by looking at Fig. 1. The particles tend to crowd at the slope leading to the extreme of a wall, leading to a smaller density right after it. At more typical locations the profile is symmetric. In Fig. 4 we show the ensemble averaged current profile at special points of a given configurations with each curve corresponding to a given value of the wall roughness W. At the extremal locations the current goes to zero at one side of the tube because of the micro roughness of an BCSOS interface. The roughness alone does not explain the current profile, as evident also from figure 2, where the values of I spread widely for small and intermediate values of W. This spread can be attributed to the dispersion of the current induced by the wall configuration. The distribution of the current (not shown) is found to be approximately gaussian.

4. Conclusions

The particle current in the two-dimensional exclusion process in a channel with rough walls was found to depend strongly on the configurations of the walls. Not only the roughness of the walls but also their overall shape determines the current. The regions close to the extrema (in the transverse direction) of the channels have a prominent effect on the transport. The channel profile in the highly dissipative case turns out to have a much greater impact on the particle flow and its dispersion than it does in the case of laminar flow previously considered in the literature. Work including interactions in both directions between fluctuating walls and the exclusion process is in progress.

References

[1] Tabeling P 2005 *Introduction to microfluidistic* (Oxford University Press)
[2] Kunert C and Harting J 2007 *Phys. Rev. Lett.* **99** 17
[3] Hyväluoma J and Harting J 2008 *Phys. Rev. Lett.* **100** 246001
[4] Harting J, Kunert C and Hyväluoma J 2010 *Microfluid Nanofluid* **8** 1
[5] Derrida B 1998 *Phys. Rep.* **301** 65
[6] Chowdhury D, Santen L and Schadschneider A 2000 *Phys. Rep.* **329** 199
[7] Singh N and Bhattacharjee S M 2009 *Physics Rev. Lett. A* **373** 3113
[8] Haus J and Kehr K 1987 *Phys. Rep.* **150** 263
[9] Barabasi A L and Stanley E H 1995 *Fractal Concepts in Surface Growth* (Cambridge University Press)
[10] Juntunen J, Pulkkinen O and Merikoski J 2007 *Phys. Rev. E.* **76** 041607
[11] Juntunen J and Merikoski J 2010 *J. Phys.: Condens. Matter* **22** 465402

Journal of Physics: Conference Series **286** (2011) 012030 doi:10.1088/1742-6596/286/1/012030

Morphological templating of metastable calcium carbonates by the amino acid leucine

S P Thompson, J E Parker, S R Street and C C Tang

Diamond Light Source, Harwell Science & Innovation Campus, Didcot, Oxon OX11 0DE, UK

E-mail: stephen.thompson@diamond.ac.uk

Abstract. The in vitro precipitation of the metastable $CaCO_3$ phases aragonite and vaterite in the presence of leucine is investigated. Under normal conditions, the production of $CaCO_3$ via the hydrolysis of urea method favours the formation of regular needle-like aragonite crystals, with very minor quantities of vaterite and calcite. However in the presence of leucine, aragonite forms highly branched structures and the vaterite yield is increased, forming flower-like clusters composed of nano-thin sheets. Both the degree of aragonite branching and the occurrence, regularity of shape and number of vaterite "petals" increases with leucine concentration. The two phases exhibit different variations in their crystallographic parameters with increasing concentration, while the molecular structure appears unaffected.

1. Introduction

Advances in technology and science are often driven by the development of new materials, or by the discovery of novel properties of existing ones. The development of various ways of tailoring the structures of existing materials to unusual and special morphologies at the sub-micron and nano scale has been particularly productive [1]. Biomimetic synthesis, inspired by the remarkable range of morphologies exhibited by biominerals, is one such approach where large macromolecules are used to control growth morphology.

Calcium carbonate ($CaCO_3$) polymorphs are important in geological [2], environmental [3] and bio-sciences [4], as well as in numerous industrial applications, where control of polymorphism, morphology and crystal size distribution are important factors. $CaCO_3$ has three anhydrous polymorphs of increasing thermodynamic stability: vaterite (hexagonal), aragonite (orthorhombic) and calcite (rhombohedral). Aragonite particle size and shape depend strongly on preparation method and conditions, with direct inorganic precipitation from soluble carbonate and Ca-salts yielding irregular crystals [5] limiting their suitability for industrial application where high-aspect ratio $CaCO_3$ needle-like crystals are in much demand for improving the mechanical properties of polymer materials, or as a filler for paints, plastics, rubber, paper or foodstuffs [6]. It is also a potential biomedical material since it is denser than calcite and could be integrated, resolved, and replaced by bone [7]. Aragonite forms only under a narrow range of physio-chemical conditions and much effort has been given to developing reliable production methods [8] such as urea hydrolysis [9]. Vaterite also has potential practical applications due to high specific surface areas, high solubility, high dispersion and low density. However it is poorly understood structurally [10] and tends to form only small crystals. The high-volume commercial availability of amino acids offers a varied source of potential templating

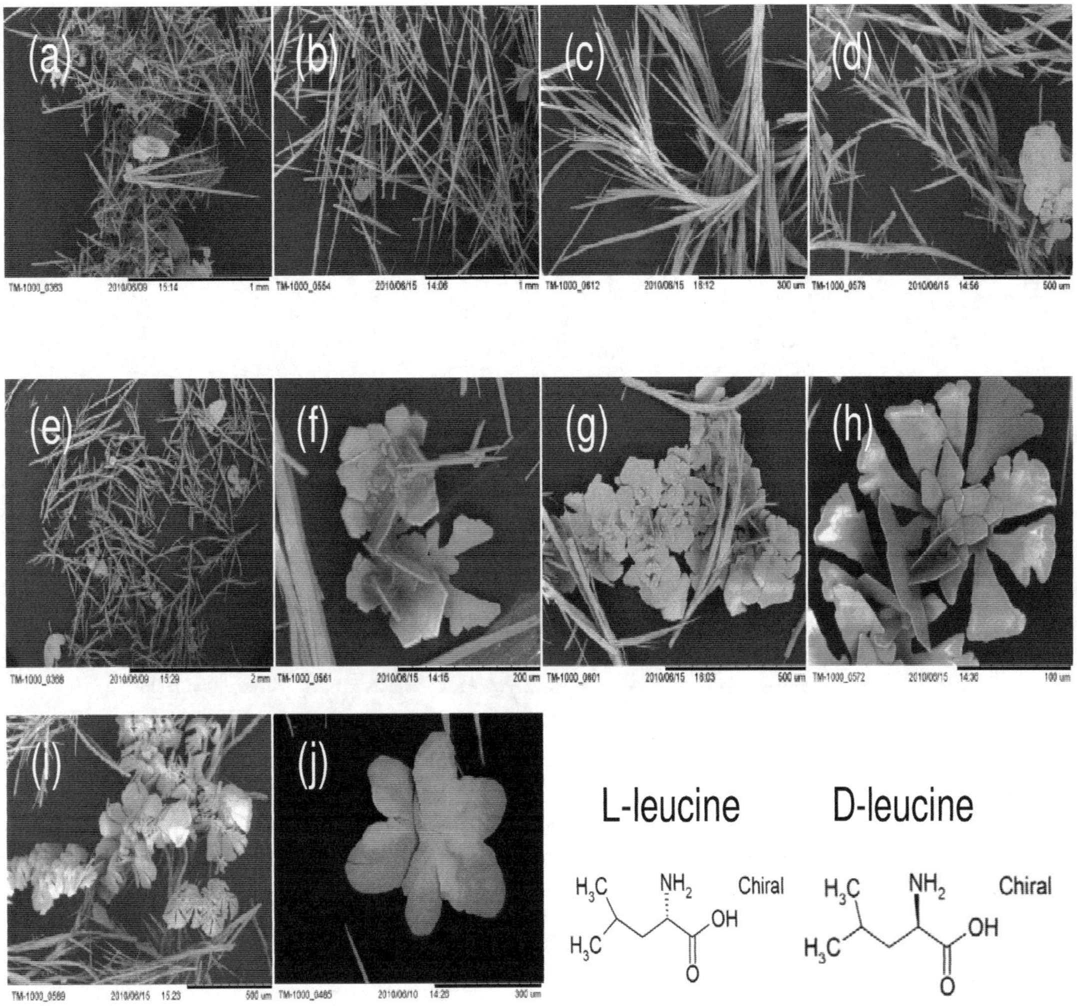

Figure 1. SEM micrographs of (a) aragonite grown in absence of amino acid; (b-e) aragonite grown in the presence of 2 g/L L-leu, 4 g/L L-leu, 6 g/L D-leu, 10 g/L L-leu; (f-j) vaterite in the presence of 2 g/L D-leu, 4 g/L L-leu, 4 g/L D-leu, 6 g/L D-leu, 10 g/L D-leu; Bottom right: molecular structure of L- and D-leucine molecules.

macromolecules for industrial production and, as part of a study investigating the templating ability of various amino acids, we report here on the effect leucine has on the precipitation of metastable $CaCO_3$. Leucine is a neutral, essential α-amino acid of the aliphatic group synthesized in plants and microorganisms and, as in nearly all α-amino acids, exists in two optical isomers denoted as L-leucine (L-leu) or D-leucine (D-leu), which are mirror images of each other. Besides practical applications, amino acids of likely extraterrestrial origin are found in certain types of meteorite [19]; while understanding the localised interaction of amino acids with the surrounding environment may provide clues to the early conditions and evolution of biomineralising creatures and the early appearance in the fossil record of protective or skeletal

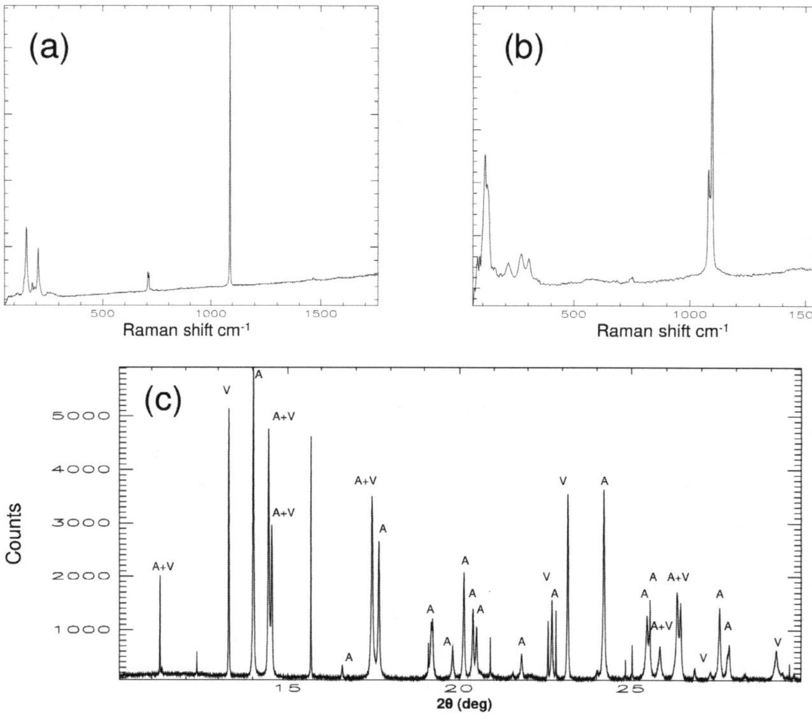

Figure 2. (a) Raman spectrum obtained from branched 4 g/L L-leu aragonite structure; (b) Raman spectrum obtained from 4 g/L D-leu vaterite flower. (c) X-ray powder diffraction pattern for $CaCO_3$ grown in presence of 4 g/L L-leu, showing the main aragonite (A) and vaterite (V) peak assignments (unmarked peaks are residual calcite phase).

structures between the Ediacaren and Early Cambrian periods 525-542 Million years ago [20].

2. Experimental details

Analytical grade $CaCl_2.H_2O$ (VWR International), urea ($(NH_2)_2CO$, Fischer Scientific) and leucine ($HO_2CCH(NH_2)CH_2CH(CH_3)_2$ L- and D- forms, Sigma Aldrich) were used to synthesize crystals of aragonite via direct precipitation in 50 ml deionised water by ageing solutions of $CaCl_2$ (0.25 mol dm^{-3}) and urea (0.75 mol dm^{-3}) at 90° C for 20 hours [9]. The precipitates were removed by filtration, rinsed with distilled water, dried overnight at 90° C and stored in a desiccator. The procedure was repeated with 0.1, 0.2, 0.3 and 0.5 g of L- and D-leu added to give 2, 4, 6 and 10 g/L concentrations. Precipitate morphologies were imaged by SEM (Hitachi TM1000) and characterised by confocal Raman spectroscopy (Horiba LabRam 800, 532 nm NdYAG laser, 600 line grating, 10x objective) and synchrotron X-ray powder diffraction (beamline I11, Diamond Light Source [16]) at 15 keV, with samples in 0.5 mm glass capillaries.

3. Results

Fig. 1 shows SEM images of the precipitates. Those grown without leucine (Fig. 1a) show the expected aragonite needle morphology [9], while increasing concentrations of leucine result in progressively split branching morphologies (Fig. 1b-e). For concentrations above 4 g/L the splitting occurs along the whole length of what would otherwise have been a single needle. Minor differences between L- and D-leu were apparent: L-leu precipitates showed a slightly

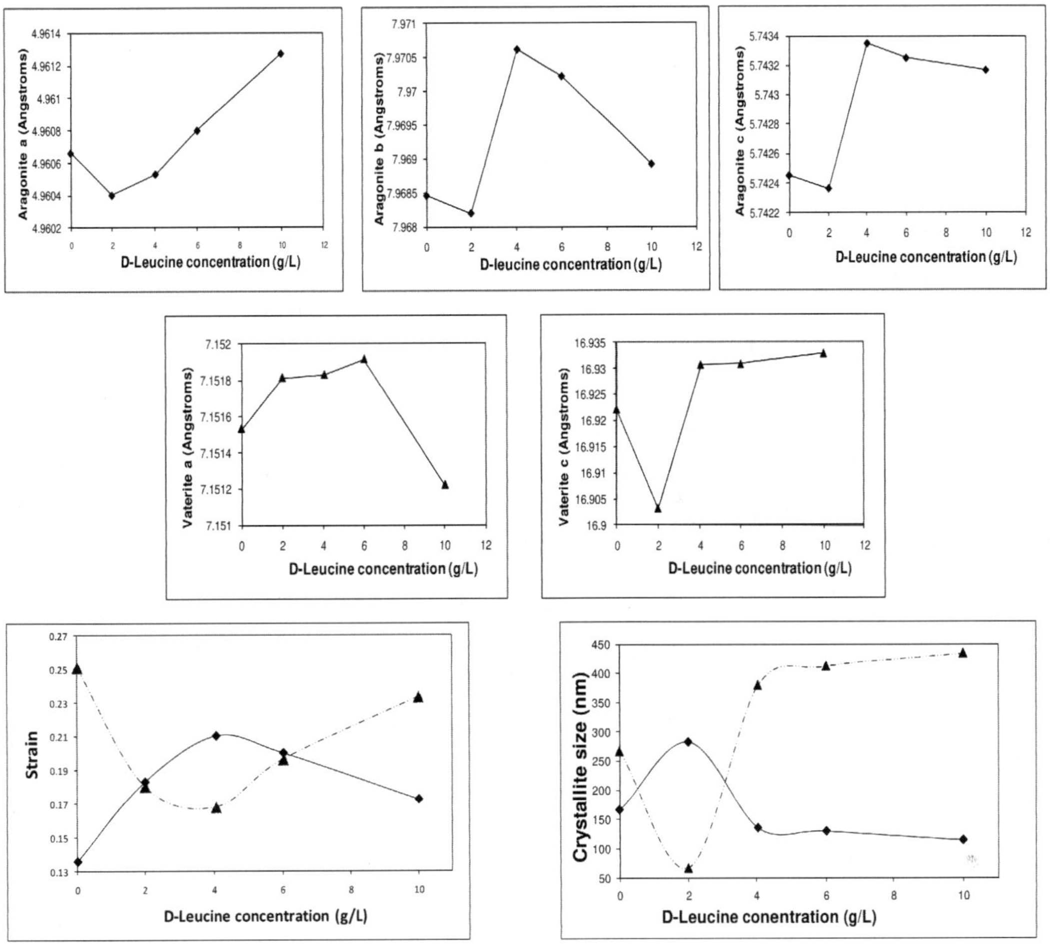

Figure 3. Variation of measured crystallographic parameters of the precipitated metastable $CaCO_3$ phases as a function of leucine content: (a to c) aragonite (◇) lattice parameters; (d, e) vaterite (△) lattice parameters; (f) lattice strain for aragonite and vaterite; (g) crystallite size for aragonite and vaterite. For aragonite a, b, c lattice parameters errors are less than $\pm 3\times10^{-5}$, $\pm 5\times10^{-5}$, $\pm 3\times10^{-5}$; For vaterite a and b are less than $\pm 8\times10^{-5}$ and $\pm 4\times10^{-4}$; errors in aragonite and vaterite strain are less than $\pm 1.3\times10^{-3}$ and $\pm 6\times10^{-3}$; errors in aragonite and vaterite crystallite sizes are ± 1.3 and ± 10 respectively across all data.

higher degree of branching and blunt ends, while the D-leu templated structures retained sharper points. In the absence of amino acid, a small number of crystals with nano-thin irregular planar shapes were also formed. L- and D-leu increased the occurrence of these and, with increasing concentration, produced complex flower-like cluster structures, with large units on the outside and smaller ones concentrated towards the middle and top (Fig. 1f-j). Again only minor differences between L- and D-leu were observed: D-leu tended to yield flowers constructed from more regularly shaped objects at lower concentrations.

Separate Raman spectra were obtained for the branched and flower structures (Fig. 2a,b). The strongest carbonate band is the internal symmetric stretch at ~1087 cm^{-1} and is independent of polytype as it depends only on O atom motions [11]. External lattice modes due to translations of the ionic centres of gravity and carbonate ion librations (restricted rotations) appear in aragonite at 143, 181, 192, 206 cm^{-1} along with a doublet at 702 and 706 cm^{-1} [13]. The branched crystals show the 700 cm^{-1} doublet, two strong features at 140 and 210 cm^{-1} and a weaker one at 180 cm^{-1}, confirming their aragonite identity. In addition to the 1087 cm^{-1} band, vaterite has an additional internal mode at $\sim$1074 cm^{-1} with external modes at 118, 270, 303 and 750 cm^{-1} [12] and confirms the flower structures as vaterite. Other than minor background differences, the spectra varied little between samples implying only growth and phase stabilisation, rather than molecular structure, are influenced by the amino acid.

From X-ray powder diffraction measurements (e.g. Fig. 2c) lattice parameters, crystallite size and microstructure strain for the two phases were obtained as a function of leucine concentration (Fig. 3) by whole pattern (150° 2θ) LeBail profile refinement. For both phases, the lattice parameters (Fig. 3a-e) are affected by the presence of leucine, with the greatest effect appearing for concentrations of 2-4 g/L and is also reflected in the precipitated crystallite sizes (Fig. 3g). Above 4 g/L crystallite sizes are approximately constant. However the presence of the amino acid results in crystal structures with very different levels of internal strain (Fig. 3f). In Aragonite, the strain increases up to 4 g/L amino acid concentration, beyond which it declines, while vaterite precipitates with reduced strain at low concentrations, which increases above 4 g/L.

4. Discussion

Although the subject of many studies, the interaction between biomolecules, calcium ions and carbonate during morphological templating and preferential phase stabilization is unclear. In biological systems, crystal lattices can maintain their stability within the microenvironments formed by organic macromolecules [14] and small amounts of acidic macromolecules exist on nearly all organic/inorganic interfaces in biomineral structures (tooth, otolith, bone, pearl and mollusc shells etc.) [15]. Early investigations established that the concentration and the conformation of polypeptides rich in acidic amino acid entrapped onto the surface of organic matrices can control polymorph selection, while morphological templating can be achieved through the preferential bonding of macromolecules to specific atomic edge-step features on the surfaces of the precipitating crystals that offer the best geometric and chemical fit [18]. This in turn changes both the shape of the steps and the edge-step free energies. Changes in step shape directly modify the macroscopic crystal shape by inducing the formation of new crystal faces resulting in macroscopic morphological modifications [18].

The in vitro hydrolysis of urea occurs in two steps [9]. First, ammonium cyanate forms: $CO(NH_2)_2 \rightarrow NH_3 + HNCO \rightarrow NH_4^+ + NCO^-$ which then rapidly hydrolyzes: $NCO^- + OH^- + H_2O \rightarrow NH_3 + CO_3^{2-}$. At elevated temperatures CO_3^{2-} and HCO_3^- ions are released and, by controlling the release rate, aragonite preferentially forms by homogeneous precipitation. Adding leucine to the urea/$CaCl_2$ solution provides additional bonding sites for Ca_2^+ ions in the form of the amino acid carboxyl group and is clearly a factor in the stabilisation of the vaterite phase, possibly through charge balancing of the amino acid's NH_2 group with the carbonate ions. The SEM results suggest preferential bonding of the amino acid via the Ca-carboxyl mechanism to the vaterite *a*-*b* plane, such that growth occurs on those faces perpendicular to the *c*-axis. At high concentrations coverage is presumably more extensive, with the macroscopic morphology (Fig. 1j) appearing to reflect the 6-fold symmetry of the hexagonal unit cell. There appears to be a strong relationship between morphology, crystallite size and internal strain in that for both phases the largest crystallites coincide with the highest strain. For aragonite, SEM imaging shows the greatest extent of both branching and branch size occurs for 4 g/L leucine concentration (Fig. 1c). At higher concentrations, the X-ray data suggests the internal strain is reduced by the precipitation of smaller crystallites and is evident in the SEM images (Fig.

1d-e). The lattice parameters of both phases in the presence of the amino acid are distorted compared to those precipitated in its absence. Similar results with respect to geologic aragonite have been reported for biomineralised aragonites in mollusc shells and coral skeletons [17, 21] and attributed to the presence of macromolecules used to template structural growth. Altered strain states are thus likely to be a universal feature of macromolecule directed templating, biological or otherwise, and will alter the functional properties such as mechanical strength [22] or thermal stability [17] of the templated material under different conditions/environments and could, for example, be the underlying reason for differences in the aragonite-to-calcite transition temperature between aragonites of biogenic and non-biogenic origin [23].

Acknowledgments
This work was carried out with the support of Diamond Light Source. The authors would like to express their thanks to Jonathan Potter (Diamond) for technical assistance.

References

[1] Lijima S 1991 *Nature* **354** 56
Braun P V, Osenar P and Stupp S I 1996 *Nature* **380** 325
Ohara P C, Heath J R and Gelbart W M 1997 *Angew. Chem. Int. Ed. Engle* **36** 1078
[2] Lippmann F 1973 *Sedimentary Carbonate Minerals vol. 6* (Springer-Verlag: Berlin)
[3] Kitamura M J 2001 *Colloid Interface Sci.* **236** 318
[4] Lowenstam H A and Weiner S 1989 *On Biomineralization* (Oxford University Press: New York)
Mann S 1988 *Nature* **332** 119
[5] Tanaka H, Horiuch H and Ohkubo T 1988 *Gypsum Lime (Sekko to Sekkai)* **216** 60
Kojima Y, Sadotomo A T and Arai Y 1992 *J. Ceram. Soc. Jpn.* **100** 1145
[6] Dalas E, Klepetsanis P and Koutsoukos P G 1999 *Langmuir 15* 8322
Stipp S L, Hochella M F, Parks G A and Leckie J O 1992 *Geochim. Cosmochim. Acta* **56** 1941
Richter A, Petzold D, Hofman H and Ullrich B 1995 *Chem. Tech.* **6** 306
Hostomsky J and Jones A G 1993 *Industrial Crystallization '93* 2-055
Garcia-Carmona J, Gomez-Morales J, Rodriguez-Clemente R 2003 *J. Cryst. Growth* **249** 561
[7] Chiroff R T, White R A, White E W, Weber J N and Roy D 1977 *J. Biomed. Mater. Res.* **11** 165
Stupp S I and Braun P V 1997 *Science* **277** 1242
[8] Wang W, Wang G, Liu Y Zheng C and Zhan Y 2001 *J. Mater. Chem.* **11** 1752
[9] Wang L, Sondi I and Matijevic E 1999 *J. Colloid Interface Sci.* **218** 545
[10] Tang C C, Thompson S P, Parker J E, Lennie A R, Azough F and Kato K 2009 *J. Appl. Cryst.* **42** 255
[11] Rutt H N and Nicola J H 1974 *J. Phys.* C **7** 4522
[12] Konoyannis C G and Vagenos N V 2000 *Analyst* **125** 251
[13] Frech R, Wang E C and Bates J B 1980 *Spectrochim. Acta* A **36** 915
Coutre L 1947 *Ann. Phys.* **2** (séerie 12) 5
[14] Manoli F and Dalas E 2000 *J. Cryst. Growth* **217** 416
Spanos N and Koutsoukos P G 1998 *J. Cryst. Growth* **191** 783
[15] Marsh M E 1989 *Biochemistry* **28** 339
Browen C E and Tang H 1996 *Comp. Biochem. Physiol.* **115A** 269
[16] Thompson S P, Parker J E, Potter J, Hill T P, Birt A, Cobb T M, Yuan F and Tang CC 2009 *Rev. Sci. Instrum.* **80** 075107
[17] Parker J E, Thompson S P, Lennie A R, Potter J and Tang CC 2010 *CrystEngComm* **12** 1590
[18] Orme C A, Noy A, Wierzbicki A. McBride M T, Grantham M, Teng H H, Dove P M and DeYoreo J J 2001 *Nature* **411** 775
[19] Botta O, Martins Z and Ehrenfreund P 2007 *Meteor. Planet. Sci.* **42** 81
[20] Porter S M 2007 *Science* **316** 1302
[21] Pokroy B, Quintana J P, Caspi E N, Berner A and Zolotoyabko E 2004 *Nat. Mater.* 3, 900
Pokroy B, Fitch A N, Lee P L, Quintana J P, Caspi E N and Zolotoyabko E 2006 *J. Struct. Biol.* 153, 145
Stolarski J, Przenioslo R, Mazur M and Brunelli M 2007 *J. Appl. Cryst.* 40, 2
[22] Eichorn S J, Scurr D J, Mummery P M, Golshan M, Thompson S P and Cernik R J 2005 *J Mater. Chem.* **15**, 947
[23] Stalport F, Coll P, Szopa C, Person A, Navarro-González R, Cabane M, Ausset P and Vaulay M J 2007 *Geophys. Res. lett.* **34**, L24102

Large-scale simulations with distributed computing: Asymptotic scaling of ballistic deposition

Bahman Farnudi
Institute for Advanced Studies in Basic Sciences (IASBS), Zanjan 45137-66731, Iran
E-mail: farnudi@iasbs.ac.ir

Dimitri D Vvedensky
The Blackett Laboratory, Imperial College London, London SW7 2AZ, United Kingdom
E-mail: d.vvedensky@imperial.ac.uk

Abstract. Extensive kinetic Monte Carlo simulations are reported for ballistic deposition (BD) in (1 + 1) dimensions. The large system sizes L observed for the onset of asymptotic scaling ($L \simeq 2^{12}$) explains the widespread discrepancies in previous reports for exponents of BD in one and likely in higher dimensions. The exponents obtained directly from our simulations, $\alpha = 0.499 \pm 0.004$ and $\beta = 0.336 \pm 0.004$, capture the exact values $\alpha = \frac{1}{2}$ and $\beta = \frac{1}{3}$ for the one-dimensional Kardar–Parisi–Zhang equation. An analysis of our simulations suggests a criterion for identifying the onset of true asymptotic scaling, which enables a more informed evaluation of exponents for BD in higher dimensions. These simulations were made possible by the Simulation through Social Networking project at the Institute for Advanced Studies in Basic Sciences in 2007, which was re-launched in November 2010.

1. Introduction

Ballistic deposition (BD) is a standard model for surfaces driven by the stochastic deposition of new material. Formulated as a model for colloidal aggregation [1, 2], BD is the prototypical model of nonconserved growth, in which the sum (or integral) of heights over the substrate is not equal to the material deposited, in this case because of void formation. The asymptotic behaviour of BD and other models of driven fluctuating interfaces, as exemplified by the variance $W(L,t)$ of the surface profile, typically exhibit dynamic scaling [3, 4]:

$$W(L,t) = \Big[\langle h^2\rangle - \langle h\rangle^2\Big]^{1/2} \sim L^\alpha f(t/L^z)\,. \tag{1}$$

Here, $h(\mathbf{x},t)$ is the surface height at position $\mathbf{x}$ and time t, L is the lateral viewing scale, α is the roughness exponent, z is the dynamic exponent, and f is a scaling function. At early times ($t \ll L^z$), $f(x) \sim x^\beta$, where $z = \alpha/\beta$, so $W \sim t^\beta$. For long times ($t \gg L^z$), $f \to$ constant, so the saturated roughness $W_{\text{sat}} \sim L^\alpha$.

In one dimension (1D), BD is generally believed to be in the same universality class as the Kardar-Parisi-Zhang equation [5],

$$\frac{\partial u}{\partial t} = \nu\nabla^2 u + \lambda(\nabla u)^2 + \xi\,, \tag{2}$$

where $u(\mathbf{x}, t)$ is the deviation of the height from its mean at position $\mathbf{x}$ and time t on a d-dimensional surface, and ξ is a Gaussian noise with mean zero and covariance

$$\langle \xi(\mathbf{x}, t) \, \xi(\mathbf{x}', t') \rangle = 2D\delta(\mathbf{x} - \mathbf{x}')\delta(t - t') \,. \tag{3}$$

Nevertheless, despite being among the first surface growth models to be studied with KMC simulations [3], the past 20 years have seen a wide scatter of estimates for the scaling exponents of BD [4, 6, 7]. In higher dimensions, the relationship between BD and the KPZ equation is considerably more problematic [8–20]. It has even been suggested [7] that BD and the KPZ equation do not belong to the same universality class on substrate dimensions greater than one.

In this paper, we report the results of extensive kinetic Monte Carlo (KMC) simulations for BD on 1D lattices with up to $2^{20} = 1,048,576$ sites, well beyond the onset of saturation for up to 10^6 independent realisations. This enables us to observe the slow approach to asymptotic scaling in this model and to obtain accurate estimates of the exponents, which agree with those of the KPZ equation. An analysis of these simulations reveals the reason behind the widespread discrepancies in previous reports of exponents for BD and provides a criterion that signifies when a system has entered the asymptotic scaling regime. The latter result enables a more informed exploration of the scaling behaviour of BD in higher dimensions, where simulations are substantially more computationally demanding.

The outline of this paper is as follows. The rules of BD are described in Sec. 2, where we also discuss some of the properties of the surfaces produced by this model. Section 3 provides a brief summary of the Simulation through Social Networking project, which made possible the extensive simulations and the highly accurate results reported in this paper. The estimates for exponents, corrections to scaling, and scaling functions are presented and discussed in Sec. 4. We summarise our findings and draw conclusions in Sec. 5.

2. Ballistic Deposition

In the classical BD model, a particle impinges on a randomly chosen lattice site and irreversibly attaches to the first vertical or lateral nearest neighbour encountered. Figure 1 illustrates how this rule is implemented on a 1D lattice for three types of incident particles. The incident particle A attaches to a *lateral* nearest neighbour, which results in a void below the attached position. As BD permits no further relaxation, for example, through surface diffusion, this void remains, so the resulting film has porosity. Hence, the volume of this film, as measured by the sum of column heights on each lattice site, exceeds that of the deposited material, i.e. growth is non-conserved. Particles B and C attach to *vertical* nearest neighbours, so no voids are produced. However, the local height configuration of the deposited position of B will result in voids upon deposition on each adjacent site. The BD rules capture the essence of the sedimentation of small

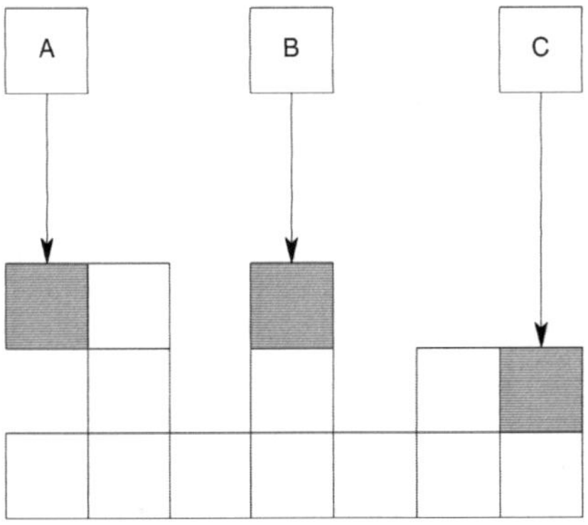

Figure 1. Schematic diagram of BD for three incident particles and their deposition sites (shown shaded). A particle impinges on a randomly chosen site and attaches to the first nearest vertical or lateral nearest neighbour encountered. Particle A attaches to a lateral nearest neighbour, while particles B and C attach to vertical nearest neighbours. As the void between the attached particle A and the surface cannot be filled by any relaxation rules within the model, the film produced by BD has porosity.

material particles under gravity in a liquid [1], or physical and chemical vapour deposition of thin films [21–23].

The updating algorithm of BD on a d-dimensional substrate based on the rules in Fig. 1 is expressed in terms of the integer height $h_i(n)$ at lattice position i after n depositions as

$$h_i(n+1) = \max(h_{i-1}(n), h_i(n)+1, h_{i+1}(n)), \tag{4}$$

for $1 = 1, 2, \ldots, L^d$, where L is the linear size of the system and $\max(x, y, z)$ yields the maximum of its three arguments. The initial conditions are usually taken as a flat substrate, with all heights set equal to zero: $h_i(0) = 0$.

3. Simulation Method

3.1. Simulation through Social Networking

The massive computational resources required for the simulations reported here relied upon unconventional "overnight office computing," which was developed into a fully-fledged social computing network. Social networking skills were employed in the development of the Simulation through Social Networking (STSN) project, in which 120-130 computers were utilised for some nine months to drive the KMC simulations of even the largest lattices deep into the saturation region.

Social computing is not a new concept. "Volunteer computing" started in the mid-1990s [24] and, by the end of that decade, expanded into the now well-known @home projects in which hundreds of thousands of personal computers work together on a single task [25]. STSN is a combination of volunteer and grid computing. Further details may be found on the project website [26] and in Ref. [27].

The success of STSN relies on exclusivity, quality, and the human touch. STSN operates on the principles characterising a family business, i.e. direct personal contact, customer care, transparency, high quality, and long-term customer retention strategy. Networks are built slowly and carefully to enhance data security through responsible volunteering and to ensure long-term participant involvement through mutual trust and benefit. To this end, a re-launch of the STSN project is currently underway with the cooperation of the students in Information and Communication Technology (ICT) at IASBS as STSN team members and team leaders. The continuity and long-term success of similar projects can be guaranteed only by all-out participation of ICT experts.

3.2. Random Number Generator

In view of the pitfalls associated with hidden correlations in random number generators [28], and the suggestion [29] that BD may be especially sensitive to such correlations, we have used the Mersenne Twister MT19937 random number generator [30, 31]. This pseudorandom number generator has a period of $2^{19937} - 1$, an output that is uniformly distributed in 623 dimensions, implying negligible serial correlation, and has passed the most stringent statistical tests [32]. These properties, and its computational efficiency, make the Mersenne Twister eminently suitable for large-scale KMC simulations.

4. Results

4.1. The roughness exponent

Figure 2 shows the determination of the roughness exponent α from the slope of a straight-line fit to a log-log plot of $W_{\rm sat}$ against L for the data in Table 1. $W_{\rm sat}$ was calculated by taking samples every 4τ to τ monolayers (MLs), depending on the lattice size, where $\tau \sim L^{3/2}$ is the relaxation time of BD [33], to ensure that the data points were statistically independent, and taking binned horizontal averages to alleviate any drift in the data over time [4]. Simulations

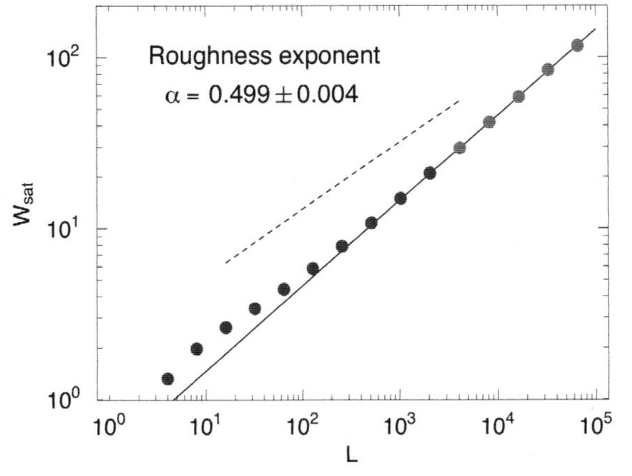

Figure 2. Log-log plot of $W_{\rm sat}$ against L for the data in Table 1. Error bars are of the order of the symbol size or smaller. The roughness exponent in (5) obtained from the slope of a linear fit over the range $L = 2^{12}$–2^{16} (red dots) is in excellent agreement with the KPZ value of $\frac{1}{2}$. The broken line indicates the approximate range of simulation data used in previous work ($L = 2^4$–2^{12}), which leads to a substantial underestimate of α.

Table 1. Saturated roughness $W_{\rm sat}$ on 1D lattices of sizes $L = 2^n$ for $n = 2, 3, \ldots, 16$ obtained from KMC simulations with the indicated number of independent realisations. The time to saturation is $t_\times$ and the simulations were continued into the saturation regime for a time t_s. Where none is indicated, the calculated error is smaller than 10^{-2}.

L	$W_{\rm sat}$	Error	Realisations	$t_\times$ (MLs)	t_s (10^6 MLs)
4	1.33	–	100,000	–	4
8	1.97	–	100,000	–	1
16	2.63	–	93,000	–	1
32	3.40	–	100,000	19.75	1
64	4.41	–	10,000	44.5	1
128	5.82	–	10,000	108	1
256	7.85	–	10,000	259	1
512	10.73	–	10,000	675	1
1,024	14.93	–	7,486	1900	1
2,048	20.95	0.03	1,010	5200	4.1
4,096	29.39	0.05	841	13700	4.1
8,192	41.58	0.1	253	38000	4.1
16,384	58.69	0.2	724	106500	10
32,768	84.09	0.2	407	297000	10
65,536	116.55	1.0	249	770000	10

for $L = 16,384$, $L = 32,768$, and $L = 65,536$ were carried out for up to 33.5×10^6 layers, but many data points were excluded from the averaging because of the stretched exponential tail in the probability distribution of $W_{\rm sat}$ [34].

Most apparent from Fig. 2 is that an accurate estimate of α necessitates using system sizes beyond $L = 2^{11}$ because of the slow approach to asymptotic scaling. For the smallest systems there is appreciable deviation from the asymptotic behaviour because of finite-size effects. But even intermediate system sizes ($L = 2^9$–2^{11}) adversely affect the estimate of α. Our estimate of

$$\alpha = 0.499 \pm 0.004 \,, \tag{5}$$

obtained from the fit between $L = 2^{12}$ and $L = 2^{16}$, captures the value of $\alpha = \frac{1}{2}$ for the 1D KPZ equation [5]. To our knowledge, this is the first time that the roughness exponent of BD has been shown to agree with the KPZ value (within the error bounds) directly from an analysis of simulation data and without any scaling corrections [4, 6, 7]. To put our results into perspective, the line from $L = 2^4$ to $L = 2^{12}$ in Fig. 2 indicates the approximate range and slope of the simulation data used in previous work for calculating α. As these data points and the fit make clear, the use of system sizes smaller than $L = 2^{12}$ results in an appreciable underestimate of α.

An alternative method of quantifying the approach to asymptotic scaling, as suggested by Reiss [6], is to first calculate L-dependent effective roughness exponents α_L defined by

$$\alpha_L = \frac{\ln[W_{\mathrm{sat}}(L)/W_{\mathrm{sat}}(\frac{1}{2}L)]}{\ln 2} \,. \tag{6}$$

These are easily determined from the data in Table 1 and enable the correction-to-scaling form

$$\alpha_L \approx \alpha + AL^{-\Delta} \,, \tag{7}$$

in which Δ is a correction-to-scaling exponent and A is a constant. These calculations are not displayed, as the data in Fig. 2 already give a clear indication of the slow approach to asymptotic scaling of BD.

4.2. The growth exponent

The slow approach of α to its asymptotic value suggests that a similar trend should be expected for β. However, in contrast to the calculation of α, which requires simulations that extend well into the saturation regime to obtain accurate values of W_{sat} (Table 1), the obstacle for determining β is not computational overhead *per se*. Rather, the difficulty lies in delineating the limits of the growth region [6] over which $W \sim t^\beta$. Our results are displayed in Fig. 3 as a plot of β against $\log(1/L)$ of the data compiled in Table 2.

The most striking trend in these data is the slow approach to the asymptotic (KPZ) value of $\beta = \frac{1}{3}$ with increasing system size, which is even more pronounced than that in Fig. 2. Lattice sizes in excess of $L = 2^{18} = 262,144$ are needed to obtain values of β to within 1% of the exact value. This observation can be quantified by fitting the data to an expression of the form

$$\beta(1/L) = \beta + a[\log(1/L)]^b \,, \tag{8}$$

where the asymptotic value is β, and a and b are fitting parameters. Optimisation over the range from $L = 2^{10}$ to $L = 2^{17}$, as shown in Fig. 4, produces

$$\beta = 0.336 \pm 0.005 \,. \tag{9}$$

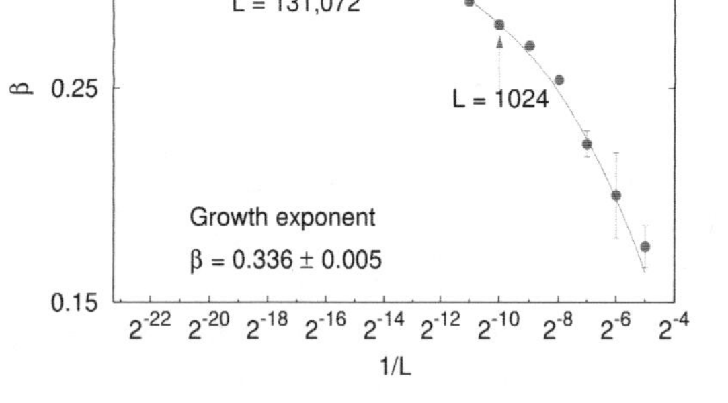

Figure 3. Log-linear plot of β against $\log(1/L)$ based on the data in Table 2. The points for the largest lattice sizes (indicated in black) have too few realisations to provide accurate estimates and are included for completeness only. The curve is a fit to the data between $L = 1024$ and $L = 131,072$ of (8). The asymptotic value $\beta = 0.336 \pm 0.005$ obtained from this fit agrees to within errors with the KPZ value of $\alpha = \frac{1}{3}$.

Table 2. The growth exponent β_L and the associated error for lattice sizes $L = 2^n$, with $n = 5, \ldots, 20$ obtained from KMC simulations with the indicated number of realisations. The corresponding growth time t_g over which β_L is calculated, and its error, are also shown.

L	β_L	Error	Realisations	t_g (MLs)	Error
32	0.176	0.01	1,000,000	2	1
64	0.20	0.02	1,000,000	4	5
128	0.224	0.006	1,000,000	10	5
256	0.254	0.002	1,000,000	20	5
512	0.270	0.002	1,000,000	80	10
1,024	0.280	0.002	100,000	284	10
2,048	0.2907	0.001	100,000	700	10
4,096	0.302	0.001	100,000	1,400	100
8,192	0.308	0.001	54,000	4,200	500
16,384	0.312	0.002	810	7,936	1,000
32,768	0.317	0.002	407	23,000	7,000
65,536	0.322	0.0005	800	61,000	4,000
131,072	0.325	0.002	105	130,000	10,000
262,144	0.323	0.01	9	260,000	50,000
524,288	0.334	0.005	2	520,000	100,000
1,048,576	0.332	0.005	2	1,040,000	200,000

Although the data at the largest lattices (indicated by black symbols in Fig. 3) lack the accuracy for an extrapolation to infinite sizes, and were not included in the fitting, the best fit lies within their error bars. Several ranges of data (including the entire range) were used to fit the above form, and in all cases the asymptotic values of β obtained for different fit ranges fluctuate around the KPZ value.

4.3. Correction to scaling

The growth times t_g and the corresponding values of β_L are compiled in Table 2. The most striking trend in these data is the slow approach to the asymptotic value of $\beta = \frac{1}{3}$ [5], which is

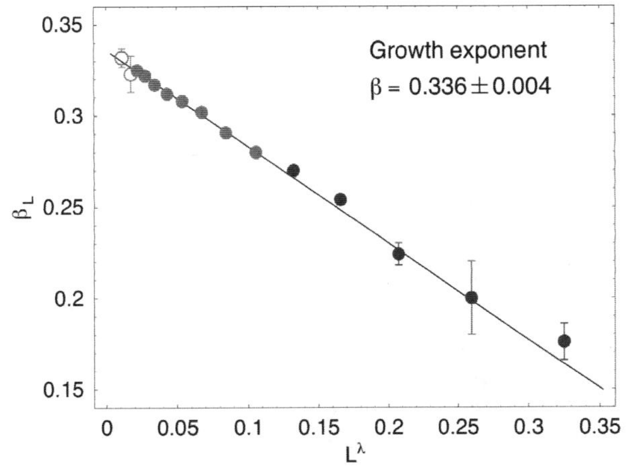

Figure 4. Effective growth exponents β_L plotted against L^λ for the data in Table 2 with optimised parameters in the scaling form (10). Only the red points, which correspond to system sizes $L = 2^{10}$–2^{17}, had large enough growth regions to be included in the fit. The empty red circles, which correspond to $L = 2^{18}$ and $L = 2^{20}$, were excluded from the fit because of insufficient accuracy. The data point corresponding to $L = 2^{19}$ has been omitted for clarity.

even more pronounced than that in Fig. 2. Lattice sizes in excess of $2^{18} = 262,144$ are needed to obtain values of β_L to within 1% of the exact value. These observations can be quantified by fitting the data to the scaling form [6]:

$$\beta_L = \beta + \frac{A}{L^\lambda}, \tag{10}$$

where β is the asymptotic value of the growth exponent, λ determines the correction to the scaling with system size, and A is a constant. The data in Table 2, shown in Fig. 4, for $L = 2^{10}$–2^{17} yields $A = -0.530 \pm 0.128$, $\lambda = 0.324 \pm 0.043$, and

$$\beta = 0.336 \pm 0.004\,. \tag{11}$$

The data for the largest lattices in Table 2 were excluded from the fitting because of insufficient accuracy, but the best fit still lies within their error bars.

4.4. The growth time

In measuring the growth times t_g in Table 2, we have expanded the criteria suggested by Reis in [6] by using floating "beginning" and "end" points for each lattice size and imposing a maximum-time criterion. Figure 5 shows a log-log plot of t_g against L with a straight-line fit of the data points from $L = 2^{16}$ to $L = 2^{20}$. The significance of this graph is that only when its slope, which we call the "growth-time exponent" and denote by γ, reaches unity can we say with confidence that we have reached the asymptotic regime. The power-law growth regime for a system of size L is influenced by two factors. (i) The transient regime, during which the system first follows random deposition, before the BD rules determine the growth characteristics. This is largely L-independent. (ii) The approach to the saturation regime, which corresponds to a deviation from power-law behaviour of the surface width. This is strongly L-dependent.

For small system sizes, the transient and saturation regimes are sufficiently close together that the growth exponent deviates significantly from the true asymptotic BD value. In effect, the system never enters a fully-developed BD growth regime. But, as L increases, the time between the transient and saturation regimes increases and eventually these are sufficiently separated that true BD behaviour can develop. It is for these values of L that we observe the slope of unity in Fig. 5.

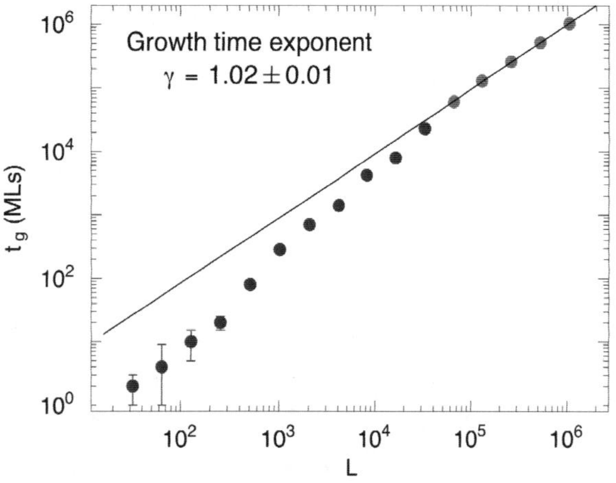

Figure 5. (Color online) Log-log plot of t_g against L for the data in Table 2. The growth time exponent $\gamma = 1.002 \pm 0.004$ obtained from the slope of a linear fit to the data points from $L = 2^{16}$–2^{20} (red dots) approaches unity very closely indicating that the system has entered the asymptotic regime.

5. Conclusion

We have used massive KMC simulations of BD onto 1D surfaces to demonstrate the slow approach of this system to the asymptotic scaling regime. The roughness and growth exponents α and β were shown to converge to the exact values obtained from the KPZ equation, with systems of up to 2^{20} sites required for a clear indication of convergence. We have used only a single random number generator, the Mersenne Twister MT19937, so our results shed no light on the reason for the slow convergence. However, given the long period of this random number generator and its other statistical and operational properties, this is likely an intrinsic property of BD.

We conclude with a few remarks about the implications of our results for simulations of BD in higher dimensions. Preliminary simulations on two-dimensional substrates suggest that the slow convergence toward asymptotic scaling behaviour persists for higher-dimensional substrates, but we have not yet determined the rate of convergence. This presents a significant computational obstacle for determining the exponents of BD and in establishing the precise relation to the KPZ equation. In this regard, Fig. 5 is vital because it indicates the whether the system has entered the asymptotic regime. This will be taken up in a future publication.

B. F. would like to express his deep gratitude to IASBS staff and students for making the STSN project possible. He is especially grateful to M. R. Kolahchi for thoughtful discussions.

References

[1] M. J. Vold, *J. Colloid Interface Sci.* **18**, 684 (1963).
[2] D. N. Sutherland, *J. Colloid Interface Sci.* **22**, 300 (1966).
[3] F. Family and T. Vicsek, *J. Phys. A: Math. Gen.* **18**, L75 (1985).
[4] A.-L. Barabási and H. E. Stanley, *Fractal Concepts in Surface Growth* (Cambridge University Press, Cambridge, 1995).
[5] M. Kardar, G. Parisi, and Y.-C. Zhang, *Phys. Rev. Lett.* **56**, 889 (1986).
[6] F. D. A. Aarão Reis, *Phys. Rev. E* **63**, 056116 (2001).
[7] E. Katzav and M. Schwartz, *Phys. Rev. E* **70**, 061608 (2004).
[8] M. Schwartz and S. F. Edwards, *Europhys. Lett.* **20**, 310 (1992).
[9] J. P. Bouchaud and M. E. Cates, *Phys. Rev. E* **47**, R1455 (1993).
[10] J. P. Doherty *et al.*, *Phys. Rev. Lett.* **72**, 2041 (1994).
[11] Y. Tu, *Phys. Rev. Lett.* **73**, 3109 (1994).
[12] M. A. Moore *et al.*, *Phys. Rev. Lett.* **74**, 4257 (1995).
[13] T. Blum and A. J. McKane, *Phys. Rev. E* **52**, 4741 (1995).
[14] M. Lässig and H. Kinzelbach, *Phys. Rev. Lett.* **78**, 903 (1997).
[15] J. K. Bhattacharjee, *J. Phys. A* **31**, L93 (1998).
[16] M. Lässig, *Phys. Rev. Lett.* **80**, 2366 (1998).
[17] C. Castellano, M. Marsili, and L. Pietronero, *Phys. Rev. Lett.* **80**, 3527 (1998).
[18] E. Marinari, A. Pagnani, and G. Parisi, *J. Phys. A: Math. Gen.* **33**, 8181 (2000).
[19] F. Colaiori and M. A. Moore, *Phys. Rev. Lett.* **86**, 3946 (2001).
[20] F. D. A. Aarão Reis, *Phys. Rev. E* **69**, 021610 (2004).
[21] H. J. Leamy, G. H. Gilmer, and A. G. Dirks, *Current Topics in Materials Science* edited by E. Katdis, page 309, (Amsterdam, North-Holland, 1980).
[22] Q. X. Jia and T. M. McCleskey and A. K. Burrell and Y. Lin and G. E. Collis and H. Wang and A. D. Q. Li and S. R. Foltyn, Polymer-assisted deposition of metal-oxide films, *Nature Materials* **3**, 529 (2006).
[23] Elshabini-Riad, A. and Barlow III, F. D., *Thin Film Technology Handbook* (McGraw-Hill, New York, 1998).
[24] D. Anderson, *ACM Queue* **3**(6), 18 (2005).
[25] Berkeley Open Infrastructure for Network Computing: `http://boinc.berkeley.edu`.
[26] `http://www.iasbs.ac.ir/els/farnudi`.
[27] B. Farnudi, *Scaling and Universality in Deposition Models of Growth*, PhD thesis, (University of London, 2010, unpublished). Available online at: `http://www.imperial.ac.uk/research/cmth/research/theses/`.
[28] A. M. Ferrenberg, D. P. Landau, and Y. J. Wong, *Phys. Rev. Lett.* **69**, 3382 (1992).
[29] R. M. D'Souza, Y. Bar-Yam, and M. Kardar, *Phys. Rev. E* **57**, 5044 (1998).
[30] M. Matsumoto and Y. Kurita, *ACM Trans. Model. Comput. Simul.* **2**, 179 (1992).

[31] M. Matsumoto and T. Nishimura, *ACM Trans. Model. Comput. Simul.* **8**, 3 (1998).
[32] P. L'Ecuyer and R. Simard, *ACM Trans. Math. Softw.* **33**, art. 22 (August 2007).
[33] F. D. A. Aarão Reis, *Physica A* **316**, 250 (2002).
[34] V. G. Miranda and F. D. A. Aarão Reis, *Phys. Rev. E.* **77**, 031134 (2008).

The effects of electron-phonon interactions on bandgaps.

J.P. Hague

Department of Physics and Astronomy, The Open University, Walton Hall, Milton Keynes MK7 6AA, UK

E-mail: J.P.Hague@open.ac.uk

Abstract. I compute the spectral functions of a 1D Holstein polaron moving in a modulated potential, to examine the effects of electron-phonon interactions on band gaps. The imaginary time Green function is computed using diagrammatic quantum Monte Carlo, which exactly sums the diagrammatic series. From the imaginary time Green function, spectral functions are computed. The electron-phonon interaction flattens the electronic dispersion and leads to an increase in the gap at momentum $k = \pi/2$. At strong coupling, polaron sidebands form in the gap. These results demonstrate the strong effect that electron-phonon interactions can have on band gaps.

1. Introduction

The size of bandgaps is a crucial feature in developing semiconductor devices, and is especially important in LED design, semiconductor lasers and influences the usefulness of semiconductors to make transistors. An open question persists about how interactions influence band gaps, with the effects of electronic correlations on band insulators remaining an important topic in condensed matter theory [1].

The Hamiltonian considered here represents a Holstein model [2] which is augmented with a modulated potential of magnitude Δ,

$$H = -t \sum_{\langle ij \rangle} c_i^\dagger c_j + \sum_i \Delta(-1)^i n_i + g\omega_0 \sum_i n_i(b_i^\dagger + b_i) + \omega_0 \sum_i (b_i^\dagger b_i + 1/2) \quad (1)$$

Here, electrons hop between sites because of an overlap integral of magnitude t. The modulated potential represents the energy shift in a superstrate due to a commensurate substrate with two types of atoms, and the Holstein electron-phonon coupling term represents an interaction of strength g with a polarisable substrate. Phonons are created on site i with the operator $b_i^\dagger$, $c_i^\dagger$ creates electrons, n_i is the number operator for electrons and ω_0 is the frequency of the Einstein phonons described by the last term in the Hamiltonian. Even indexed sites with energy offset $+\Delta$ will be denoted as A sites and odd index sites with energy offset $-\Delta$ will be denoted as B sites. This model is similar to the ionic Hubbard model [1] with the Coulomb repulsion replaced by an electron-phonon interaction. It is common to introduce a dimensionless electron-phonon coupling, which is defined here as $\lambda = g^2\omega_0/2t$.

In the absence of electron-phonon interaction, the Hamiltonian leads to energy bands of the form $E_{A,B}(k) = \pm\sqrt{\gamma^2(k) + \Delta^2}$ where $\gamma(k) = -2t\cos(k)$ i.e. there is an energy gap with magnitude 2Δ at momentum $k = \pi/2$ as illustrated in Figure 1. It is the aim of this paper to demonstrate the influence of electron-phonon interactions on this band gap.

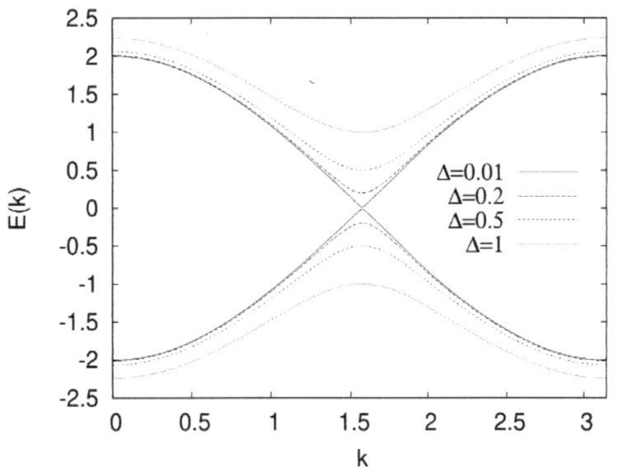

Figure 1. Bare dispersion at $\lambda = 0$ showing various modulated potential sizes, Δ.

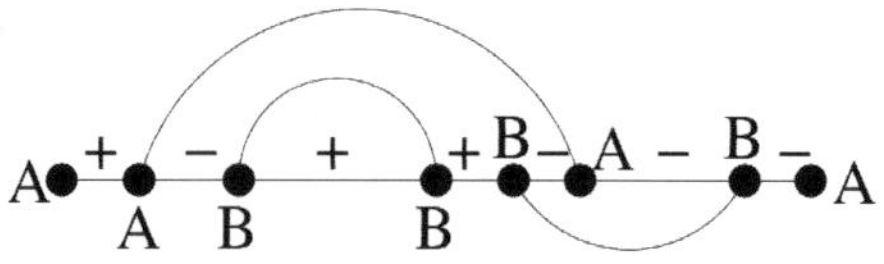

Figure 2. Example of a diagram in the expansion. Curved lines represent phonons and the straight line the electron. Filled circles represent interaction vertices. Also shown are the signs of the constituent propagators. It can be seen that any possible minus sign problems associated with the anomalous propagators cancel because there are always an even number of such propagators.

2. Method

I use a form of diagrammatic quantum Monte Carlo to compute the exact imaginary time Green functions for the polaron [3, 4]. A small modification is required to deal with a binary basis. In the presence of a modulated potential, it is appropriate to represent non-interacting Green functions as a matrix. For imaginary time, $\tau > 0$ and at absolute zero, the Green functions are defined as follows:

$$\mathbf{G} = \begin{pmatrix} G_{AA} & G_{AB} \\ G_{BA} & G_{BB} \end{pmatrix} = \begin{pmatrix} -\langle c_A(\tau)c_A^\dagger(0)\rangle & -\langle c_A(\tau)c_B^\dagger(0)\rangle \\ -\langle c_B(\tau)c_A^\dagger(0)\rangle & -\langle c_B(\tau)c_B^\dagger(0)\rangle \end{pmatrix}, \tag{2}$$

where,

$$G_{AA} = \frac{1}{2}\left[\exp(-E_A\tau) + \exp(-E_B\tau)\right] + \frac{1}{2}\frac{\Delta}{\sqrt{\gamma^2+\Delta^2}}\left[\exp(-E_A\tau) - \exp(-E_B\tau)\right] \tag{3}$$

$$G_{BB} = \frac{1}{2}\left[\exp(-E_B\tau) + \exp(-E_A\tau)\right] + \frac{1}{2}\frac{\Delta}{\sqrt{\gamma^2+\Delta^2}}\left[\exp(-E_B\tau) - \exp(-E_A\tau)\right] \tag{4}$$

$$G_{AB} = G_{BA} = \frac{1}{2}\frac{\gamma}{\sqrt{\gamma^2+\Delta^2}}\left[\exp(-E_B\tau) - \exp(-E_A\tau)\right]. \tag{5}$$

For $\tau < 0$ and absolute zero, all Green functions are zero valued because the polaron only contains a single electron. In its most basic form, the algorithm proceeds by inserting and removing interaction lines into or from the electron propagator. Since $G(\tau < 0) = 0$, vertices are time ordered between times 0 and τ, where τ is the length of the diagram. This time ordering averts any major sign problems. The propagators of the Holstein interaction represent local phonons with a local interaction term, so the interaction lines have the form $\exp(-\omega_0\tau)\delta_{XX}$ where $X \in \{A, B\}$ represents sub-lattice type and δ_{XX} is the Kronecker delta function (i.e. phonons on A sites can only interact with electrons on site A). By fixing the lattice type of the vertices at the ends of the diagram and inserting multiple lines with both A and B end vertices, the full interacting Green function can be calculated.

It is worth noting that there is a unusual symmetry between the two non-interacting bands in the anomalous Green functions, G_{AB} and G_{BA}, such that reassigning A and B sites introduces a minus sign. This is not a problem, as there are always an even number of anomalous non-interacting Green functions in the full diagonal Green functions G_{AA} and G_{BB}. Therefore any negative signs arising from reassigning the bands cancel to give a net positive sign, as demonstrated by the example in Figure 2. The factor $\gamma(k)$ can be negative or positive depending on the momentum, however the average sign has been found to be $\langle s\rangle > 0.9999$ for all cases discussed here. The τ dependent Green function is measured here using the estimator given in Ref. [4] which is free of systematic errors.

3. Results

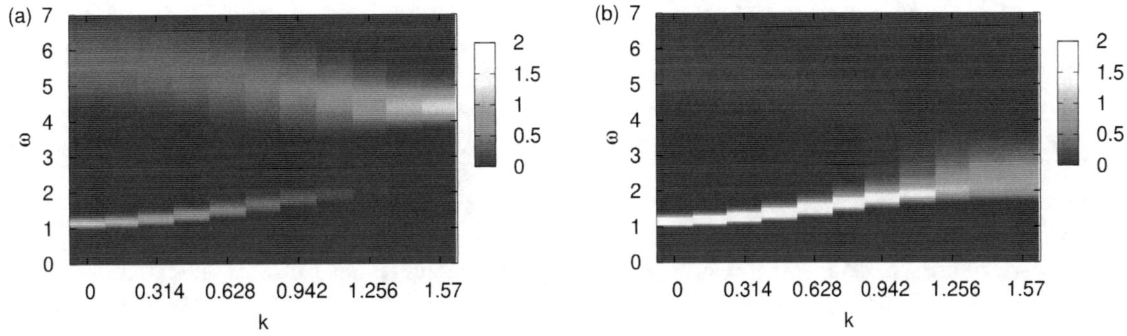

Figure 3. Momentum dependent electronic spectrum when $\lambda = 0.5$ for (a) band A electrons (b) B band electrons.

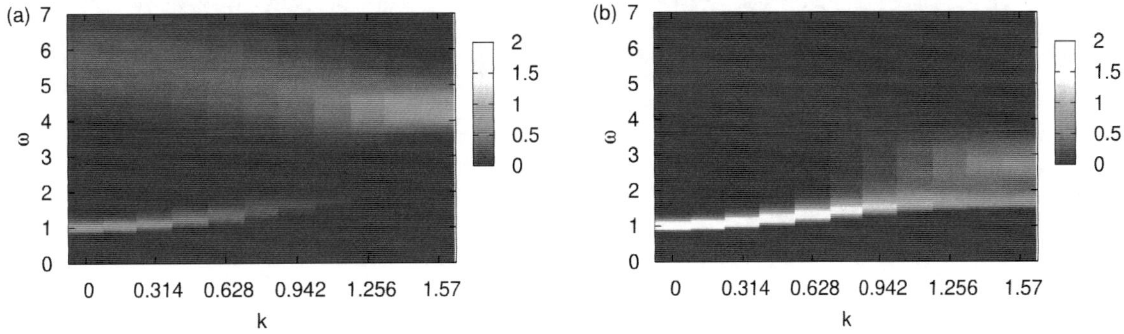

Figure 4. Momentum dependent electronic spectrum when $\lambda = 1.0$ for (a) band A electrons (b) B band electrons.

I briefly discuss some of the results obtained for this model. For illustration purposes, a large non-interacting half band gap $\Delta = 1$ is chosen so that the bands are clearly resolved. The Green functions G_{AA} and G_{BB} are computed for a range of momenta and λ values by making 10^9 measurements to obtain an accuracy of greater than 0.5% at 500 imaginary time points ranging from $\tau = 0$ to $\tau = 5$. To obtain spectral functions, a numerical procedure is used to invert the integral transform relating the τ-dependent Green function and the spectral function,

$$G(\tau) = \int_0^\infty \exp(-E\tau)A(E), \tag{6}$$

where $A(E)$ is the spectral function. In this way, the analytic continuation of the Green function is carried out. A continuous-frequency Monte Carlo approach is used to determine spectral functions, with sets of several thousand delta functions free to change frequency and magnitude. These are sampled according to the weight, $\exp(-\chi^2/\alpha)$. α is decreased until the spectrum meets the historic maximum entropy condition of $\chi^2 = N_d$ where N_d is the number of data points. It is anticipated that the more sophisticated stochastic analytic inference will be used for a future improvement in spectral function resolution [5].

I start by showing spectral functions for A and B type electrons at $\lambda = 0.5$ (Fig. 3) and $\lambda = 1.0$ (Fig. 4). For all figures, $\omega_0 = t = 1$. Two bands can be seen, with spectral weight at high energies for A type electrons, and at lower energies for B type electrons. There is a gap at $k = \pi/2$. Both spectra have the same basic form as shown in figure 1. As the electron-phonon interaction is increased from $\lambda = 0.5$ to $\lambda = 1.0$, there is a slight increase in the band gap, and some band flattening occurs. This band flattening has also been seen for polarons in the absence of a modulated potential [6]. There is an increase in the width of the excitation peaks indicating a decrease in quasi-particle lifetime, and a polaron side band can be seen forming above the electron band in figure 4(b) around $k = \pi/2$ which is the signature of an excitation consisting of an electron and a single phonon.

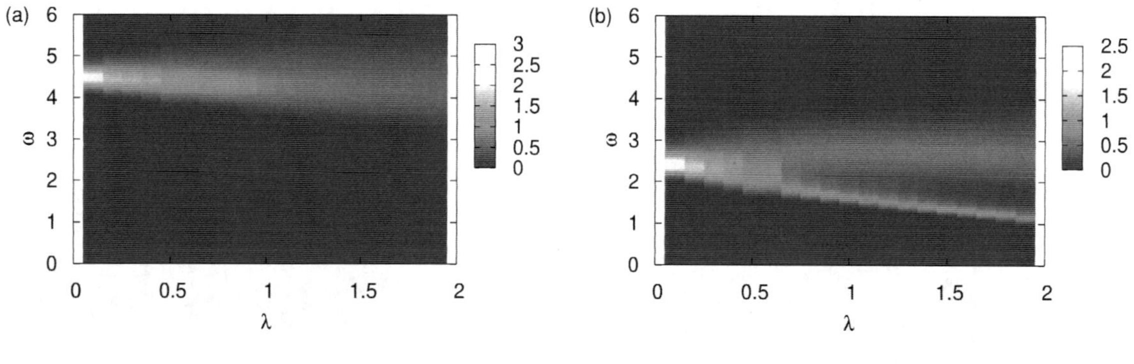

Figure 5. (a) Evolution of the spectral weight of A site electrons as the electron-phonon interaction λ is increased. (b) As panel (a) for B site electrons.

To examine the λ dependence of the gap more clearly, I show the evolution of the spectral functions at $k = \pi/2$ in figure 5. As the electron-phonon interaction increases, the energy of the B type band decreases more rapidly than the energy of the A type band, which leads to a band gap increase. The polaron side band can also be seen. Since a phonon needs to be created for decays from the top band into the polaron side band, it is anticipated that such transitions are rarer that those into the ground state. To show the gap more clearly, both A and B type spectra are shown on a waterfall plot in figure 6.

4. Summary and conclusions

In this paper, I computed spectral functions of a polaron in a 1D modulated potential, demonstrating that in one dimensional systems, electron-phonon interactions could be used

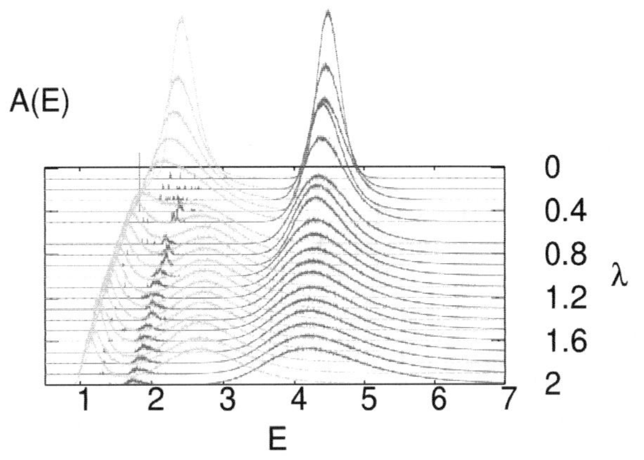

Figure 6. Spectral functions of A and B type electrons at $k = \pi/2a$. Note the increase in the band gap as the electron-phonon coupling is increased, and the polaron side band.

to modify substrate induced bandgaps. By analysing the spectral function, it was found that the electron-phonon interaction initially leads to an increase in the gap, although with stronger interactions polaron side bands form in the gap.

Photoemission results suggest that a bandgap of the order of 0.25eV is formed in graphene placed on a substrate of SiC [7] and a bandgap has been predicted for graphene on hexagonal boron nitride [8]. This raises the possibility that a similar electron-phonon mediated enhancement might be accessible for graphene. Work to explore this possibility is currently in progress.

Acknowledgments
I am pleased to acknowledge EPSRC grant no. EP/H015655/1 and useful discussions with Pavel Kornilovitch, Sasha Alexandrov, Mervyn Roy, Peter Maksym, Ed McCann, Nigel Mason and Nick Braithwaite.

[1] Garg A, Krishnamurthy H R and Randeria M 2006 *Phys. Rev. Lett.* **97**
[2] Holstein T 1959 *Ann. Phys.* **8** 325
[3] Prokof'ev N V and Svistunov B V 1988 *Phys. Rev. Lett.* **81** 2514
[4] Mishchenko A S, Prokof'ev N V, Sakamoto A and Svistunov B V 2000 *Phys. Rev. B* **62** 6317
[5] Fuchs S, Pruschke T and Jarrell M 2010 *Phys. Rev. E* **81** 056701
[6] Kornilovitch P E 1999 *Phys. Rev. B* **60** 3237
[7] Zhou S Y, Gweon G H, Fedorov A V, First P N, de Heer W A, Lee D H, Guinea F, Neto A H C and Lanzara A 2007 *Nature Materials* **6** 770
[8] Giovannetti G, Khomyakov P A, Brocks G, Kelly P J and van den Brink J 2007 *Phys. Rev. B* **76** 073103

Condensed Matter and Materials Physics Conference (CMMP10) IOP Publishing
Journal of Physics: Conference Series **286** (2011) 012033 doi:10.1088/1742-6596/286/1/012033

Disorder in materials with complex crystal structures: the Non-Local Coherent Potential Approximation for compounds with multiple sublattices

,

A Marmodoro and J B Staunton

Department of Physics, University of Warwick, Coventry CV4 7AL, UK

E-mail: A.Marmodoro@warwick.ac.uk

Abstract. Over the last few years the Non-Local Coherent Potential Approximation (NL-CPA) has been shown to provide an effective way to describe the electronic structure and related properties of disordered systems, where short-range order (SRO) and other local environment effects are important. Here we present its generalization to materials with multi-atom per unit cell lattices. The method is described using a Green function formalism and illustrated by an implementation for a simplified one-dimensional tight-binding model with substitutional disorder. This development paves the way for a natural reimplementation of the Korringa-Kohn-Rostoker (KKR) multiple scattering solution of Kohn-Sham equations for ab-initio calculations of real materials.

1. Introduction

A certain degree of disorder is a common feature in all real materials, and may have a very significant impact on their electronic structure and related properties. With the loss of translational invariance an analytical, exact solution becomes rapidly inaccessible even for simplified, small size model systems, and the typical challenges of a fully *ab-initio* density functional theory -based approach are also strongly enhanced.

The Coherent Potential Approximation (CPA) [1] has historically provided a powerful, fully self-consistent mean field approximation framework for the systematic evaluation of local disorder effects across a wide variety of solid state physics applications. In combination with the Korringa-Kohn-Rostoker multiple scattering method for solving the electronic problem of the Kohn-Sham equations [2],[3] it is possible to fully account for the impact of on-site deviations from periodicity. Bloch states, on the average, are correctly predicted to acquire a finite lifetime, and the conceptual tool of the Bloch spectral function is used to generalize the original band structure description in the presence of randomly distributed impurities.

A recent development of the method (NL-CPA) [4, 5, 6, 7, 8, 9] has enhanced its capabilities to also cover the effective evaluation of off-diagonal disorder necessary for an accurate treatment of short-range order (SRO) effects or dopant clustering, at a comparatively negligible computational cost with respect to similarly accurate supercell -based calculations [10],[11],[12].

This extension is based on a reinterpretation of the original CPA self-consistency conditions as the first order term only in a coupled expansion over a coarse grained set of $N_c \geq 1$ sites $\boldsymbol{R}_I$ and associated discrete cluster momenta $\boldsymbol{K}_n$ in reciprocal space [4]. These are selected to preserve the original point group symmetries of the underlying geometry [13], and have so far allowed successful investigations at a higher than single-site resolution of a range of simple crystal structures derived from a cubic lattice (*Strukturbericht* designation A_h, $A1$, $A2$), with application to metallic alloys [5],[14] and electrical conductivity studies [9],[15] where macroscopic observables are found to be strongly affected by local environment effects and formation of complexes.

On the other hand, the problem of how to handle the associated Brillouin zone coarsening for a system with many atom per unit cell has so far made the idea incompatible with an earlier development, designed to address multi-sublattice systems (MS-CPA) [16]. In such an approach the same original limitations of a local-only single site approximation still persist, but it is possible to investigate a broad range of selective chemical disorder scenarios, such as hydrogen absorption for energy storage[17], cation defects in transition metal oxides [18], as well as cases where the local order parameter may be for instance represented by the orientation of a local moment in a magnet[19],[20] or by strain and polarization in a ferroelectric [21].

We propose here an unification of the two extensions (MS-NL-CPA) to combine their respective merits for the efficient treatment of disorder and SRO effects in arbitrary complex structures. In order to aid the interpretation of investigations for the first real materials from an ab-initio DFT-KKR adaption, for which preliminary calculations are currently under evaluation, we present here the general method in a flexible Green function formalism and illustrate for better clarity the type of physics it can describe by an implementation for a tight-binding (TB) model for substitutional disorder.

The setup of this simplified, 1D representation of the problem is introduced in section 2. The new general formalism and a practical algorithm for its numerical solution are described in section 3. Results from this new framework are first checked in the special test case of a degenerate complex unit cell system, in which the NL-CPA treatment for a single atom per unit cell approach can be recovered from the MS-NL-CPA (section 4.1). We then discuss a selection of representative instances of the full model, illustrating the topical examples of selective doping with SRO, and order-disorder transitions (sections 4.2.1, 4.2.2). The final section draws some overall conclusions including implications for studies of realistic systems.

2. Model description and Green function solution

A basic model for the problem of substitutional disorder in a complex unit cell scenario can be given by the tight-binding hamiltonian:

$$H = \sum_{i,j}^{N} \sum_{s,s'}^{N_{sub}} H_{(i,s);(j,s')} \qquad H_{(i,s);(j,s')} = W_{(i,s);(j,s')} + \epsilon_{(i,s)} \delta_{i,j} \delta_{s,s'}. \tag{1}$$

where the general i, j indices refer to the $N \to \infty$ superlattice vectors of an effectively infinite disordered sample, and the $s, s' = 1, \ldots, N_{sub}$ indices identify different sublattices within a complex unit cell of N_{sub} atomic sites.

In our 1D TB model we restrict kinetic effects to nearest neighbour hopping processes in the form:

$$W_{(i,s);(j,s')} = \underbrace{t\, \delta_{s,s'=s\pm 1} \delta_{ij}}_{\substack{\text{intra-cell} \\ \text{hopping}}} + t\, \underbrace{\delta_{i,j=i\pm 1} \left(\delta_{s,1} \delta_{s',N_{sub}} + \delta_{s',1} \delta_{s,N_{sub}} \right)}_{\substack{\text{inter-cell} \\ \text{hopping}}} \tag{2}$$

where the positive hopping parameter t describes lowering of total energy through electronic delocalization among equidistant ions.

Local substitutions are represented by the second term $\epsilon_{(i,s)}$, which in a general treatment may acquire different values among $N_{alt.,s}$ alternative ones, selected from the species available to each specific sublattice s. Alternatively, the method can also be appropriate to systems with off-diagonal disorder, involving instead different site to site hopping parameters.

In this illustrative 1D example unit cells are repeated at $|\boldsymbol{R}_i - \boldsymbol{R}_j| = 0, L, 2L, \ldots$, with periodic boundary conditions, and we make the further convenience-only assumption of equal separation between the N_{sub} sublattices at $|\boldsymbol{a}_s - \boldsymbol{a}_{s'}| = 0, a, 2a, \ldots, (N_{sub} - 1)a$, with $L = N_{sub}\ a$ (figure 1). We apply this description to the example of a $N_{sub} = 2$ unit cell binary compound of the form: $(A_{c_1} B_{1-c_1})(C_{c_2} D_{1-c_2})$.

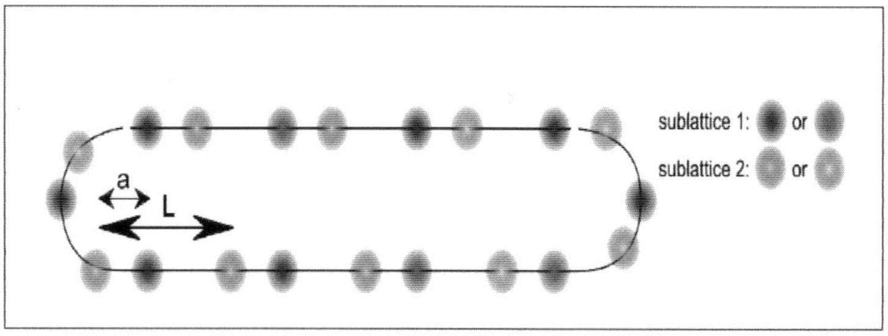

Figure 1. Schematic illustration of a 1D tight-binding model for substitutional disorder in a complex unit cell, here evaluated under periodic boundary conditions for $N_{sub} = 2$, $L = 2a$.

An exact solution to the model of eq. 1 would in principle require accurate knowledge of the atomic species on each site of the effectively infinite N unit cells in the bulk, and the ability to numerically handle the associated $(N \times N_{sub}) \times (N \times N_{sub})$, almost tridiagonal sparse matrix $\underline{H}$:

$$\underline{H} = \begin{pmatrix} \epsilon_{(1,1)} & -t & 0 & \ldots & -t \\ -t & \epsilon_{(1,2)} & -t & \ldots & 0 \\ \ldots & \ldots & \ldots & \ldots & \ldots \\ -t & \ldots & \ldots & -t & \epsilon_{(N,N_{sub})} \end{pmatrix} \tag{3}$$

to construct the direct space Green function [22]:

$$(z\mathbb{1} - \underline{H})\,\underline{\mathcal{G}}(z) \quad = \quad \mathbb{1} \qquad \text{evaluated at the complex energy:} \qquad z = \epsilon + i\eta \tag{4}$$

Considering however that self-averaging effects over a great number of possible configurations will always dominate actual measurements even in a microscopic sample, we can resort instead to an effective medium description. Translational invariance is reintroduced as an *a priori* condition, following the motivations of the original CPA development. With the definition of a mean-field self-energy $\underline{\Xi}$, it is then possible to directly compute the bulk-averaged Green function $\langle \underline{\mathcal{G}}(z) \rangle$ as:

$$\langle \underline{\mathcal{G}}(z) \rangle = \langle (z - \underline{H})^{-1} \rangle = (z - \underline{W} - \underline{\Xi}(z))^{-1} \tag{5}$$

from which observables such as the density of states (DOS) can be immediately determined:

$$n(\epsilon) = \sum_{s=1}^{N_{sub}} n_s(\epsilon) \qquad \text{with sublattice contributions:} \quad n_s(\epsilon) = \lim_{\eta \to 0^+} -\frac{1}{N\pi} \mathrm{Im} \sum_{i=1}^{N} \langle \mathcal{G}_{(i,s);(i,s'=s)}(z) \rangle \tag{6}$$

The problem is therefore transformed into the search for a properly sublattice-resolved self-energy $\underline{\Xi}$. This shall in particular apply to the general case $N_{sub} \geq 1$, and keep track of possible SRO correlations at multiple length scales beyond a single site $N_c = 1$ CPA description, recovering the exact solution in the limit $N_c \rightarrow N$.

3. The Multi-Sublattice generalization of the NL-CPA

We follow and generalize the derivation of the single atom per unit cell NL-CPA as provided by Jarrell and Krishnamurthy [13] and Rowlands *et al.* [4]. With the introduction of translational invariance at the effective medium level, averaged quantities must be periodic at the appropriate length scales across every complex unit cell, but retain sublattice dependence. This condition can be enforced by introducing a (super)lattice Fourier transform[1] satisfying the idempotence relationship:

$$\frac{1}{N_c}\sum_{\boldsymbol{K}_n}^{N_c} e^{i\boldsymbol{K}_n\cdot(\boldsymbol{R}_I-\boldsymbol{R}_J)} = \delta_{I,J} \tag{7}$$

for $\boldsymbol{K}_n$, $\boldsymbol{R}_I, \boldsymbol{R}_J$ general vectors identifying different coarse-graining regions respectively in reciprocal and direct space [4], to express the self-energy as:

$$\Xi_{s,s'}(\boldsymbol{K}_n, z) = \sum_{J=1}^{N_c} \Xi_{(I,s);(J,s')}(z) e^{-i\boldsymbol{K}_n\cdot(\boldsymbol{R}_I+\boldsymbol{a}_s-\boldsymbol{R}_J-\boldsymbol{a}_{s'})} \tag{8a}$$

$$\Xi_{(I,s);(J,s')}(z) = \frac{1}{N_c}\sum_{\boldsymbol{K}_n}^{N_c} \Xi_{s,s'}(\boldsymbol{K}_n, z) e^{+i\boldsymbol{K}_n\cdot(\boldsymbol{R}_I+\boldsymbol{a}_s-\boldsymbol{R}_J-\boldsymbol{a}_{s'})}. \tag{8b}$$

Correspondingly, the bulk-averaged Green function of eq. 5 is:

$$\mathcal{G}_{(I,s);(J,s')}(z) = \sum_{J=1}^{N_c} \mathcal{G}_{s,s'}(\boldsymbol{K}_n, z) e^{+i\boldsymbol{K}_n\cdot(\boldsymbol{R}_I+\boldsymbol{a}_s-\boldsymbol{R}_J-\boldsymbol{a}_{s'})} \tag{9}$$

with:

$$\mathcal{G}_{s,s'}(\boldsymbol{K}_n, z) = \frac{N_c}{\Omega}\int_{\Omega_{\boldsymbol{K}_n}} \left[z\underline{\mathbf{1}} - \underline{W}(\boldsymbol{k}) - \underline{\Xi}(\boldsymbol{K}_n, z)\right]^{-1}_{s,s'} d\boldsymbol{k} \tag{10}$$

where the integration extends over the original Brillouin zone Ω, without the reciprocal space rescaling that would be involved in a supercell-based description of the problem [6]. The Brillouin zone of the lattice is instead partitioned into N_c equal "tiles" $\Omega_{\boldsymbol{K}_n}$ of extension Ω/N_c, centered on each cluster momentum $\boldsymbol{K}_n$ [4], to give the solution $\mathcal{G}_{s,s'}(\boldsymbol{K}_n, z)$ as a coarse-grained approximation of the implicit equation:

$$\mathcal{G}_{s,s'}(\boldsymbol{k}, z) = \mathcal{G}^0_{s,s'}(\boldsymbol{k}, z) + \sum_{s_1,s_2}^{N_{sub}} \mathcal{G}^0_{s,s_1}(\boldsymbol{k}, z)\Xi_{s_1,s_2}(\boldsymbol{k}, z)\mathcal{G}_{s_2,s'}(\boldsymbol{k}, z) \tag{11}$$

for $\mathcal{G}^0_{s,s'}(\boldsymbol{k}, z) = \left(z - W_{s,s'}(\boldsymbol{k})\right)^{-1}$ the free space propagator, and $\boldsymbol{k}$ the originally continuous momentum defined for $N \rightarrow \infty$ within the whole Ω.

Furthermore by introducing in direct space a cavity Green function

$$\mathcal{G}^{cav}_{(I,s);(J,s')}(z) = \left(\underline{\mathcal{G}}^{-1}(z) + \underline{\Xi}(z)\right)^{-1}_{(I,s);(J,s')} \tag{12}$$

[1] This definition of the cluster momenta $\boldsymbol{K}_n$'s is not unique, but fixed up to a phase choice [7]. Here we limit ourselves to the single set with $\phi = 0$, but remark, following Rowlands *et al*, how a multi-phase reformulation involving averages over possibly other sets can be developed straightforwardly.

we can write down an impurity Green function appropriate to each configuration $\boldsymbol{\gamma}$ of the $N_{sub} \times N_c$ atomic energies at positions $\boldsymbol{R}_I + \boldsymbol{a}_s$ within a multi-site cluster:

$$\mathcal{G}^{\gamma}_{(I,s);(J,s')}(z) = \left(\underline{\mathcal{G}}^{cav-1}(z) - \underline{\epsilon}^{\gamma}\right)^{-1}_{(I,s);(J,s')} \quad \text{for} \quad \underline{\epsilon}^{\gamma} = \epsilon^{\gamma}_{(I,s)}\delta_{I,J}\delta_{s,s'} \tag{13}$$

Summing over all of these with an appropriate probability distribution $P(\boldsymbol{\gamma})$ gives an improved expression for the bulk-averaged propagator:

$$\overline{\mathcal{G}}_{(I,s);(J,s')}(z) = \sum_{\gamma}^{N_{tot}} P(\gamma)\mathcal{G}^{\gamma}_{(I,s);(J,s')}(z) \text{ with } N_{tot} = \prod_{s=1}^{N_{sub}} N^{N_c}_{alt.,s} \tag{14}$$

enforcing at this stage the possible inclusion of SRO and/or sublattice-specific effects, as illustrated in the examples of section 4.

This result provides a new starting point for the iterative determination of a better effective medium description. The algorithm is applied at each energy beginning from an arbitrary Ansatz for the self-energy in eq. 10, and computing a new $\underline{\Xi}(z)$ as:

$$\Xi_{(I,s);(J,s')}(z) = \left(\underline{\mathcal{G}}^{cav-1}(z) - \underline{\overline{\mathcal{G}}}(z)^{-1}_{(I,s);(J,s')}\right) \tag{15}$$

to re-evaluate the lattice Fourier transform of eq. 8 and following steps until an arbitrary degree of self-consistency is achieved.

4. Numerical results and discussion

4.1. Validation studies

As a check, the method can be examined for its ability to reproduce the CPA results in the special case $N_{sub} = 1$, $N_c = 1$. This is illustrated by comparing the DOS obtained from an independent implementation of the original algorithm [22], with a study of the same single sublattice system for varying cavity sizes in the new method (figure 2).

We consider then the case of a possibly complex material with $N_{sub} \geq 1$ sites per unit cell. Under the special constraint of every sublattice hosting only the same alternative on-site energies $\epsilon_{(i,s)} = \epsilon_{(i,s')}\ \forall s = 1, \ldots, N_{sub}$ (degenerate scenario), the effective medium has the same translational invariance as the lattice spanned by vectors $\boldsymbol{R}_{(i,s)} = \boldsymbol{R}_i + \boldsymbol{a}_s$, and the self-energy can be written as:

$$\Xi_{s,s'}(\boldsymbol{K}_n, z) = \frac{1}{N_{sub}} \sum_{\boldsymbol{g}_l}^{N_{sub}} \Xi(\boldsymbol{K}_n + \boldsymbol{g}_l, z)e^{-i\boldsymbol{g}_l(\boldsymbol{a}_s - \boldsymbol{a}_{s'})} \tag{16}$$

for $\boldsymbol{g}_l$ reciprocal space (sub)lattice vectors. Combining this with eq. 8 gives:

$$\Xi_{(I,s);(J,s')}(z) = \frac{1}{N_c N_{sub}} \sum_{\boldsymbol{K}_n}^{N_c} \sum_{\boldsymbol{g}_l}^{N_{sub}} \Xi(\boldsymbol{K}_n + \boldsymbol{g}_l, z)e^{+i(\boldsymbol{K}_n + \boldsymbol{g}_l)(\boldsymbol{R}_I + \boldsymbol{a}_s - \boldsymbol{R}_J - \boldsymbol{a}_{s'})} \tag{17}$$

and similar expressions are appropriate for $\mathcal{G}_{(I,s);(J,s')}(z)$ or $\mathcal{G}_{s,s'}(\boldsymbol{K}_n, z)$, showing how in this limit a MS-NL-CPA N_{sub} sites per unit cell $N_c = 1$ single tile calculation can be equivalently decomposed into a NL-CPA study with N_c tiles over a simple $N_{sub} = 1$ lattice, or intermediate cases (table 1). This is illustrated in figure 3, comparing three possibilities for a cavity of 4 elements.

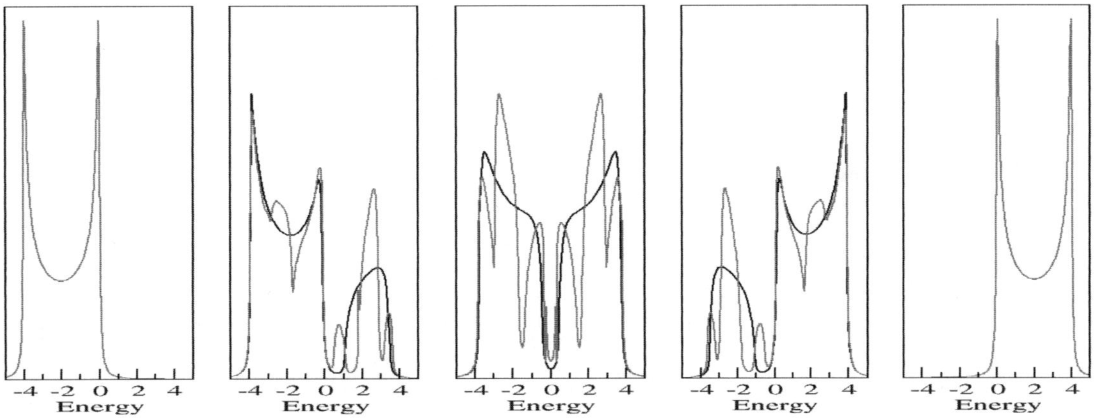

Figure 2. The density of states vs. energy (evaluated at $z = \epsilon + i\eta$ for $\eta = 0.05t$) for the model of eq. 1 in the simple unit cell case $N_{sub} = 1$, with on-site energies $\epsilon_A = -2t$, $\epsilon_B = +2t$ for a 1D compound A_cB_{1-c}. Calculations at varying concentrations $c = 0.0, 0.25, 0.5, 0.75, 1.0$ with the original CPA and NL-CPA algorithms, in an equivalent $SRO = 0$ regime (see text) and for different cavity sizes. The CPA results are perfectly reproduced for $N_c = 1$ (black). Going beyond the single-site treatment to $N_c = 2$ (red) and above allows to pick up additional features in the presence of disorder.

NL-CPA	MS-NL-CPA
$N_c = 2$, $N_{sub} = 1$	$N_c = 1$, $N_{sub} = 2$
$N_c = 4$, $N_{sub} = 1$	$N_c = 2, N_{sub} = 2$ $N_c = 1, N_{sub} = 4$
$N_c = 8$, $N_{sub} = 1$	$N_c = 4, N_{sub} = 2$ $N_c = 2, N_{sub} = 4$ $N_c = 1, N_{sub} = 8$

Table 1. Equivalent MS-NL-CPA descriptions in the particular case of a degenerate scenario, explicitly listed up to a total computational budget of $N_{sub} \times N_c = 8$ sites.

4.2. Complex unit cell, multi-site treatment

Following the previous validation tests in the degenerate limit, we now investigate sublattice dependent disorder effects, with specific on-site energies on the different sublattices. The extended MS-NL-CPA treatment enables an accurate study of the effect of selective disorder, together with SRO; specific examples are discussed in section 4.2.1.

As a bi-product of the development we point out that it is also possible to trade unit cell complexity (larger N_{sub}) for a simplification in the Brillouin zone tiling procedure (lower N_c). A sample application of this approach in the context of disorder-to-order phase transitions is illustrated in section 4.2.2.

4.2.1. Sublattice-selective doping with SRO For a first fully representative test system for the MS-NL-CPA we consider a complex unit cell material, with different on-site energies ϵ_A, ϵ_B and ϵ_C, ϵ_D for atoms belonging to two sublattices.

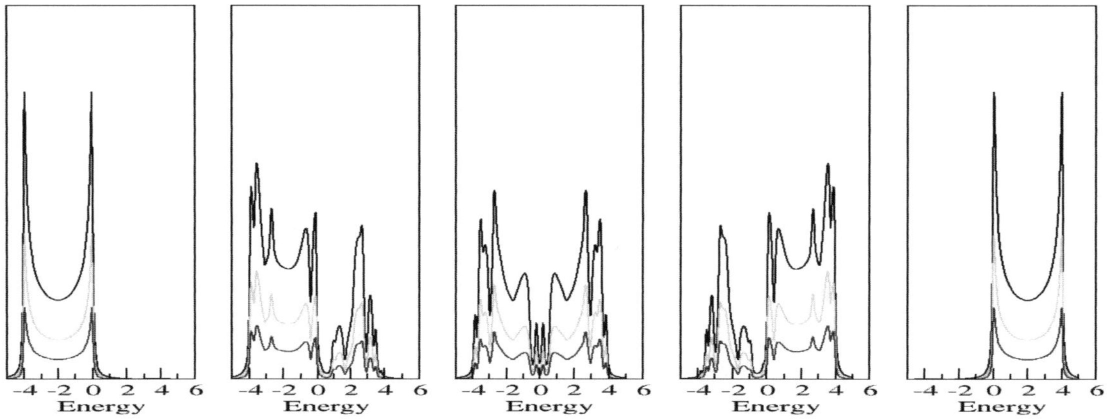

Figure 3. NL-CPA ($N_c = 4$, $N_{sub} = 1$, black), MS-NL-CPA ($N_c = 2$, $N_{sub} = 2$, green) and MS-NL-CPA (($N_c = 1$, $N_{sub} = 4$, blue) DOS for the same model setup as figure 2, evaluated for a degenerate test case $\epsilon_{(i,s)} = \epsilon_{(i,s')} \, \forall s = 1, \ldots, N_{sub}$. In the simple unit cell treatment with $N_{sub} = 1$ the larger cavity size allows to resolve additional short length scale features. The same results are obtained from the equivalent decomposition in $N_{sub} = 2$ or $N_{sub} = 4$ sublattices, lying on top of each other for this degenerate scenario, as the total cavity size is kept constant.

Figures 4-6 illustrate the impact of various SRO regimes, evaluated for a multi-site cavity size of $N_c = 2$ elements, and in the particular example of a selectively doped compound. Sublattice $s = 1$ hosts a regular array of species A, with on-site energies $\epsilon_A = -2t$. A disordered distribution of impurities (C_{c_2}, D_{1-c_2}), with $\epsilon_C = +1t$, $\epsilon_D = +3t$, occurs instead with varying concentration $c_2 = 0, 0.25, .5, .75, 1$ on sublattice $s = 2$ alone.

The non-local nature of the method allows to investigate the effect of different biases in the average dispersion of the dopant. We consider in particular two extremal cases from the whole series of physical possibilities.

In a total clustering regime ($SRO = +1$ [5] or multi-site probability distribution $P(\gamma) = c_2$ if $\epsilon_{(I,2)} = \epsilon_C$, or $P(\gamma) = 1 - c_2$ if $\epsilon_{(I,2)} = \epsilon_D \, \forall I = 1, \ldots, N_c$, and $P(\gamma) = 0$ otherwise, figure 4) similar atomic species show a tendency towards phase segregation into local regions of homogeneous elements. This creates the conditions for a DOS dominated by superposition of the typical 1D Van Hove singularities for individually pure systems, centered on each constituent on-site energy at $\epsilon = -2t$ (ordered sublattice, in red) and $\epsilon = +1t$ or $\epsilon = +3t$ (disordered sublattice, in green), with a relative contribution proportional to the concentration c_2 or $1 - c_2$. The initial gap is progressively narrowed down from $\Delta\epsilon \simeq 5t$ for a pure $(A), (D)$ system, to $\Delta\epsilon \simeq 3t$ for the pure $(A), (C)$ one. This takes place through the growth of intermediate states that carry the features associated with unperturbed 1D hopping between like neighbors on sublattice 2 (all C or all D, in the homogeneous background of interleaved A elements on the other sublattice).

The opposite case of a total ordering regime ($SRO = -1$ [5] or $P(\gamma) = c_2$ if $\epsilon_{(I,2)} = \epsilon_C, \epsilon_{(I+1,2)} = \epsilon_D, \epsilon_{(I+2,2)} = \epsilon_C, \ldots$, and $P(\gamma) = 1 - c_2$ if $\epsilon_{(I,2)} = \epsilon_D, \epsilon_{(I+1,2)} = \epsilon_C, \epsilon_{(I+2,2)} = \epsilon_D, \ldots$, and $P(\gamma) = 0$ otherwise, figure 5) provides a qualitatively different outlook. Here contiguity between unlike elements is favored. Since the surrounding sites around the minority impurity (C for concentration $c_2 < 50\%$, D for $c_2 > 50\%$) are always occupied by different neighbors, hopping processes respectively associated with its on-site energy ϵ_C or ϵ_D are strongly

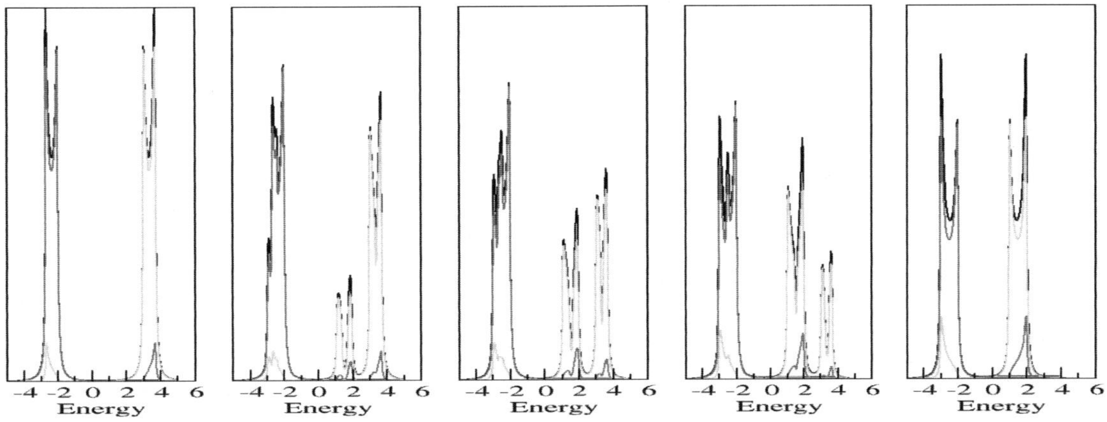

Figure 4. Selective doping with clustering of impurities ($SRO = +1$): DOS for a $N_{sub} = 2$, $N_c = 2$ study of a complex unit cell compound with distinct on-site energies and disorder (sublattice $s = 1$: $\epsilon_{(i,s=1)} = \epsilon_A = -2t\ \forall i$, red; sublattice $s = 2$: $\epsilon_{(i,s=2)} = \epsilon_C = +1t$ or $\epsilon_D = +3t$ with concentration $c_2 = 0.0, 0.25, 0.5, 0.75, 1.0$, green), favoring contiguous alike elements.

suppressed, leading to a DOS contribution in the form of an impurity -like single peak around its energy $\epsilon = +1t$ or $\epsilon = +3t$. The same argument however is reversed for the majority species D or C, giving rise to an almost pure, Van Hove -like double feature around that energy. The transition takes place through the special concentration value $c_2 = 50\%$, for which both species of sublattice $s = 2$ are equally localized, and correspondingly two single peaks appear.

We also note that in both cases of the series in figure 4 and 5 some SRO -induced differences also extend to the DOS contributions from the ordered sublattice $s = 1$ (red). In the presence of impurity clustering on sublattice $s = 2$, atoms belonging to the orderly distributed species A find themselves on average surrounded by extended domains of homogeneously different background $\epsilon_{(I,2)} = \epsilon_C$ or $\epsilon_{(I,2)} = \epsilon_D$. This can be expected to impact hopping processes at the effective medium level, giving raise to the peculiar superposition of localized/delocalized features around the original on-site energy $\epsilon = -2t$ (figure 4). The effect is different in nature, and on the whole less appreciable, in a $SRO = -1$ regime, where as remarked an atom of species A tends to always appear within a similar environment of species C on one side, D on the other one, and vice-versa. This affects only subtly delocalization, which remains the dominant process on the sublattice, and results in a contribution mainly sensitive to the 1D nature of the model. Such developments may prove useful in some materials cases, in order to gain a measure of control over the contribution from the otherwise ordered and more stable constituents of a complex compound, by means of annealing or cold-working procedures having a larger impact on the disordered sublattice.

These calculations can be contrasted with a totally uncorrelated setup ($SRO = 0$ [5] or $P(\gamma) = \prod_{I=1}^{N_c}(c_2)$ if $\epsilon_{(I,2)} = \epsilon_C$, $(1 - c_2)$ if $\epsilon_{(I,2)} = \epsilon_D$), mimicking the single-site design of the MS-CPA. In the absence of any particular short range ordering on the doped sublattice $s = 2$, the higher resolution physical features resolved in both figure 4 and 5 appear to be contained *in nuce* and superimposed within the DOS of figure 6. As discussed above, their relative importance follows the bias towards total clustering or total ordering. These differences are enhanced by the unidimensional nature of the TB model adopted in this exposition, but further studies from

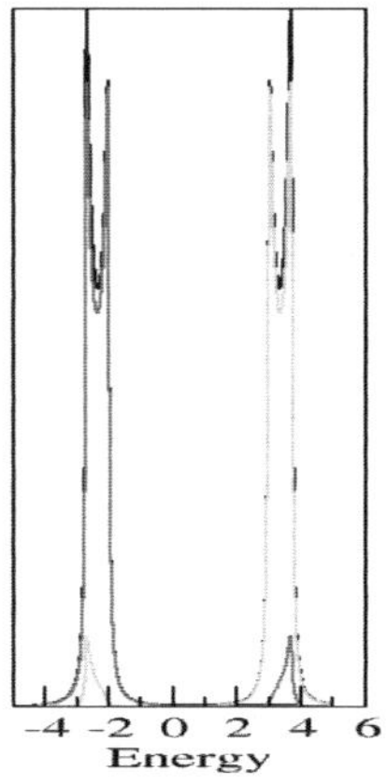

Figure 5. Selective doping with local ordering of impurities ($SRO = -1$): DOS for a $N_{sub} = 2$, $N_c = 2$ study of a complex unit cell compound with distinct on-site energies and disorder (sublattice $s = 1$: $\epsilon_{(i,s=1)} = \epsilon_A = -2t\ \forall i$, red; sublattice $s = 2$: $\epsilon_{(i,s=2)} = \epsilon_C = +1t$ or $\epsilon_D = +3t$ with concentration $c_2 = 0.0, 0.25, 0.5, 0.75, 1.0$, green), favoring contiguous dissimilar elements.

a fully 3D reimplementation of the theory in a multiple scattering formalism for first-principles investigations also confirm their persistence in the case of realistic materials.

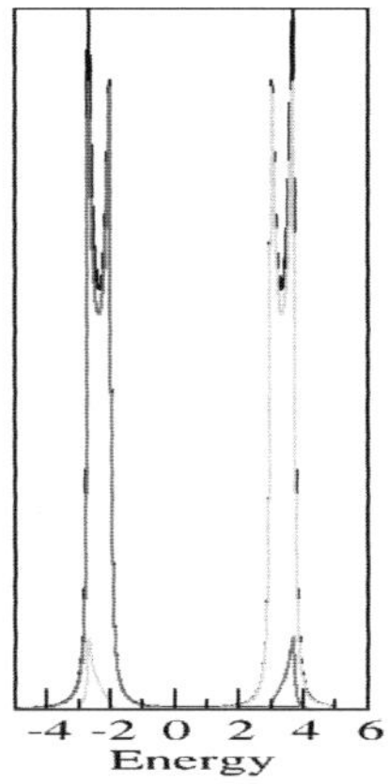

Figure 6. Selective doping with totally uncorrelated impurities ($SRO = 0$): DOS for a $N_{sub} = 2$, $N_c = 2$ study of a complex unit cell compound with distinct on-site energies and disorder (sublattice $s = 1$: $\epsilon_{(i,s=1)} = \epsilon_A = -2t\ \forall i$, red; sublattice $s = 2$: $\epsilon_{(i,s=2)} = \epsilon_C = +1t$ or $\epsilon_D = +3t$ with concentration $c_2 = 0.0, 0.25, 0.5, 0.75, 1.0$, green)

4.2.2. Order-disorder transition scenario If a homogeneously disordered system is cooled through a phase transition, the underlying symmetry is broken and long range order (LRO) develops and grows as the temperature is lowered towards 0K. In our second example we use the model of eq. 1 to examine such scenario, considering again similar on-site energies on both sublattices $\epsilon_A = \epsilon_C = -2t$, $\epsilon_B = \epsilon_D = +2t$ as in the degenerate test cases of figure 3, but adopting here constituents probability distributions $P(\gamma)$ governed by the coupled concentrations: $c_2 = 1 - c_1 = (1 - \eta)/2$.

In this setup η is a LRO parameter, ranging from 0 to 1 as a fixed $1 : 1$ ratio (A, D) binary alloy stabilizes into more and more regular structures, up to the totally ordered limit. Figure 7 illustrates the effect of such modifications on the DOS, starting from the same totally uncorrelated results also obtained in the central panel of figure 3 for a $c = 0.5$, simple unit cell $N_c = 4$ NL-CPA calculation. Using now an additional resolution into two sublattices as afforded by the MS-NL-CPA, it is possible to investigate this disorder-to-order transition through the $N_c = 2$, $N_{sub} = 2$ decomposition suggested in table 1.

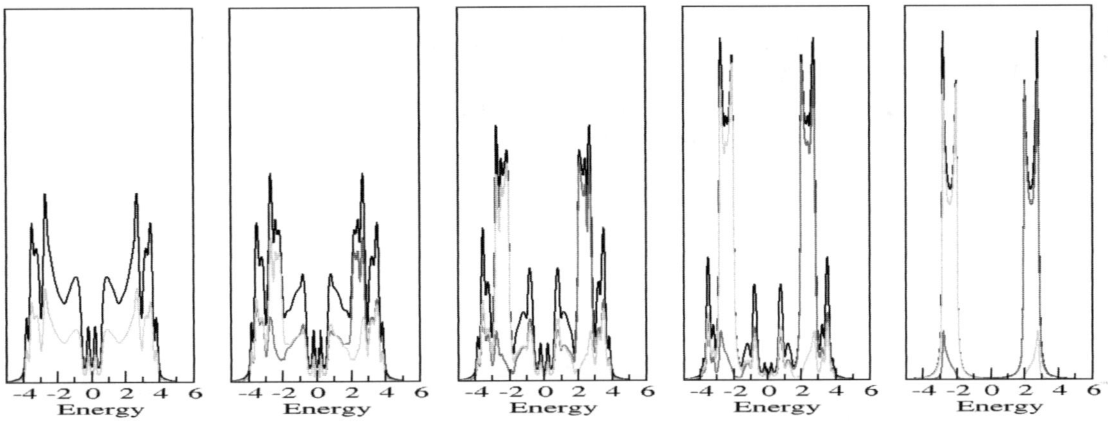

Figure 7. Disorder-to-order transition: DOS for a $1 : 1$ binary alloy, evaluated for $N_c = 2$ and $N_{sub} = 2$ with the same on-site energies $\epsilon_A = \epsilon_C = -2t$ and $\epsilon_B = \epsilon_D = +2t$ on both sublattices. Coupled concentrations: $c_1 = 1 - c_2 = (1 - \eta)/2 = 0.5, 0.375, 0.25, 0.125, 0$ describe an increasing LRO parameter $\eta = 0, 0.25, 0.5, 0.75, 1$. Non-local treatment with $SRO = 0$.

As the order parameter increases, we can follow the evolution of the most significant DOS features associated with on-site energies ϵ_A, ϵ_D. In the leftmost panel interactions with totally uncorrelated random neighbhours are responsible for a pronounced smearing and broadening of the states around $\epsilon = \pm 2t$. Both sublattices present an exactly equivalent environment, and give therefore contributions lying again on top of each others.

A more and more regular distribution of the same ratio of species A and D into well defined atomic sites at $\boldsymbol{a}_s$ and $\boldsymbol{a}_{s'}$ shows how this picture develops towards the pure scenario (rightmost panel). Disorder tends to gradually fill the initially empty gap between distinct Van Hove singularities, through the insurgence of impurity -like states similarly to the situation observed for the cases of figures 4-6. Here however the peculiarity of coupled concentrations and same on-site energies for both sublattices gives raise to such structures in the form of resonant peaks, occurring approximately at $\epsilon = \pm 4t$ and 0. Considering in particular the intermediate regimes, we note how these differ qualitatively from the on-site states, which remain down to $\eta = 0$

dominated by different individual sublattices contributions (either red or green). The additional features are instead characterized by finding their origin equally on both sites (green and red lines lying almost on top of each others, panels 2-4), and growing at the expense of the local "matching" species (green for sublattice $s = 1$ at $\epsilon = -2t$, red for $s = 2$ at $\epsilon = +2t$), while the importance of the alternate one (red for sublattice $s = 1$ at $\epsilon = -2t$, and vice-versa) remains relatively unchanged.

This can be qualitatively explained by comparing the ordered $\eta = 1$ sequence: $\ldots, [\epsilon_A, \epsilon_D], [\epsilon_A, \epsilon_D], [\epsilon_A, \epsilon_D], [\epsilon_A, \epsilon_D], \ldots$ with a possible realization of the central panel case $c_1 = 0.25$ or $\eta = 0.5$: $\ldots, [\epsilon_A, \epsilon_D], [\epsilon_D, \epsilon_A], [\epsilon_D, \epsilon_A], [\epsilon_D, \epsilon_A], \ldots$, adopting here the [] notation to outline the choice of a $N_{sub} = 2$ reference unit cell.

The instances of unperturbed intra-sublattice propagation between equivalent species A at $\boldsymbol{a}_s$ or D at $\boldsymbol{a}_{s'}$, which are the main factors responsible for the Van Hove double peaks probing the model dimensionality around $\epsilon = \pm 2t$, get reduced on each sublattice proportionally to the anti-site defects concentration c_1 or c_2, until acquiring an almost localized -only aspect (leftmost panel). In the same energy subregion the contribution from the other site is instead always impurity -like, reflecting the interleaved nature of this particular model setup. As opposed to the example of figure 3, here in fact free propagation is always hindered by a different species even in the "pure" limit. This results in the superposition of a similar dominant symmetric structure for each sublattice, on top of an asymmetric one from the other, giving the skewed aspect also seen in the previous series (figures 4-6, leftmost and rightmost panels).

On the other hand, the progressive collapse of LRO in a system of equidistant sites at $0, L = 2a, 2L = 4a, \ldots$, with the same possible on-site energies for both sublattices, can only take place by opening up more and more nearest neighbour propagation channels at the new halved wavelength a. Replacing any atom A with its alternative D will always result in having two equivalent elements next to each other across the arbitrary unit cell twice, somewhere in the bulk; and the translational invariant effective medium treatment is insensitive to the actual different placement of such $\ldots, A], [A, \ldots$ and $\ldots, D][D, \ldots$ regions, which therefore appear in the DOS equally from both sublattices. Going beyond to $N_c = 4$ and above in the non-local expansion of the investigation also confirms the propagation -like nature of these extra peaks, already hinted in figure 7. As the calculation becomes even more resolved, these in fact split again following the expected Van Hove behaviour, while the main on-site features at $\epsilon = \pm 2t$ retain their single sublattice origin.

5. Conclusions

The recently proposed NL-CPA method for describing disordered electron systems with short-range order can be straightforwardly extended to describe materials with multi-atom per unit cell lattice structures. In many crystalline systems the distribution of impurities on a subset of the sublattices can modify properties nominally associated with the chemical composition of the others. In this paper we have shown how to approach a quantitative analysis of such effects.

The method has been introduced through the language of Green function theory, discussing in particular numerical results from an illustrative 1D implementation. Topical examples from the context of selectively doped materials and the case of disorder-to-order phase transitions show how subtle modifications to an originally single-site effective medium description become accessible, and can be resolved into their specific origin at the level of average local alterations to the microscopic environment in the sample.

Work is in progress on the application of the theory to a multiple scattering (KKR), fully ab-initio treatment of realistic materials.

Acknowledgments

The authors wish to thank Dr. Martin Lüders and Dr. Arthur Ernst for insightful discussions throughout the ongoing development of this work.

Technical support from the Centre for Scientific Computing at the University of Warwick is gratefully acknowledged.

References

[1] Soven P 1967 *Phys. Rev.* **156** 809–813
[2] Györffy B and Stocks G 1978 First principles band theory for random metallic alloys *Electrons in disordered metals and at metallic surfaces* NATO advanced study institutes series ed Phariseau P, Gyorffy B and Scheire L NATO Scientific Affairs Division (Plenum Press)
[3] Faulkner J S and Stocks G M 1980 *Phys. Rev. B* **21** 3222–3244
[4] Rowlands D A, Staunton J B and Györffy B L 2003 *Phys. Rev. B* **67** ISSN 1098-0121
[5] Rowlands D A, Staunton J B, Györffy B L, Bruno E and Ginatempo B 2005 *Phys. Rev. B* **72** 045101 ISSN 1098-0121
[6] Rowlands D A 2006 *Psi-k newsletter* **77**
[7] Rowlands D A, Zhang X G and Gonis A 2008 *Phys. Rev. B* **78** 115119
[8] Tulip P R, Staunton J B, Rowlands D A, Györffy B L, Bruno E and Ginatempo B 2006 *Phys. Rev. B* **73** 205109
[9] Tulip P R, Staunton J B, Lowitzer S, Ködderitzsch D and Ebert H 2008 *Phys. Rev. B* **77** 165116
[10] Leslie M and Gillan N J 1985 *J. Phys. C: Solid State Physics* **18** 973
[11] Makov G and Payne M C 1995 *Phys. Rev. B* **51** 4014–4022
[12] Freysoldt C, Neugebauer J and Van de Walle C G 2009 *Phys. Rev. Lett.* **102** 016402
[13] Jarrell M and Krishnamurthy H 2001 *Phys. Rev. B* **63** 125102
[14] Rowlands D A, Ernst A, Györffy B L and Staunton J B 2006 *Phys. Rev. B* **73** 165122
[15] Lowitzer S, Ködderitzsch D, Ebert H, Tulip P R, Marmodoro A and Staunton J 2010 *EPL* **92**
[16] Pindor A J, Temmerman W M and Györffy B L 1983 *J. Phys. F: Met. Phys.* **13** 1627–1644
[17] Temmerman W M and Pindor A J *J. Phys. F: Met. Phys.* **13** 1869
[18] Däne M, Lüders M, Ernst A, Ködderitzsch D, Temmerman W M, Szotek Z and Hergert W 2009 *J. Phys.: Cond. Mat.* **21** 045604
[19] Györffy B L, Pindor A, Staunton J B, Stocks G and Winter H 1985 *J. Phys.: Cond. Mat.* **15** 1337
[20] Kudrnovský J, Drchal V and Bruno P 2008 *Phys. Rev. B* **77** 224422
[21] Bellaiche L, Garcia A and Vanderbilt D 2000 *Phys. Rev. Lett.* **84** 5247
[22] Gonis A 1992 *Green functions for ordered and disordered systems* (North-Holland)

Condensed Matter and Materials Physics Conference (CMMP10) IOP Publishing
Journal of Physics: Conference Series **286** (2011) 012034 doi:10.1088/1742-6596/286/1/012034

A fast, stable method for density functional simulations of nanostructures

P J Hasnip

Dept of Physics, University of York, York YO10 5DD, U.K.

E-mail: phil.hasnip@york.ac.uk

Abstract. The accurate simulation of nanostructures by density functional theory techniques is often hampered by numerical instabilities, which lead to poor convergence or even divergence. We discuss the physics behind these instabilities, and propose a fast, stable method for suppressing them. The performance of this method is demonstrated for a variety of simulation systems.

Density Functional Theory[1] (DFT), as proposed by Kohn and Sham[2], has become the *ab initio* method of choice for many scientists worldwide. Advances in growth and electron microscopy techniques have enabled experiments on smaller and smaller length scales, whilst the development of improved algorithms and more powerful computer hardware has enabled DFT calculations to be performed on larger and larger length scales. At the nanoscale the length scales of the smallest experiments and largest calculations coincide, enabling coupled experimental and theoretical studies that are at the forefront of their respective disciplines.

DFT calculations have been crucial to our understanding of much of the behaviour of materials at the nanoscale, providing interpretations of electron microscope images as well as a host of spectroscopic techniques. Nevertheless many of the conventional density mixing methods[3][4] employed to solve the Kohn-Sham equations show poor convergence, or even fail to converge at all, when employed to study large anisotropic simulation cells such as those required for many nanoscale studies.

The root cause of these instabilities is the ill-behaviour of the Hartree potential, which in reciprocal space may be expressed as

$$V_H(\mathbf{G}) = \frac{4\pi\rho(\mathbf{G})}{\Omega|\mathbf{G}|^2}, \tag{1}$$

where $\mathbf{G}$ is a reciprocal lattice vector, ρ is the electronic charge density and Ω is the volume of the simulation cell. As the size of simulation cell is increased, the size of the smallest $\mathbf{G}$-vector decreases and a small error in ρ can lead to a large error in $V_H[\rho]$.

We propose the use of a model auxiliary density functional to model the system's behaviour, accounting exactly for the dependence of the Hartree potential on the density and only approximating the kinetic and non-local contributions. The convergence of this method is not only superior to many conventional methods (see Fig. 1), but invariant with respect to the size and shape of the simulated material. This method has been implemented in the CASTEP computer program[5], and its performance will be demonstrated for a range of simulations.

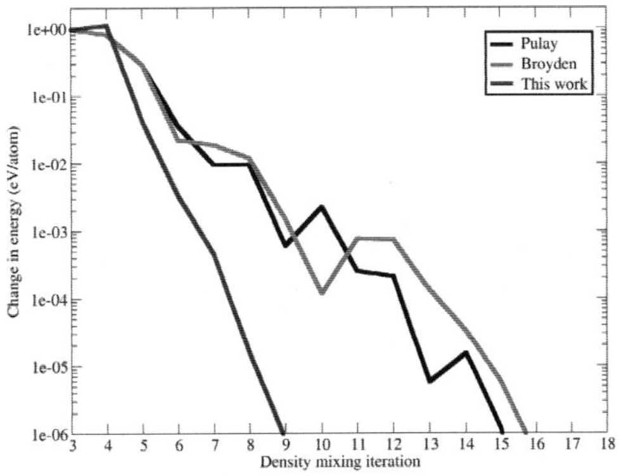

Figure 1. Convergence of the auxiliary DFT scheme (blue) for a 2D periodic MgO slab, separated from its periodic image by 10Å. The convergence of the conventional preconditioned Pulay (black) and Broyden (red) schemes are shown for comparison.

[1] P. Hohenberg and W. Kohn, *Phys. Rev.* **136** B864 (1964)
[2] W. Kohn and L.J. Shamm, *Phys. Rev.* **140** A1133 (1965)
[3] P. Pulay, *Chem. Phys. Lett.* **73** 393-398 (1980)
[4] C.G. Broyden, *Math. Comput.* **19** 577 (1965)
[5] S. J. Clark, M. D. Segall, C. J. Pickard, P. J. Hasnip, M. J. Probert, K. Refson and M. C. Payne, *Z. Kristallogr.* **220**(5-6) pp.567-570 (2005)

Numerical Modeling of the Transient Response of Metal-Semiconductor-Metal Photodetector Using Discrete Fourier Transform Method

Atefeh Habibpoor[1], Hamid R. Mashayekhi[2]

[1, 2] Physics Group, Faculty of Science, The University of Guilan, Rasht, Iran.

(Corresponding Author: mashhr@guilan.ac.ir)

Abstract: A one dimensional (1-D) simulation program based on the drift-diffusion model and Discrete Fourier Transform method (DFT) is developed. This program numerically solves the time-dependent continuity equations for electrons and holes in a semiconductor device. This model simulates carrier concentrations and the impulse response of a GaAs metal-semiconductor-metal (MSM) photodetector at a constant bias voltage. From this simulation, we found that for a smaller value of carrier lifetime, the response fall time decreases without significantly reducing the responsivity of the device. The simulation results are well in agreement with the experimental finding and our results match the work of other people who have done the same simulation but with different approach.

Key Words: Simulation; Discrete Fourier Transform (DFT) Method; Impulse Response; Metal-Semiconductor-Metal (MSM) Photodetector.

1. Introduction

The planar Metal-Semiconductor-Metal (MSM) photodetector has become very attractive for application in optoelectronic communication systems [1]. This device consists of a semiconductor absorbing layer on which two electrodes have been deposited to form back-to-back Schottky diodes. The two top electrodes can be as simple as two parallel electrodes or designed in an interdigitated [2] or circular pattern [3]. It has several advantages over traditional p-i-n photodiodes. MSM photodetectors have a lower intrinsic capacitance than p-i-n photodetectors, which results in higher speeds for high bit rate, high sensitivity optoelectronic integrated receivers [4]. In addition, since MSM structures have a lateral design, they have, for the same absorbing layer thickness, a larger surface area than vertical p-i-n photodetectors for similar capacitance and bandwidth [5]. This larger area results in higher alignment tolerance, and thus, reduced packaging cost. Also MSM structure has process compatibility with field effect transistor (FET). In fact, the monolithic integration of MSM structure with a FET is very simple. This is because the electrodes of the MSM photodetector and the gate fingers of a FET can be defined with the same photolithographic step and then deposited with the same metallization [6].

The impulse response of the MSM photodetector often exhibits two different components, a fast rise time, and a long fall time. In high speed digital communications, the long fall time results in intersymbol interference. This is typically caused by the holes since they have much lower mobility than the electrons [7]. One of the reported approaches for reducing the fall time of the transient response of the GaAs MSM photodetector is to use intermediate temperature grown GaAs (ITG-GaAs) for which the growth temperature of the semiconductor composite is intermediate between the standard growth temperature of GaAs and that for low temperature grown (LTG) GaAs. The outcome is that the carrier lifetime in the light absorption region is only slightly longer than the transit-time between the electrodes. Thus, the speed of the photodetector is improved without significantly reducing the responsivity [8].

In this paper, we use the Discrete Fourier Transform (DFT) method to simulate the transient response of the GaAs MSM photodetector and to investigate the effect of the lifetime parameter value on the device impulse response. The simulation is based on a one-dimensional (1-D) numerical solution of the time–dependent continuity equations for electrons and holes.

2. Modeling and Simulation

The basic planar MSM photodetector is schematically shown in Fig. 1. It consists of interdigitated metal fingers formed on the top surface of a semiconductor layer. These electrodes are made to form Schottky contacts with the semiconductor. A voltage is applied to the metal fingers to create sufficient electrical field within the semiconductor layer. Light incident on the top surface of the photodetector is absorbed to generate electron-hole pairs. The electric field sweeps the photogenerated carriers out of the device to produce a photocurrent. This photocurrent can be modeled in a one-dimensional regime [9-13].

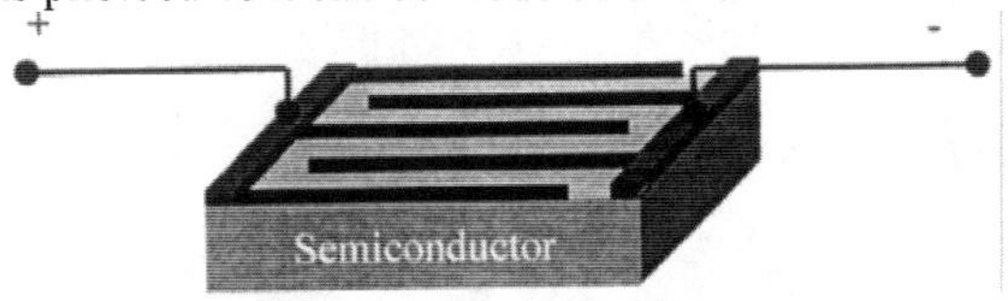

Fig. 1 Schematic MSM structure.

To understand the behavior of photocarriers under the influence of the drift-diffusion mechanism, the time dependent continuity equations for electrons and holes are numerically solved on the basis of a one dimensional approximation. The time-dependent continuity equations for electrons and holes are given by [13]

$$\frac{\partial n}{\partial t} = \frac{1}{q}\frac{\partial J_n}{\partial x} - R + G_{opt} \quad (1)$$

$$\frac{\partial p}{\partial t} = -\frac{1}{q}\frac{\partial J_p}{\partial x} - R + G_{opt} \quad (2)$$

$$J_n = q\mu_n nE + qD_n\frac{\partial n}{\partial x} \quad (3)$$

$$J_p = q\mu_p pE - qD_p\frac{\partial p}{\partial x} \quad (4)$$

$$n = n_0 + \delta n \quad (5)$$

$$p = p_0 + \delta p \quad (6)$$

Where $J_{n,p}$ are the electron and hole currents, R is the net carrier recombination rate and G_{opt} is the generation rate due to illumination, $\mu_{n,p}$ and $D_{n,p}$ are the electron/hole mobilities and diffusivities, E and q are the electric field and the elementary charge, respectively. A constant electric field is assumed across the device in accordance with ref. [13]. n_0 and p_0 are the thermal equilibrium electron and hole concentrations and are independent from the time and space coordinate (assuming a homogeneous active region), δn and δp are photogenerated (excess) electron and hole concentrations. The recombination rates are chosen to be n/τ_n and p/τ_p, where $\tau_{n,p}$ are the carriers lifetime.

We use the field-dependent expression for the carriers' mobilities μ_n and μ_p as follows [14]

$$\mu_{n,p} = \frac{\mu_{0n,0p}}{1 + (\mu_{0n,0p}E/v_{sn,sp})} \quad (7)$$

Where $\mu_{0n,0p}$ are the low field electron and hole mobilities, and $v_{sn,sp}$ are the electron and hole saturation velocities, respectively. The parameters which have been used for the GaAs MSM photodetector are μ_{on}=8500 cm^2/V.s, μ_{op}=400 cm^2/V.s, $E_0=4\times10^3 V.cm^{-1}$, $v_{sn}=8.5\times10^6 cm.s^{-1}$ and $v_{sp}=1.5\times10^7 cm.s^{-1}$.

The carriers' diffusion constants are determined by use of the Einstein relation [15-16] as follows

$$D_{n,p} = \frac{kT}{e}\mu_{n,p} \tag{8}$$

Where k is Boltzmann's constant and T is the carrier temperature (room temperature).

From the symmetrical geometry of the interdigitated electrodes, it is sufficient to treat only a single unit cell whose boundaries are at two adjacent electrical metal fingers. We take the fingers to be located at x=0 and x=L. With the assumption of ideal Ohmic contacts with an infinite surface recombination velocity, the densities of electrons and holes at the fingers are taken to be at thermal equilibrium, i.e., their excess densities are zero [12].

The DFT method is used to numerically solve continuity equations for excess electrons and holes. Once the carrier concentrations have been calculated, the photocurrents can be found afterwards.

3. Simulation Method

In this simulation, the illumination is assumed to be uniform over the entire active region with an arbitrary band-to-band generation rate of 1×10^{20} (electron-hole).m^{-3} at t=0. It has been assumed that no generation of carriers happens in latter time steps.

We have numerically solved the continuity equations for electrons and holes by the use of Fourier transform technique. By substituting equations (3-6) into the equations (1) and (2), we obtain:

$$\frac{\partial(\delta n(x,t))}{\partial t} = D_n\frac{\partial^2(\delta n(x,t))}{\partial x^2} + \mu_n E\frac{\partial(\delta n(x,t))}{\partial x} - \frac{\delta n(x,t)}{\tau_n} \tag{9}$$

$$\frac{\partial(\delta p(x,t))}{\partial t} = D_p\frac{\partial^2(\delta p(x,t))}{\partial x^2} - \mu_p E\frac{\partial(\delta p(x,t))}{\partial x} - \frac{\delta p(x,t)}{\tau_p} \tag{10}$$

The Fourier transforms of the entire equations (9) and (10) are as follows [17]

$$F[\frac{\partial(\delta n(x,t))}{\partial t}] = D_n F[\frac{\partial^2(\delta n(x,t))}{\partial x^2}] + \mu_n EF[\frac{\partial(\delta n(x,t))}{\partial x}] - \frac{1}{\tau_n}F[\delta n(x,t)] \tag{11}$$

$$F[\frac{\partial(\delta p(x,t))}{\partial t}] = D_p F[\frac{\partial^2(\delta p(x,t))}{\partial x^2}] - \mu_p EF[\frac{\partial(\delta p(x,t))}{\partial x}] - \frac{1}{\tau_p}F[\delta p(x,t)] \tag{12}$$

To solve the above two equations, we have used the Fourier transform properties [17]. The Fourier transform of the first and second-order spatial derivatives and the Fourier transform of the time derivative of the photogenerated electron density, respectively, are equal to

$$F[\frac{\partial(\delta n(x,t))}{\partial x}] = ikF[\delta n(x,t)] \tag{13}$$

$$F[\frac{\partial^2(\delta n(x,t))}{\partial x^2}] = (ik)^2 F[\delta n(x,t)] \tag{14}$$

$$F[\frac{\partial(\delta n(x,t))}{\partial t}] = \frac{\partial}{\partial t}F[n(x,t)] \tag{15}$$

The same expressions can be found for the photogenerated hole density. Substituting these expressions into equations (11) and (12) results in

$$(-k^2 D_n + ikE\mu_n - \frac{1}{\tau_n})F[\delta n(x,t)] = \frac{\partial}{\partial t}F[\delta n(x,t)] \tag{16}$$

$$(-k^2D_p - ikE\mu_p - \frac{1}{\tau_p})F[\delta p(x,t)] = \frac{\partial}{\partial t}F[\delta p(x,t)] \quad (17)$$

Equations (16) and (17) are first-order constant coefficient differential equations. Their solutions are as follows

$$F[\delta n(x,t)] = F[\delta n(x,0)]\exp((-k^2D_n + ikE\mu_n - \frac{1}{\tau_n})t) \quad (18)$$

$$F[\delta p(x,t)] = F[\delta p(x,0)]\exp((-k^2D_p - ikE\mu_p - \frac{1}{\tau_p})t) \quad (19)$$

Where δn(x,0) and δp(x,0) are the initial photogenerated electrons' and holes' densities in x-space. Equations (18) and (19) give the electron and hole densities at any time in k-space, respectively. We have used DFT method to numerically solve these two equations [18-21]. We have arbitrary chosen 2N=512 sample points in x and k-space. The values of x and k satisfy the following formulas

$$x_k = \frac{kL}{2N}, \qquad 0 \le k \le (2N-1) \quad (20)$$

$$k_j = \frac{2\pi j}{L}, \qquad 0 \le j \le (2N-1) \quad (21)$$

The Discrete Fourier Transforms of the excess electron and hole concentrations at any time have the following formulas

$$F[\delta n(x_k,t)] = \delta N(k_j,t) = \frac{1}{2N}\sum_{k=0}^{2N-1}\delta n(x_k,t)e^{-ik_jx_k}, \qquad 0 \le j \le (2N-1) \quad (22)$$

$$F[\delta p(x_k,t)] = \delta P(k_j,t) = \frac{1}{2N}\sum_{k=0}^{2N-1}\delta p(x_k,t)e^{-ik_jx_k}, \qquad 0 \le j \le (2N-1) \quad (23)$$

To obtain the carriers' concentrations, in the first step, we take Discrete Fourier Transform of the excess electron and hole concentrations at t=0 by using equations (22) and (23). In the second step, equations (18) and (19) are numerically solved at any time step to obtain the excess carrier's concentrations in k-space, δN(k,t) and δP(k,t). Finally, we take the Inverse Discrete Fourier Transform of δN(k,t) and δP(k,t) to obtain the excess carrier's concentrations in x-space, δn(x,t) and δp(x,t), using the following formulas

$$\delta n(x_k,t) = \sum_{j=0}^{2N-1}\delta N(k_j,t)e^{ik_jx_k}, \quad 0 \le k \le (2N-1) \quad (24)$$

$$\delta p(x_k,t) = \sum_{j=0}^{2N-1}\delta P(k_j,t)e^{ik_jx_k}, \quad 0 \le k \le (2N-1) \quad (25)$$

After calculating the excess carriers' concentrations, the electron and hole currents are calculated by using the following formulas, respectively

$$I_e(t) = \frac{qE}{L}\sum_{k=0}^{2N-1}\delta n(x_k,t)\mu_n\Delta x \quad (26)$$

$$I_p(t) = \frac{qE}{L}\sum_{k=0}^{2N-1}\delta p(x_k,t)\mu_p\Delta x \quad (27)$$

The total current passing through the device is the summation of the electron and hole currents given in equations (26) and (27).

4. Simulation Results

Plots of the electron, hole and the total photocurrents are shown in Fig. 3 at 10 volts bias voltage and 6 μm fingers spacing. In each subplot, the upper curve is for electrons and the lower one is for holes. Since electrons have a greater mobility than the holes, they are swept out of the device much more quickly than the holes and this results in a strong increase in the total photocurrent. As can be seen from Fig. 3(a), the electron and hole carriers with the same lifetime have the same FWHM and fall time in their response currents. In Fig. 3(b) holes have a larger lifetime than the electrons and this results in a long tail in the total photocurrent.

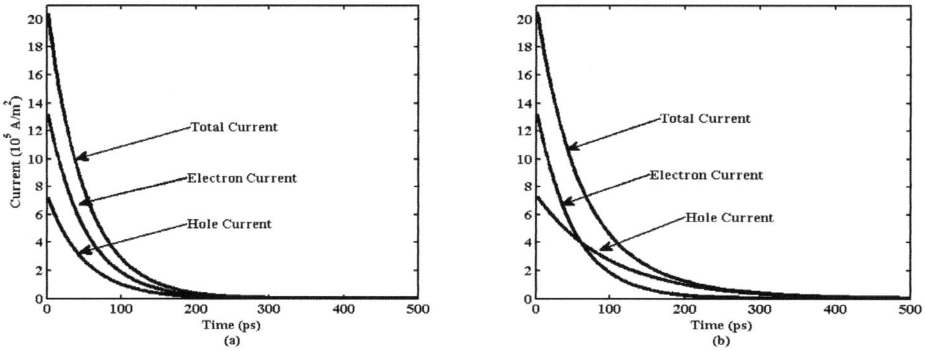

Fig. 2 The electron, hole and the total photocurrent in an MSM photodetector versus time for the cases (a) carriers with the same lifetime (τ_n and τ_p equal to 50 ps), (b) carriers with different lifetime (τ_n=50 ps and τ_p=100 ps). The fingers spacing is 6 μm and the applied voltage is 10 V.

A plot of the hole response current for various lifetimes is shown in Fig. 3.

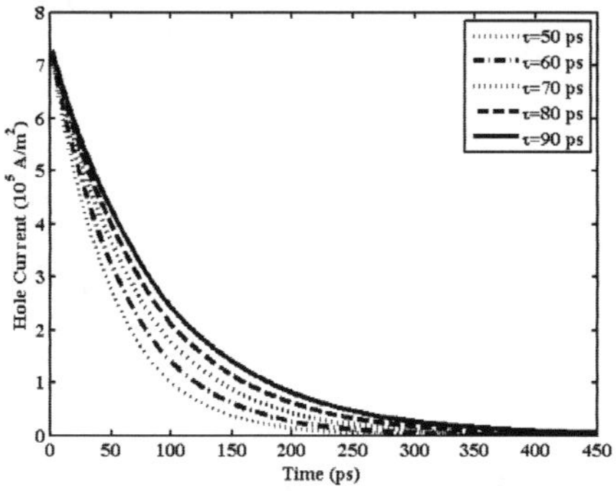

Fig. 3 The hole current of an MSM photodetector for various values of lifetime parameter of 50, 60, 70, 80 and 90 ps at 10 volts applied voltage and 6 μm finger spacing.

As can be seen from Fig. 2 and Fig. 3, hole carriers with shorter lifetime give a faster response time or shorter tail in the response pulse without significantly reducing the device responsivity. These results show good correspondence to the experimental finding of Fig. 4 in ref. [8] and are in agreement with simulation results of the effect of lifetime parameter on the hole impulse response of Fig. 4 and Fig. 5 in ref. [22].

Conclusion

In this paper, by using DFT method, we have numerically solved the continuity equations for the electron and hole carriers with the inclusion of drift, diffusion and bulk recombination terms in one-dimension. Transient photocarrier concentrations and photocurrents are numerically calculated in a low applied field regime. Also the effect of variation of carriers' lifetime on the impulse response of the photodetector is investigated. The simulation results are comparable with the reported experimental finding and published simulation results.

Acknowledgment
The authors would like to thank Adrian Boland-Thoms for his kindness in reviewing and editing this article.

References

[1] S. Y. Chou and M. Y. Liu, "Nanoscale tera-hertz MSM photodetectors," IEEE J. Quant. Electron., Vol. 28, pp. 2358-2368, 1992.
[2] M. Y. Liu, S. Y. Chou, et al., "Nanoscale metal-semiconductor-metal photodetectors with subpicosecond response time fabricated using electron beam lithography," J. Vac. Sci. Technol., B 10(6), p. 2932, 1992.
[3] J. Burn, et al., "Optimization of high-speed metal-semiconductor-metal photodetectors," IEEE Photonics Technology. Letters, Vol. 6, No. 6, p. 722, 1994.
[4] E. John and M. B. Das, " Design and performance analysis of InP-based high speed and high sensitivity optoelectronic integrated reciviers," IEEE Trans, Electron Devices, Vol. 41, No. 2, pp. 162-171, 1994.
[5] Dennis L. Roger, "Integrated optical reciviers using MSM photodetectors," J. Lightwave Technol., Vol. 9, pp. 1635-1638, 1991.
[6] Paul R. Berger, "MSM photodiodes: a new, faster way to ensure lightwaves get their messages across," IEEE Potentials, p. 28, 1996.
[7] S. W. Seo, N. M. Jokerst, "Suppressing long tail in impulse response of MSM photodetectors using slow hole current subtraction," Electronics Letters, Vol. 43, No. 15, , 2007.
[8] V. Krishnamurthy, M. C. Hargis, and M. R. Melloch, "A 4-GHz large-area (160000 μm2) MSM-PD on ITG-GaAs," IEEE photonics technology letters, Vol. 12, No. 1, pp. 71-72, 2000.
[9] Ch. S. Harder, B. J. Van Zeghbroeck, M. P. Kesler, H, P, Meier, P. Vettiger, D. J. Webb, and P. Wolf, "High speed GaAs/AlGaAs optoelectronic devices for computer applications," IBM J. Res. Devel., Vol. 34, No. 4, pp. 568-584, 1990.
[10] S. M. Sze, D. J. Coleman, and A. Loya , "Current transport in metal-semiconductor-metal (MSM) structures," Solid-State Electron., Vol. 14, pp. 1209-1218, 1971.
[11] W. C. Koscielniak, J. L. Pelouard, and M. A. Littlejohn, "Dynamic behavior of photocarriers in a GaAs metal-semiconductor-metal photodetector with sub-half-micron electrode pattern", Appl. Phys, Lett., Vol. 54, pp. 567-569, 1989.
[12] A. E. Iverson and D. L. Smith, "Mathematical modeling of photodetector transient response," IEEE Transactions on Electron Devices. Vol. ED-34, No. 10, p. 2098, p. 2100. 1987.
[13] A. W. Sarto, B. J. Van Zeghbroeck, "Photocurrents in a metal-semiconductor-metal photodetector," IEEE Quantum electronics, Vol. 33, No. 12, p. 2188-2189, 1997.
[14] S. M. Sze, "Physics of semiconductor devices," New York: Wiley, pp. 324-325, 1981.
[15] K. Seeger, "Semiconductor physics," 2nd ed., New York: Springer, 1982.
[16] W. Fichtner , D. J. Rose, and R. E. Bank, "Semiconductor device simulation," IEEE Transaction on Electron Devices, Vol. ED-30, No. 9, p. 1020, 1983.
[17] R. Haberman, "Elementary applied partial differential equations with Fourier series and boundary value problems," 2nd ed., Prentice-Hall Inc., pp. 352-354, 1987.
[18] E. O. Brigham, "The fast Fourier transform and its applications," Prentice-Hall, Inc., p. 22, pp. 131-132, 1988.
[19] J. F. James, "A student's guide to Fourier transforms with applications in physics and engineering," 2nd ed., Cambridge University Press, pp.117-119, 2002.
[20] A. Quarteroni, F. Saleri, "Scientific computing with Matlab and Octave," 2nd ed., Springer-Verlag Italian, pp. 81-83, 2006.
[21] H. B. Wilson, L. H. Turcotte, D. Halpern, "Advanced mathematics and mechanics applications using Matlab," 3nd ed., Chapman & Hall/CRC, ch. 6, 2003.
[22] S. Wu, S. M. Kang, "Modeling and simulation of metal-semiconductor-metal photodetector using VHDL-AMS," IEEE, pp. 61-62, 2004.

Journal of Physics: Conference Series **286** (2011) 012036 doi:10.1088/1742-6596/286/1/012036

Monte Carlo simulation of the magnetic exchange spring system $DyFe_2(1)$ /$YFe_2(4)$

S. Djedai[(1),(2)], **P.E. Berche**[(1)]

[(1)]Groupe de Physique des Matériaux - UMR CNRS 6634, Université de Rouen, BP 12, 76801 Saint Etienne du Rouvray, France

[(2)]Laboratoire de Physique Appliquée et Théorique, Université de Tébessa, BP 12002, Tébessa, Algérie

E-mail: saoussen.djedai@inv-univ-rouen.fr

Abstract. We investigate the magnetic properties of exchange coupled super-lattices constituted by a hard ferrimagnet $DyFe_2$ and a soft ferrimagnet YFe_2. The interface is characterized by a strong antiferrromagnet due to the negative exchange couplings between the Dy and Fe magnetic moments. In this work, we reproduce hysteresis loops for the net and compound-specific magnetisations at different temperatures, by means of Monte Carlo simulations, and assess the quality of our results by a direct comparison to experimental hysteresis loops.

1. Introduction

Lave phase superlattices such as $DyFe_2$ / YFe_2 are interesting materials since they exhibit exchange spring properties [1]. They are exchange coupled systems constituted by a hard ferrimagnet ($DyFe_2$) whose magnetisation is dominated by the Dy moments, and a soft ferrimagnet alloy (YFe_2) in which Y carries a small moment antiparallel to that of iron [2]. The interface antiferromagnetic coupling (due to the negative exchange coupling between Dy and Fe moments) permits us to tailor extremely rich behaviours, such as exchange spring materials [1, 3]. In these hard/soft heterostructures, the classical magnetisation reversal process is expected to occur via the development of exchange springs in the soft part before the irreversible switch of the hard one. The magnetisation of the hard material remains tightly stuck by anisotropy, whereas the magnetic moments of the soft layer, locally pinned by exchange at the interfaces, may be rotated freely by the external magnetic field far from the pinning center. This simple description of magnetic exchange springs sometimes permits us to infer the magnetic profile from macroscopic magnetisation measurements [4]. The technological relevance of these materials ranges from applications as permanent magnets [5], as giant magnetoresistance (GMR) spin devices [6], to usage for advanced recording media [7]. In this paper, we report on the wide thermal evolution of the magnetisation reversal process in a super-lattice $DyFe_2$(5nm)/YFe_2(20nm), and we compare the experimental results to the simulation's one by using the Monte Carlo method. These results have especially revealed an unexpected high temperature situation, where magnetisation reversal first occurs in the harder $DyFe_2$ layers, and the magnetic springs are partly translated from inside the soft YFe_2 layers to the hard $DyFe_2$ ones.

2. Model and numerical simulations

Due to the many degrees of freedom of the spin system, we have chosen to use numerical methods in order to give a detailed microscopic description of the system. Since we are especially interested in the magnetisation reversal process we use Monte Carlo simulations in the canonical ensemble with time step quantification [8]. The model chosen is the classical Heisenberg model [9],

$$H = -\sum_{<i,j>} J_{ij}\mathbf{S_i}.\mathbf{S_j} - D(\mathrm{Dy})\sum_{i\in \mathrm{Dy}}(\mathbf{S_i}.\mathbf{x})^2 - D(\mathrm{Y})\sum_{i\in \mathrm{Y}}(\mathbf{S_i}.\mathbf{x})^2 - \mathbf{B}.\sum_i \mathbf{m_i},$$

where $\mathbf{S_i}$ are spin vectors at site i. The first term represents the exchange energy over nearest neighbours, the second and third sums represent an uniaxial anisotropy which favors the x axis as the easy axis of the magnetisation and the last sum corresponds to the Zeeman interaction with the applied magnetic field. The values of the magnetic parameters have been estimated from experimental data concerning the Curie temperatures of the different species.

3. Simulated versus experimental results at low temperature

$DyFe_2$ / YFe_2 superlattices are characterizd by a high Curie temperature ($T \sim 600$K) [10] due to the strong Fe-Fe exchange interactions. Moreover, the Dy-Fe exchange interactions give rise to antiferromagnetic couplings between the Dy and Fe moments which lead to ferrimagnetic structures when there is no magnetic applied field. At $T = 10$K, the experimental and numerical hysteresis loops are compared in Figure 1 for several superlattices of different thicknesses which exhibit three different behaviours [11, 12].

- The first behaviour is a $DyFe_2$ magnetically dominated behaviour obtained when the thickness of the hard layer ($DyFe_2$) is larger than the soft layer's one ((1) and (4)). In this case, the loop is square as for a single hard phase, there is no signature of magnetic walls in the YFe_2 layers and the magnetisation reversal is completely governed by the anisotropy in the $DyFe_2$ layers. The giant ferromagnetic superlattice behaves as unique block with a one step magnetisation reversal, where the hard and soft magnetic layers switch as a unit.

- The second case corresponds to an almost magnetically compensated behaviour when the thicknesses of the layers are comparable ((2) and (5)). The measured hysteresis loops display a characteristic feature of magnetic exchange spring with

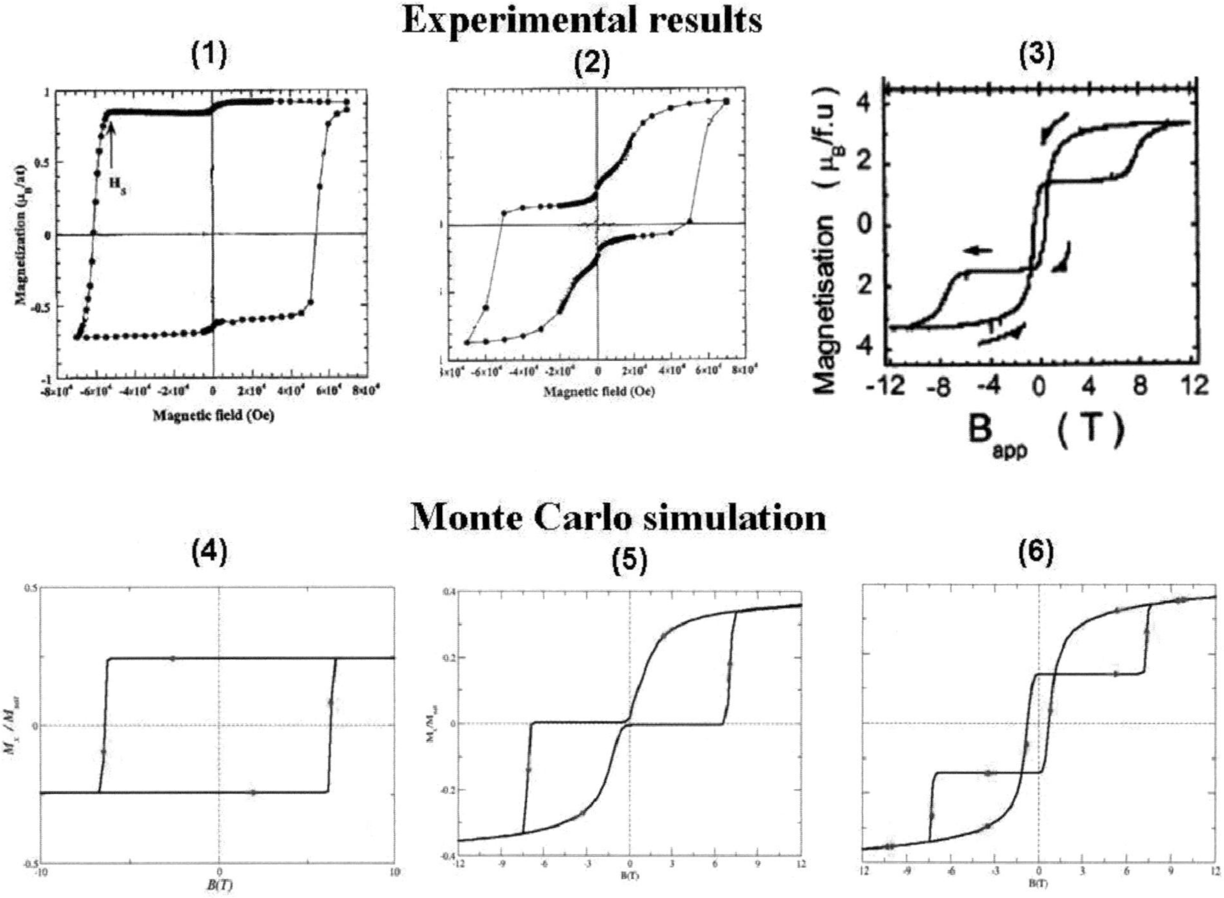

Figure 1. Experimental ((1), (2), (3)) [11, 12] and simulated ((4), (5), (6)) hysteresis loops of $DyFe_2$ / YFe_2 superlattices measured at low temperature (T = 10K). The magnetic behaviour of the samples is strongly related to the thickness of the layers: (1) - $DyFe_2$ (10nm)/YFe_2 (5nm); (2) - $DyFe_2$ (10nm)/YFe_2 (13nm); (3) - $DyFe_2$ (5.5nm)/YFe_2 (22nm). The figures with the experimental data ((1), (2), (3)) are shown with the permission of the authors.

the rise of an interface wall in the soft YFe_2 layer. The reversal of the magnetisation then occurs in two steps: firstly the rise of the interface wall, secondly the abrupt reversal of the hard $DyFe_2$ layer.

- The third behaviour is an YFe_2 magnetically dominated behaviour when the soft layer thickness is much larger than the hard layer one ((3) and (6)). In this case, various schematic spin configurations can be considered, as a function of the applied field. The thickness of the soft magnetic layer YFe_2 is so large relative to the hard layer that the reversal of the magnetisation occurs with a characteristic negative coercivity since the net magnetisation becomes negative even though the applied magnetic field is still positive.

4. Influence of the temperature for the $DyFe_2(1)/YFe_2(4)$ superlattice

The experimental hysteresis loop of the sample $DyFe_2$(5nm)/YFe_2(20nm) at $T = 200$K is shown in Figure 2 ((1)) from XMCD and macroscopic magnetisation measurements [4, 13]. As a comparison, we have also simulated the hysteresis loop of the $DyFe_2(1)/YFe_2(4)$ superlattice ((6) in Figure 1) at $T = 200$K ((2) in Figure 2). We observe that in both cases (experimental or simulated) the magnetisation reversal then occurs in a three step process and that the coercive field is positive contrary to the case at low temperature (Figure 1) or even at $T = 100$K.

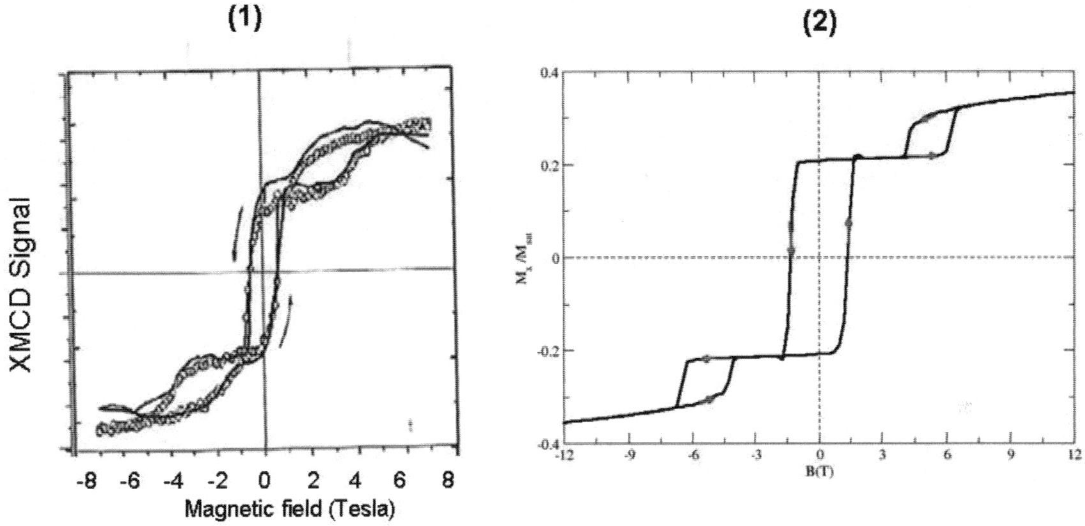

Figure 2. Experimental hysteresis loop (1) of a $DyFe_2$(5nm)/YFe_2(20nm) superlattice obtained from XMCD and macroscopic magnetisation measurements [4, 13] and numerical hysteresis loop (2) measured at $T = 200$K. The figure with the experimental data ((1)) is shown with the permission of the authors.

Then, these results suggest a strong temperature dependence of the magnetisation reversal. In this sample, the YFe_2 magnetisation is dominant at all temperature but its influence becomes stronger when the temperature increases since the Dy magnetic moments become disordered. Thus, the reversal of the magnetisation evolves from a soft first mechanism below 200K (Figure 1) to a very unusual mechanism above this temperature, where the magnetisation reversal first affects the hard $DyFe_2$ layers, while the soft YFe_2 magnetisation remains aligned with the applied field direction. A sharp transition between these two processes has been observed, that account for the thermal evolution of coercivity from a negative to a positive value.

5. Conclusion

In this article, it was shown that the Monte Carlo method, adapted to include magnetic exchange springs, can be used to provide a good semi-quantative interpretation of $DyFe_2$ /YFe_2 multilayers. In paticular, we have been able to reproduce hysteresis loops with characteristics matching those of experimental work at different temperatures. The strong thermal dependence of anisotropy plays a key role for the reversal processes, and the interplay with exchange and Zeeman energy is understood on a qualitative level.

Acknowledgments

The simulations were performed at the Centre de Ressources Informatiques de Haute Normandie (CRIHAN) under the project No. 2008001. Moreover, the authors are indebted to Dr Catherine Dufour and Dr Karine Dumesnil from the Institut Jean Lamour of Nancy University for valuable discussions on the experimental results.

References

[1] Dumesnil K, Dutheil M, Dufour C and Mangin Ph 2000, *Phys. Rev.* B **62** 1136
[2] Dumesnil K, Dufour C, Mangin Ph, Fitzsimmons M R, Park S, Rhyne J J, Rogalev A and Borchers J A 2005 , *J. Appl. Phys.* **97** 10K108
[3] Sawiki M, Bowden G J, De Groot P A J, Rainford B D, Beaujour J M L, Ward R C C and Wells M R 2000, *Phys. Rev.* B **62** 5817
[4] Dumesnil K, Dufour C, Mangin Ph, Wilhelm F and Rogalev A 2004, *J. Appl. Phys.* **95** 6843
[5] Kneller E F and Hawig R 1991, *IEEE Trans. Magn.* **27** 3588
[6] Gordeev S N, Beaujour J M L, Bowden G J, Rainford B D, De Groot P A J, Ward R C C, Wells M R and Jansen A G M 2001, *Phys. Rev. Lett.* **87** 186808
[7] Ando T and Nishihara T 1997 *IEEE Trans. Magn.* **33** 2983
[8] Nowak U, Chantrell R W and Kennedy E C 2000, *Phys. Rev. Lett.* **84** 163
[9] Hinzke D and Nowak U 1998, *Phys. Rev.* B **58** 265
[10] Beaujour J M L, Gordeev S N, Bowden G J, De Groot P A J, Rainford B D, Ward R C C and Wells M R 2001, *Appl. Phys. Lett.* **78** 964
[11] Dumesnil K, Dufour C, Mangin Ph and Rogalev A 2002, *Phys. Rev.* B **65** 094401
[12] Bowden G J, Beaujour J M L, Zhukov A A, Rainford B D, De Groot P A J, Ward R C C and Wells M R 2003, *J. Appl. Phys.* **93** 6480
[13] Dumesnil K, Dufour C, Mangin Ph, Rogalev A and Wilhelm F 2005, *J. Phys.: Cond. Matt.* **17** L215

Journal of Physics: Conference Series **286** (2011) 012037 doi:10.1088/1742-6596/286/1/012037

Study of Porous Materials Acoustic Signatures Behaviour in Dark Field

S Bouhedja[1], F Hamdi[1], A Doghmane[2] and Z Hadjoub[2]
[1]Laboratoires des Hyperfrequences et Semi-conducteurs, Université de Constantine ; B.P. 125, DZ-25000 Algeria
[2]Laboratoire des Semi-Conducteurs, Département de Physique, Faculté des Sciences, Université Badji-Mokhtar, BP 12, Annaba, DZ-23000, Algeria.

E-mail: bouhedja_samia@yahoo.fr ; a_doghmane@yahoo.fr

Abstract. Several kinds of lens-transducer system exist in the scanning acoustic microscope. In this work, annular lenses are chosen in order to quantify the occultation limiting angle to suppress Rayleigh mode generation. Hence, we have numerically simulated, through variable occultation of generated rays at the lens center, the porous silicon acoustic signatures at an operating frequency of 142 MHz. In non destructive control, this investigation is of a great importance in the measurement of the surface waves attenuation. The obtained results enabled us to evaluate the maximum relative occultation at Rayleigh waves.

1. Introduction

Scanning acoustic microscope (SAM) is a key technique not only in qualitative micro characterizations, but also in quantitative measures [1, 2]. In micro-analysis, it can be used to directly determine the mechanical properties of materials via acoustic signatures, V (Z); this signature is obtained by recording the change in the reflected acoustic signal with the defocusing distance, z, as the sample is moved towards the acoustic lens [3]. Hence, acoustic microscopy is applied in fundamental research as well as in industry. It can successfully be used in the investigations of porous materials in general and porous silicon, PSi, in particular.

Several types of acoustic microscopes can be designed with various forms of transducer lens systems such as linear acoustic microscopy, spherical concave transducer, flat lens microscope, low opening lens-angle and annular lenses etc. [4, 5]. For annular lens acoustic microscope, the incident energy on the central part of the transducer is blocked by using an absorbing stop that blocks the transmission of the central ultrasound beam. This type of lens was mainly used in a qualitative way in the dark field imaging to examine objects with high transmission and low diffraction [6, 7]. On the other hand their applications in quantitative control remain still a subject of investigations.

The dimensions of these stops could be varied in order to study the influence of the occultation of the central energy in acoustic materials signatures. This leads to a better understanding of the interaction incident and reflected beams. Hence, this would be of great importance in optimizing the characteristics of transducer-lens system as well as improving the generation of surface modes.

In this context, we proceed via a program that simulates the oscillations of the acoustic signature, in dark field. We first show the effects of porosity on the Rayleigh mode and the amplitude of the acoustic signature. Then, we evaluate the angle of maximum decrease and therefore the attenuation of the signal V (z) at this critical angle.

2. Methodology

In the theory of acoustic microscopy, the angular spectral model [8] is the most widely used to calculate the acoustic materials signature which is written as:

$$V(z) = \int_0^{\pi/2} P(\theta)^2 R(\theta)^2 \sin\theta \cos\theta \exp(2ikz\cos\theta)d\theta \quad (1)$$

$R(\theta)$ is a complex function which allows amplitude and phase is given by:

$$R(\theta) = \frac{(Z_L Cos^2 2\theta_T + Z_T Sin^2 2\theta_T - \rho_{Liq} V_{Liq} / Cos\theta)}{(Z_L Cos^2 2\theta_T + Z_T Sin^2 2\theta_T + \rho_{Liq} V_{Liq} / Cos\theta)} \quad (2)$$

In this work we focus on the effect of the dark field on the attenuation of Rayleigh waves propagating in PSi. We have therefore used the relations given by the theory of elastic media [9, 10] to write the dependence of velocity, V, and porosity, P, as:

$$V_{RP} = V_R(1-P)^{1.095} \quad (3)$$

Porosity is, by definition, the ratio of the volume of the vacuums to the total volume of the body considered. Its value remains unchanged by dilation/contraction of all dimensions of the structure) [11]. The density of *PSi* is also given by the following expression:

$$\rho_{PSi} = \rho_{Si}(1-P) + \rho_{Liq} P \quad (4)$$

3. Condition of simulation

The removal of surface modes has been applied to lenses with a large half-opening angle (50°) whose center was occulted by stops of variable diameters. The methodology consists of covering, by an absorbing material of circular form, the central part of the lens [5] then calculating R(θ) and V(z) for each occulted angle, θ_{occ}. In order to compare some results, we considered both crystalline and porous. It should be noted the PSi possess good photoluminescent properties [10.12] that are applied in many modern technologies and the industrial fields.

Calculations are performed under normal operating condition of simulation of a scanning acoustic microscope operated in the reflection mode: the working frequency is 142 MHz, a half-opening angle of 50 ° and water as coupling fluid (velocity = 1500 m/s and density (ρ) = 1000 kg/m^3) and silicon (V_L = 9160 m/s, V_T = 5085 m/s, ρ_{Si} = 2300 kg/m^3). Note that these conditions favour the excitation of the Rayleigh mode which represents the most dominant mode.

4. Influence of illumination intensity on V(z): results and discussion

Figure 1 illustrates the amplitude (– – –) and the phase (——) of the reflection coefficient as functions of the incidence angle for water/silicon interface. R(θ) provides important information on the angles of different excited modes. It is clear that as θ increases, the longitudinal mode appears first at θ_L = 9.43° and the transverse mode occurs at θ_T = 17.16°. Beyond it, all the energy is reflected, this is due to the absence of transmission in the solid and, therefore, the module R (θ) becomes constant and equal to unity. However, the biggest fluctuation in phase occurs a few degrees after θ_T; this corresponds to the excitation of Rayleigh waves at θ_R = 18.54 °. Note that the R(θ) phase changes with

2π for a very small change in θ. The shaded area corresponds to the different diameters of stop absorbing.

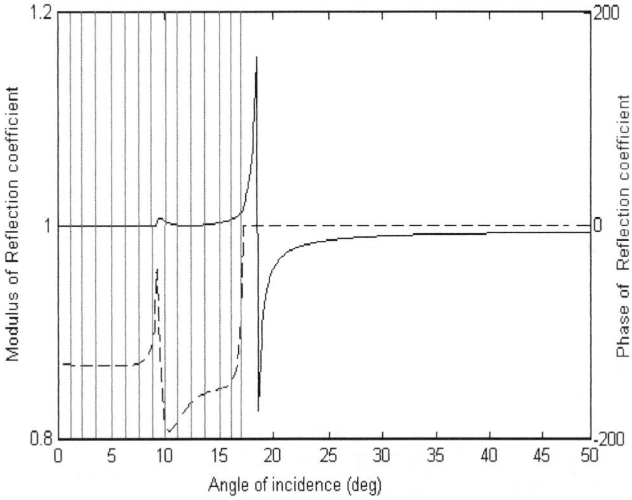

Figure 1. Amplitude (– – –) and phase (——) of the reflection coefficient as functions of incidence angle for water/silicon interface.

4.1. Bulk Silicon

Figure 2 shows V(z) curves at different occulted angles for the structure water/Silicon. It is clear that as θ_{occ}, increases the apparent period increase and the oscillation amplitude decreases These results are in agreement with those recently reported [13, 14]. For the largest occulted angle, the amplitude of the oscillations of the acoustic signatures tends towards an exponential decay of type: $Ae^{-\alpha z}$ where A is the amplitude of V(Z) at the focal distance ($Z = 0\ \mu m$) and α is the total attenuation of the wave of Rayleigh in material considered.

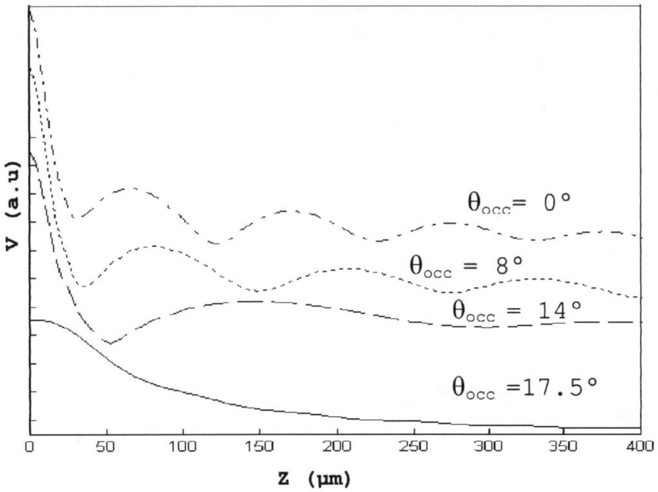

Figure 2. Effects of occultation angles on acoustic signatures for Water/Silicon.

This attenuation can be evaluated by using an optimization method for fitting of the curve V(Z) by a mathematical function; it is found to be 0.041 μm^{-1} in the case of water/Silicon structure.

4.2. Porosity effects

Porosity effects on V(z) curves are shown in Figure 3 for water/PSi (at $P = 20\%$) and in Figure 4 for water/PSi (at $P = 47\%$). Similar observations, as above, can be formulated.

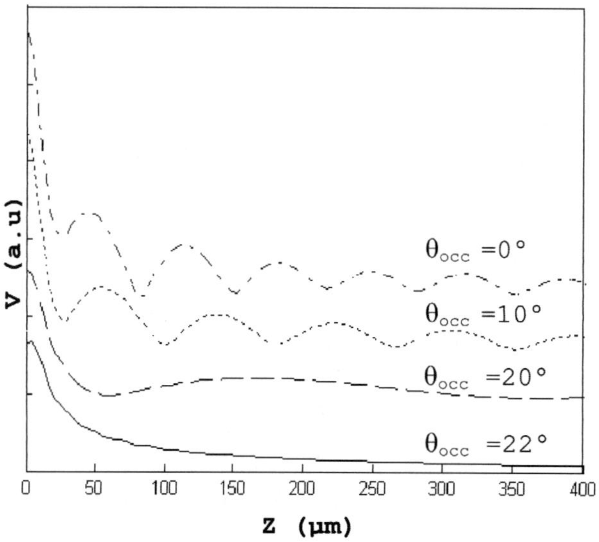

Figure 3. Effects of occultation angles on acoustic signatures for Water/PSi at P=20%.

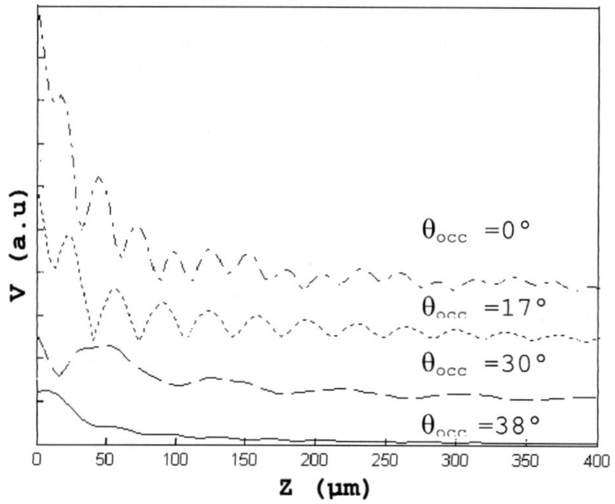

Figure 4. Effects of occultation angles on acoustic signatures for Water/PSi at P=47%.

The decrease in amplitude of signatures V(Z) as well with focal distance ($Z = 0\ \mu m$) as for the other distances from defocusing is explained by the fact that the distribution of energy at the transducer decreases. For a porosity of 20%, we note that with $\theta_{occ.} = 17.5°$, the oscillations persist and disappear only at $\theta_{occ.} = 22°$. Consequently we find the value of the total attenuation equalizes with $0.038\mu m^{-1}$. For 47% porosity, the oscillations disappear for $\theta_{occ} = 38°$ accompanied by a total attenuation of about $0.023\ \mu m^{-1}$.

In order to quantify the effects of porosity, we plot on Figure 5 the shift the critical angle of excitation of Rayleigh, θ_R, as a function of porosity. Two important observations are deduced from this study: the first is the shift of the critical angle of excitation of Rayleigh, θ_R, to large values when the porosity increases. This is explained by the presence of air in the pores.

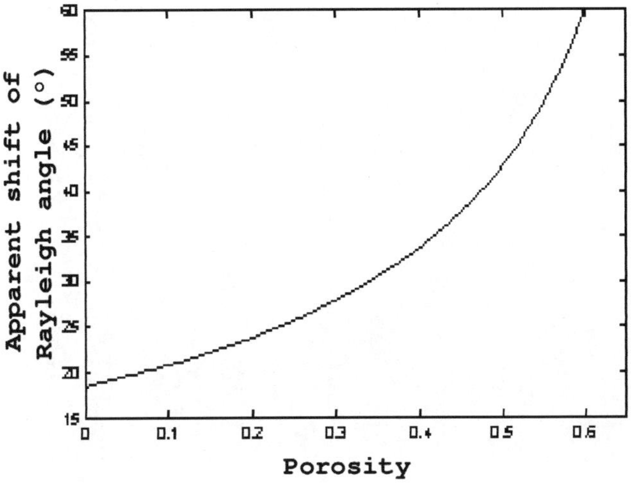

Figure 5. Effects Variation of the apparent shift of Rayleigh Angle with Porosity.

The second consequence concerns the wave attenuation. We found that the value of the attenuation decreases as the porosity increases. This result is physically acceptable since the Rayleigh velocity of PSi decreases with porosity. In the same context, we calculated the values of the attenuation of the wave function of the porosity at the same angle occulted and we found that the attenuation values increased gradually as the porosity increases. This is due the presence of air in the pores that leads to ultrasonic attenuation.

5. Conclusion

In quantitative dark field investigation, we found that the decrease of the energy distribution at the transducer strongly influences both the amplitude of acoustic signatures as well as their periods.This threshold angle of total disappearance of V(z) oscillations shift to large values when the porosity increases. The decay in V(z) curves with porosity is related to Rayleigh total attenuation. This parameter is found to be much lower for porous Silicon. Note that the attenuation at the same angle of occultation increases when the porosity increases.

6. References

[1] Briggs A 1995 *Advances in Acoustic Microscopy* Ed (Plenum Press, New York Technical).
[2] Zinin P V 2001 *Handbook of Elastic Properties of Solids, Liquids and Gases*, Ed Levy, Bass and Stern (Academic Press) **vol 1**. p.187
[3] Achenbach J D 1993 Ed Evaluation of Materials and Structures by Quantitative Ultrasonic, (Springer, Verlag)
[4] Hadjoub Z, Alami K, Doghmane A, Saurel J M, Attal J, 1991 *Electron.Lett.* **27** 11 981
[5] Doghmane A, Hadjoub Z, 1997 *Appl. Phys.* **30** 2777
[6] Smith I R, Wickramasingh, H K e. 1982 *Electron. Lett,* **18** 22 955.
[7] Al-Surayhi I, Doghmane A, Hadjoub Z 2009, *Damage and Fracture Mechanics* (Springer Verlag) p 415
[8] Sheppard C G R, Wilson T. 1981 *Appl. Phys. Lett.* **38** 858.

[9] Da Fonsesa R J M Y, Saurel J M and Despaux G. 1994, *Appli, Surf,* **16**, 21
[10] Boumaiza Y, Hadjoub A, Doghmane A, Deboub L 1999; *J. Mat. Scie. Lett* **18** 295
[11] Bourbié T, Coussy O, Zinszner B 1986, Ed TEHNIP.
[12] Da Fonsesa R J M, Saurel J M, Foucaran A 1995 *Appli, Surf, Scie,* **255** 155
[13] Bouhedja S, Hadjoub I, Doghmane A & Hadjoub Z 2005, *Phys.Stat. Soli. (a)* **202** 1025-1032
[14] Bouhedja S, Hadjoub I, Doghmane A and Hadjoub Z 2009 *J. optoelec. Advanced. mat. Symp,* **1** 420-423

Optical Design and Fabrication of Fully Sputtered CdTe/CdS Solar Cells

R. E. Treharne, A. Seymour-Pierce, K. Durose

Department of Physics, Durham University, South Road, Durham. DH1 3LE

E-mail: r.e.treharne@durham.ac.uk

K. Hutchings, S. Roncallo, D. Lane

Department of Applied Science, Security and Resilience, Cranfield University, Shrivenham, Swindon. SN6 8LA

Abstract. A model is presented for optimising the integrated intensity of light passing into the absorber layers of thin-film solar cells. Dispersion data for each of the transparent layers in the cells was obtained from variable angle spectroscopic ellipsometry. This data was then used in a transfer matrix model of multi-layer optics to determine the intensity spectrum passing over the interface from the transparent to the absorbing parts of the cell. Application of the method to the case of CdTe/CdS/ZnO/ITO/glass solar cells is described in detail, including the fitting of oscillator models to extract the complex dielectric function from the ellipsometry data. It was found that light coupling into the absorber was particularly sensitive to the CdS thickness but that tuning the ZnO (highly resistive transparent) layer was also influential. All-sputtered cells of this typed were demonstrated as having efficiencies of 12.5% for 5×5 mm^2 contacts.

1. Introduction

There are some significant opportunities to further increase the conversion efficiencies of thin-film solar cell devices; the present day best efficiencies are 16.5% for sublimation-grown [1] and 14.5% for sputtered [2] devices. These achievements fall short of the $\sim 30\%$ limit predicted for single junction devices. It is therefore likely that gains may be made, for example by optimising doping or by reducing the influence of polycrystallininty. It is the object of this paper to explore the *optical* gains in efficiency that may arise from optimising the thicknesses of both the window and transparent conducting oxide (TCO) layers of thin film solar cells. While the measurement and modelling methodology used in this work has general applicability, it is developed and tested here for the CdTe/CdS/ZnO/ITO/glass device as an exemplar. Industrially, SnO_2 is used as the transparent conductor, but here ITO (indium tin oxide) was chosen for its compatibility with an in-house all-sputtered cell fabrication process. The CdS is n-type, and is a window layer admitting light into the p-type CdTe absorber. It is widely reported [1, 3, 4] that inclusion of a so-called 'high resistive transparent' (HRT) layer between the TCO and the window layer acts to enhance efficiency, although there is no consensus on the mechanism by which this occurs. In this work the HRT is ZnO.

Here, optical optimisation of the design of combined window and oxide multi-layers is investigated using the following methodology: Optical dispersion (or the refractive index and extinction coefficient) are obtained from ellipsometry [5, 6, 7]. This allows optimisation of the layer thicknesses, the results of which inform cell designs for experimental fabrication. In ellipsometry, the change in components of the reflected light are measured and related to the two standard ellipsometric parameters Ψ and Δ according to

$$\frac{R_p}{R_s} = \tan \Psi e^{i\Delta} \tag{1}$$

In the variant of this technique known as variable angle spectroscopic ellipsometry (VASE), both the angle and wavelength give additional experimental variables that contribute to the robustness of the data interpretation. The experimental spectra of Ψ and Δ are fitted using models of the complex frequency dependant dielectric function $\varepsilon(\omega)$. Physical models that are available to describe the dielectric function include:

(a) Drude model [8]. Used for conductive materials in which there is a cut-off in transmission (i.e high reflectivity) in the near infra-red.

(b) Lorentz oscillators [9, 10]. Apropriate for insulators and accounts for atom-dipole interactions.

(c) PSEMI-M0 oscillators. Based on a model by Johs [11], and being appropriate for direct-gap semiconductors.

The usual procedure is to fit the measured spectra of Ψ and Δ by invoking appropriate oscillators. $\varepsilon(\omega)$ data obtained in this way gives the dispersion of both refractive index and of extinction coefficient (n and κ) directly. It is this data that is carried forward for modelling the optical transmission through multi-layer structures. The approach is to write the electromagnetic boundary conditions as simultaneous equations that are amenable to matrix algebra. This so-called 'transfer matrix method' allows the transmission and reflection spectra for multi-layers of thin films to be calculated and is described in detail in book-length reviews [12].

2. Experimental

Solar cells composed of a CdTe/CdS/ZnO/ITO/glass structure and samples of each of the individual layers were fabricated by radio-frequency sputtering. An AJA International Orion twin chamber system having 3" targets (obtained from Pi-Kem Ltd) was used, one chamber being dedicated to growth of the oxides and a vacuum transfer between the two being possible. The sputter targets were of the compounds themselves, with the ITO comprising 90% In_2O_3 and 10% SnO_2 by weight. All substrates were low iron soda-lime glass (Pilkington OptiWhite^{TM}).

Variable angle spectroscopic ellipsometry (VASE) was undertaken using a Woollam M2000UI instrument operating in reflection mode in the range 250-1500 nm, and at angles of 50, 60 and 70°. Fitting of the spectra was achieved using the manufacturer's software. Transmission of the layers and stacks was measured using a Shimadzu Solid Spec UV-VIS-IR 3600 spectrophotometer. Transmission spectra were modelled using the transfer matrix method incorporating both dispersion of refractive index and extinction coefficient. Coding was done in FORTRAN95.

3. Results and Discussion

Ellipsometry data was obtained for each of the individual layers ITO, ZnO, CdS and CdTe. As an example, figure 1 shows the spectra of both Ψ and Δ (250-1500 nm) as function of angle. Fits obtained using appropriate oscillator models are shown as solid lines. For the ITO, which is considered to be highly conductive, it was necessary to use one Drude and two Lorentz oscillators. However, for the undoped ZnO, a single PSEMI-MO oscillator was sufficient, while both CdS and CdTe required two. Full details of the models chosen, and of the fitting procedures shall be presented elsewhere. The ITO fitting procedure generated a range of materials parameters that are compared to experimental and literature values in table 1. It is a confirmation of the model's validity that the thickness, carrier concentration and mobility, effective mass, band gap and permittivity all conform to the expected measured values.

Figure 2 shows the dispersion of n and κ in the wavelength range 250-1500 nm derived from the ellipsometry data for ITO, ZnO, CdS and CdTe. It is this data that was used in the transfer matrix model to simulate experimental transmission spectra for each of the individual layers in order to verify the procedure. Figure 3 shows examples of the simulated transmission spectra, and the corresponding experimental data. In figure 3a, a close correspondence is demonstrated for both ITO and for OptiWhite glass. Figure 3b shows how the modelling was able to predict the transmission spectrum of a test structure that was grown by sputtering and measured by spectrophotometry. The structure was CdS(120 nm)/ZnO(115 nm)/ITO(200 nm), with the experimental points

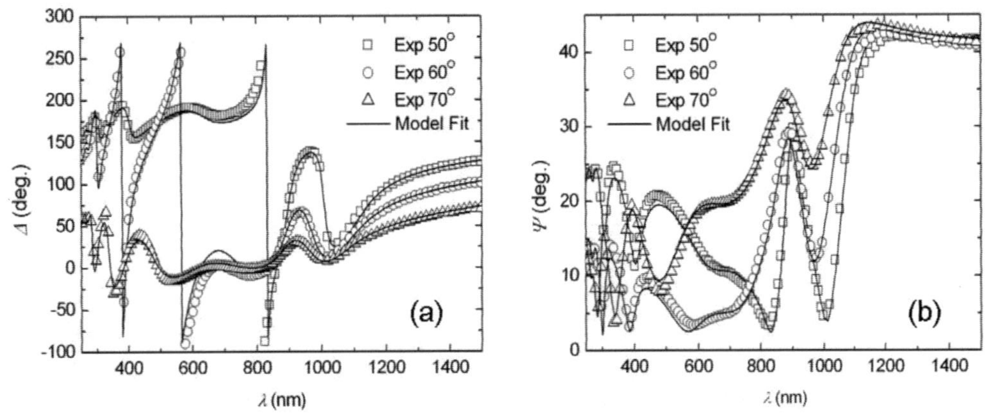

Figure 1. Experimental plots of a) Ψ and b) Δ of ITO film and the corresponding model fits

Table 1. Key parameters extracted from ellipsometric model for ITO and a comparison with measured or reference values.

	From Model	Comparison
d (nm)	237	230[1]
n_e ($\times 10^{21}$ cm^{-3})	1.3	1.2[2]
μ_e (cm^{2}V^{-1}s^{-1})	42	40[2]
m_e^*	0.4	0.35 [13]
E_G (eV)	4.1[3]	3.75-4.3 [14, 15]
ε_∞	3.0	3.95 [13]

[1] Surface profilometry measurement
[2] Hall measurement
[3] Extrapolated from a $(h\nu\alpha)^2$ vs $h\nu$ plot

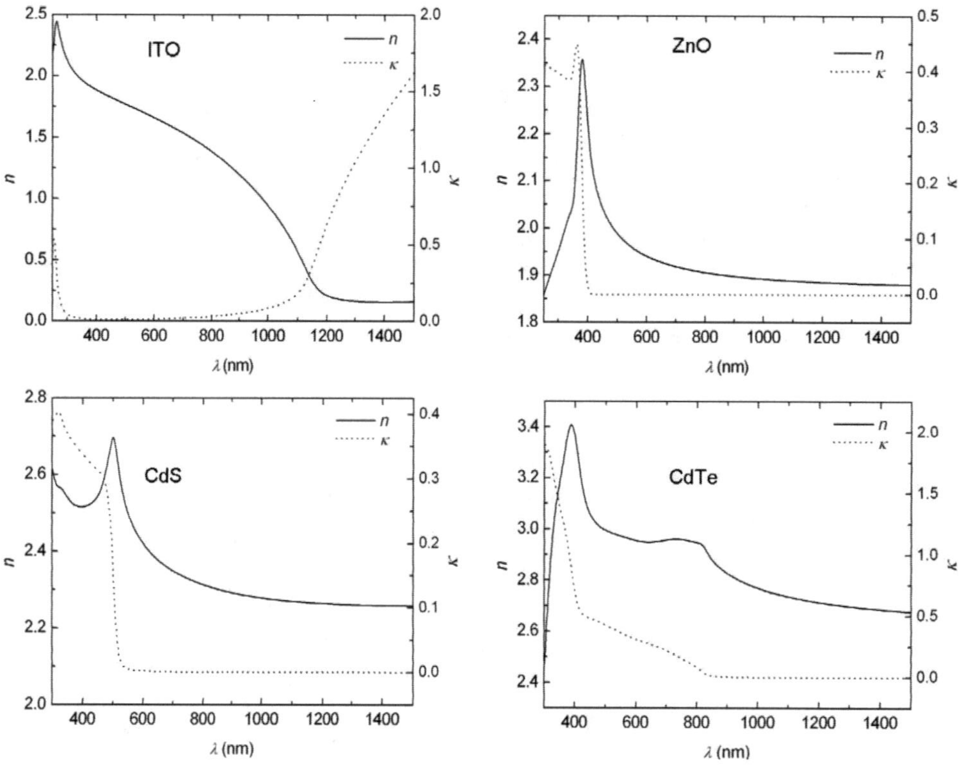

Figure 2. Extracted dispersion (n and κ) for ITO, ZnO, CdS and CdTe

in the figure conforming closely to the output from the model. Hence it was established with some confidence that the model is appropriate for use in estimating the transmission into the CdTe through the full CdS/ZnO/ITO/glass combination. The following results

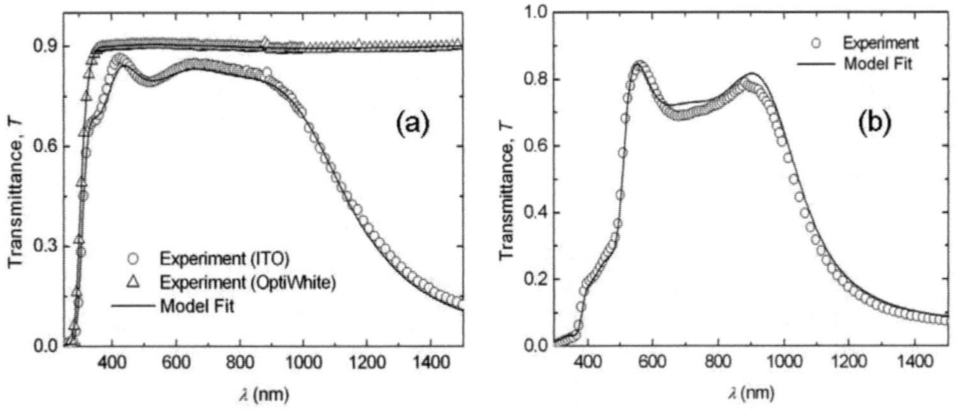

Figure 3. (a) Experimental transmission curves for ITO and OptiWhite substrate, and corresponding model fits. (b) Multi-layer model fit to the transmission curve for a CdS/ZnO/ITO/glass (200/115/120 nm) stack. Note that the exit medium in this case is air and not CdTe.

and discussion are for spectra calculated for light passing over the CdS-CdTe interface (whereas those in figure 3a assumed that transmission was into air rather than into CdTe). Note that the CdTe layer is assumed to have an infinite thickness (i.e all light is absorbed in this layer). A selection of the structures investigated are described in table 2. For each a full transmission spectrum was calculated, and from this the normalised integrated transmission into the CdTe absorber was evaluated as follows:

$$\tilde{T} = \frac{1}{b-a}\int_a^b T d\lambda \tag{2}$$

where $\tilde{T}$ is the average fractional transmission over the range a-b specified. For CdTe/CdS devices, the range 300-850 nm was selected.

Table 2. A selection of ITO/ZnO/CdS stack structures showing the potential for increased transmittance in structures with thinner layers.

	d_{ITO} (nm)	d_{CdS} (nm)	d_{ZnO} (m)	$\tilde{T}$ (300-850 nm)	% change[1]
A	230	200	100	0.53	-
B	230	50	50	0.65	+22.6
C	115	50	50	0.67	+26.4
D	115	50	25	0.70	+32.1
E	115	25	25	0.71	+33.9
F	80	25	25	0.72	+35.3

[1] Relative to current baseline structure, A

Generally it was found that small changes in the thicknesses of CdS and ZnO had a greater influence on $\tilde{T}$ than did changing the ITO thickness. This may be seen by comparing the baseline structure A to structure B (Table 2), in which the CdS and ZnO are thinned, which results in an increase in $\tilde{T}$ of +22.6%. Thinning the ITO (C) causes a further +3.8% increase in transmission. Further reduction in the CdS and ZnO thickness (D and E) allow for further gains. This is explored further in figure 4, in which the ITO thickness is fixed at 115 nm and $\tilde{T}$ is shown as a function of both the CdS and ZnO thicknesses. It is seen in the figure that changing the CdS thickness for a given ZnO thickness has a greater effect than *vice-versa*. For example, for a constant 50 nm of ZnO, reduction of the CdS from 200 to 50 nm increased $\tilde{T}$ from 53 to 73%. However, for 50 nm of CdS, the same reduction in ZnO thickness acted to increase $\tilde{T}$ from 64 to 72%. This behaviour may be accounted for a by examining the transmission spectra in figure 5 in which the absorption loss in CdS (below $\sim$ 500 nm) is revealed in the progression $F \rightarrow C \rightarrow A$. Nevertheless, reduction in the ZnO thickness has a significant effect on $\tilde{T}$, and by extension, possibly on cell efficiency. Overall, gains from optical optimisation of $> 30\%$ may be expected over the performance of the baseline structure A in table 2.

It must however be emphasised that these findings are based upon optical considerations only. The electrical impact of changing the layer thickness is not accounted for here - optical optimisation may well require different thicknesses than for electrical optimisation. Moreover, the materials data used in the modelling here was extracted from the layers in their as-grown state. Therefore the effects of processing, such as inter-diffusion, are not accounted for. In optimising solar cells, it is therefore essential to make real devices, and to use the results of this kind of optical study to inform the choice of layers and thicknesses for use in the experiments.

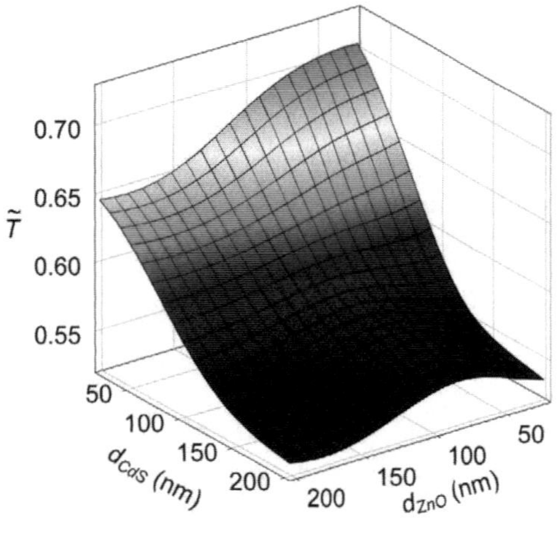

Figure 4. The variation in $\tilde{T}$ shown as a function of both ITO and CdS thickness while maintaining a constant ITO thickness of 115 nm.

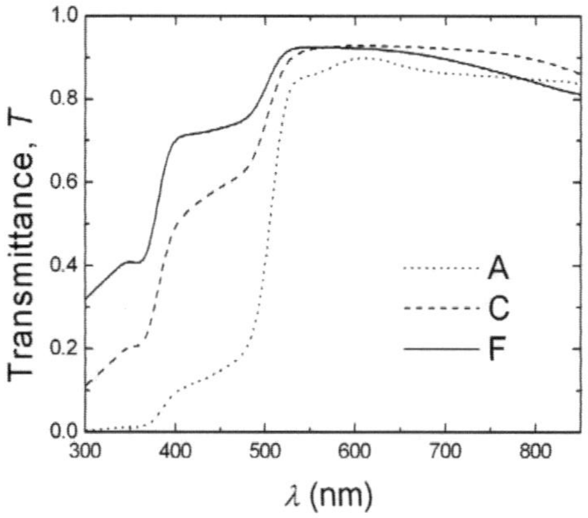

Figure 5. A selection of ITO/ZnO/CdS stack structures showing the potential for increased transmittance in structures with thinner layers.

Preliminary cells were therefore fabricated in preparation for the next stage of the investigation. The film thicknesses used were CdTe(3.4 μm)/CdS(200 nm)/ZnO(50 nm)/ITO(200 nm)/glass. Cells having 5×5 cm^2 gold contacts were tested under AM1.5 spectral conditions at 1000 W.m^{-2}. The best working parameters obtained were V_{OC} (open circuit voltage) = 820 mV, J_{SC} (short circuit current density) = 24.9 mA.cm^{-2} and FF (fill factor) = 61% yielding an efficiency of 12.4%. This compares well with the best reported device of this kind (14.5%) [2], and hence provides the basis for continuing experiments. In further work a systematic study of varying the window and buffer layer design in real devices shall be undertaken, with the choices being informed by the results of the optical modelling.

4. Conclusions

Optical coupling into the absorber layers of thin film solar cells - after transmission through the TCO, high resistive transparent, and window layers - has been investigated. It was evaluated by a transfer matrix model that employed experimentally derived dispersion data for both the refractive indices and extinction coefficients of each material involved. The methodology was applied to cells comprising CdTe/CdS/ZnO/ITO, with dispersion data being extracted from as-grown sputtered films using variable angle spectroscopic ellipsometry. This involved modelling the dielectric dispersion using physical models that were appropriate for each layer, which then yielded optical dispersion data directly. Spectra for optical transmission into the absorber layer itself were derived for a series of test structures having systematically varied layer thicknesses. It was found that varying the ITO thickness had a relatively weak effect on optical transmission into the absorber, while the CdS and ZnO were more influential. The greatest effect was achieved by thinning the CdS to 25 nm, with gains of $\sim$ 20% being possible with respect to a test structure with 100 nm thick CdS - this is expected from the blue absorption of CdS. Nevertheless, optical optimisation of the ZnO thickness may be expected to yield further gains of $\sim$ 5%.

Since the model is based on data for as-grown layers only, materials changes that occur during processing may be expected to influence the cell behaviour in practice. Moreover, optimisation based on optical criteria alone may compromise electrical performance. It is therefore essential to carry out optimisation of cells by experiment, using the results of this type of study as a starting point. Preliminary all-sputtered cells fabricated during this work achieved a maximum efficiency of 12.5%, this being an excellent starting point for optically informed device optimisation.

Acknowledgements

The authors would like to thank Neil McSporran and Paul Warren of Pilkington for useful discussions. They are also grateful to Michelle Sestak of Toledo University for helpful discussion on elliposmetry. Funding for the PV-21 project was provided by the EPSRC SUPERGEN programme.

References

[1] Wu X, Keane J C, Dhere R G, DeHart C, Albin D S, Dudam A, Gessert T A, Asher S, Levi D H and Sheldon P 2001 *Proc. 17th Eur. PVSEC (Munich)* 995

[2] Gupta A and Compaan A 2004 *App. Phys. Lett.* **85** 684
[3] Bosio A, Romeo N, Mazzamuto S and Canevari V 2006 *Prog. Cryst. Growth Charact. Mater.* **52** 247
[4] Ferekides C S, Marinskiy D, Viswanathan V, Tetali B, Palekis V, Selvaraj P and Morel D L 2000 *Thin Solid Films* **361** 520
[5] Durose K, Asher S, Jaegermann W, Levi D, McCandless B, Metzger W, Moutinho H, Paulson P, Perkins C, Sites J, Teeter G and Terheggen M 2004 *Prog. Photovoltaics: Res. App.* **12** 177
[6] Paulson P and Mathew X 2004 *Sol. Energ. Mat. Sol. Cells* **82** 279
[7] Li J, Podraza N and Collins R 2008 *Phys. Stat. Sol. A* **205** 901
[8] Edwards P P, Porch A, Jones M O, Morgan D V and Perks R M 2004 *Dalt. Trans.* 2995
[9] Synowicki R 1998 *Thin Solid Films* **313** 394
[10] Jung Y 2004 *Thin Solid Films* **467** 36
[11] Johs B, Herzinger C, Dinan J, Cornfeld A and Benson J 1998 *Thin Solid Films* **313** 137
[12] Macleod H A 1986 *Thin-film Optical Filters* (Bristol: Adam Hilger Ltd)
[13] Hamberg I and Granqvist C 1986 *J. Appl. Phys.* **60** R123
[14] George J and Menon C 2000 *Surf. Coat. Tech.* **132** 45
[15] Terzini E, Thilakan P and Minarini C 2000 *Materials Science and Engineering B* **77** 110

Evaporation and decomposition of acrylic acid grafted luminescent silicon quantum dots in ultrahigh vacuum

Yimin Chao[1], Qi Wang[1], Paul Coxon[1] and Alex Walton[2]

[1] Energy Materials Lab, School of Chemistry, University of East Anglia, Norwich NR4 7TJ, United Kingdom
[2] School of Physics and Astronomy, University of Leeds, Leeds LS2 9JT, United Kingdom

E-mail: y.chao@uea.ac.uk

Abstract. A thin film of silicon quantum dots (Si-QDs) is a potential thermoelectric material with a high figure of merit. Evaporation and deposition in ultrahigh vacuum is a novel method to produce such a thin film of high quality. Acrylic acid grafted luminescent Si-QDs have been synthesized by a simplified method. Their surface electronic core levels were investigated by X-ray photoelectron spectroscopy (XPS) at various temperatures. At room temperature, three components at 100.4, 102, and 104.4 eV were identified, corresponding to Si core, Si-C, and silicon oxide on surface, respectively. The appearance of Si-C component showed that acrylic acid has been successfully grafted onto Si-QDs surface. At 200° C, intact evaporation took place and the ratios of silicon to carbon in core levels remained constant as the annealing time increrased. Decomposition occured at 340°C, where the ratios of Si2*p* to C1*s* began to change. The XPS results are in agreement with thermogravimetric analysis (TGA) measurements which showed a sharp weight loss of 80% at 200° C, which is the intact evaporation temperature. Another slow weight loss of 14% took place between 300°C and 500°C, which is a footprint of surface Si-C decomposition.

1. Introduction

Owing to its electronic and optical properties, functionalised Si-QDs have attracted enormous interest in the past decades [1-7]. Since smaller QDs were preferred to evaporate at lower temperatures, the intact evaporation of quantum dots could provide a unique method to process thin films of size-selected functionalised QDs [8], which could open the door for applications in memory devices and thermoelectric modules. With the conventional oxide layer replaced by Si-QDs in memory devices, the retention time and lifetime of the devices will be enhanced enormously [9, 10]. Since the nanostructured silicon size would be smaller than the phonon mean free path, the thermal conductivity would be reduced by 100 fold [1, 2]. Such thin films of Si-QDs could be next generation of energy materials converting waste heat or solar energy to electricity. Owing to phonon blockage the QDs with low thermal conductivity could possess high figure of merit thus convert thermal energy to electricity more efficiently. Furthermore, current resources of thermoelectric materials, such as Bi_2Te_3 and PbTe, are running out rapidly. Silicon is one of the most abundant resources on the earth, is non-toxic and ecologically friendly. If it could be used to replace current thermoelectric materials, it would mark a significant breakthrough in fundamental sciences and energy conversion technologies.

QDs have been used as fluorescent labels for bio-molecules. They could be future alternatives to the traditional organic dyes in case of bio-imaging [7, 11-13]. Moreover, QDs have several advantages over the traditional organic dyes [7]: resistance to photobleaching which permits the QDs to be imaged for a longer period to time devoid of any loss of luminescence; QDs could be designed with an emission range from the visible to the IR region; they have narrow, symmetrical emission peaks and a broad absorbance band; relatively high quantum yields; different QDs of different emission characteristics can be excited simultaneously at a single wavelength; and QDs are 20 times bright than traditional organic dyes. Since Si-QDs surface can be modified with water dispersible ligands, and they emit visible red light, water dispersible Si-QDs are desired by biological and medical applications. Acrylic acid grafted luminescent Si-QDs are among bio-compatible quantum dots that are water dispersible, non-toxic, and emitting bright red light [14-16]. Therefore their thermal stability with temperatures is important for future applications.

The evaporation and decomposition of alkylated Si-QDs has been investigated [8], but such a study on acrylic acid grafted Si-QDs remains to be performed. Here we report an XPS study on the evaporation and decomposition of acrylic acid grafted Si-QDs, with confirmation via TGA results. This is the first report comparing the XPS results with TGA analysis and proved these results are in agreement.

2. Experimental

The simplified method of synthesising acrylic acid grafted luminescent Si-QDs has been published elsewhere [14]; here we provide a brief description of the sample preparation. Photoluminescent silicon layers were formed by galvanostatic anodisation of boron-doped p-Si (100) oriented wafer (10 Ω·cm resistivity, Compart Technology, Peterborough, UK) in a 1:1 v/v solution of 48% aqueous hydrofluoric acid (HF) and ethanol solution. The circular electrochemical etching cell (1 cm diameter) was machined from polytetrafluoroethylene (PTFE). The silicon wafer was sealed to the base using a Viton^{TM} O-ring. The counter electrode was a piece of platinum wire coiled into loop to improve the uniformity of the current distribution, and the etching was carried out using KEITHLEY 2601 in constant current source mode. A layer of luminescent porous silicon was made at high current density (5 minutes at 550 mA/cm^2). The solution was decanted and the fluorescing porous silicon layer on the chip was transferred to a Schlenk flask and dried under the vacuum of a rotary pump for 2 hours. The air within a 1:99 v/v solution of acrylic acid and ethanol aqueous was purged by N_2 bubbling for 2 hours. The solution was then added to the Schlenk flask containing Si chips (working under N_2 protection.) The Schlenk flask was tightly closed and subjected to ultrasonic dispersion for 50 minutes at 40 °C. The mixture was then poured into a polyethylene bottle and treated with N_2 bubbling for 10 minutes to purge the air. Afterwards the mixture was inserted into an UV reactor and stirred at 50 °C for 5 h. This procedure changes the Si-QDs surface termintation from hydrogen (H-) to poly acrylic acid(PAA-) group. After filtration the luminescent solution was collected. A photoluminescence (PL) spectrum was acquired with a PerkinElmer LS55 Fluorescence Spectrometer with the excitation from

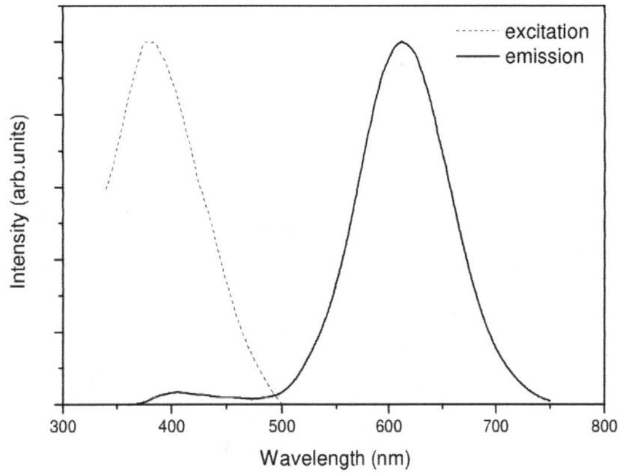

Figure 1. Solid line: PL emission spectrum from PAAc grafted Si-QDs with UV excitation wavelength 250 nm and filter at 290 nm. The main peak position is at 610 nm and the minor peak at 403 nm; dash line: excitation spectrum at emission 610 nm, the peak wavelength is at 378 nm.

Journal of Physics: Conference Series **286** (2011) 012039 doi:10.1088/1742-6596/286/1/012039

a Xenon lamp at 250 nm and a filter at 290 nm. The excitation spectrum was obtained by collecting the emission at 610 nm.

A few drops of the prepared sample were cast onto a gold foil substrate prior to insertion inside the XPS system preparation chamber. High resolution XPS core level measurements were performedwith a Scienta ECSA300 at NCESS, Daresbury Laboratory. Evaporation and decomposition experiments were carried out in a VG Escalab 250 in LENNF, Leeds University, equipped with a conventional hemispherical sector analyser and controlled by a VGX900 data system. XPS experiments were carried out using a high intensity monochromated Al Kα source (1486.6 eV) operated at 15 kV and 20 mA. O1*s*, C1*s* and Si2*p* spectra were recorded using a pass energy of 20 eV. The energy scale of the spectrometer was calibrated to the Ag $3d_{5/2}$ peak at 368.3 eV. The binding energy scale was calibrated to the $Au4f_{7/2}$ at 84.0 eV. High-resolution peak fitting was performed using the manufacturers' software.

Thermogravimetric analysis (TGA) was performed with a METTLER TOLEDO TGA system from ambient environment conditions up to 650° C, with the rate of temperature increase set at 20° C per minute and using a 40 µl aluminium pan under dry nitrogen gas flow protection.

3. Results and discussion:

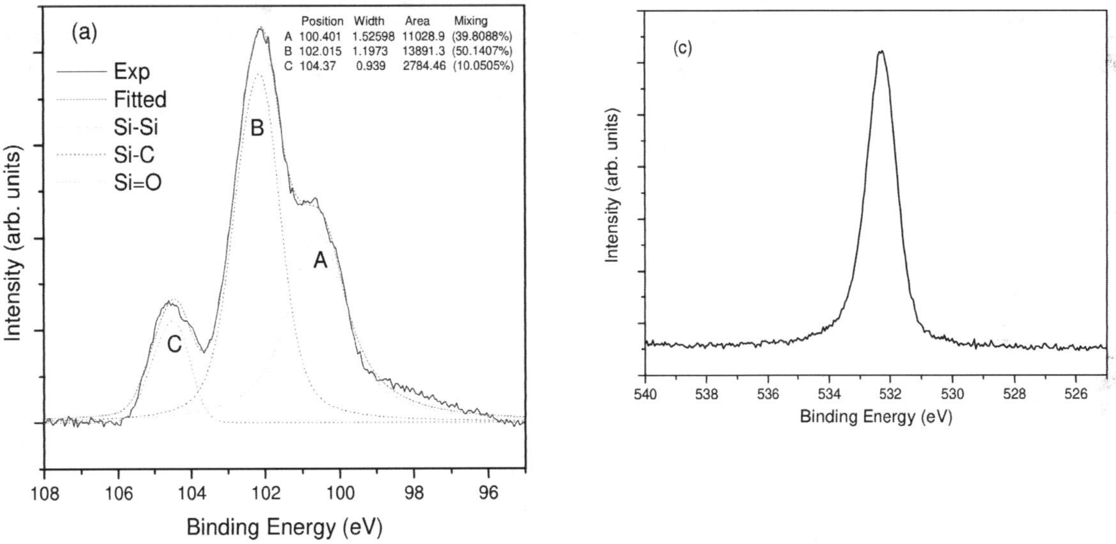

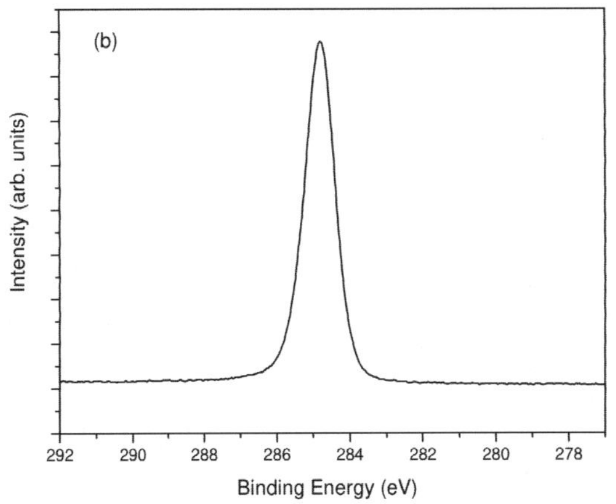

Figure 2. XPS core levels at room temperature in region (a) Si2*p*, (b) C1*s*, and (c) O1*s*. The Si2*p* was fitted with a Shirley background and three doublets to demonstrate there are different species on Si-QDs surface. Three components at 100.4, 102, and 104.4 eV, respectively, were ascribed to Si core, Si-C, and Si-O related species.

The prepared samples show a strong luminescent emission at peak 610 nm (orange), equivalent to 2.03 eV, and the excitation spectrum shows a peak at 378 nm (or 3.28eV), see figure 1. The separation between these peaks is 232 nm (or 1.25 eV). There are two mechanisms proposed for PL from Si-QDs [17]: one is sized effect based quantum confinement model, i.e. the excitation process occurs between the quantized levels of QDs, and subsequent radiative recombination processes may take place directly via these levels [18, 19], or alternatively via surface/interfacial states [20, 21]; another mechanism is due to particular species irrelevant to the quantum dots, i.e. both the processes take place in certain kind of species (e.g. a surface-bound oxyhydride-like emitter or SiO_x-related defects act as luminescent centres like those found in a-SiO_2) [22-25]. The orange peak from our acrylic acid grafted Si-QDs follows the first mechanism proposed above. Based on the calculation by Wolkin [21], this indicates the size of Si-QDs to be approximately 5 nm in diameter, which is in consistent with our previous DLS, HRTEM and AFM results [14]. The minor peak at 403 nm (blue) is attributable to Si-O related species following the second mechanism above. PL bands located in near ranges have been found in a-Si:H:O films formed by CVD and Si^+ implanted SiO_2 films [23, 26].

That Si-O species exist on the Si-QDs surface can be demonstrated by the XPS core level spectra. Figure 2(a) shows a core level spectrum at room temperature in Si2*p* region, which can be fitted with a Shirley background and three doublets. Three components at 100.4, 102, and 104.4 eV, respectively, are ascribed to Si core, Si-C, and Si-O related species. Since the QDs were grafted by acrylic acid, Si-C species was the majority in this fitting. This is similar to the results from alkylated Si-QDs [27]. The core levels at room temperature in C1*s* and O1*s* regions, shown in figure 2(b) and (c), confirmed the possibility of existence of Si-C and Si-O related species.

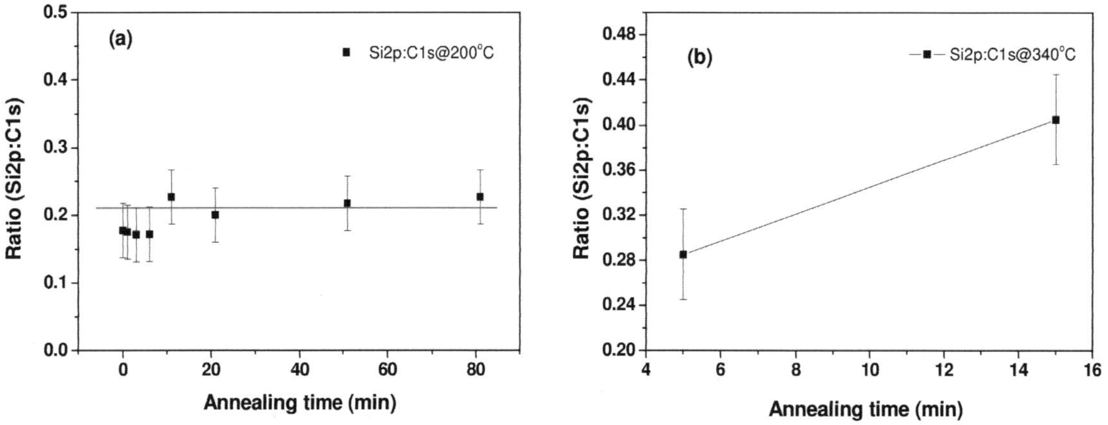

Figure 3. The analysis of ratio of Si:C in the core level spectra at different annealing temperatures over a series of annealing times: (a) 200° C; (b) 340° C. The results show that there is no significant Si:C ratio change when the samples were annealed for 1, 3, 6, 11, 21, 51, and 81min at 200°C, which is the intact evaporation temperature; when samples were annealed at 340° C, the ratio of Si:C increased with annealing time, owing to decomposition at this temperature.

In 2007 it was unveiled that evaporation of intact alkylated Si-QDs in UHV at relatively low temperature of ~ 200°C, compared to the normal silicon boiling point of 1414°C [8]. The intact sublimation is enabled by a combination of anomalously weak inter-particle interactions and the extremely high thermal stability of the alkyl-coated silicon nanoparticles. Hence, the exploration of the evaporation behaviour of acrylic acid grafted Si-QDs in vacuum is interesting. The measurements were carried out by monitoring the Si2*p* and C1*s* core levels over several annealing cycles of of 1, 3, 6, 11, 21, 51, and 81 minutes at 200° C, which is the temperature at which they are expected to evaporate intact. The results (figure 3a) show that there is no significant Si: C ratio change. The constant ratio of Si to C implies that the PAAc Si-QDs evaporated intact without undergoing decompostion. It is worth noting the desorbing temperature for non-covalently anchored monolayers such as the alkanethiol self-assembled monolayers is about 175 °C [28]. This proves that the carboxyl groups were covalently anchored to Si-QDs surface, which is consistent with the core level Si2*p* at room temperature which showed the existence of Si-C component.

Interestingly, once the samples were annealed at 340°C, the ratio of Si to C rose (figure 3b). At this temperature it is believed that decomposition of the covalent anchored monolayer on Si took place [29]. Decomposed carboxyl groups from Si-QDs surface were burned and pumped away in UHV, thus the Si:C ratio rose.

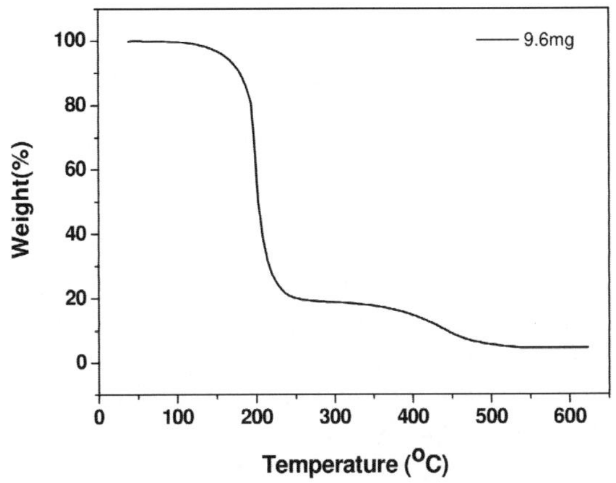

Figure 4. TGA analyses from ambient environment to 650°C under the protection of dry nitrogen gas flow at 20° C per minute temperature rate using 40 μl aluminium pan: a sharp weight loss, 80%, at 200° C which is intact evaporation temperature; further 14% weight lost between 300° C and 500° C during the decomposition.

To investigate the thermal properties of poly acrylic acid grafted Si-QDs further, TGA analyses were carried out with dry samples. Figure 4 shows a scan from the ambient environment to 650° C under the protection of dry nitrogen gas flow at a rise of 20° C per minute using a 40μl aluminium pan: a sharp weight loss, 80%, at 200° C which is intact evaporation temperature. Further 14% weight lost between 300° C and 500° C was observed, which is the footprint of surface decomposition.

4. Conclusions:

Acrylic acid grafted luminescent Si-QDs were synthesized by a simplified method. Their surface electronic core levels were investigated by XPS at various temperatures. At room temperature, the core level in Si2p region was fitted with three doublets and a Shirley background. Three components at 100.4, 102, and 104.4 eV were corresponding to the Si core, Si-C, and silicon oxide on the surface, respectively. The appearance of a Si-C component shows that acrylic acid had been successfully grafted onto the Si-QDs surface. At 200° C, the Si-QDs evaporated intact and the ratios of silicon to carbon core levels were constant with the increasing annealing time. Decomposition at 340° C was observed with the change in Si2*p* to C1*s* ratio with the annealing time. The XPS results were in agreement with TGA. TGA measurement showed a sharp weight loss of 80% at 200°C which is the intact evaporation temperature. Another slow weight loss happened between 300° C and 500° C, a further loss of 14% weight, which is a footprint of surface Si-C decomposition.

Acknowledgement:

QW is grateful to the International Scholarship Funding panel of the University of East Anglia for awarding an international scholarship. YC is grateful to the Royal Society for awarding a Research Grant 2007/R2 and the EPSRC for financial support under the project EP/G01664X/1. NCESS and LENNF are acknowledged for awarding XPS beamtime.

References:

[1] Boukai A I, Bunimovich Y, Tahir-Kheli J, Yu J K, Goddard W A and Heath J R 2008 Silicon nanowires as efficient thermoelectric materials *Nature* **451** 168-71

[2] Hochbaum A I, Chen R K, Delgado R D, Liang W J, Garnett E C, Najarian M, Majumdar A and Yang P D 2008 Enhanced thermoelectric performance of rough silicon nanowires *Nature* **451** 163-7

[3] Lehtonen O and Sundholm D 2006 Bright luminescence from silane substituted and bridged silicon nanoclusters *PCCP* **8** 4228-32

[4] Sundholm D 2004 Density functional studies of the luminescence Of Si29H36 *PCCP* **6** 2044-7

[5] Garoufalis C S and Zdetsis A D 2006 High accuracy calculations of the optical gap and absorption spectrum of oxygen contaminated Si nanocrystals *PCCP* **8** 808-13

[6] Hodes G 2007 Semiconductor and ceramic nanoparticle films deposited by chemical bath deposition *PCCP* **9** 2181-96

[7] Michalet X, Pinaud F F, Bentolila L A, Tsay J M, Doose S, Li J J, Sundaresan G, Wu A M, Gambhir S S and Weiss S 2005 Quantum dots for live cells, in vivo imaging, and diagnostics *Science* **307** 538-44

[8] Chao Y, Šiller L, Krishnamurthy S, Coxon P R, Bangert U, Gass M, Kjeldgaard L, PatoleO S N, Lie L H, O'Farrell N, Alsop T A, Houlton A and Horrocks B R 2007 Evaporation and deposition of alkyl-capped silicon nanocrystals in ultrahigh vacuum *Natur. Nanotech.* **2** 486-9

[9] Feng T, Yu H B, Dicken M, Heath J R and Atwater H A 2005 Probing the size and density of silicon nanocrystals in nanocrystal memory device applications *Appl. Phys. Lett.* **86** 033103

[10] Rao R A, Gasquet H P, Steimle R F, Rinkenberger G, Straub S, Muralidhar R, Anderson S G H, Yater J A, Ledezma J C, Hamilton J, Acred B, Swift C T, Hradsky B, Peschke J, Sadd M, Prinz E J, Chang K M and White B E 2005 Influence of silicon nanocrystal size and density on the performance of non-volatile memory arrays *Solid-State Electron.* **49** 1722-7

[11] Loumaigne M, Praho R, Nutarelli D, Werts M H V and Debarre A Fluorescence correlation spectroscopy reveals strong fluorescence quenching of FITC adducts on PEGylated gold nanoparticles in water and the presence of fluorescent aggregates of desorbed thiolate ligands *PCCP* **12** 11004-14

[12] Cooper D R, Suffern D, Carlini L, Clarke S J, Parbhoo R, Bradforth S E and Nadeau J L 2009 Photoenhancement of lifetimes in CdSe/ZnS and CdTe quantum dot-dopamine conjugates *PCCP* **11** 4298-310

[13] Medintz I L and Mattoussi H 2009 Quantum dot-based resonance energy transfer and its growing application in biology *PCCP* **11** 17-45

[14] Wang Q, Pietzsch A, Hennies F, Ni H and Chao Y 2010 Synthesis of Water dispersible Photoluminescent Silicon Nanoparticles and Their Use in Biological Fluorescent Imaging *J. Nanopart. Res.* in press

[15] Li Z F and Ruckenstein E 2004 Water-soluble poly(acrylic acid) grafted luminescent silicon nanoparticles and their use as fluorescent biological staining labels *Nano Lett.* **4** 1463-7

[16] Sato S and Swihart M T 2006 Propionic-acid-terminated silicon nanoparticles: Synthesis and optical characterization *Chem. Mater.* **18** 4083-8

[17] Zhu X P, Suematsu H, Jiang W H and Yatsui K 2008 Structures and photoluminescence properties of silicon thin films prepared by pulsed ion-beam evaporation *Materials Science and Engineering B-Advanced Functional Solid-State Materials* **149** 105-10

[18] Ledoux G, Gong J and Huisken F 2001 Effect of passivation and aging on the photoluminescence of silicon nanocrystals *Appl. Phys. Lett.* **79** 4028-30

[19] Patrone L, Nelson D, Safarov V I, Giorgio S, Sentis M and Marine W 1999 Synthesis and properties of Si and Ge nanoclusters produced by pulsed laser ablation *Appl. Phys. A* **69** S217-S21

[20] Dinh L N, Chase L L, Balooch M, Siekhaus W J and Wooten F 1996 Optical properties of passivated Si nanocrystals and SiOx nanostructures *Phys. Rev. B* **54** 5029-37

[21] Wolkin M V, Jorne J, Fauchet P M, Allan G and Delerue C 1999 Electronic states and luminescence in porous silicon quantum dots: The role of oxygen *Phys. Rev. Lett.* **82** 197-200

[22] Gole J L, Dudel F P, Grantier D and Dixon D A 1997 Origin of porous silicon photoluminescence: Evidence for a surface bound oxyhydride-like emitter *Phys. Rev. B* **56** 2137-53

[23] Liao L S, Bao X M, Zheng X Q, Li N S and Min N B 1996 Blue luminescence from Si+-implanted SiO2 films thermally grown on crystalline silicon *Appl. Phys. Lett.* **68** 850-2

[24] Nishikawa H, Watanabe E, Ito D, Sakurai Y, Nagasawa K and Ohki Y 1996 Visible photoluminescence from Si clusters in gamma-irradiated amorphous SiO2 *J. Appl. Phys.* **80** 3513-7

[25] Prokes S M, Carlos W E, Seals L and Gole J L 2000 Defect study of light-emitting HCl-treated porous silicon *Phys. Rev. B* **62** 1878-82

[26] Tong S, Liu X N, Gao T and Bao X M 1997 Intense violet-blue photoluminescence in as-deposited amorphous Si:H:O films *Appl. Phys. Lett.* **71** 698-700

[27] M'Gaieth R, Maaref H, Mihalcescu I and Vial J C 1999 Porous silicon: photoluminescence decay in the nanosecond range *Microelectron. J.* **30** 695-8

[28] Nishida N, Hara M, Sasabe H and Knoll W 1996 Dimerization process in alkanethiol self-assembled monolayer on Au(111) *Jpn. J. Appl. Phys., Part 2* **35** L799-L802

[29] Sung M M, Kluth G J, Yauw O W and Maboudian R 1997 Thermal behavior of alkyl monolayers on silicon surfaces *Langmuir* **13** 6164-8

Electrodeposition and characterisation of ZnTe layers for application in CdTe based multi-layer graded bandgap solar cells

D G Diso, F Fauzi, O K Echendu, A R Weerasinghe and I M Dharmadasa
Electronic Materials & Sensors Group, Materials and Engineering Research Institute, Sheffield Hallam University, Sheffield S1 1WB, United Kingdom.

E-mail: *dgdiso@yahoo.co.uk*

Abstract

Zinc Telluride (ZnTe) thin films have been deposited on glass/conducting glass substrates using low-cost aqueous electrodeposition (ED) method. The structural, optical and morphological properties of the resulting films have been characterized using X-ray diffraction (XRD), spectrophotometry and Scanning Electron Microscopy (SEM). It has been confirmed by XRD technique that the deposited layers are nano- and polycrystalline mixture for as deposited layers. Photoelectrochemical (PEC) studies revealed that the layers are p-type in electrical conduction. Optical absorption measurement has been used for the bandgap determination of the deposited layers. The bandgap of the polycrystalline ZnTe layers are in the range (2.65 – 2.75) eV for the $ZnSO_4$ precursor, and (2.70 – 2.87) eV for the $ZnCl_2$ precursor instead of (2.21 – 2.26) eV reported for crystalline ZnTe. This increase may be due to the quantum effect which confirmed by the absence of strong XRD peaks from these layers and/or inclusion of ZnO in the deposited layers.

1. Introduction

The family of group II–VI semiconductors has been widely studied due to their stability of fundamental structural, electrical and optical properties for use in thin film solar cells, photodetectors and light emitting diodes (LEDs) [1-3]. Crystalline ZnTe has a direct bandgap of 2.21 to 2.26 eV at room temperature with low electron affinity of 3.53 eV [4, 5]. The development of thin film solar cells makes use of at least two kinds of semiconducting layers; a wide bandgap window material and a narrow bandgap absorber material. The two most researched absorber materials are CdTe and $CuInGaSe_2$, and in both cases the majority of researchers use an n-CdS layer as a window material.

Recent work on multi-layer graded bandgap solar cells based on AlGaAs/GaAs structures with p-type window materials has demonstrated their potential by producing high V_{oc} (1100 - 1175 mV) and excellent FF (0.80 - 0.87) values [6, 7]. The use of p-type materials instead of n-type layers (CdS) as wide bandgap windows produce higher potential barriers for electron transport, yielding higher V_{oc} values. The use of ZnTe as a window layer in solar cells can also reduce the toxic nature by replacing chemical bath deposited CdS layers currently used in thin film solar cells.

Various methods have been employed to deposit ZnTe thin films such as molecular beam epitaxy (MBE) [8], metal-organic chemical vapour deposition (MOCVD) [9], thermal evaporation [10], metal organic vapour phase epitaxy (MOVPE) [11], r.f and dc sputtering [12], isothermal close space sublimation (ICSS) [13], pulsed laser ablation [14], brush plating [15] and electrodeposition [5, 16-18]. This paper presents the results of attempts made to electrodeposit ZnTe layers from an aqueous solution, for use in thin film solar cells based on CdTe.

2. Experimental

2.1. Electrodeposition of ZnTe films

ZnTe thin films were cathodically electrodeposited onto glass/conducting glass substrates using two different chemical baths (i.e. using $ZnSO_4$ and $ZnCl_2$ precursors). The chemical compositions of the two baths are (0.15 M $ZnSO_4.7H_2O$, 0.5 M Na_2SO_4 and 3 ml TeO_2) for $ZnSO_4$ and (0.15 M $ZnCl_2$ and 3 ml TeO_2) for $ZnCl_2$ baths respectively. In both electrolytes the growth temperature and pH were ~78°C and 3.50 ± 0.01 respectively. Fluorine doped tin oxide (FTO) coated conducting glass substrates with the sheet resistance of 15 Ω/□ were cleaned in organic solvents (methanol and acetone), dilute HNO_3 and finally rinsed in glacial acetic acid. In between different solvents, the glass/FTO substrates were rinsed in de-ionized water. All chemicals used for electrodeposition were analytical reagent grade of purity 5N (99.999%). A two-electrode system was used in this project in order to simplify the process and to eliminate an impurity source; the reference electrode. A graphite anode and a glass/FTO cathode were used to pass an electric current through the electrolyte. The study of the electrodeposition mechanism was carried out using cyclic voltammogram.

2.2. Measurement Techniques

Cyclic voltammetry (CV) was performed using a Gill AC computeriser potentiostat (S/N 1313 ACM Instruments). With the information obtained from the voltammogram, ZnTe thin film deposition was carried out around -1600 mV for both chemical baths. After the deposition, the films were rinsed in distilled water and subsequently dried using nitrogen gas.

The samples were studied using a Cary 50 Scan UV-Visible spectrophotometer (Varian) for investigating the optical properties and for estimation of bandgap values. The bandgap energy, E_g was derived from Stern relation [4, 5, 15, 18] using:

$$\alpha = \frac{k\left(h\upsilon - E_g\right)^{\frac{n}{2}}}{h\upsilon} \qquad (1)$$

where, υ is the frequency, *h* is the Planck's constant and *n* takes the value of either 1 or 4. The value of *n* is 1 and 4 for direct and indirect transitions respectively.

Photoelectrochemical (PEC) measurements using an electrolyte, 0.10 M $Na_2S_2O_3$, were carried out to determine the electrical conductivity type of the ED-ZnTe layers. A solid/liquid junction was formed by immersing glass/FTO/ZnTe in the electrolyte solution. The PEC signal was determined by the difference between voltages observed under illumination and under dark conditions. The magnitude of the PEC signal depends on the doping condition of the solid layer and hence, the width of the depletion region formed. The sign of the PEC signal determines the electrical conduction type.

X-ray diffraction measurements (XRD) were carried out using a Philips PW 3710 X-ray diffractometre using CuK_α radiation (λ = 1.5418 Å) in the range of 2θ = (10 – 70)° for the structural analysis of the deposited film. The X-ray generator tension and current were 40 kV and 40 mA respectively. A Philips XL 30 ESEM-Field Emission Gun with accelerating voltage of 20 kV was used to study the surface morphology of the thin film layers.

3. Results and Discussion

3.1. Voltammogram

Typical cyclic voltammogram related to the electrodeposition of ZnTe (ED-ZnTe) thin film are shown in Figure 1 (a) and (b). These figures indicate that the formation of ZnTe compound takes place beyond 1620 mV, cathodic voltage. Two peaks were observed in the reverse direction which are attributed to the removal of Zinc and Tellurium at (-1078 & -300) mV for $ZnCl_2$ precursor and (-993 & -391) mV for $ZnSO_4$ precursor respectively. pH is among the key parameters necessary to be considered for the deposition of ZnTe layers. According to references [5, 17], at lower pH (3.50 ± 0.01), free tellurium was deposited and at comparatively higher pH (5.50 ± 0.01), the solution became cloudy due to the precipitation of TeO_2 or $Cd(OH)_2$. Furthermore, the influence of deposition current densities is important on the properties of deposited ZnTe thin films. It is found that [5], higher deposition current densities (>400 μAcm^{-2}) led to the films with a rough surfaces, whereas films deposited at lower current densities (≈100 μAcm^{-2}) had a smooth surfaces.

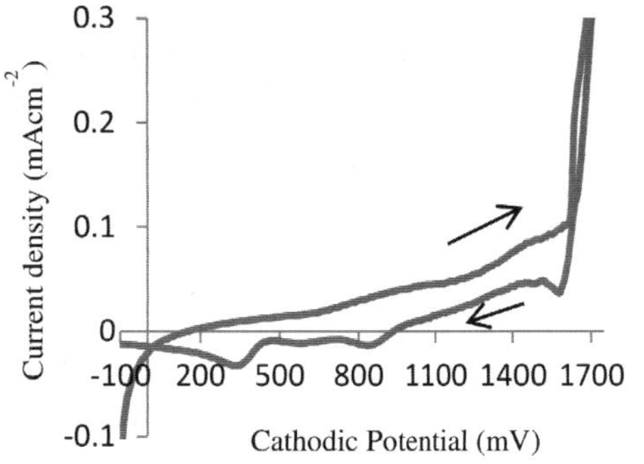

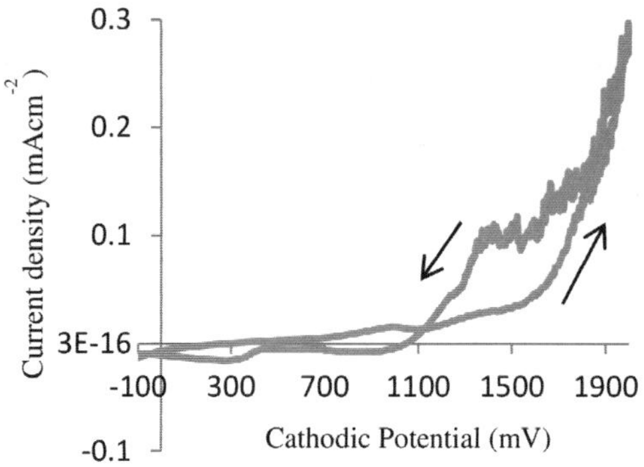

(a) *(b)*

Figure 1: *Typical voltammograms of electrolytic baths used for deposition of ED-ZnTe for (a) $ZnSO_4$ and (b) $ZnCl_2$ baths.*

3.2 Visual appearance

ZnTe films of about 0.4 μm thick were deposited in about 40 minutes while the solution is moderately stirred. Figure 2 (a) and (b) shows the visual appearance of ZnTe layers deposited as a function of growth voltage (V_g) for $ZnSO_4$ and $ZnCl_2$ baths respectively. As shown in the Figures, at higher and lower cathodic voltages, the samples become dark which are attributed to the Zn- and Te- richness. In the intermediate voltages, the layers appear reddish brown, similar to that of bulk ZnTe material.

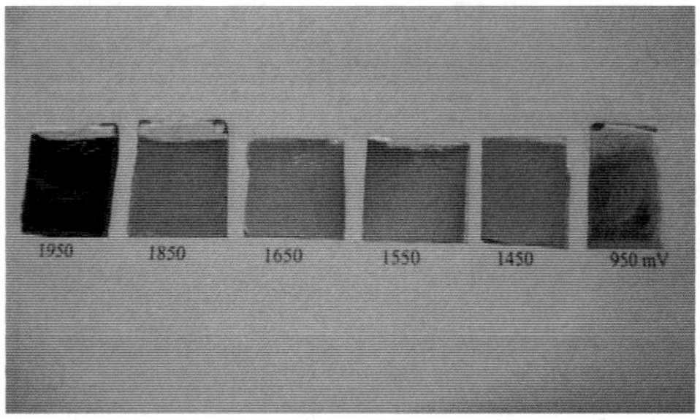

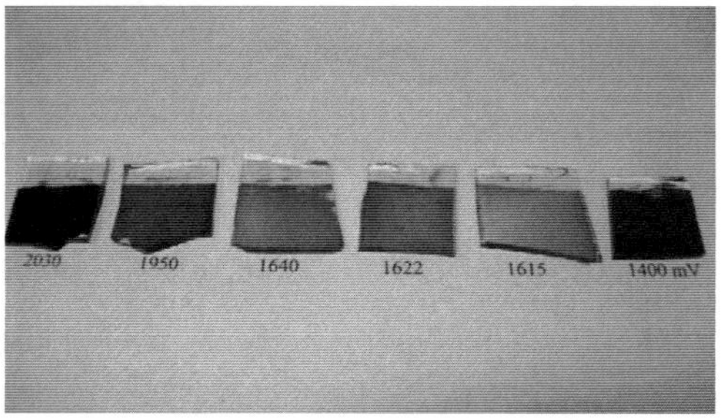

(a) *(b)*

Figure 2: *The appearance of as-deposited ED-ZnTe layers grown on glass/FTO substrate, at different growth voltages from (a) $ZnSO_4$ and (b) $ZnCl_2$ baths.*

3.3. Thickness measurement

The thickness of ED-ZnTe layers deposited for 40 minutes shows a value of ~0.40 μm when measured using the Dektak technique. The theoretical estimate using Faraday's relation given by equation (2), gave the value ~0.35 μm.

$$T = \frac{1}{nFA} \times \frac{ItM}{\rho} \qquad (2)$$

where, T is the thickness of the ZnTe film in cm, I is the deposition current density in $mAcm^{-2}$, t is the deposition time in seconds, M is the molecular weight in grams, n is the number of electrons transferred, F is the Faradays constant = 96485, A is the area of the deposited layers in cm^2, and ρ is the density of ZnTe in gcm^{-3}.

The thickness estimated using the two methods gives similar values of the same order of magnitude for the electrodeposited films.

3.4. Optical absorption

Optical absorption experiments were carried out using three different ZnTe layers grown at different cathodic voltages as shown in Figure 3 (a) and (b). The values of E_g were determined by plotting the square of absorbance (α^2) against photon energy (hv). The bandgap of crystalline ZnTe layer is (2.21 – 2.26) eV [4-5], but the bandgap obtained for ED-ZnTe layers is higher than those reported values. The larger values in the range (2.65 – 2.75) eV for $ZnSO_4$ precursor and (2.70 – 2.87) eV for $ZnCl_2$ precursor, indicates that the layers formed consist of nano-crystals, producing larger bandgap values due to quantum effects. This increase could also be due to incorporation of ZnO (E_g = 3.08 – 3.26) eV [19, 20].

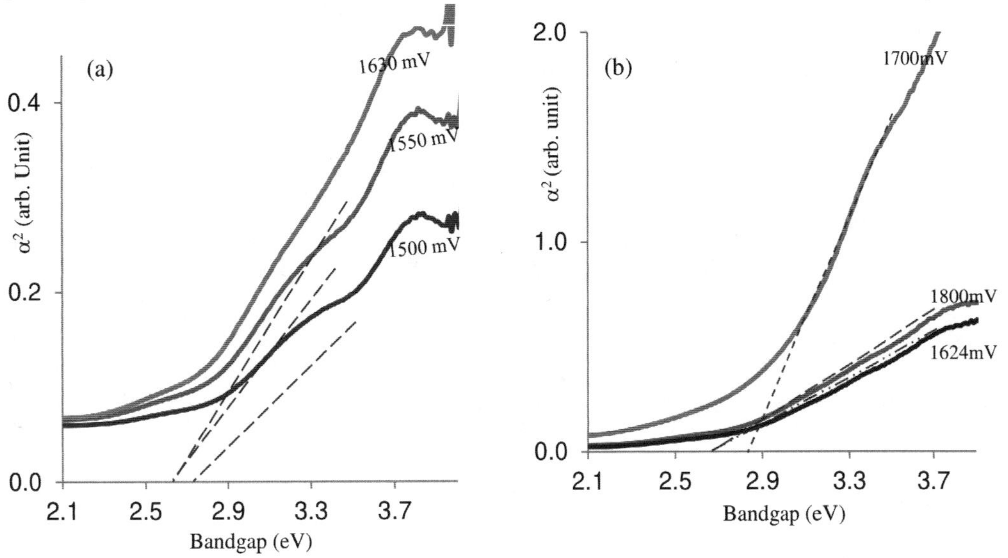

Figure 3: *Optical absorption edges for as-deposited ED-ZnTe grown at different voltages on glass/FTO substrates from (a)* $ZnSO_4$ *and (b)* $ZnCl_2$ *containing baths.*

3.5. Photoelectrochemical (PEC) studies

The results of PEC investigation for ED-ZnTe layers are shown in Figure 4 (a) and (b). These results indicate that all ED-ZnTe layers grown at shown voltages were p-type in electrical conduction. PEC studies provide vital information on electrical conductivity type and hence help in fabrication of electronic devices using these films. It should be noted that Na_2SO_4 has been used as a supporting electrolyte to grow ZnTe in the $ZnSO_4$ bath. The presence of sodium in the electrolyte could act as a p-type dopant in ZnTe producing p-type ZnTe layers always. Furthermore, the presence of chloride ions in the $ZnCl_2$ bath did not change the electrical conductivity type even though chlorine could act as an n-type dopant.

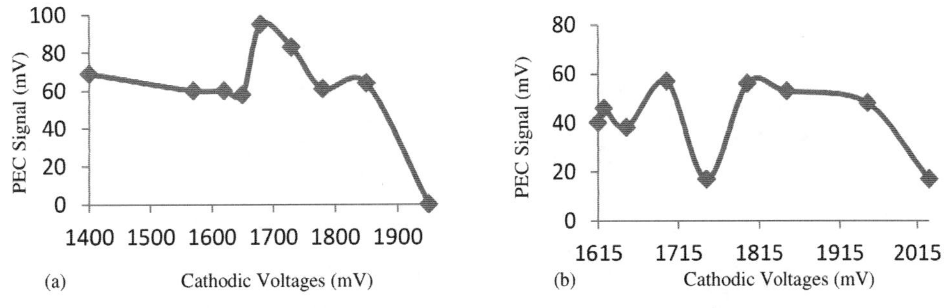

Figure 4: *PEC signal as a function of growth voltage for as-deposited ED-ZnTe layers for (a)* $ZnSO_4$ *bath and (b)* $ZnCl_2$ *bath.*

3.6. X-ray diffraction

XRD studies were carried out for the best samples which show uniform reddish colour and the sharpest optical absorption edge. Figure 5 shows the XRD diffractogram obtained for glass/FTO substrate and as-deposited ZnTe layers from $ZnSO_4$ and $ZnCl_2$ baths. It is observed that there are no strong XRD peaks arising from both ZnTe layers. The only sign of peaks due to ZnTe are the ones appearing at $2\theta = 35^o$ and 47° corresponds to (102) and (200) planes of hexagonal structure respectively. This confirms the existence of nano- or poly-crystalline ZnTe materials in these layers. In addition a broad hump has developed in the region of $(15 - 30)^o$, indicating that there is an amorphous material present in these layers. Therefore, these as-deposited layers seems to be a mixture of amorphous (nano-) and poy-crystalline ZnTe material

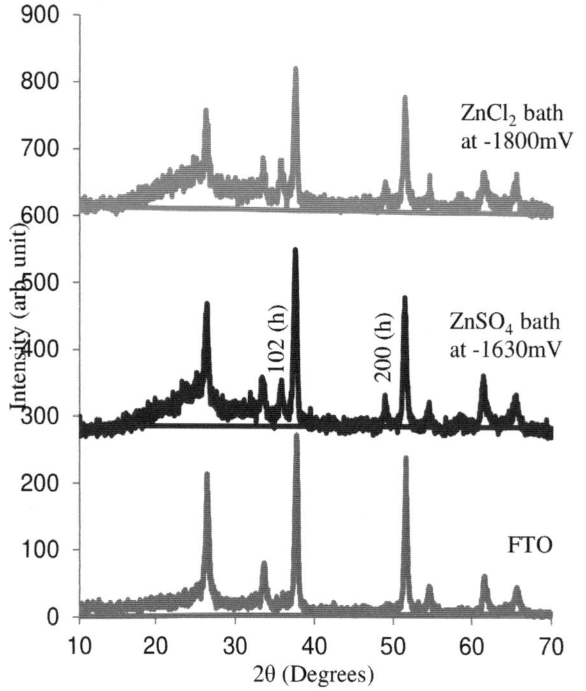

Figure 5: *X-ray diffraction patterns of glass/FTO, and as-deposited ZnTe layers from $ZnSO_4$ and $ZnCl_2$ containing baths.*

3.7. Scanning Electron Microscopy (SEM) studies

SEM studies were carried out to investigate the surface morphology and uniformity of ED-ZnTe layers. The SEM pictures of as-deposited ED-ZnTe layers are shown in Figure 6 (a) and (b) from two different baths. The appearance indicates the uniform distribution of grains size of ~10 nm from $ZnSO_4$ bath and ~15 nm from $ZnCl_2$ bath.

Figure 6: *SEM picture of as-deposited ED-ZnTe layers from (a) $ZnSO_4$ and (b) $ZnCl_2$ containing baths.*

4. Conclusions

This initial work demonstrates that ZnTe layers can be electrodeposited using an aqueous electrolyte and a two-electrode system. The layers obtained have excellent adhesion properties with a uniform reddish appearance. Optical absorption measurement provides bandgap of (2.65 – 2.75) eV for $ZnSO_4$ precursor and (2.70 – 2.87) eV for $ZnCl_2$ precursor instead of reported bandgap of (2.21 – 2.26) eV [4, 5] for crystalline ZnTe material, indicating that the layers consist of nanoparticles introducing quantum effects which increase the bandgap. The same effect could also be introduced by inclusion of possible component ZnO, in these layers. XRD studies, further confirms the nano- and polycrystalline nature of ZnTe layers grown by this method. PEC studies indicate the p-type electrical conduction for electrodeposited ZnTe layers. The work is continuing to further develop these material layers and incorporate them in multi-layer graded bandgap solar cell devices [6, 7] based on electrodeposited CdTe materials.

Acknowledgements

Pilkington Group Limited is thanked for providing glass/FTO substrates for this work and Vinay Patel is thanked for his contribution in carrying out SEM measurements. The main author (D. G. Diso) wishes to acknowledge the financial support given by Kano University of Science & Technology, Wudil – Nigeria.

References

[1] Merchant J D and Cocievera M 1996 J. Electrochemical Soc. **143** 4054
[2] Ota T and Takahashi K 1973 Solid State Electronic **16** 1089
[3] Bellakhder H, Outzourhit A and Ameziane E L 2001 Thin Solid Films **382** 30
[4] Mahalingam T, John V S, Ravi G and Sebastian P J 2002 Crystal Research Technol. **37** (4) 329
[5] Neumann-Spallart M and Konigstein C 1995 Thin Solid Films **265** 33
[6] Dharmadasa I M 2009 Curr. Appl. Phys. doi: 10.1016/j.cap.2008.12.0121
[7] Dharmadasa I M, Roberts J S and Hill G 2005 Sol. Energy Mater. & Sol. Cells **88** 413
[8] Franta D, Ohlidal I, Klapetek P, Montaigne-Ramil A, Bonanni A, Stifter D and Sitter H 2004 Thin Solid Films **468** 193
[9] Mora-Ser I, Tena-Zaera R, Gonzalēz J and Munōz-Sanjosé V 2004 J. Crystal Growth **262** 19
[10] Bacaksiz E, Aksu S, Ozer N, Tomakin M and Özcékik A 2009 Appl. Surface Sci. **256** 1566
[11] Nishio N, Guo Q and Ogawa H 1999 Thin Solid Films **343–344** 512
[12] Bellakhader H, Outzourhit A and Ameziane E L 2001 Thin Solid Films **382** 30
[13] De melo O, Climent-Front A, Martinez Duart J M and Hernández Vēlez M 2009 Thin Solid Films **517** 1559
[14] Rouleau C M and Lowandes D H 1998 Appl. Surface Sci. **127–129** 418
[15] Murali K R, Ziaudeen M and Jayaprakash N 2006 Solid-State Electronics **50** 1692
[16] Islam O A B M, Chaure N B, Wellings J, Tollan G and Dharmadasa I M 2009 Materials Characterisation **60** 160
[17] Mahalingam T, John V S, Rajendran S and Sebastian P J 2002 Semicond. Sci. Technol. **17** 465
[18] John V S, Mahalingam T and Chu J P 2005 Solid-State Electronics **49** 3
[19] Fahoume M, Magfoul O, Aggour M, Hartiti B, Chraïbi F and Ennaoui A 2006 Sol. Energy Mat. & Sol. Cells **90** 1437
[20] Chander R and Raychaudhuri A K 2008 Solid State Communication **145** 81

Journal of Physics: Conference Series **286** (2011) 012041 doi:10.1088/1742-6596/286/1/012041

Solar Cells Active in Complete Darkness

I. M. Dharmadasa, O. Elsherif and G. J. Tolan
Materials & Engineering Research Institute
Sheffield Hallam University, Sheffield S1 1WB, United Kingdom.
E-mail: Dharme@shu.ac.uk Tel: +44 (0)114 225 6910 Fax: +44 (0)114 225 6930

ABSTRACT: A graded bandgap multi-layer solar cell device structure was designed to absorb UV, visible and IR radiation, and to incorporate impact ionisation and impurity photovoltaic effects within one device. The design was experimentally tested with a well researched material system, MOVPE grown GaAs/AlGaAs. Open circuit voltages of ~1175 mV with highest possible FF values (0.83-0.87) and J_{sc}~12 mAcm^{-2} have been observed [1,3]. These parameters were independently verified by measuring in five different laboratories in Europe and United States including NREL. While the work is continuing to increase short circuit current density values, these devices were tested to explore the experimental evidence of impurity PV effect, as expected from this design. Responsivity measurements and PV activity in dark conditions have been carried out to investigate impurity PV effect in these devices. Responsivity measurements indicate current collection in the infra-red region confirming the contribution from IR photons. The I-V measurements in dark conditions produce open circuit voltages exceeding 750 mV confirming the contribution from surrounding heat radiation. The new features of graded bandgap devices enable impurity PV effect to dominate and create useful charge carriers, suppressing detrimental recombination process. These experimental results will be presented in this paper.
Keywords: GaAs/AlGaAs, graded bandgap solar cells, impurity PV effect, thin film solar cells.

1 INTRODUCTION

Traditionally, solar radiation is exploited using two different methods; Infra-red (IR) radiation in "Solar Thermal Technology" and visible and ultra violet (UV) radiation in "Photovoltaic Technology". In order to be cost effective, it is attractive to combine both together in one device. In addition, if the incorporation of phenomena such as "impact ionisation" and "impurity photovoltaics" can be built into the same device, the performance of a solar cell can be enhanced. With these aims in mind, a graded bandgap multi-layer device structure has been designed and experimentally tested with well researched GaAs/AlGaAs system. This paper summarises the results obtained to date for graded bandgap solar cells, and the experimental evidence observed for the impurity photovoltaic effect in these device structures.

2 SUMMARY OF EXPERIMENTAL RESULTS

The device design proposed for combination of all above features is shown in Figure 1, and the details are published elsewhere [1-3]. The three most important steps for an efficient solar cell are satisfied for this structure according to solid state physics principles; namely (i) absorption of a major part of the solar spectrum, (ii) creation of e-h pairs in the presence of an internal electric field and (iii) effective separation of charge carriers towards two electrical contacts and transport through an external electric circuit. Because of the starting from a large bandgap p-type semiconductor and completing the device with a narrow bandgap n-type semiconductor, the probability of impact ionisation and the impurity PV effect has been increased. The detailed working of this device is illustrated with an animation placed in the website:
http://www.apsl.org.uk/dharme/default.aspx for readers use.

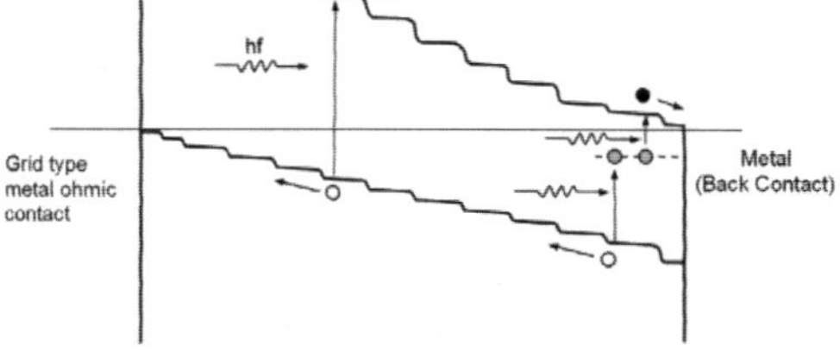

Figure 1. New design proposed for PV solar cells to maximise optical absorption, minimise thermalisation & transmission losses combining impurity photo-voltaics and impact ionisation.

The first attempt to grow this device structure was made using metal organic vapour phase epitaxy (MOVPE) - GaAs/AlGaAs materials. The growth started from an n-type GaAs substrate and the bandgap was gradually increased by increasing the aluminium content in the layers. The narrow bandgap end (back of the solar cell) was doped n-type with silicon (Si) and the wide bandgap end (front of the solar cell) was doped p-type with background carbon (C) in the reactor. The thickness of the device was set to ~3.2 μm and Si-doping was gradually decreased towards the middle of the device. Then, the carbon incorporation was increased gradually by adjusting the growth temperature. This process was adopted expecting to create the device structure as shown in Figure 1, and the devices were fabricated by forming an ohmic contact to the GaAs substrate and a grid type contact to the front surface of the solar cell. Growth and processing details are given in one of our previous publications [2].

Different sizes of solar cells were fabricated as shown in Figure 2 for detailed characterisation of their PV activity.

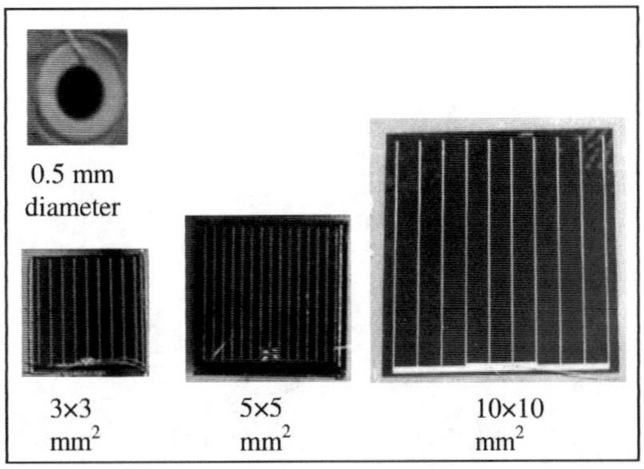

Figure 2. Photograph of solar cells fabricated with different dimensions to test the effect of the scaling up process.

The current-voltage (I-V) curves under AM1.5 illumination were measured in five different laboratories for comparison, and to confirm accuracy of measurements. A typical I-V curve measured at NREL in USA is shown in Figure 3, and the detail results are summarised in Table I. It is clear that these devices produce V_{oc} values higher than the highest reported [4,5] to date, together with highest possible FF values in the mid 80%. However, J_{sc} values were low in the range (10-12) mA cm^{-2}. Detailed studies carried out with electron beam induced current (EBIC) and secondary ion mass spectroscopy (SIMS) depth profiling showed that the first growth did not produced the expected device structure as shown in Figure 1. It was found that nearly flat band regions exist at both ends due to heavy doping of Si and C. Both Si and C doping has risen to ~10^{18} cm^{-3} levels towards the end of the solar cell during this first attempt.

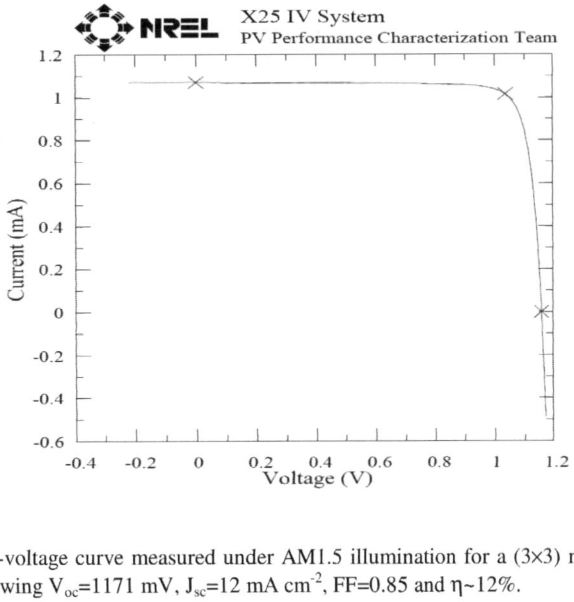

Figure 3. A typical current-voltage curve measured under AM1.5 illumination for a (3×3) mm^2 graded bandgap multilayer GaAs/AlGaAs solar cell showing V_{oc}=1171 mV, J_{sc}=12 mA cm^{-2}, FF=0.85 and η~12%.

The second growth with reduced Si doping to achieve ~10^{15} cm^{-3}, increased the depletion region at rear of the solar cell and the J_{sc} value rose to ~24 mA cm^{-2} as expected. But the reduction of C doping in the front of the device cannot be achieved using MOVPE technique. The work is continuing to grow the device structures using molecular beam epitaxy (MBE) technique in order to achieve these doping requirements.

Table I: Summary of solar cell parameters measured at five different laboratories for (3×3) mm^2 solar cells processed in one batch.

Device	V_{oc} (mV)	FF	J_{sc} (mA cm^{-2})	R_s (Ω cm^{-2})	R_p (Ω cm^{-2})	Place of Assessment
1	1170	0.87	----	----	----	SHU Labs
2	1160	0.86	----	----	----	Dharmadasa et al
3	1148	0.86	10.7	2.5	10,400	Zürich Labs
4	1141	0.86	10.3	4.0	5,100	Tiwari et al
5	1169	0.85	11.5	----	----	SBU Labs
6	1149	0.86	10.0	----	----	Reehal et al
7	1150	0.85	12.1	3.8	----	EPFL Labs Grätzel et al
8	1159	0.85	12.3	----	----	NREL - USA
9	1171	0.85	12.0	----	----	Emery et al

3 SEARCH FOR EXPERIMENTAL EVIDENCE OF IMPURITY PV EFFECT

The device structure shown in Figure 1 was designed to incorporate impurity PV effect using high concentrations of defects naturally available closer to the GaAs substrate at the back of the device. The experiments were therefore performed in order to test the existence of this effect in these devices. The most relevant technique is the quantum efficiency (QE) measurement to test whether there is a collection of current in the infra-red region. The authors' laboratory has no such facility to carry out these experiments. Therefore these devices were sent to several different laboratories equipped with this technique. There are two types of QE measurements which are used in solar cell assessments; external QE and internal QE. The external QE measurements include the effects of optical losses such as transmission through the cell and reflection from the surface. The internal QE refers to the efficiency with which photons that are not reflected or transmitted from the cell can generate collectable charge carriers. Internal QE is also referred to as incident photon to charge carrier efficiency (IPCE), and this is the method used to measure the cells produced in this project.

Figure 4 shows the IPCE measurements carried out for two devices in two different laboratories. Their features are similar, and the losses at UV and IR ends are due to high doping of the materials at both front and the back of the cell. These issues are dealt with separately in order to achieve full features shown in Figure 1 and increase the performance of the device structure. The aim of this work is to observe the impurity photovoltaic effect in these devices. All these IPCE measurements show a sharp cut-off at ~740 nm, and there is no collection in the IR region. According to these measurements, there is no contribution from the impurity PV effect in these devices.

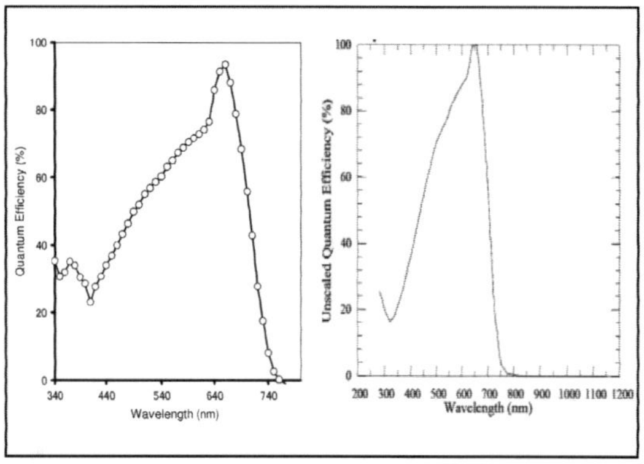

Figure 4. Typical IPCE% curves measured for two (3×3) mm^2 solar cells. Note the sharp cut-off wavelength at ~740 nm indicating nil collection in the I-R region.

4 RESPONSIVITY MEASUREMENTS

The above measurements clearly raised an issue of disagreement between the theoretically expected impurity PV effect and the experimentally observed null results. This encouraged us to explore the ways of data collection, analysis and presentation of IPCE results and found that there are grey areas which are not clear to the experimentalists. For this reason, a simple measurement of responsivity as a function of wavelength was explored. Responsivity of a device is defined in the simplest way as

Responsivity (A/W) = Electrical out put of the solar cell in (A)/Light in put to the solar cell in (W).

This method is fully transparent to the experimentalist and removes all complications in data collection, analysis and graphical presentation. The devices were sent to measure responsivity against the wavelength to explore further.

Figure 5 shows a typical measurement and the main spectrum shows similar losses in UV and IR ends as observed in IPCE measurements. However, the collection in the IR region, from 800 - 1200 nm, is clearly observed in these measurements. This is a strong indication that during IPCE data collection, analysis and graphical presentations, this useful information is completely lost.

5 CURRENT-VOLTAGE (I-V) MEASUREMENTS UNDER DARK CONDITIONS

The positive indication of existence of impurity PV effect encouraged us to further investigate this effect using other techniques.

5.1 Current-voltage (I-V) as a function of light intensity

In order to test the behaviour of these devices, in diffused light the device parameters were measured as a function of light intensity. Table II summarises the device parameters observed under different illumination conditions. It is an excellent feature of this device that the V_{oc} and FF values saturate to their highest values even at 1% of the one-sun intensity, of 1.2 mW cm^{-2}. In fact, this is almost dark, and the saturation of V_{oc} is a unique feature enabling the use of these devices in diffused light conditions. The production of V_{oc}=980 mV and FF=0.81 at 1% of the one-sun intensity indicates that there is a positive contribution from IR radiation towards the PV activity of this device.

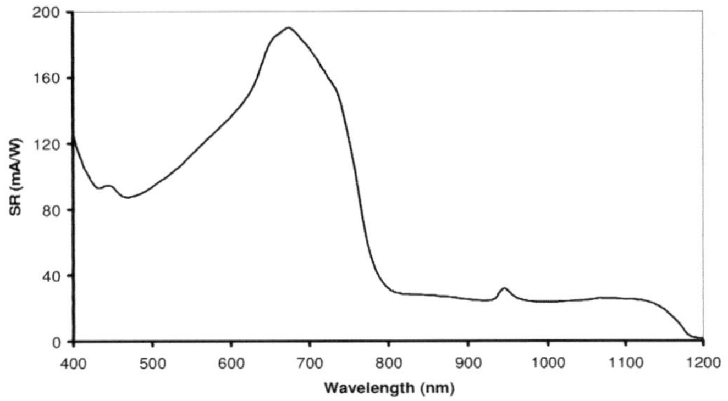

Figure 5. A typical responsivity versus wavelength curve measured for graded bandgap devices. Note the positive response in the IR region up to ~1200 nm.

Table II: Summary of device parameters observed for (3x3) mm^2 solar cells, when measured individually and two cells connected in series, as a function of light intensity.

Light Intensity (mW cm^{-2})	One (3x3) mm^2 cell			Two (3x3) mm^2 cells connected in series		
	V_{oc}(mV)	FF	J_{sc}(mA cm^{-2})	V_{oc}(mV)	FF	J_{sc}(mA cm^{-2})
1.2	980	0.81	0.10	1960	0.81	0.12
10.5	1080	0.83	1.20	2160	0.83	1.26
53.0	1140	0.85	6.19	2270	0.85	6.49
100.0	1150	0.85	12.12	2310	0.85	12.14

5.2 Current-voltage (I-V) measurements under complete darkness

The above observation of production of high values for V_{oc} and FF at lowest light intensity guided us to perform I-V characteristics under complete darkness and as a function of temperature. The devices were mounted inside a metal cryostat eliminating background light completely and the I-V characteristics were measured as a function of temperature. A typical set of linear-linear I-V curves observed under dark conditions are shown in Figure 6.

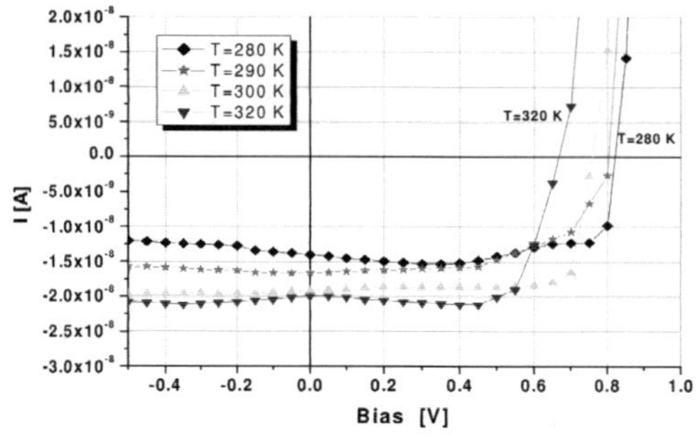

Figure 6. Linear-linear I-V curves measured for graded bandgap devices under dark conditions at different temperatures.

The observation of J_{sc} and the production of V_{oc} values in the region of (650 - 850) mV, under complete darkness confirms the PV activity created by thermal energy from the surrounding, and hence the existence of impurity PV activity in these devices.

In a separate I-V measurement system, diodes were measured under dark and illuminated conditions for further confirmation. Two such plots are shown in Figure 7, in both linear-linear and log-linear forms. Again the device exhibit ~1175 mV and ~950 mV open circuit voltage, under AM1.5 and in dark conditions respectively, re-confirming the activity due to IR radiation.

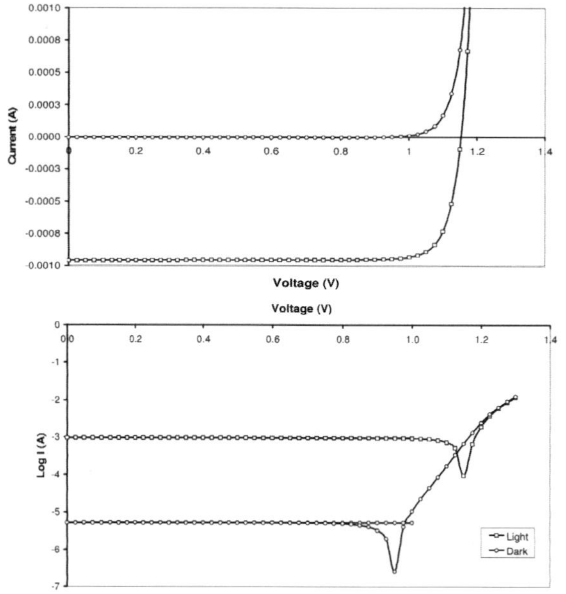

Figure 7. Linear-linear and log-linear I-V curves measured for graded bandgap devices under dark and AM1.5 illuminated conditions. Note the V_{oc}~950 mV at complete darkness, and V_{oc}~1175 mV under illumination.

6 DISCUSSION

The above observations highlight the existence of two power inputs to these graded bandgap multi-layer solar cells as shown in Figure 8; (i) the normal solar radiation when illuminated and (ii) the environment of the solar cell acting as an infinite heat reservoir. The heat from the surroundings also creates charge carriers in the device, especially at the back end and enhances the combined PV effect of the device. Production of charge carriers using I-R photons depends on naturally occurring defect levels specially occurring close to the growth substrate at the rear of the solar cell.

Table 11.2. Experimentally observed defect levels in MOVPE $Al_xGa_{(1-x)}As$ layers[6,7].

Al content (x)	Activation Energy, E_a (eV)	Approximate concentration (cm^{-3})
0.05 - 0.11	0.41	10^{14} - 10^{16}
0.15	0.50	10^{15}
0.20	0.55	10^{15}
(0.00 - 0.25)	(0.80 - 0.90)	The varying EL2 level according to value of x

It is appropriate to look at the deep energy levels present in MOVPE grown $Al_xGa_{(1-x)}As$ layers. This system is the most researched semiconductors next to silicon. The deep levels present are thoroughly studied and documented [6,7] in the literature. Some of these are shown in Table III and therefore it is clear that there exists a ladder of deep levels in the bandgap. The concentration of these defects must be higher closer to the substrate. These levels therefore must be helping to create charge carriers using multi-step generation (or impurity PV effect), utilising IR radiation from surroundings.

The detrimental R&G process also takes place in parallel, but the shape of this device allows impurity PV effect to dominate and positively contribute to overall PV effect. Therefore, when fully optimised, this device has a huge potential to increase J_{sc} and hence the conversion efficiency.

7 CONCLUSIONS

The work presented in this paper leads to make several important conclusions.

(i) The responsivity measurements as a function of wavelength is simple, and experimentally confirms existence of impurity PV effect in graded bandgap multi-

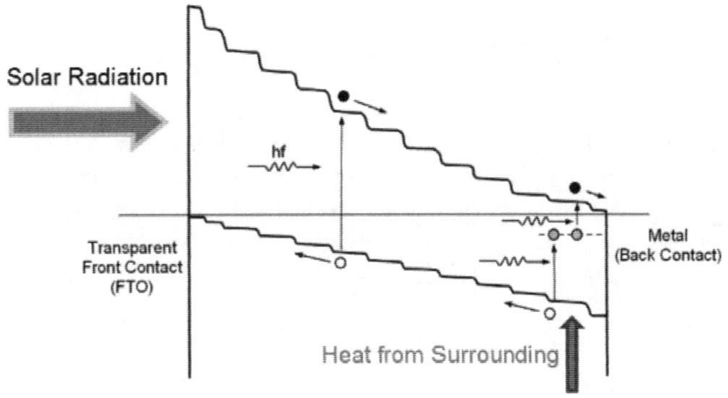

Figure 8. Schematic diagram showing the two energy inputs available to this device structure. When placed in complete darkness, the device still produces PV activity using heat energy from the surrounding due to impurity PV effect.

layer devices. These observations are re-confirmed by the production of high V_{oc} values together with other PV parameters in complete darkness. The theoretical expectation of existence of impurity PV effect is therefore experimentally confirmed for these new device structures.

(ii) The graded bandgap multi-layer solar cell structure has two power inputs; normal solar radiation and the surrounding heat energy. Heat energy creates charge carriers and this process will enhance the overall PV effect of the device when illuminated. Therefore when the doping concentrations are optimised to achieve fully depleted device, high J_{sc} values and hence high efficiencies are expected from this device structure.

(iii) The current QE measurement systems miss-out some important information, most probably due to methods used for data collection, analysis and/or presentations of the results. This needs appropriate actions to improve this important PV characterisation technique.

ACKNOWLEDGEMENTS

Authors would like to acknowledge the contributions made by Drs. J. S. Roberts and G. Hill for MOVPE growth and device processing. All the collaborating colleagues are gratefully thanked for their contributions to this work by carrying out some experimental measurements.

References

[1] I. M. Dharmadasa, 2005 *Sol. Energy Mater. Sol. Cells* 85 293.
[2] I. M. Dharmadasa, J. S. Roberts and G. Hill, 2005 *Sol. Energy Mater. Sol. Cells* 88 413.
[3] I. M. Dharmadasa, G. J. Tolan, J. S. Roberts, G. Hill, S. Ito, P. Liska and M. Grätzel, 2006, *Proc. of 21st EU photovoltaic conference, Dresden Germany*, 257.
[4] M. A. Green, K. Emery, D. L. King, S. Igari and W. Warta, 2005 *Prog. Photovolt: Res. Appl.* 13 49.
[5] K. Takahashi, S. Yamada and T. Unno, U.D.C. 621.383.51:523.9-7:[546.681'62'19: 546. 681'19].
[6] *Properties of Gallium Arsenide*, 2nd Edition, EMIS Data review series No.2, An INSPEC publication, 1990.
[7] S. Adachi (Edited), *Properties of Aluminium Gallium Arsenide*, EMIS Data reviews Series No. 7, An INSPEC Publication, 1993.

Hypervelocity impacts into graphite

S Latunde-Dada, C Cheesman, D Day, W Harrison and S Price

AWE Plc, Aldermaston, Reading RG7 4PR, UK

E-mail: seyi.latundedada@awe.co.uk

Abstract. Studies have been conducted into the characterisation of the behaviour of commercial graphite (brittle) when subjected to hypervelocity impacts by a range of projectiles. The experiments were conducted with a two-stage gas gun capable of launching projectiles of differing density and strength to speeds of about 6kms^{-1} at right angles into target plates. The damage caused is quantified by measurements of the crater depth and diameters. From the experimental data collected, scaling laws were derived which correlate the crater dimensions to the velocity and the density of the projectile. It was found that for moderate projectile densities the crater dimensions obey the '2/3 power law' which applies to ductile materials.

1. Introduction

The normal impact of spheres into semi-infinte targets is a problem that has been studied extensively in the laboratory and through the use of numerical hydrocodes with in-built material models. Most of these studies have been done with ductile targets examples of which can be found in [1, 2] and a few with brittle targets, an example of which is [3]. In these studies, the dimensions of the craters formed have been used to characterize the damage to the target and scaling laws with impact velocity have been derived. Typically, due to the large deformations prevalent in the hypervelocity regime, detailed mathematical analysis of this problem has proved too difficult to derive scaling laws theoretically. Therefore, simple power laws relating the crater dimensions to the governing parameters of the problem have been developed by fitting experimental or numerical data.

In this paper, an attempt will be made to derive scaling laws for the dimensions of the craters formed from experimental data generated by normal hypervelocity impacts of spherical projectiles into a graphite target.

In section 2, the cratering process is discussed and contrasted with the morphology of craters formed in ductile and brittle targets such as graphite. Next, in section 3, the impact experiment is described and the experimental data presented in appendix A. From this data, in section 4, the variation of crater depth against impact velocity is studied and compared with the 2/3 power law characteristic of impacts into ductile targets. In section 5, the correlation of the crater diameter with velocity is studied and hence, a power law obtained for the variation of the crater volume with impact velocity. In section 6, the effects of projectile density on the correlations are studied and predictions made for the crater dimensions formed by polypropylene and cellulose acetate projectiles which are compared to data. Finally in section 7, a summary and some conclusions are reported.

2. Cratering process and morphology for normal impacts

The process of crater formation from normal hypervelocity impacts can be divided into four phases: shock, steady state, cavitation and recovery [4]. At the instant of impact, the particles at the impact surface of the projectile are instantaneously brought to rest by the target particles. Shock waves are produced at this surface which propagate into both the target and back into the projectile, causing large scale plastic deformations accompanied by melting or vaporization. This is the **shock phase** which ends when the rarefraction waves relieve the shock pressures. These rarefraction waves originate from reflection of the shock waves from free surfaces where there is a zero-pressure boundary condition. In semi-infinite targets where the lateral and rear surfaces are too far away to generate the rarefraction waves, the shock is primarily released by the rarefraction waves generated from the free surfaces of the projectile. In thinner targets, on the other hand, both the projectile and target rarefraction waves relieve the shock wave in the target. Accompanying this release is the ejection of target particles at high speeds.

During the shock phase, the projectile, being subjected to pressures much greater than its yield stress, begins to erode and effectively becomes a fluid which is now the primary cause of crater formation. This is the **steady state phase** during which the governing parameters of crater formation are the density and velocity of the projectile. The duration of this phase depends on the dimensions of the projectile. For spherical projectiles travelling at hypervelocities, this phase can be very short.

After the projectile has completely eroded, the crater continues to expand under its inertia until it is unable to overcome its own intrinsic resistance. This is the **cavitation phase**.

Finally, once the crater has reached its maximum extent, its dimensions may diminish slightly due to an elastic rebound which produces tensile stresses. This is the **recovery phase** which signals the end of cratering.

Typically, in brittle materials, due to their low tensile strength, cratering lasts longer than in ductile materials and hence the damage is more extensive. This because the shock and rarefraction waves last longer during the shock phase due to the low intrinsic resistance. In addition, due to their relative tensile strengths, the target material ejected during this phase forms the characteristic raised lip in ductile materials which is absent in impacts on brittle targets.

Likewise, during the cavitation phase, the low resistance of brittle materials causes sub-surface cracks to develop which can result in surface cracks and spallation around the initial crater unlike in ductile materials. This results in relatively wider craters which are therefore shallower in terms of the ratio of the crater depth to its diameter.

Figures 1 (a) and (b) show the morphologies of ductile and brittle craters labelled with the crater dimensions. In figure 2, a photograph of the brittle crater formed by a 4mm diameter alumina projectile impacting a graphite target at 3.02kms^{-1} is shown.

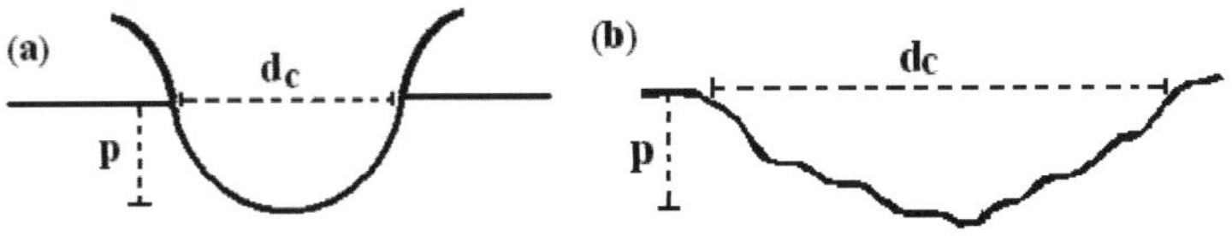

Figure 1. Typical morphologies of crater formed by hypervelocity impacts into (a) ductile and (b) brittle targets.

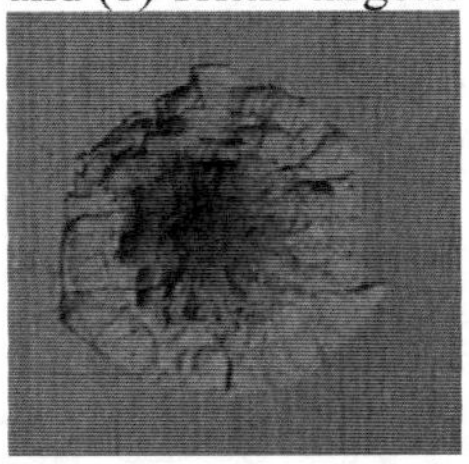

Figure 2. Photograph of a brittle crater

3. The experiment

The experiments were conducted with a two-stage light gas gun which is used to accelerate small spherical projectiles to hypervelocity speeds (approximately 1–7kms^{-1}) into target plates. It operates by the ignition of a propellant charge to force a piston forward into a breech filled with hydrogen gas at pressures of 200–500psi. This pressure rapidly increases until a bursting disc 'petals' and the sabot containing the projectile is launched forward. The sabot is stripped either by a metal obstruction or gas pressure and the speed of the projectile is measured by lasers just prior to impact into the target plate.

The spherical projectile materials used include HDPE, alumina, aluminium, soda glass and steel with diameters ranging from 1-4.76mm. The thickness of the blocks used varied from 25mm to 50mm to 100mm. The target blocks were made of commercial graphite (Graphite 25) which is a brittle and porous material with a porous density of 1.83gcm^{-3} and a compact density of 2.13gcm^{-3}.

The primary method of quantitative damage comparison between each impact is the measurement of the crater depth, diameter and volume using a laser scanner which generates profiles of the craters. Due to the irregular nature of the craters, the depth and diameter are taken in two perpendicular planes and the averages of each are found. The crater volumes are also calculated. The experimental data obtained from these experiments are tabulated in appendix A.

4. Correlation of crater depth *p* with impact velocity *v*

In previous studies of hypervelocity impacts, the dimensionless form $p/d_p = \Phi(v)$ has been used for the correlation where p is the crater depth, d_p is the diameter of the projectile, v is its impact velocity and Φ is the correlation to be determined [5]. This convention will be adopted here since as can be seen in the data tables in appendix A, d_p has been varied for certain projectile-target configurations.

Previous studies on normal hypervelocity impacts on semi-infinite ductile targets have demonstrated two fundamental concepts about the correlation of crater depth with impact velocity [1, 2]. Firstly, as the impact velocity is increased, it has been observed that the crater tends to a hemispherical shape i.e. the ratio of the crater depth p to crater diameter d_c approaches 1/2. The direction of the approach to this regime was found to be dependent on the ratio of the density of the projectile to the target. For projectiles denser than the target, p/d_p starts off greater than 1/2 and vice versa for projectiles of a lower density than the target. In addition, the greater the difference between the densities of the projectile and the target, the slower is the approach to the hypervelocity regime and the greater the velocity at which hypervelocity phenomena dominate the cratering process.

Secondly, a number of studies have suggested that the crater volume becomes proportional to the kinetic energy of the projectile in the hypervelocity regime. Coupling these two observations together, the following relations are arrived at:

$$\begin{aligned} V_c &= \frac{4}{3}\pi p^3 \\ E_p &= \frac{2}{3}\pi \rho d_p^{\,3} v^2 \\ V_c &\propto E_p \\ \therefore \frac{p}{d_p} &= K_p v^{\frac{2}{3}} \end{aligned} \tag{1}$$

where V_c is the crater volume, E_p is the kinetic energy of the projectile, ρ is its density and K_p is a constant of proportionality. This is the 2/3 power law which has been verified for a variety of ductile materials. It is worth noting that some fits to experimental and numerical data have provided velocity exponents between 1/3 and 2/3 which is indicative of a mixture between energy scaling and momentum scaling i.e. scaling in which the momentum rather than the kinetic energy is proportional to the crater volume [1].

Impacts in brittle materials like graphite have undergone much less investigation. However, as discussed in section 2, most of the differences between the brittle and ductile cratering processes are related to the crater width and not the depth. Hence, a similar scaling law is expected for the depth of craters in graphite.

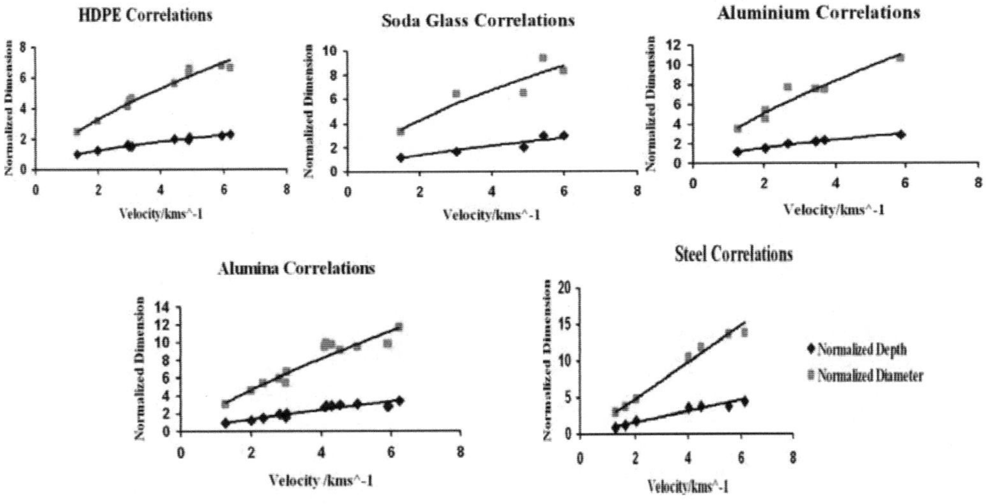

Figure 3. Correlation fits from experimental data.

Figure 3 shows the correlation fits obtained from the experimental data recorded in tables A1-A6 in appendix A. In table 1, the velocity exponents n_p obtained for the depth correlations are presented for the various projectiles under study. The values of the proportionality constants K_p and the correlation coefficients R^2, of the global fits are also presented. R^2 is a measure of the goodness of fits of the global trend lines.

Table 1. Correlation parameters for crater depth with impact velocity.

Projectile Material	ρ/gcm^{-3}	n_p	K_p
HDPE	0.969	0.6972	0.7837
Soda Glass	2.48	0.6319	0.9292
Aluminium	2.71	0.6059	1.0754
Alumina	3.69	0.6022	1.0512
Steel	7.8	0.87	0.9434

From the table, it appears that except for steel projectiles, the velocity exponents obtained agree with the 2/3 power law if some momentum scaling is allowed for. The higher value of n_p observed for steel projectiles may be attributed to the inclusion in the fits, of data from outside the hypervelocity regime since, as mentioned earlier, the speed of approach to the hypervelocity regime is expected to be slower for denser materials. This will be discussed further in appendix B. It should be noted also that this trend should not be surprising as a previous study into hypervelocity impacts on glass targets gave a velocity exponent of 0.8 [6].

5. Correlation of crater diameter d_c with impact velocity v

As discussed in section 4, for ductile targets, the crater formed from a hypervelocity impact tends to a hemispherical shape. This implies that, the crater diameter d_c scales as the penetration depth p does with velocity i.e.

$$\frac{d_c}{d_p} \propto v^{\frac{2}{3}} \tag{2}$$

In brittle targets however, as explained in section 2, the correlation is not so straightforward due to the shallow craters formed by the subsurface cracks and spallations characteristic of such materials and hence the craters formed are not hemispherical. To illustrate this, scatter plots of the ratio of the crater diameter to its depth d_c/p as a function of velocity as shown in figure 4.

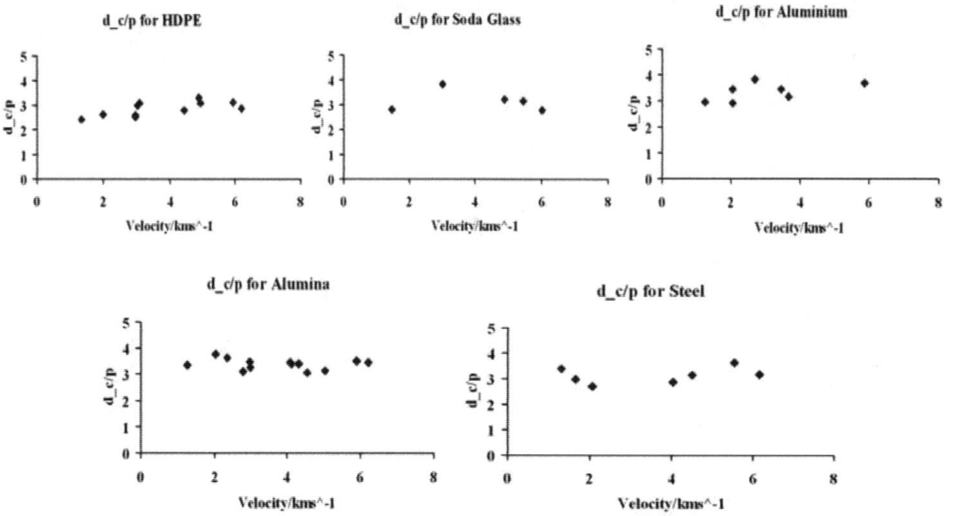

Figure 4. Ratios of crater diameter to crater depth.

It is important to note though that despite the irregularities of the cracking process, d_c/p stays relatively constant with velocity and ranges from an average value of about 2.8 for HDPE to 3.4 for alumina. The average values $\langle d_c/p \rangle$ and standard deviations σ are recorded in table 2.

Table 2. Means and standard deviations of the ratios of crater diameter to crater depth.

Projectile Material	$\langle d_c/p \rangle$	σ
HDPE	2.86	0.29
Soda Glass	3.15	0.43
Aluminium	3.34	0.35
Alumina	3.39	0.2
Steel	3.13	0.32

This suggests that crater diameter should have a similar variation with impact velocity as crater depth. This is demonstrated in table 3 where n_d represents the velocity exponents for the normalized crater diameter correlations (see figure 3) and K_d and R are the proportionality constant and correlation coefficient respectively.

Table 3. Correlation parameters for crater diameter with impact velocity.

Projectile Material	ρ/gcm^{-3}	n_d	K_d
HDPE	0.969	0.718	1.9498
Soda Glass	2.48	0.5818	3.0603
Aluminium	2.71	0.6655	3.1472
Alumina	3.69	0.6686	3.2256
Steel	7.8	0.9308	2.7355

The values of n_p for the depth correlations obtained in section 4 can now be compared to n_d. As mentioned in section 2, during the cratering process in brittle targets, an initial crater is formed which then subsequently expands in diameter as the tensile strength of the target is overcome and cracks propagate. The scaling of d_c and hence the crack length, l, therefore suggests that the energy absorbed by crack propagation (as well as the spallation of the target material above the crack) is proportional to the projectile energy as well as volume of target material spalled which scales roughly as l^3.

As an aside, it is interesting to note that for impact velocities above 2.5kms^{-1}, the velocity exponents for steel approach 2/3 as shown in table 4.

Table 4. Correlation parameters for impact velocities above 2.5kms^{-1}.

Projectile Material	n_p	K_p	n_d	K_d
Steel ($v < 2.5$kms^{-1})	0.6468	1.3449	0.6265	4.5084

Therefore it appears that, for steel projectiles, the hypervelocity regime includes speeds of above about 2.5kms^{-1}, although this can not be confirmed with conviction because of the relative paucity of data used in these new fits.

In the data tables of appendix A, the crater volume is also recorded. During the measurements, it was observed that the craters formed were roughly conical in shape. This can be verified by comparing the volume of a cone of height p and diameter d_c which is given by $\pi d_c^2 p/12$ with the measured values. Figure 5 shows scatter plots of the volume ratio $V_c^{\text{cone}}/V_c^{\text{measured}}$ from which it can be seen that a cone does give a good description of the crater shape. The mean of this distribution is 0.97 with a standard deviation of 0.12.

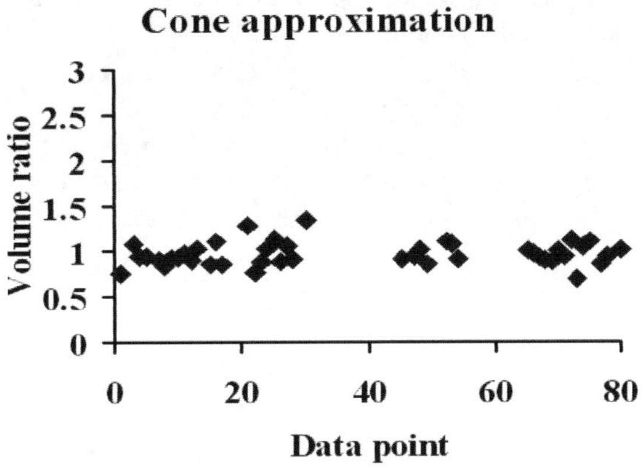

Figure 5. Scatter plot for conical volume ratio

6. Projectile density correlations

The differences in the velocity exponents and proportionality constants obtained for the different projectiles may be attributed to their varying densities since as discussed in section 4, the density of the projectile relative to the target's determines the speed at which hypervelocity phenomena becomes important.

Looking at tables 1 and 3, it can be noted that in general as the density increases, so do the proportionality constants K_p and K_d with the velocity exponents (excepting steel's) remaining relatively constant. Assuming a power law variation for the exponents n and constants K with ρ and ignoring the values for steel, the following correlations are obtained for a projectile of density ρ;

$$\frac{p}{d_p} = 0.791\rho^{0.2339} v^{0.6936\rho^{-0.1146}}; \rho < 3.69\,gcm^{-3},$$
$$\frac{d_c}{d_p} = 2.02\rho^{0.4061} v^{0.6974\rho^{-0.0758}}; \rho < 3.69\,gcm^{-3}. \tag{3}$$

With these equations, predictions were made for four experiments which were excluded from the fits. The results are compared to the experimental measurements in table 5 and both can be seen to agree to within at least 90% accuracy.

Table 5. Experimental and predicted crater dimensions.

Projectile Material	ρ/gcm^{-3}	v/kms^{-1}	$\frac{p}{d_p}^{exp}$	$\frac{p}{d_p}^{pred}$	$\frac{d_c}{d_p}^{exp}$	$\frac{d_c}{d_p}^{pred}$
Cellulose acetate	1.273	3.015	1.72	1.76	5.27	4.74
Cellulose acetate	1.273	5.928	2.53	2.78	7.07	7.54
Polypropylene	0.881	3.204	1.64	1.74	4.26	4.36
Polypropylene	0.881	1.415	0.98	0.99	2.45	2.43

7. Summary and conclusions

In summary, this study into the hypervelocity impact of projectiles into graphite targets has:

- Developed power laws for the correlation of crater dimensions with the impact velocity for a wide range of projectiles.
- Confirmed that projectiles up to the density of alumina (3.69gcm^{-3}) obey the 2/3 power law approximately. Investigation of the data from the steel impacts also hint at an obedience of the 2/3 power law at impact velocities greater than 2.5kms^{-1}.
- Confirmed that the craters were roughly conical in shape with a height and diameter equalling the crater depth p and diameter d_c.
- Calculated the density variation of the velocity exponent and proportionality constant and tested this against experiments with success.

The hypervelocity impact data has provided a valuable insight into the laws of cratering in graphite targets. However, the study will benefit from the inclusion of more data in the fits from normal impacts of steel at velocities above 2.5kms^{-1} to determine the onset of the hypervelocity regime.

Acknowledgements

The author would like to acknowledge the valuable help received from Nigel Thomas during the compilation of this report. This work was funded by AWE Plc, Aldermaston, Reading, Berkshire RG7 4PR.

Appendix A. Experimental data

The experimental data obtained from the impact experiments are presented in tables A1-A6. The notation below has been applied for the parameters.

Table A1. Notation for data tables.

v	Impact velocity of projectile
t	Thickness of the graphite block
d_p	Diameter of spherical projectile
p	Average penetration depth of the crater
d_c	Average diameter of the crater (normal impact)
V_c	Crater volume

Table A2. Experimental data for normal impacts of HDPE projectiles into graphite.

v/kms^{-1}	t/mm	d_P/mm	p/mm	d_c/mm	V_c/mm^3
0.729	50	4.75	2.54	7.13	45.06
1.076	50	4.75	4.36	9.68	87.91
1.342	50	4.76	4.85	11.78	334.1
1.995	50	3.15	3.83	10.00	103.7
2.97	50	3.15	5.03	13.08	237.9
2.97	50	3.15	5.17	13.03	249.2
3.02	100	3.15	4.76	14.25	334.1
3.09	100	3.15	4.75	14.73	294
4.01	50	3.15	5.49	14.6	389.2
4.44	50	3.15	6.32	17.68	531.9
4.88	100	3.15	5.97	19.75	554
4.92	100	3.15	6.67	20.69	795.7
5.94	50	3.15	6.85	21.38	818.9
6.21	50	3.15	7.27	20.88	889.2

Table A3. Experimental data for normal impacts of soda glass projectiles into graphite.

v/kms−1	t/mm	d_p/mm	p/mm	d_c/mm	V_c/mm^3
1.46	50	4	4.74	13.28	509.8
3.03	100	3.95	6.65	25.49	1199
4.89	100	3.95	7.99	25.57	1626
5.44	50	2	5.98	18.74	268.6
6.01	100	2	5.96	16.55	499.1
0.794	50	3.95	3.6	12.1	167.8
1.048	50	3.95	3.3	11.35	137.9
3.963	50	3.95	10.8	29.25	2382

Table A4. Experimental data for normal impacts of aluminium projectiles into graphite.

v/kms−1	t/mm	d_p/mm	p/mm	d_c/mm	V_c/mm^3
1.25	50	3.95	4.66	13.78	332.4
2.04	100	2.05	3.2	9.3	80.65
2.04	100	2.05	3.2	11.06	102.2
2.7	50	3.95	7.93	30.38	2248
3.45	50	2.05	4.5	15.48	253.2
3.68	100	2.05	4.86	15.3	329.1
5.88	50	2.05	5.88	21.67	814.4
0.78	50	3.94	3.75	10.61	112.2
1.03	50	3.95	4.47	12.25	184.5
3.968	50	2.04	5.83	14.8	335.4
4.391	50	2.05	5.7	15.2	370.9
5.076	50	2.05	6	16.95	538.7

Table A5. Experimental data for normal impacts of alumina projectiles into graphite.

v/kms−1	t/mm	d_p/mm	p/mm	d_c/mm	V_c/mm^3
0.797	50	4	4.1	11.75	164.1
0.994	50	4	6	16.15	366.3
1.27	50	4	3.7	12.38	166.9
2.02	50	4	4.93	18.58	496.2
2.35	50	3	4.45	16.16	285.6
2.82	50	4	7.65	23.8	1003
3	50	2	3.18	11.07	97.32
3.02	100	4	8.2	26.73	1404
4.1	100	4	10.94	37.71	4321
4.13	100	4	11.63	39.49	4442
4.31	50	3	8.55	29.14	1852
4.54	50	4	11.95	36.5	4729
5.03	100	2	6.07	18.97	566.7
5.89	50	2	5.64	19.67	752.5
5.92	50	2	5.61	19.58	746.7
6.24	50	2	6.78	23.34	1019

Table A6. Experimental data for normal impacts of steel projectiles into graphite.

v/kms−1	t/mm	d_p/mm	p/mm	d_c/mm	V_c/mm^3
0.774	50	3.99	3.2	9.69	90.5
0.964	50	3.99	4.4	11.55	130.4
1.3	50	3	2.65	8.98	54.32
1.65	50	3	3.75	11.18	96.24
2.06	50	4	6.99	18.87	589.5
2.515	50	3.01	7	20.6	750.6
3.394	50	3	8.9	29.45	1850
4.032	50	4	7.21	20.77	830.2
4.51	50	3	11.25	35.43	4361
5.54	50	2	7.47	27.17	1695
6.15	50	2	18.7	27.5	1924

Appendix B. Low velocity impacts

Since alumina and steel are much denser than graphite, the approach to the hypervelocity regime is expected to be slower than less dense materials which might explain why the velocity exponents obtained for these projectiles fall out of the range $1/3 <\psi n_p{}^{c}\psi n_d <\psi 2/3$ predicted by momentum and energy scaling. To understand this further, the velocity dependence of craters formed by low velocity impacts will be investigated.

At low velocities, the equation of motion of the indenting sphere of mass m_p and diameter d_p is given empirically by,

$$m_p \frac{d^2 p}{dt^2} = m_p v' \frac{dv'}{dp} = -Y\pi \frac{d_c^2}{4} \approx -Y\pi p d_p \tag{B.1}$$

where t is the time, v≃←is the velocity, and Y is the mean pressure of resistance of the target to penetration. We have assumed that d_c^2 ♡←$4pd_p$ which is true when p⊲d_p is small. Solving equation (10), the following correlation for the depth p with impact velocity v is obtained,

$$\frac{p}{d_p} \propto v, \tag{B.2}$$

which suggests a velocity exponent of 1 for low velocity impacts. Thus, the large velocity exponents obtained from the steel projectile correlations might be due to the influence of low velocity data in the correlation fits.

References

[1] Shanbing Y, Gengchen S and Qingming T 1994 *Int. J. Impact Eng* **15** 1 67-77
[2] Cour-Palais B G 1987 *Int. J. Impact Eng* **5** 221-237
[3] Taylor E, Kay L and Shrine N 1997 *Adv. Space Res.* **20** 8 1437-1440
[4] Hermann W and Wilbeck J S 1987 *Int. J. Impact Eng* **5** 307-322
[5] Holsapple K 1987 *Int. J. Impact Eng* **5** 343-355
[6] Burchell M J and Grey I D S 2001 *Material Science and Engineering* **A303** 1340141

Journal of Physics: Conference Series **286** (2011) 012043 doi:10.1088/1742-6596/286/1/012043

An Equation of State for Titanium

Geoffrey Cox

AWE, Aldermaston, Reading, RG7 4PR, United Kingdom

geoffrey.cox@awe.co.uk

Abstract. In hydrocode simulations the equations that govern fluid motion are solved. The variation of density of a material with pressure and temperature is needed. This is called the equation of state (EoS). In this work the development of an EoS for titanium is discussed, and comparisons with the theoretical model and experimental data are shown. The EoS produced is a thermodynamically consistent multiphase EoS where solid-solid, as well as, solid-liquid phase transitions are modelled. The Helmholtz free energy as a function of density and temperature is calculated for each phase. The stable phase is the phase with the lowest Gibbs free energy, and phase boundaries are where the Gibbs free energy is the same for two phases. For the solid phase the Helmholtz free energy is decomposed into three terms, the cold curve contribution, the thermal electron contribution, and the thermal contribution from lattice vibrations, and for the liquid phase just the thermal electron and ion contribution is needed. In most cases a classical approach is used to model these contributions, but for one of the solid phases density functional theory is used to calculate the cold curve contribution. The final model covers a large range of phase space and accurately reproduces regions of shock instability, caused by one of the solid-solid phase transitions.

1. Multiphase Equation of State

With a change in pressure and temperature many materials undergo solid-solid phase transitions where the crystal structure changes. Melting of materials is also another phase transition. Across a phase boundary discontinuities in energy, density and properties of a material are seen. Even small changes can have a significant effect such as wave splitting under shock loading. The changes will affect the EoS and should be included in an accurate model. A thermodynamically consistent multiphase EoS is needed. In a multiphase EoS an EoS is generated for each phase and at any pressure-temperature point the phase with the lowest Gibbs free energy is the phase that exists. Phase boundaries are the locus in the pressure-temperature plane where the Gibbs free energy of two phases are equal. The EoS in this region may be determined by assuming that along the phase boundary the material consists of two phases in pressure equilibrium, temperature equilibrium and thermodynamic equilibrium, and using the method of mixtures.

Titanium has a complex phase diagram, shown in figure 1. The multiphase EoS generated for titanium consists of three solid phases (α, ω and β) and a liquid phase. All thermodynamic quantities are obtained from the Helmholtz free energy. For the solid phases the Helmholtz free energy, is decomposed into the contribution from the cold curve, the contribution from lattice vibrations, and the thermal electron contribution. For the liquid only the thermal contribution of the electrons and ions is needed.

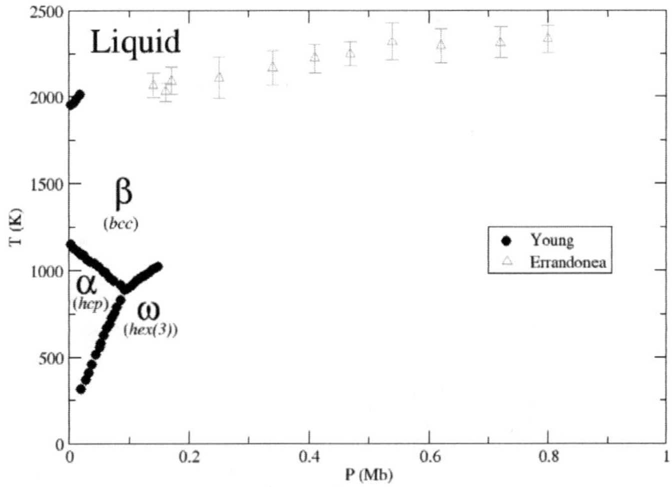

Figure 1: Phase diagram data[1,2] for titanium.

For the α and ω phases the cold cure and thermal ion contribution were taken from Kerley[3]. For the β phase the projector augmented wave approach to DFT, as implemented in VASP[4,5], was used to calculate a cold curve. A mean field model was used for β phase thermal ion contribution, where the pairwise interactions were extracted from the cold curve. For thermal ions in the liquid phase a fluid perturbation model was used, where the pair interaction term was obtained from the β phase cold curve. For all of the phases a Thomas-Fermi model was used for the thermal electron contribution.

A constant shift was added to the entropy and Helmholtz free energy of each phase to obtain the desired phase boundaries. The phase boundaries were chosen to give the best agreement with the Hugoniot and isothermal compression data. The final EoS covers a large range of phase space and compares well with experimental data, although the latent heat change at the α-β boundary is too large. On the principal Hugoniot the model shows wave splitting due to α→ω phase transition, but no wave splitting for the ω→β phase transition. This is consistent with experimental data. Trunin et al.[6] performed shock experiments for porous and non-porous samples. No wave splitting was observed for any of the porous samples. Calculations using the multiphase EoS with a snowplough porosity model offer an explanation of these anomalous experimental measurements. In the calculations the energy from closing the pores raises the temperature such that the β phase transition occurs at lower pressures, and the shock remains stable through this solid-solid phase transition.

References

[1] D. A. Young, Phase Diagrams of the Elements (University of California, USA, 1991)
[2] D. Errandonea, B. Schwager, R. Ditz, C. Gessmann, R. Boehler, and M. Ross, Phys. Rev. B, **63**, 132104-1 (2001)
[3] G. I. Kerley, SANDIA REPORT, SAND 2003-3785 (2003)
[4] G. Kresse and J. Joubert, Phys. Rev. B, **59**, 1758 (1999)
[5] P. E. Bloch, Phys. Rev. B, **50**. 17953 (1994)
[6] R. F. Trunin, G. V. Simakov, and A. B. Medvedev, High Temperature, **36**(6), 851 (1999)

A Solid State Ion Collider with Transient Current Pulses

C. D. Papageorgiou[1], T. E. Raptis[2]

[1]Dept. of Electrical & Electronic Engineering,
National Technical University of Athens, Greece
E-mail: chrpapa@central.ntua.gr
[2]National Centre for Science and Research "Demokritos",
E-mail: rtheo@dat.demokritos.gr

Abstract

We present experimental evidence for the appearance of strong forces inside the lattice of metallic conductors in the phase of transient excitation by a sequence of high power, high voltage but low energy electrical pulses.

1. Introduction

In a series of previous publications [1, 2] we analyzed the problem of wire fragmentation, which is associated with a long-standing controversy in Maxwellian electrodynamics. The first experiments in wire explosions under very high currents and voltages were made by Nasilowsky during the 60s. At the 80s, Pappas, Graneau and Rambaut and Vigier [3-6] repeated similar experiments and argued on the existence of longitudinal forces beyond the ordinary Maxwellian electrodynamics. Other attempts [7, 8] for a correct explanation in our opinion fail to provide a clear picture of the phenomena in terms of the Maxwell stress tensor. A corrected picture within the classical Maxwell theory has been presented in [1, 2] which is amenable to experimental verification based on the instantaneous forces between radiating dipoles created by the action of a transient high power current pulse.
Experimental results are reported in section 2 with a specially prepared pulse generator of very low internal resistance and low capacitor energy. An additional ongoing project for a microscopic study via simulations is explained in section 3. In the last section we discuss the possibility of using similar effects in order to create a solid state ion collider. Arguments are presented for the treatment of special metals and/or unstable isotopes by this method for low energy nuclear transmutations and possibly for neutralization of radio wastes.

2. Experimental investigation of bipolar forces

According to the macroscopic analysis presented in [1, 2] when a abrupt transient electrical pulse of an exponential shape akin to a capacitor discharge excites a long thin linear conductor (heretofore referred as the "wire"), the conductor during the transient excitation time interval behaves as a long transmission line. After the first moment of excitation, propagation of the electrical disturbance across the wire causes the temporary establishment of a system of stationary waves. This sets into motion additional secondary excitations inside the wire that cause the subsequent polarization of the metallic lattice in a time-dependent manner. In a simplified form, one may analyze this situation with the aid of Fig. 1 where a system of transient dipoles is shown along the wire.
This system was analyzed in [1, 2] in terms of classical Maxwell theory and the instantaneous forces between the dipoles were found to be given by the following formula

$$F(I,\lambda) = -180\frac{I^2}{c}\phi(1/\lambda), \lambda = L/2\pi D \qquad (1)$$

where D stands for the wire diameter, L is the wire total length and the function $\phi(1/\lambda)$ is a special integral from the dipole radiation formula that can be evaluated numerically.

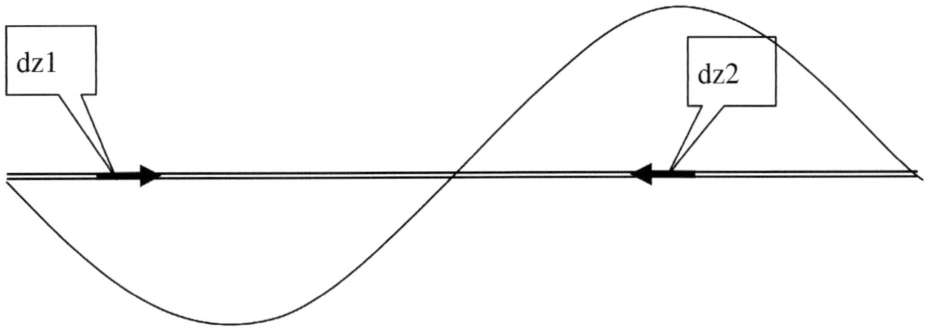

Fig. 1

The overall function force factor $\phi(1/\lambda)$ is shown in Fig. 2

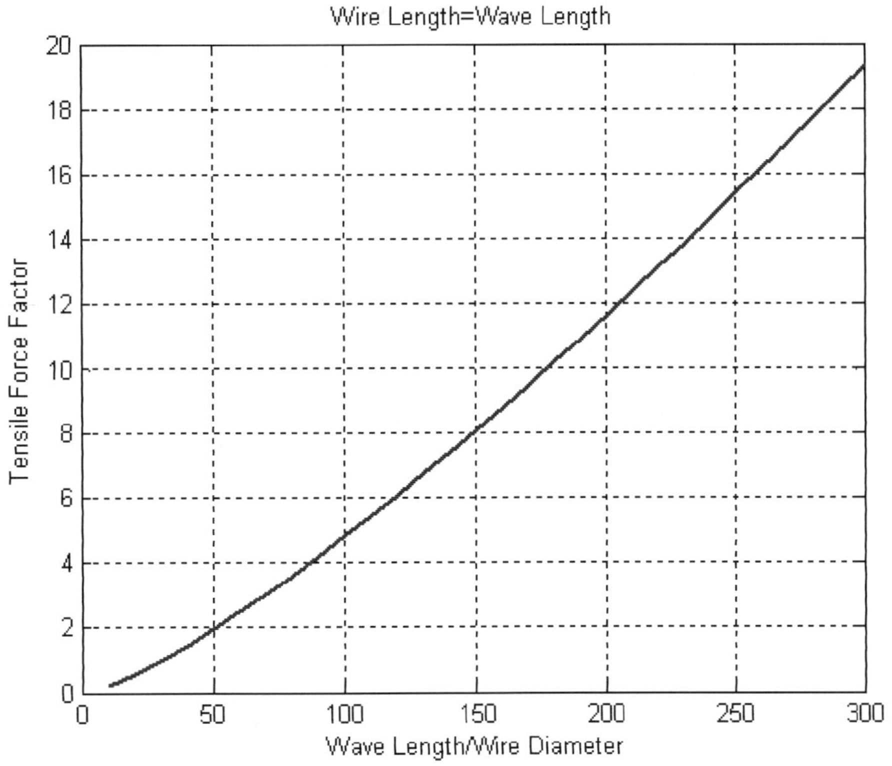

Fig. 2

In order to prove the existence of strong ionic collisions inside thin wires a special pulse generator was designed. The basic circuit is shown in Fig. 3. The aim behind this design was to have the overall internal resistance of this circuit to lie in a range of some mΩ's. For this, a special set of capacitors and a thyristor of very low internal resistance were chosen.

The current pulses were created by the discharge of the capacitor bank composed of 4 capacitors in parallel, each of 3 μF with 12 μF total capacitance (4.3), with the aid of a special digital controller circuit (4.6) that allowed both a single pulse or a sequence of consecutive pulses (see Fig. 3). The duration between subsequent pulses in the sequential mode was of the order of 3 seconds. These were then applied to a variety of wires of variable diameters and a constant length of approximately 27 cm.

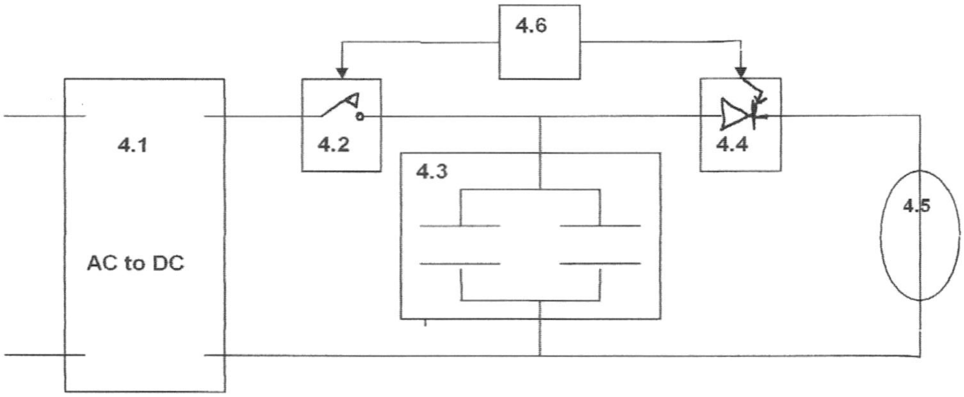

Fig. 3

The exercised voltage of the capacitors was nearly 1000 Volts and the discharge resistances including the circuit and the additional wire resistance were from 20 mΩ to 200 mΩ. The time constants of the occurring discharges were thus in the range of several μsecs.

The values of the instantaneous currents exerted on the wires from these current pulses were estimated in the range of 5000A-30000A. It was not possible to measure the magnitude of the fundamental harmonic but its existence could be detected by the appearance of longitudinal vibration of the thin wires that could be detected visually and easily could be captured by a high resolution camera.

Direct inspection showed that during the first pulse, a cloud of metallic dust gets violently expelled from the wire surface. This is speculated to be the result of the bipolar forces on defects and dislocations in the metallic crystal lattice. After several pulses depending on the wire diameter some of the wires get cut not necessarily near the center of the total length. This apparent discrepancy with the theoretical prediction of section 2 is probably due to the simultaneous coexistence and action of several different wavelengths due to the continuous spectrum of the discharge curve.

A number of wires with a steel core and a conductive metallic layer deposited with an electrolytic bath were also tried to show that it is possible to increase the tensile strength to avoid fragmentation. Indeed, the type of fatigue of the conductive metallic layer remained the same although the wire did not get cut despite a large number of pulses applied sequentially for a long time interval of several minutes up to half an hour.

In particular, the specific type of fatigue appearing in a set of experiments with copper wires of 27cm length and an average resistance of 100 mΩ was shown to be characterized by three major effects.

a) A number of almost periodic burns on the external surface of the wire that is probably associated with the wavelengths of higher harmonics.
b) An overall zig-zag shape obtained by a loose wire and a subsequent length shortening that is strongly suggestive of the opposite local dipole polarities ,due to the action of several harmonics, at certain sections of the wire.
c) At some of the experiments sparks appeared to emanate from points of the surface totally unrelated with the final break point. Some of these were captured with the

video camera and may be related with local charge concentrations on previously created cracks of the metal surface.

Observation of the condition of the wires surface took place at the Laboratory of Metallurgical Microscopy of the National Technical University of Athens. Results were photographed and are shown in the pictures of Fig. 4 Examination of these samples revealed both vertical as well as helical cracks on the metal surface.

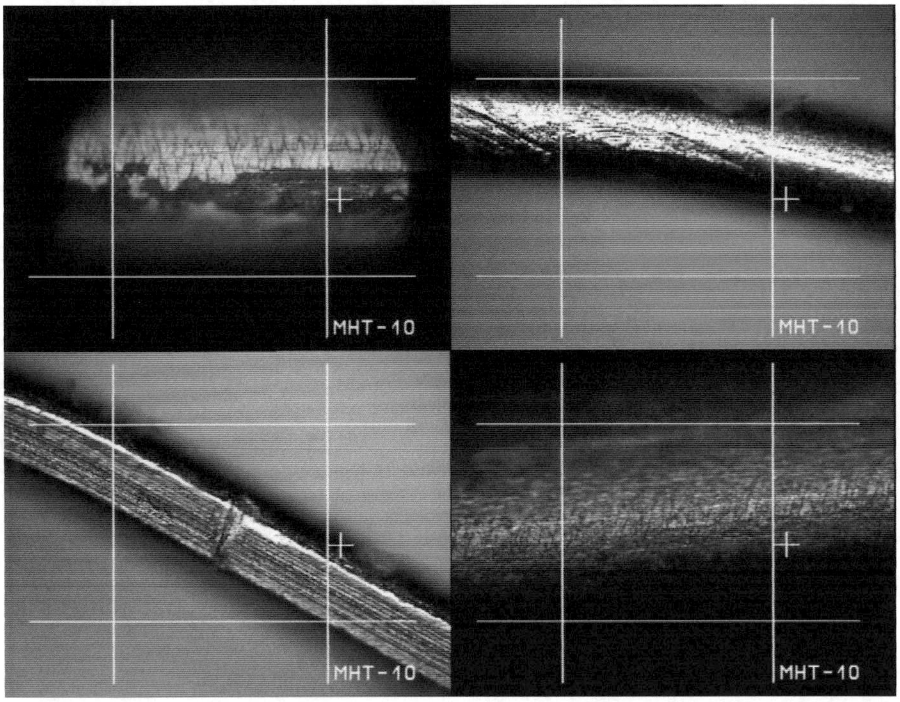

Fig 4. Condition of a wire surface after a succession of high power current pulses.

Another question concerning the appearance of the characteristic "neck"-like configuration of a metal near the break point due to increased metal plasticity as usually predicted was not verified. The final breaking mechanism may require separate study but our purpose in these experiments was rather to prevent the wires from reaching their tensile strength threshold in order to create a sustainable mechanism for exerting bipolar forces on a metallic lattice without breaking.

3. Conclusions

Both theoretical analysis and experimental evidence suggests the appearance of very strong forces inside the crystal lattice of metallic conductors which cause certain fatigue to the surface of the conductors. Although their macroscopic analysis in the present study was based on the abstract notion of ideal Hertzian dipoles, it is more than obvious that in the microscopic reality of the lattice, these forces will be of a purely Coulmobic nature due to the local strong polarization phenomena. This implies that in the short transient phase that the lattice moves strongly away from equilibrium, there will be a severe increase in the kinetic energies of the associated phonon-phonon and electron-phonon scattering events that could be utilized in cases where one would like to study the effects of strong ionic collisions without the use of an accelerator. It is also possible to use this technique in order to study the influence of such phenomena in the case of unstable or even radioactive metals of heavy Actinides.

References

[1] C. D. Papageorgiou. T. E. Raptis, Eur. Phys. J. Appl. Phys. **48**, 31002 (2009)

[2] C. D. Papageorgiou. T. E. Raptis, 7th Conf. Balkan Physics Union, AIP Conf. Proceedings, (2009) 955 - 960

[3] P. T. Pappas, P. G. Moyssides, Phys. Let. A, **111**(4), (1985) 193 – 198

[4] P. Graneau, Phys. Let. A, **107**(5), (1985) 235 – 237

[5] P. Graneau, N. Graneau, Phys. Let. A, **174**(5-6), (1993) 421 – 427

[6] M. Rambaut, J. P. Vigier, Phys. Let. A, **148**(5), (1990) 229 – 238

[7] S. Molokov, D.P. Wall, J.E. Allen, J. Phys. D: Appl. Phys. **36**, (2003) 2757

[8] F. Winterberg, Annales de la Fondation Louis de Broglie **31**, (2006) 363

[9] J. D. Jackson, "*Classical Electrodynamics*", J. Wiley (1975)

[10] W. Lochte-Holtgreven, Atomkernenergie **28** (1976) 150

[11] S. K. Handel, O. Johnson, Atomkernenerg. Kerntech. **36** (1980) 170

[12] P. Graneau, N. Graneau, Phys. Lett. A, **165**(1), (1992) 1-13

[13] J. D. Sethian *et al.*, Phys. Rev. Lett. **59** (1987) 892

[14] A. Shyam, R. K. Rout, IEEE Trans. Plasma Sci. **27**(4), (1999) 1210 - 1213

[15] W. Jones, N. H. March, *"Theoretical Solid State Physics"*, J. Wiley (1973)

[16] P. L. Taylor, *"A Quantum Approach to the Solid State"*, Prentice-Hall (1970)

Calculation of the Casimir-Polder interaction between Bose-Einstein condensates and microengineered surfaces: a pairwise-summation approach

M N A Halif[1], R Messina[2] and T M Fromhold[1]

[1] Midlands Ultracold Atom Research Centre, School of Physics & Astronomy, University of Nottingham, NG7 2RD, Nottingham, United Kingdom.

[2] SYRTE - Observatoire de Paris 61, Avenue de l'Observatoire, F-75014 Paris, France.

E-mail: ppxmna@nottingham.ac.uk

Abstract. Understanding the Casimir-Polder (CP) attraction between ultracold atoms and dielectric surfaces is of fundamental importance to the design and miniaturization of atom chips [1] and for understanding quantum reflection from, and sticking to, the surface [2 - 6]. So far, calculations have focussed on planar dielectrics, either semi-infinite or of finite width [7-9], and non-planar geometries including corrugated surfaces [10], spheres and carbon nanotubes [11-12]. Here, we present a simple scheme, based on the Pairwise-Summation (PWS) approximation, for estimating the CP interaction between a ground-state alkali atom and a dielectric volume of arbitrary shape. The PWS approximation integrates the van der Waals interaction between two ground-state atoms over the many atoms in the surface volume. We relate the microscopic polarizability of atoms in the surface to the bulk dielectric constant via the Clausius-Mossotti relation within the Lorentz local-field theory. We find that the CP potential depends strongly on both the geometry and density of the surface, which modify the fluctuation spectrum of the electromagnetic field that couples the alkali atoms to the dielectric. We compare our results with known analytical expressions for the CP potential near planar surfaces and use them to investigate how BECs interact with, and may quantum reflect from, curvilinear surface structures.

1. Introduction

The ability to manipulate ultra-cold atoms using current-carrying wires microfabricated on the surface of atom chips has fuelled interest in interactions between the atoms and the chip. Understanding and reducing the intrinsic attractive interaction between an atom and a surface is crucial for trapping ultra-cold atoms within a micron of the surface. The reason for this is that the stability and lifetime of coherent atom clouds depend mainly on this atom-surface potential, called the Casimir-Polder (CP) potential. Classically, the CP potential can be estimated by summing the van der Waals (vdW) interaction between an atom above the surface and each atom within the surface. A full description can, however, only be explained within the framework of Quantum Electrodynamics (QED). In this approach, the CP potential results from the interaction of a single atom (in the ground or an excited state) close to a microscopic body in a quantum vacuum. The quantum vacuum fluctuations generate virtual particles that interact with the atom and the surface as predicted theoretically by Casimir and

Polder in 1948 [13] as an extension to the vdW-London interaction between two ground-state atoms. This CP regime, also known as the vdW retarded regime, occurs when the separation between two bodies greatly exceeds the atomic transition wavelength. The CP potential of a *single* atom with a macroscopic body has been used to approximate the interaction between a Bose-Einstein condensate (BEC) and a surface, e.g. for quantum reflection (QR) studies [4,5]. Theoretical investigation shows that the QR probability and dynamics are both qualitatively the same for an abrupt-step potential and the exact power-law attractive CP potential [4]. Motivated by experimental studies of QR from microfabricated pillar arrays [3], here we use our PWS approximation to estimate the interaction potential between an atom and a non-planar surface, in this case a cylinder. This simple approach agrees reasonably well with exact calculations for planar surfaces and, moreover, offers the flexibility required to estimate the CP attraction to engineered surface structures of arbitrary shape.

2. Pairwise-Summation approach

We now introduce the pairwise-summation (PWS) approach to estimating the CP interaction between a single atom and a surface. There are a number of ways of understanding how this CP potential arises. The most common interpretations of the CP potential are: (i) a many-body vdW potential resulting from additive interactions between the approaching atom and each atom in the surface [12], (ii) an attraction originating from the exchange of virtual photons [8,9] between the approaching atom and the surface, and (iii) a Stark shift of the atomic energy levels [7].

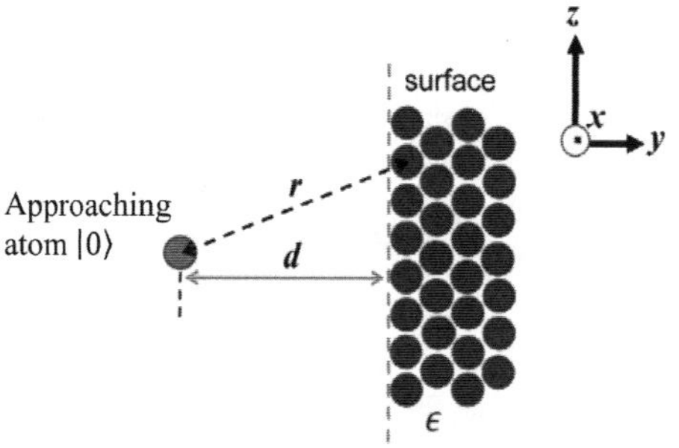

Figure 1. Schematic illustration of the PWS approximation for the attractive CP interaction potential between a ground-state atom (red) at distance *d* from a planar bulk surface of relative permittivity ϵ.

The PWS approximation neglects the many-body vdW interactions between atoms *within* the surface. Then, the CP attraction between the approaching atom and the surface is the summation of the vdW potentials between the approaching atom and each atom in the surface, as illustrated in Figure 1. Specifically, we write the attractive CP potential in the form

$$\mathcal{E}_{CP} = -\mathcal{C}_m \bar{n} \int_V \frac{dV}{r^m} \quad (1)$$

In equation (1), the interaction constant, $\mathcal{C}_m$, is taken from the retarded ($m = 7$) and non-retarded ($m = 6$) limits of the vdW atom-atom interaction [7], $\bar{n}$ is the volume (V) density of surface atoms and r is the atom-atom separation. The polarizability, α, of the approaching atom and the dielectric constant, ϵ, of the surface satisfy the Clausius-Mossotti (CM) relation $(\bar{n}\alpha/3\varepsilon_0) = (\epsilon - 1)/(\epsilon + 2)$ derived from Lorentz local-field theory. Comparing our PWS approach with standard expressions [1, 9] for an ^{87}Rb atom approaching a planar Si dielectric, which is either semi-infinite or of finite width, we find, respectively, that equation (1) gives a 38% overestimate and a 66% underestimate of the exact results.

2.1. *Non-planar surface: Cylinder*

Figure 2 illustrates the interaction of an atom with a cylindrical surface. As the cylinder's radius $\mathcal{R}_C \to \infty$, the effect of the cylinder on the atom approaches that of a semi-infinite dielectric whose left-hand edge is shown by the bold line in Figure 2. Using equation (1), we estimate the CP potential in the retarded (*Rt*) and non-retarded (*Non-Rt*) regimes for an atom at a distance $d \leq \mathcal{R}_C$ from an infinitely long solid dielectric cylinder as [14]

$$\mathcal{E}_{Rt} = -\mathcal{C}_4\left[\frac{\mathcal{R}_C^2(\mathcal{R}_C^2 + 2y^2)}{(y^2 - R_C^2)^4}\right] \; ; \; \mathcal{E}_{Non-Rt} = -\mathcal{C}_3\frac{\mathcal{R}_C^2}{y^5}\mathcal{F}\left(\frac{5}{2},\frac{5}{2};2;\frac{\mathcal{R}_C^2}{y^2}\right), \tag{2}$$

where $y = d + \mathcal{R}_C$, $\mathcal{F}$ is the hypergeometric function, and the interaction constants are

$$\mathcal{C}_4 = \frac{23\hbar c\alpha(0)}{40\pi^2\varepsilon_0}\left(\frac{\epsilon - 1}{\epsilon + 2}\right) \; ; \; \mathcal{C}_3 = \frac{27\hbar}{128\pi\varepsilon_0}\int_0^\infty d\omega\, \alpha(i\omega)\frac{\epsilon(i\omega) - 1}{\epsilon(i\omega) + 2} \tag{3}$$

In the limit $\mathcal{R}_C \to \infty$ and $d \to 0$, $\mathcal{E}_{Non-Rt}$ reduces to the CP potential for an atom near a half-plate.

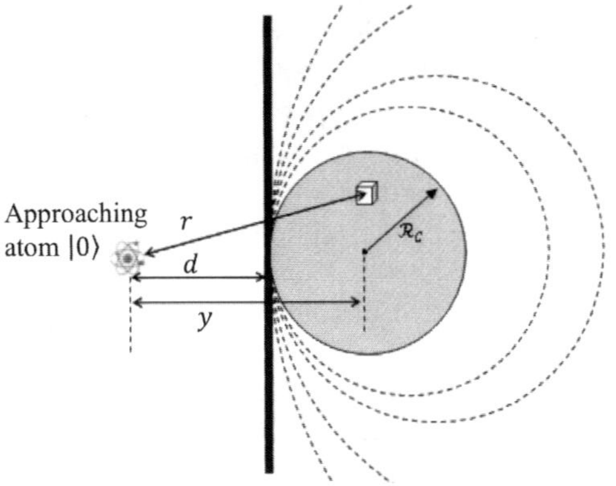

Figure 2. Schematic interaction of an atom at distance *d* [*y*] from the left-hand edge [centre] of an infinitely long cylinder (cross-section shown grey). Dashed curves show various radii of the cylinder as $\mathcal{R}_C$ increases towards ∞. We calculate the CP potential by integrating the vdW potential over the surface volume: cube shows a voxel at distance *r* from the atom.

Comparing CP potential values obtained from equation (2) with exact expressions (see for example Ref.[15]) for a perfectly conducting cylinder with $\epsilon = \infty$, gives errors ranging from ~ 20% to 80% depending on the system parameters and geometry (a more detailed analysis will be presented in [16]). Here, we present PWS calculations of the retarded CP potential between a ground-state ^{87}Rb atom near either a Si slab of finite width or a solid cylinder [Figures 3(a) and (b) respectively)]. When the slab width and cylinder radius are 100 and 10 microns [top two curves in Figure 3(a,b)], these two dielectric geometries produce almost identical CP potentials when *d* is between 1 and 10 microns. For thinner slabs and cylinders with widths or radii of 0.1 microns and 1 nm [bottom two curves in Figure 3(a,b)], the CP potential for the cylinder is significantly lower than that for the planar slab – between approximately three and four orders of magnitude lower when the slab width/cylinder radius is only 1 nm. Detailed comparison between the PWS and exact scattering approaches now needs to be done for a range of dielectric geometries and parameters in order to confirm our initial calculations.

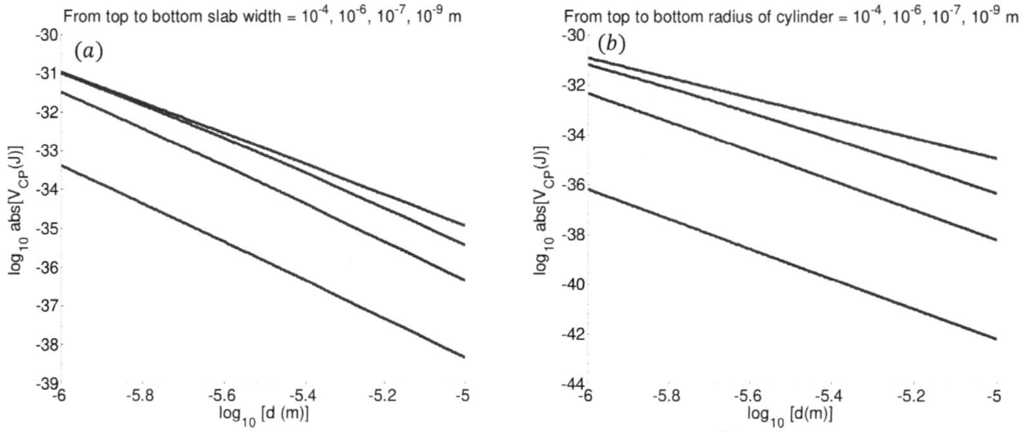

Figure 3. PWS approximation for the attractive CP potential of an ^{87}Rb atom versus distance d from (a) a dielectric slab of finite width and (b) a dielectric cylinder.

3. Conclusion

We have used the PWS approach to calculate the CP potential for an atom near a bulk dielectric with either a planar or cylindrical geometry. We find that when the atom-surface separation is comparable to the width of a dielectric slab or the radius of a cylinder, the magnitude of the CP potential can be orders of magnitude smaller for the cylinder than for the slab. This suggests that thin or curved substrates could provide a route to trapping atoms within a micron of an atom-chip surface [4]. Although for the slab and cylinder, the PWS approach gives CP potential values that can differ by up to 100% from exact values obtained from a scattering method, it gives a fast and simple method for estimating the values and spatial variation of CP potentials for any surface geometry, e.g. MEMS devices. Our preliminary calculations show that the quantum reflection probability for a single atom approaching a surface is higher for a cylindrical dielectric than for a dielectric slab, due to the weaker CP potential of the former. We are now undertaking numerical calculations of the dynamics of Bose-Einstein condensates undergoing quantum reflection from solid and hollow cylinders with CP potentials calculated using both the PWS method and the exact scattering approach [16].

Acknowledgement
We thank C Eberlein and R Zietal for helpful discussions. This work is supported by EPSRC, HEFCE, MOHE Malaysia and UNIMAP.

References

[1] R Salem, Y Japha, J Chabé, B Hadad, M Keil, K A Milton and R Folman, *New J. Phys.* **12**, 023039 (2010).
[2] J X de Carvalho, M S Hussein and Weibin Li, *Phys. Rev. A* **78**, 032906 (2008).
[3] T A Pasquini, Y Shin, C Sanner, M Saba, A Schirotzek, D E Pritchard and W Ketterle, *Phys. Rev. Lett.* **93**, 223201 (2004); T A Pasquini, M Saba, G B Jo, Y Shin, W Ketterle and D E Pritchard, *Phys. Rev. Lett.* **97**, 093201 (2006).
[4] R G Scott, A M Martin, T M Fromhold and F W Sheard, *Phys. Rev. Lett.* **95**, 073201 (2005); T E Judd, R G Scott, G Sinuco, T W A Montgomery, A M Martin, P Krüger and T M Fromhold, *New J. Phys.* **12**, 063033 (2010); G Sinuco-León, B Kaczmarek, P Krüger and T M Fromhold, arXiv:1007.5339.
[5] S L Cornish, N G Parker, A M Martin, T E Judd, R G Scott, T M Fromhold and C S Adams, *Physica D* **238**, 1299 (2009).
[6] F Arnecke, H Friedrich and J Madronero, *J. Phys.: Conf. Series* **88**, 012041 (2007).

[7] S Y Buhmann, H Safari, Ho Trung Dung and D G Welsch, *Optics & Spectroscopy* **103**, 374, (2007).
[8] C Eberlein and R Zietal, *Phys. Rev. A* **80**, 012504 (2009).
[9] A M Contreras Reyes and C Eberlein, *Phys. Rev. A* **80**, 032901, (2009).
[10] Diego A R Dalvit, Paulo A Maia Neto, Astrid Lambrecht and Serge Reynaud, *J. Phys. A: Math. Theor.* **41**, 164028, (2008).
[11] R Fermani, S Scheel and P L Knight, *Phys. Rev. A* **75**, 062905 (2007).
[12] E V Blagov, G L Klimchitskaya and V M Mostepanenko, *Phys. Rev. B* **71**, 235413 (2005).
[13] H B Casimir and H Polder, *Phys. Rev.* **73**, 360 (1948).
[14] V A Kirsch, *Adv. in Colloid and Interface Sci.* **104**, 311 (2003).
[15] S A Ellingsen, S Y Buhmann and S Scheel, *Phys. Rev. A* **82**, 032516 (2010).
[16] R Messina *et al.*, to be published.

Luminescence patterns in photoexcited quantum wells: diffusion of the Coulomb plasma versus exciton superfluidity

S. E. Savel'ev and A.S. Alexandrov

Department of Physics, Loughborough University, Loughborough LE11 3TU, United Kingdom

E-mail: S.Saveliev@lboro.ac.uk, a.s.alexandrov@lboro.ac.uk

Abstract.
Inspired by experimental observations of ring-like luminescence patterns outside the laser spot in double [1] and single [2] quantum wells (QWs), we performed molecular dynamics (MD) simulations of the electron-hole plasma. Electron and hole density profiles generally follow the hydrodynamic two-fluid kinetic equations with non-trivial power laws for hole and electron velocities. While the MD or its kinetic approximation explain the inner ring formation, they fail to account for the macroscopic outer luminesce rings observed both in double and single QWs. Solving the Gross-Pitaevskii equation we show that in contrast to previous theoretical models describing the outer ring formation via conventional (normal state) diffusion, this phenomenon is a sequence of superfluidity of excitons in a dark state.

To describe the dynamics of hot electrons and holes, we use classical equations of motion [3]

$$m_e^*\ddot{\mathbf{r}}_e^i + \gamma_e\dot{\mathbf{r}}_e^i = \sum_{j\neq i}\frac{e^2(\mathbf{r}_e^i - \mathbf{r}_e^j)}{\left|\mathbf{r}_e^i - \mathbf{r}_e^j\right|^3} - \sum_k \frac{e^2\left(\mathbf{r}_e^i - \mathbf{r}_h^k\right)}{\left[\left(\mathbf{r}_e^i - \mathbf{r}_h^k\right)^2 + d^2\right]^{3/2}} + \xi_e, \tag{1}$$

$$m_h^*\ddot{\mathbf{r}}_h^i + \gamma_h^*\dot{\mathbf{r}}_h^i = \sum_{j\neq i}\frac{e^2(\mathbf{r}_h^i - \mathbf{r}_h^j)}{\left|\mathbf{r}_h^i - \mathbf{r}_h^j\right|^3} - \sum_k \frac{e^2(\mathbf{r}_h^i - \mathbf{r}_e^k)}{[(\mathbf{r}_h^i - \mathbf{r}_e^k)^2 + d^2]^{3/2}} + \xi_h,$$

where vectors $\mathbf{r}_e^i$ and $\mathbf{r}_h^j$ are in-plane positions of i-th electron and j-th hole, e is the electron charge, and d is the interlayer distance. The left-hand side of Eqs.(1) accounts for inertia of electrons and holes with their effective masses m_e^* and m_h^* respectively as well as for the momentum damping (described by constants γ_e and γ_h) due to interaction with acoustic phonons. The right-hand sides account for the in-layer electron-electron and hole-hole Coulomb repulsions, the inter-layer electron-hole attraction, and random forces, $\xi_{e,h}$ describing electron and hole diffusion. Here $\langle\xi_e\rangle = \langle\xi_h\rangle = 0$ and $\langle\xi_e(0)\xi_e(t)\rangle = \langle\xi_h(0)\xi_h(t)\rangle = \delta(t)$. In simulations we normalized time by $t_s = m_e^*\mu_e/e$ and all distances by $r_s = \sqrt[3]{m_e^*\mu_e^2}$ with electron mobility μ_e. The optical excitation of electrons and holes was modeled by generating the particles in random positions inside the excitation spot (i.e., within interval $-x_0 < x < x_0$) with a rate p. Initial velocities of electrons were also chosen randomly. An exciton formation happened if an electron and a hole were close enough to each other, $|\mathbf{r}_e - \mathbf{r}_h| < a$, where $a(d)$ is a phenomenological

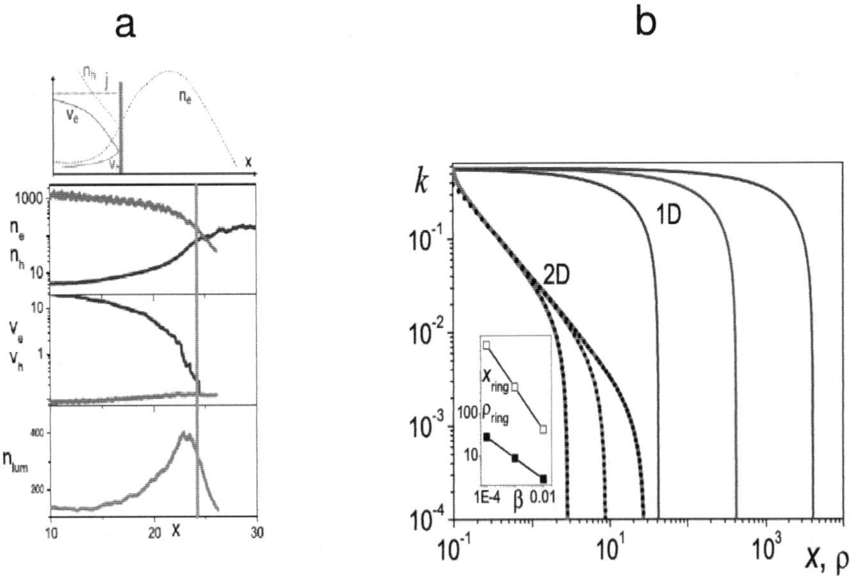

Figure 1. (a): Qualitative comparison of distributions of the electron and hole velocities and densities predicted by the hydrodynamical model (top panel) with the simulated distributions (the distance from the spot x, densities, n_e, n_h, and velocities, v_e, v_h, as well as density of exciton formation events, n_{lum}, are given in arbitrary units). (b): The exciton superfluid momentum as a function of the distance (arb. units) from the laser spot obtained using Eq.(3) for three different values of the decay rate $\beta = \gamma/2\mu = 10^{-2},\ 10^{-3},\ 10^{-4}$ in 1D and 2D geometries [4].

in-layer exciton radius, and their relative velocity $V = |\dot{\mathbf{r}}_e - \dot{\mathbf{r}}_h|$ was smaller than some critical value V_c. Here d is the characteristic separation between electron and hole wells.

Some of our 1D results obtained by simulating the above classical equations of motion without the random forces ($\xi_{e,h} = 0$) are shown in Fig.1a. As soon as we begin simulations switching on the laser pumping, the number N of excited electrons and holes starts to increase and finally saturates after a short time. We plot the spatial distributions of electrons, holes and exciton formation events by averaging the number of particles (events) within a certain short spatial interval. One can see in Fig.1a that electrons and holes are spatially separated: holes are accumulated close to the excitation spot while electrons are mostly located far away from the spot. In the region of the overlap of electron and hole densities the luminescence peak occurs at a certain macroscopic distance from the spot. Qualitatively this picture is somewhat similar to the earlier explanations of luminescence patterns within purely diffusion model, but here we did not make any ad-hoc assumptions (e.g. assuming an electron flux from external parts of the sample, and/or neglecting the Coulomb forces). As seen from Fig. 1a, the electron average velocity decreases with x monotonically while the hole velocity surprisingly increases with x.

As shown in Fig.1a most of our MD simulation results can be reproduced in a simplified hydrodynamic model derived by truncating the Bogoliubov - Born - Green - Kirkwood - Yvon

(BBGKY) hierarchy of many-particle distribution functions to one-particle distributions,

$$J^2 \frac{dn_e^{-1}}{dx} + \gamma_e J = \frac{n_e F_e^C}{m_e^*}$$
$$J^2 \frac{dn_h^{-1}}{dx} + \gamma_h J = \frac{n_h F_h^C}{m_h^*}. \quad (2)$$

with a constant J, which is the current for both electrons and holes, and the mean-field Coulomb forces, $F_{e,h}^C$.

We have also studied how electron, hole and luminescence distributions depend on the laser (pump) intensity by changing the e-h generation rate p. It turns out that the position of the luminesce peak depends very weakly on p: slightly increasing at low p and then starting to decrease. Thus, our *normal state* MD of the electron-hole plasma fails to explain the experimental observations [1, 2] of a strong dependence of the outer (external) ring radius on the laser intensity, but it is compatible with the properties of the inner ring observed in double QWs [1]. So we can conclude that the ring observed in our simulations can explain the innner ring where excitons with a finite momentum are formed.

To explain the *external* macroscopic rings with the radius of about a few hundreds microns [1, 2] we propose [4] that excitons condense in a coherent superfluid with a finite momentum from electron-hole plasma at the inner ring since the electron-hole plasma expands inside the spot due to the Coulomb repulsions. If the lattice temperature is below the Berezinskii-Kosterlitz-Thouless (BKT) transition temperature of 2D excitons, this new state of matter allows excitons to travel in a dark state on a macroscopic distance from the laser spot producing photo-luminesces (PL) rings far away from the photo-excited region as observed in Refs.[1, 2]. To describe a free superflow of the 2D Bose-liquid of decaying excitons we apply a generalized Gross-Pitaevskii (GP) equation [5] for the order-parameter, $\psi(\mathbf{R})$ taking into account the exciton recombination rate $\gamma/\hbar$, as originally proposed by Keldysh [6]:

$$-\left(\frac{\hbar^2}{2m_{ex}}\Delta + \mu\right)\psi(\mathbf{R}) + V|\psi(\mathbf{R})|^2\psi(\mathbf{R}) - i\frac{\gamma}{2}\psi(\mathbf{R}) = 0. \quad (3)$$

Here μ and V conveniently parameterize the average superfluid density and the short-range repulsion, respectively. To determine location of luminescence rings, we use that excitons can radiate in 1D and 2D by resonant emission of photons only if their momentum is inside the photon cone, $K < K_c$. The solutions of the GP equation are shown in Fig.1b for 1D and 2D geometries.

In dimensional units the PL external ring radius is found as $R_{ring} \approx \sqrt{\hbar v \tau_{ex}(16\pi d a n/m_{ex})^{1/2}/\gamma}$. Importantly γ in this estimate is the *nonradiative* recombination rate since the coherent excitons have their momentum outside the photon cone, d is the screening radius in the electron-hole plasma, τ_{ex} is the exciton formation time, v is the characteristic speed of photocarries, m_{ex} is the exciton mass and n is the photocarrier density. Remarkably, using the realistic material parameters yields R_{ring} about 300 μm for the photoelectron density $n = 10^{10}$ cm^{-2}, explaining the external PL rings [4].

References

[1] Butov L V *et al.*, Nature **418** (2002) 751
[2] Snoke D *et al.*, Nature **418** (2002) 754
[3] Paraskevov A V, Savel'ev S E, Phys. Rev. B **81** (2010) 193403
[4] Alexandrov A S and Savel'ev S E, arXiv:1008.2778
[5] Gross E P, Nuovo Cimento **20** (1961) 454; Pitaevskii L P, Zh. Eksp. Teor. Fiz. **40** (1961) 646 [Soviet Phys. JETP **13** (1961) 451]
[6] Keldysh L V, *Problems of Theoretical Physics* (Nauka, Moscow, 1972)

The peculiarities of δ-plutonium electronic structure and magnetic susceptibility

A N Filanovich, A A Povzner, A G Volkov

Department Of Physics, Ural Federal University, Mira. St., 19, Ekaterinburg, Russia

E-mail: fil@kf.ustu.ru, povz@kf.ustu.ru

Abstract. The calculations of electronic density of states, exchange-enhanced spin and orbital magnetic susceptibility of plutonium in its δ–phaze existing region are performed in terms of a generalized s(p)df-model, which enables to take the intra-site Hubbard ff-, dd-interactions and inter-site df-exchange interaction into account along with s,p,f and d-electrons band movement and orbital splitting of f-sates. It is shown that the splitting of the electronic spectra induced by fluctuating exchange fields leads to appearing of the temperature-induced local magnetic moments, to changing of d,f-bands occupation numbers with temperature and to appearing of charge fluctuations induced by spin fluctuations. A good agreement with the experimental data on temperature dependencies of magnetic susceptibility of $Pu_{0.95}Ga_{0.05}$ and $Pu_{0.957}Ga_{0.043}$ alloys is obtained.

1. Introduction

Elucidation of the nature of plutonium and its alloys anomalous electronic properties is extremely important both from scientific and practical points of view. On the on hand it is necessary for the definition of type and concentrations of defects, which appears as a consequence of plutonium's radioactive decay. On another hand the question remains open on the fundamental difference among plutonium and other f-metals, which connected with Kondo or Kondo-like behavior of its susceptibility and resistivity [1,2].

In [3] in terms of ab initio LDA+U+SO method it is shown that magnetic moments in the case of plutonium's ground state disappear due to strong (comparing to exchange) spin-orbit interaction. However the experiments reveal both orbital and spin magnetic susceptibility already in the low-temperature region [4,5]. In the recent article [6] in terms of combination of DMFT method and density functional theory (DFT) a transformation from Pauli-like local magnetic susceptibility to Curie-Weiss susceptibility is demonstrated. Calculated in such a framework density of states demonstrates a considerable overlap between 7/2 and 5/2 multiplets comparing to [3], while the compensation of the local magnetic moments due to spin-orbit coupling gets broken. However, in terms of this model only a qualitative agreement with experimental data can be achieved and only in the range of relatively high temperatures (T>400 K).

It is worth noting that plutonium can be attributed to the group of nearly magnetic metals, which electronic system is unstable to the formation of magnetic ordering [7]. Previously in terms of the Hubbard model for d-electrons and a generalized Hubbard sd-model [8,9] it was shown that the fluctuating inner exchange fields (ξ_v) which appear in nearly magnetic systems, lead to the fluctuations of charge fields (η_v) and to the splitting of electronic spectrum, which increases with increasing temperature. In the range of low temperatures exchange fields fluctuate both in direction and magnitude (transverse and longitudinal fluctuations). For the temperatures higher than some temperature T^* the longitudinal spin and charge fluctuations disappear and the transverse spin fluctuations lead to Curie-Weiss law for magnetic susceptibility, which is analogous to the picture of appearing of local magnetic moments (LMM). The last in terms of the fluctuation theory is interpreted as formation of the temperature-induced LMM.

In the present study we develop a self-consistent procedure which combines an ab initio calculation of electronic structure using LDA+U+SO method (within FP-LAPW) with model representations on spin and charge fluctuations in the d,f-electrons system. A generalized s(p)df-model is considered, where along with s,p,f and d-electrons band movement and orbital splitting of f-sates the intra-site Hubbard ff-, dd-interactions and interstitial exchange df-interaction are taken into account.

2. Model description

In order to calculate the free energy and electronic densities of states we use an approach, which was previously developed in [8,9] with regard to the two band Hubbard sd-model. In the present self-consistent model calculation combined with ab initio calculation along with spin magnetic moments the orbital magnetic moments are taken into account. Therefore in the expression for magnetic susceptibility along with spin susceptibility a Van Vleck-type terms appears:

$$\chi = \chi_{\mathrm{s}} + \chi_{\mathrm{orb}} \tag{1}$$

Here

$$\chi_{\mathrm{orb}} = 2\sum_{\alpha,\alpha'} n_l \left[N_l - n_l\right]\cdot\left[N_l\left(\Delta_l + U^{(l)} m^{(l)}(\alpha - \alpha')\right)\right]^{-1}, \tag{2}$$

$$\chi_{\mathrm{s}} = \chi^{(f)} + \chi^{(d)} + \chi^{(sp)} + I\chi^{(f)}\chi^{(d)}, \tag{3}$$

where N_l is the value of orbital degeneracy of the band with an l quantum number (l=f,d), $2n_l$ is the averaged energetic distance between multiplets (which are spitted due to coulomb and spin-orbital interaction), $U^{(l)}$ is the parameter of the Hubbard's repulsion of electrons of the band with an l quantum number, I is the uniform part of the inter-site f-d exchange interaction,

$$\chi^{(sp)} = 2g_0^{(sp)}(\mu), \tag{4}$$

$$\chi^{(l)} = 2\left(\frac{2}{3}\chi_{\perp}^{(l)} + \frac{1}{3}\chi_{//}^{(l)}\right)D^{(l)} \tag{5}$$

are paramagnetic susceptibilities of s,p- (4) and d,f- (5) electrons,

$$D^{(l)-1} = 1 - \frac{2}{3}U^{(l)}\chi_{\perp}^{(l)} - \frac{1}{3}U^{(l)}\chi_{//}^{(l)} \tag{6}$$

is the susceptibility exchange-enhancement factor of d- or f-electrons,

$$\chi_{\perp}^{(l)} = \sum\nolimits_{\alpha=\pm1}\alpha\int d\varepsilon g_0^{(l)}(\varepsilon) f(\varepsilon - \mu - U^{(l)}\eta_0^{(l)} - \alpha U^{(l)} m^{(l)})/\left(2U^{(l)} m^{(l)}\right) \tag{7}$$

corresponds to the transverse, and

$$\chi_{//}^{(d)} = 2\left(\prod\nolimits_{\alpha=\pm1} g_0^{(d)}(\mu + U\eta_0 + \alpha U m)\right)/\sum\nolimits_{\alpha=\pm1} g_0^{(d)}(\mu + U\eta_0 + \alpha U m) \tag{8}$$

to the longitudinal components of the paramagnetic susceptibility of non-interacting f- or d-electrons (l=f,d), the mean-squared amplitude of spin fluctuations (square root from the spin-spin correlator on a site) is

$$m^{(l)2} = B^{(l)}\left(D^{(l)}T/U^{(l)}\right)^2\left\{1 + \left[4\left(I^2 D^{(l)}/U^{(f)}U^{(d)}\right)/\left(D^{(l')-1} - \left(B^{(l')}/B^{(l)}\right)D^{(l)-1}\right)\right]\times\right. \tag{9}$$
$$\left.\times\left(2 - \left(B^{(l')}/B^{(l)}\right)D^{(l')2}\left(D^{(l')-2} - \left(B^{(l')}/B^{(l)}\right)D^{(l)-2}\right)/D^{(l')-1} - \left(B^{(l')}/B^{(l)}\right)D^{(l)-1}\right)\right\},$$

$$\eta_0^{(l)} = n_l(0) + \Delta n_l,\ \Delta n_l = (1/2)\sum_{\alpha}\int_{-\infty}^{\infty} g_0^{(l)}(\varepsilon) f(\varepsilon - \mu - \alpha U^{(l)} m^{(l)} + U^{(l)}\Delta n_l)d\varepsilon - n_l(0),$$

while the amplitude of the charge fluctuations is defined as follows

$$\delta\eta_{v,l} \approx \left(\xi_{vl}^2 - \left(U^{(l)} m_l\right)/m\right)\left(\sum_{\alpha=\pm1}\alpha g_0^{(l)}(\mu - U\Delta n_l + \alpha U m_l)\Big/\sum_{\alpha=\pm1} g_0^{(l)}(\mu - U\Delta n_l + \alpha U m_l)\right). \tag{10}$$

$B^{(l)}$ and $X^{(l)}$ parameters are the coefficients of the expansion of Pauli susceptibility in the effective mass approximation, $n_l(0)$ is the occupancy of electrons belonging to the band with an l quantum number in the ground state, $g_0^{(l)}(\varepsilon)$ is the uniform density of states of electrons belonging to the band with an l quantum number (l=f,d) which is calculated using LDA+U+SO approximation (in the FP-LAPW basis). The calculation of the ground state was performed with FP-LAPW Elk code [10]. We use experimental values [11] for the lattice parameters of considered dilute alloys, a 32x32x32 k-point grid, an LSDA exchange-correlation potential as parameterized by Perdew-Wang [12] and a interpolation scheme between around mean field and fully localized limits for LDA+U calculation (see [13]). The intra-site f-f exchange parameter J_H was determined automatically by implementing method described in [10]. Other parameters of the band calculations have been checked for the stable convergence.

While performing the self-consistent calculation along with magnetic susceptibility we have defined the electronic spectra, splitting of which is accounted by the replacement $\varepsilon_{\mathbf{k},l,j} \to \varepsilon_{\mathbf{k},l,j} + \alpha U^{(l)} m^{(l)}(T)$. Along with this a formation of the temperature-induced local magnetic moments is observed, which is accompanied by filling of the d,f-states analogously to how it occurs in Kondo effect. The states with energy $\varepsilon_{\mathbf{k},l,j} - U^{(l)} m^{(l)}(T)$ are filled both by states of another bands (with $l \neq l'$) and by states of the same band with energy $\varepsilon_{\mathbf{k},l,j} + U^{(l)} m^{(l)}(T)$. Thus appearing difference of occupancies corresponds to the effective concentration of electrons in the expression for spin magnetic susceptibility, which appears due to the transverse spin fluctuations.

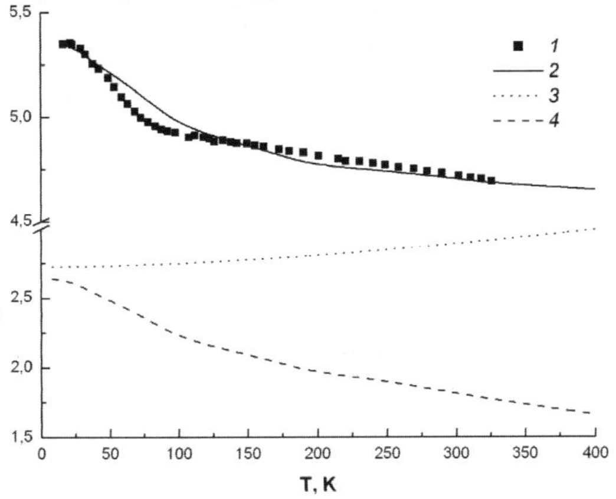

Figure 1. Temperature dependence of $Pu_{0.95}Ga_{0.05}$ alloy magnetic susceptibility: 1 – experimental data [4]; 2 – calculation in the present study using the equation (1) and the DOS (figure 3) with Fermi energy equals to –0.02 eV; 3 – Van Vleck contribution (calculated with (2)); 4 – spin contribution (calculated with (3))

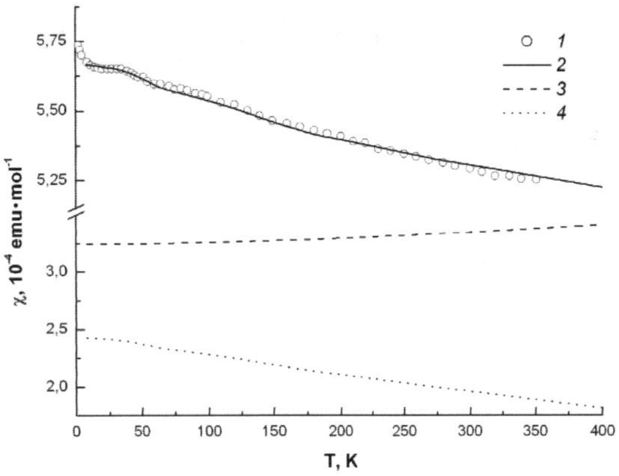

Figure 2. Temperature dependence of $Pu_{0.957}Ga_{0.043}$ alloy magnetic susceptibility: 1 – experimental data [5]; 2 – calculation in the present study using the equation (1) and the DOS (figure 3) with Fermi energy equals to 0.02 eV; 3 – Van Vleck contribution (calculated with (2)); 4 – spin contribution (calculated with (3))

3. Results and discussion

The results of the calculations of temperature dependencies of magnetic susceptibility are presented in figures 1, 2. Agreement with the experimental data [1,2] is obtained by taking two different values of the Fermi energy (see figures captions) and with single values of electrons interaction: $U^{(f)}$=2.67 eV, J_H=0.5 eV (in the ab initio calculations [3], where exchange-enhancement effect is not taken into account, $U^{(f)}$=2.5 eV). The averaged distance between multiplets' energies is Δ_l=2 eV, which corresponds to the half of energy distance between centers of gravity of the bands which correspond do 5/2 and 7/2 multiplets and the value of spin-orbit coupling is 0.35 eV, as in [3]. Also while performing the self-consistent calculation of densities of states (figure 3) and of magnetic susceptibility (figure 1, 2) we used the value of I=0.1$U^{(f)}$ for the inter-site fd-exchange interaction and the following values for imaginary part of Pauli susceptibility of non-interacting ($\mathrm{Im}U^{(l)}\chi_0^{(l)}(\mathbf{q},\omega)=B^{(l)}\omega q_c/|\mathbf{q}|$) f- and d electrons: $B^{(f)}$=1 and $B^{(d)}=B^{(f)}U^{(d)}g_0^{(d)}(\varepsilon_F)/(U^{(f)}g_0^{(f)}(\varepsilon_F))$, respectively.

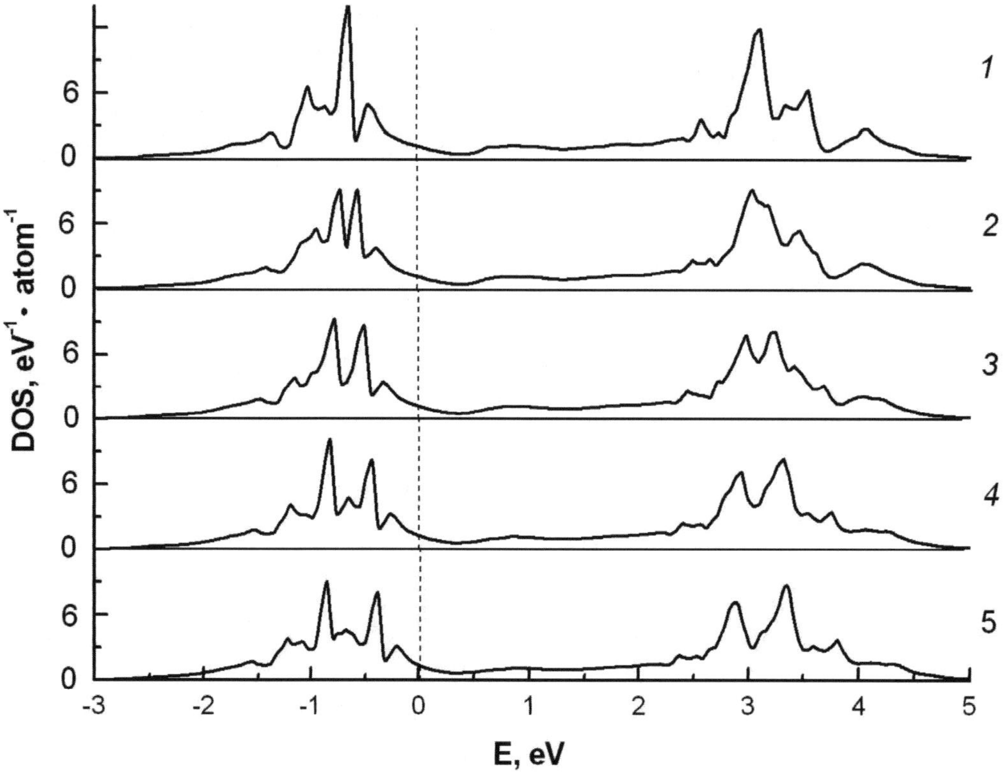

Figure 3. Temperature dependence of δ-plutonium f-electrons DOS: 1 – 0 K, 2 – 100 K, 3 – 200 K, 4 – 300 K, 5 – 400 K; the dashed curve shows the position of chemical potential

Additionally let us note that the orbital susceptibility renormalized by spin fluctuations accounts for somewhat greater than 70% of the value of full magnetic susceptibility, which is in agreement with the experimental estimations [4]. Moreover, the developed model of plutonium's electronic structure describes instability to the formation of magnetic ordering. By shifting the Fermi level to the left on energy scale by amount of 0.1 eV, we obtain a ferromagnetic solution (divergence of spin susceptibility). With temperature change in the deltha-phaze existing range the factor of exchange enhancement changes from 11 to 7 for f-electrons and from 1.6 to 1.5 for d-electrons.

In the T>T* temperature range (for $Pu_{0.95}Ga_{0.05}$ T*=200K and for $Pu_{0.957}Ga_{0.043}$ T*=150K) the spin-fluctuation renormalization of δ-plutonium electronic spectrum leads to disappearance of longitudinal spin fluctuations ($\chi_{\perp}^{(f)} >> \chi_{//}^{(f)}$) and a state with temperature-induced local spin magnetic moments appears (figure 1,2).

The densities of states (DOS) at different temperatures are shown in figure 3. Changing of DOS with temperature leads to decrease of the averaged occupation numbers of f,d electrons while the occupancy of sp-bands increases. For example at T=400 K we obtain Δn_f = -0.2 electrons/site and Δn_d = -0.013 electrons/site. The obtained change of valence with temperature is accompanied by charge fluctuations, specifically the maximal mean-squared value of fluctuation of on-site electrons number is $\sqrt{\sum_\nu \delta\eta_{\nu,f}^2/N_0}$ ~ 0.01 electrons/site for f-electrons, and is smaller by an order of magnitude for d-electrons. However, the obtained changes of DOS with temperature are only in qualitative agreement with the photoemission experiments [14], because maximum of the calculated spectral function is shifted by 0.4 eV from the experimental one.

Thus when the temperature-induced LMM are emerging in δ-plutonium a temperature dependence of spin and orbital susceptibility is appeared which is consistent with the experimental data. Thus appearing temperature-dependent renormalization of the electronic DOS leads to the effects of charge fluctuations and variable valence.

Other mechanisms for LMM formation are defects and impurities. However, in the present study we discussed experimental data [4,5] on magnetic susceptibility of specimens, which have undergone a special heat processing, which eliminates defects produced by radioactive decay and leads to conserving of electrons concentration.

References

[1] Clementyev E S and Mirmelstein A V 2009 *JETP* **109** 128-139
[2] Tsiovkin Yu Yu, Korotin M A, Shorikov A O, Anisimov V I, Voloshinskii A N, Lukoyanov A V, Koneva E S, Povzner A A, Surin M A 2007 *Phys. Rev. B* **76** 075119
[3] Shorikov A O, Lukoyanov A V, Korotin M A, Anisimov V I 2005 *Phys. Rev. B* **72** 024458.
[4] Verkhovski S V, Arkhipov V E, Zuev Yu N, Piskunov Yu V, Mikhalev K N, Korolev A V, Svyatov I L, Pogudin A V, Ogloblichev V V, Buzlukov A L 2005 *JETP Letters* **82** 139-14
[5] McCall S K, Fluss M J, Chung B W, McElfresh M W, Jackson D D, Chapline G F 2006 *PNAS* **103** 17179–17183
[6] Marianetti C A, Haule K, Kotliar G, Fluss M J 2008 *Phys. Rev. Lett.* **101** 056403
[7] Moore K T, van der Laan G 2009 *Rev. Mod. Phys.* **81** 235
[8] Kudasov Yu B et al. 1999 *JETP* **89** 960-965
[9] Volkov A G, Povzner A A, Kryuk V V, Bayankin P V 1999 *Physics Of The Solid State* **41** 960-962
[10] Bultmark F, Cricchio F, Granas O, Nordstrom L 2009 *Phys. Rev. B* **80** 035121
[11] Lawson A C, Roberts J A, Martinez B, Richardson J W 2006 *Phil. Mag.* **86** 2713
[12] Perdew J P, Wang Y 1992 *Phys. Rev. B* **45** 13244
[13] Petukhov A G, Mazin I I, Chioncel L, Lichtenstein A I 2003 *Phys. Rev. B* **67** 153106
[14] Arko A J, Joyce J J, Morales L, Wills J, Lashley J 2000 *Phys. Rev. B* **62** 1773

Approximation of the entanglement in quantum dot chains using Hubbard models

J P Coe[1], V V França[2], I D'Amico[1]

[1] The Department of Physics, The University of York, Heslington, York, UK
[2] Physikalisches Institut, Albert-Ludwigs-Universität, Hermann-Herder-Straße 3, D-79104 Freiburg, Germany

E-mail: jpc503@york.ac.uk, vivian.franca@physik.uni-freiburg.de, ida500@york.ac.uk

Abstract. We investigate the Hubbard model, and the extended Hubbard model with and without correlated hopping, as an approximation to the average single-site (or local) entanglement of two-electron one-dimensional quantum dot chains with *long range interactions*. We focus on the effect of the chain length on the approximation accuracy. We also consider the particle-particle spatial entanglement of the three Hubbard models using the technique of Ref. [3] and compare the accuracy of this approximation to the numerically exact results of the nanostructure system.

1. Introduction

Systems of quantum dots are considered a feasible basis for quantum information processing and for investigating fundamental quantum mechanical properties such as entanglement. Modelling these many electron systems can by computationally difficult so approximations such as the use of lattice models are often employed. If these models could be seen to be valid for the entanglement, at least for certain parameter regions, they could be used for investigations of numerically-challenging many-electron nanostructures composed of longer chains of quantum dots. This could also pave the way for the use of density-functional techniques for lattice systems [1] thereby allowing an even more efficient approach to approximate the entanglement and appraise the suitability of quantum dot chains as components for quantum technologies.

The accuracy of the one-dimensional Hubbard model (HM) as an approximation to the average single-site entanglement of two-electrons trapped in a chain of quantum dots (QDs) was considered in Ref. [2]. This considered a contact interaction between particles and found the HM to be accurate for single-site entanglement calculations. Later, Ref. [3] considered a similar system with a more realistic long-range particle-particle interaction and a chain of four quantum dots. The accuracy of the average single-site entanglement of one-dimensional Hubbard model variants as an approximation to that of the QD structure was then appraised. The next largest term—at least for 3d transition metals—in the Coulomb interaction expansion [4] was incorporated to include nearest neighbour interactions: the extended Hubbard model (EHM). The third largest term [4], the correlated hopping (CH) term, was also considered to give the EHM+CH. In addition, in Ref. [3] a method to calculate the particle-particle spatial entanglement of the HM and EHM was also proposed. This resulted in the spatial degrees of freedom being severely reduced to just the number of sites, yet worked relatively well as an approximation to the spatial entanglement of the QD system.

In this paper we consider the effect of the chain length on the accuracy with which the HM, EHM and EHM+CH approximates the average single-site and spatial entanglement of a one-dimensional long-range interacting quantum dot chain.

2. Models

The Hamiltonian for the quantum dot nanostructure system is

$$H_{QD} = \sum_{i=1,2} \left(-\frac{1}{2} \frac{d^2}{dx_i^2} + v(x_i) \right) + C_U f(|x_1 - x_2|), \tag{1}$$

where $v(x)$ models an array of identical, regularly spaced wells, each well representing a QD. The array is symmetric around the origin, and defined by the quantities: M the number of wells, d the barrier width between two consecutive wells, and w and v_0 the width and depth of each well respectively. The long range interaction is of the form (see, for example, refs. [5, 6])

$$f(|x_1 - x_2|) = \frac{1}{\sqrt{(l^2 + (x_1 - x_2)^2)}}. \tag{2}$$

Here we use $v_0 = 10$ Hartrees, $d = w = 2\ a_0$ and $l = 1\ a_0$ where a_0 is the (effective) Bohr radius. We define the interaction strength as C_U and vary this to enable a useful comparison of the QD system with a variant of the Hubbard model. We solve the system using 'exact' diagonalization with a basis formed from 20 or 30 of the wave functions of the non-interacting system.

The Hamiltonian for the EHM+CH lattice model with dimensionless interaction strengths is

$$\begin{aligned} H = -\sum_{i,\sigma} \left(\hat{c}^{\dagger}_{i,\sigma} \hat{c}_{i+1,\sigma} + \hat{c}^{\dagger}_{i+1,\sigma} \hat{c}_{i,\sigma} \right) &+ U \sum_{i} \hat{n}_{i,\uparrow} \hat{n}_{i,\downarrow} + U' \sum_{i,\sigma,\sigma'} \hat{n}_{i,\sigma} \hat{n}_{i+1,\sigma'} \\ &+ t' \sum_{i,\sigma} \left(\hat{c}^{\dagger}_{i,\sigma} \hat{c}_{i+1,\sigma} + \hat{c}^{\dagger}_{i+1,\sigma} \hat{c}_{i,\sigma} \right) \left(\hat{n}_{i,-\sigma} + \hat{n}_{i+1,-\sigma} \right). \end{aligned} \tag{3}$$

With $U' = t' = 0$ this is the HM, while for $t' = 0$ it reduces to the EHM. We use the procedures of Ref. [2, 3] to calculate the Hubbard parameters corresponding to C_U in the QD system and find $U = 17865 C_U$, $U' = 0.28U$ and $t' = 2 \times 10^{-5} U$. We solve the system with open boundary conditions for two particles using exact diagonalization.

3. Entanglement results

To quantify the entanglement we use the von Neumann entropy of the reduced density matrix, $S = -Tr[\rho_{\text{red}} \log_2 \rho_{\text{red}}]$. To find the reduced density matrix we consider the system as bipartite and trace out one of the subsystems: $\rho_{\text{red}} = Tr_B \left|\psi\right\rangle \left\langle\psi\right|$.

Firstly, we consider the single-site or local entanglement S_i. This can be thought as quantifying the information gained about the quantum state of the other sites by measurement of a single-site i. For the lattice system the ρ_{red} at site i is diagonal in the occupation basis [7, 8], $\rho_{\text{red,i}} = \text{diag}\left[P_i(\uparrow\downarrow), P_i(\uparrow), P_i(\downarrow), P_i(0)\right]$. Here $P_i(\uparrow\downarrow)$ is the probability of double particle occupation of site i, $P_i(\uparrow)$ and $P_i(\downarrow)$ are those of single occupation, and $P_i(0)$ is that of empty occupation. For the nanostructure system, QDs and their surrounding barrier regions are assigned to sites in such a way that all of space is accounted for. This again results in a diagonal $\rho_{\text{red,i}}$ (see Ref. [2]). We calculate the average single-site entanglement (S_{site}) which, for a system of M sites or QDs, is defined by the value of S_i averaged over all sites.

We see in Fig. 1 that for attractive interactions $U < 0$ all of the approximations work well for the average single-site entanglement of the six quantum dot system, with the EHM and EHM+CH being a small improvement over the HM for $-50 \lesssim U \lesssim 0$. For repulsive interactions the approximation is reasonable for $U \lesssim 10$, but then the QD entanglement drops much more sharply than that of the lattice system. The use of the EHM or EHM+CH is only a small improvement over the HM but does little to reconcile the difference which is over 100% of the QD entanglement by $U \sim 300$. For eight quantum dots, Fig. 2, the situation is very similar except the entanglement of the QD system has not dropped by as much with respect to the lattice system's entanglement by $U \sim 300$, but still the error is around 75%.

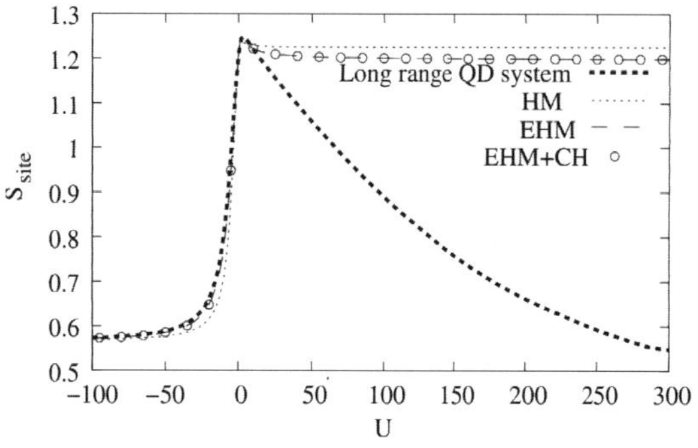

Figure 1. Average single-site entanglement quantified using S_{site} plotted against the interaction strength U for $M = 6$. QD system with long-range interaction, $d = w = 2\,a_0$, $l = 1\,a_0$, $v_0 = 10$ Hartree, $U = 17865C_U$ (thick dashed line); HM (dotted line); EHM (thin dashed line) with $U' = 0.28U$ and EHM+CH (circles) with $t' = 2 \times 10^{-5}U$.

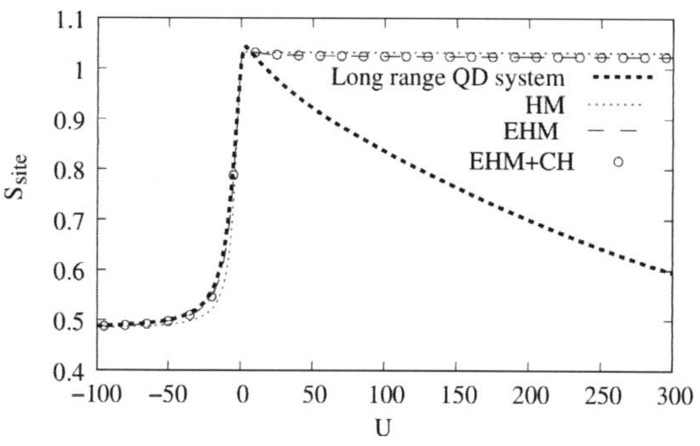

Figure 2. Average single-site entanglement quantified using S_{site} plotted against the interaction strength U for $M = 8$. QD system with long-range interaction, $d = w = 2\,a_0$, $l = 1\,a_0$, $v_0 = 10$ Hartree, $U = 17865C_U$ (thick dashed line); HM (dotted line); EHM (thin dashed line) with $U' = 0.28U$ and EHM+CH (circles) with $t' = 2 \times 10^{-5}U$.

Let us now consider the particle-particle spatial entanglement [9], S_{spatial}, which can be thought of as quantifying the spatial information gained about one particle by measuring the other particle's position. Using the procedure of Ref. [3] we can calculate the spatial entanglement for the lattice system. Here the degrees of spatial freedom have been reduced with respect to the system from, in principle, an infinite amount to simply the number of sites.

We see in Figs. 3 and 4 that for $U < 0$ the spatial entanglement tends to *increase* with increasing magnitude of U while the average single site entanglement (Figs. 1 and 2) generally decreased. For six wells we see in Fig. 3 that, similarly to the average single site entanglement, the lattice system approximation to the spatial entanglement is good for $U < 0$ with the EHM and EHM+CH offering a small improvement over the HM for $-50 \lesssim U \lesssim 0$. For $U > 0$ after an initial increase the entanglement of the QD system displays a maximum then decreases and saturates. The approximation from the Hubbard models for $U > 0$ is not as good as for $U < 0$, but better than that of the average single site entanglement particularly for large U. Interestingly, for $U \gtrsim 10$ the EHM and EHM+CH are a *worse* approximation to the spatial entanglement than the HM. Similar behaviour can be observed in the case of eight quantum dots (Fig. 4). We saw this behaviour for four quantum dots in Ref. [3] where we proposed that the spatial entanglement stemmed from the competition between the Coulomb interaction increasing the spatial correlation of the particles, and hence the entanglement, and the density being forced into the outer QDs resulting in a triplet-type state [6] with a lower entanglement. The EHM achieves the former better than the HM but neither model does much of the latter, so the overestimation of the spatial entanglement of the QD system is worse when using the EHM.

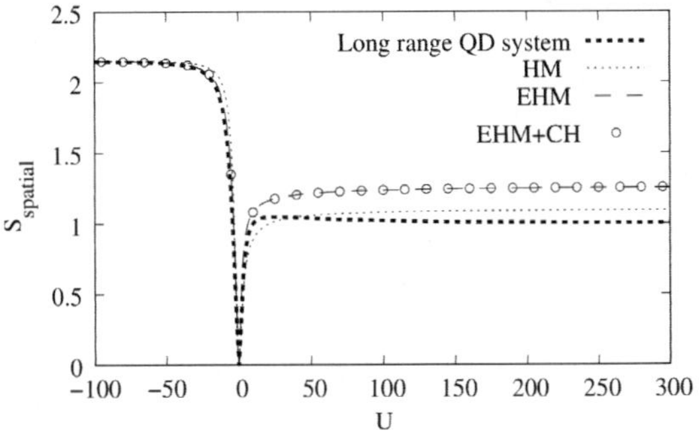

Figure 3. Particle-particle spatial entanglement quantified using S_{spatial} plotted against the interaction strength U for $M = 6$. QD system with long-range interaction, $d = w = 2\ a_0$, $l = 1\ a_0$, $v_0 = 10$ Hartree, $U = 17865C_U$ (thick dashed line); HM (dotted line); EHM (thin dashed line) with $U' = 0.28U$ and EHM+CH (circles) with $t' = 2 \times 10^{-5}U$.

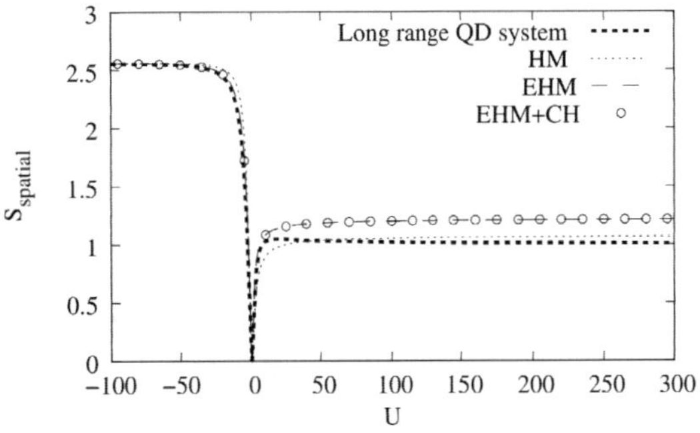

Figure 4. Particle-particle spatial entanglement quantified using S_{spatial} plotted against the interaction strength U for $M = 8$. QD system with long-range interaction, $d = w = 2\ a_0$, $l = 1\ a_0$, $v_0 = 10$ Hartree, $U = 17865C_U$ (thick dashed line); HM (dotted line); EHM (thin dashed line) with $U' = 0.28U$ and EHM+CH (circles) with $t' = 2 \times 10^{-5}U$.

We noted in Ref. [3] that values of U as large as ~ 5700 could correspond to experimentally viable quantum dots for the parameters chosen. Table 1 shows that around this value of U the approximate average single-site entanglement using the EHM+CH is very different to that of the QD system and *this percentage difference increases with the number of sites.* In fact we can expect that, at very large U, the QD system for two particles approaches a situation where double occupation is excluded and almost all of the density resides in the outer QDs. This gives rise to an entanglement of $2/M$ which is indeed what we see for large enough U in the cases of four [3], six and eight QDs. The EHM (+CH) are unable to alter the density to such an extent even for very large U, although they succeed in excluding double occupation. For EHM (+CH) we then expect to be able to approximate the large U entanglement by that of a homogeneous density with zero double occupation [2], i.e.

$$S^{\text{th}}_{\text{site},U\to\infty} = \frac{2}{M}\log_2(M) + \left(\frac{2}{M} - 1\right)\log_2\left(1 - \frac{2}{M}\right). \tag{4}$$

This gives 1.25 and 1.06 for six and eight QDs respectively, to be compared with the values in table. 1. Eq. 4 is then a reasonable approximation to the entanglement in the EHM (+CH) at large U and supports our suggestions on the behaviour of the EHM (+CH) density.

The approximation to the spatial entanglement at this very large U is better, although it still has around 20% error. Even this error though is surprisingly small when compared with that of the average single- site entanglement and bearing in mind the huge reduction in spatial degrees of freedom when moving to a lattice system.

For both the spatial and average single-site entanglement we have seen that for the parameters considered in this contribution the addition of the correlated hopping term to the EHM makes negligible difference to the accuracy of the approximation to the entanglement for six and eight QDs.

Table 1. Table showing entanglement (average single-site and spatial) at $U = 5700$ for the EHM+CH and the nanostructure system with six and eight QDs.

	Six QDs	EHM+CH six sites	Eight QDs	EHM+CH eight sites
Average single-site	0.34	1.19	0.25	1.01
Spatial	1.00	1.26	1.00	1.22

4. Summary

We have investigated the approximation to the entanglement in a long-range interacting six or eight quantum dot chain by the Hubbard model and extended Hubbard model with or without correlated hopping. For the average single site entanglement we found that, for small values of U, the approximation was good. However the Hubbard models severely overestimated the entanglement as U increased. This overestimation became worse at large U as the number of dots increased. We note that even these very large U values could correspond to realistic quantum dot models. The spatial entanglement was much better approximated even for large U values which was especially interesting given the large number of spatial degrees of freedom discarded when approximating the QD system by a lattice model. The discrepancy between the QD system and the entanglement from the Hubbard model does not display in this case a marked dependency on the QD chain length, decreasing only slightly as the chain length increases.

Acknowledgments

JPC and IDA acknowledge funding from EPSRC grant EP/F016719/1. VVF is supported by Brazilian funding from CAPES(4101/09-0).

References

[1] França V V and Capelle K 2008 *Phys. Rev. Lett.* **100** 070403
[2] Coe J P, França V V and D'Amico I 2010 *Phys. Rev. A* **81** 052321
[3] Coe J P, França V V and D'Amico I 2010 arXiv:1007.5441v1 [cond–mat.str–el]
[4] Hubbard J 1963 *Proc. R. Soc London. Series A* **276** 238
[5] Javanainen J, Eberly J H and Su Q 1988 *Phys. Rev. A* **38** 3430
[6] Abdullah S, Coe J P and D'Amico I 2009 *Phys. Rev. B* **80** 235302
[7] Zanardi P 2002 *Phys. Rev. A* **65** 042101
[8] Larsson D and Johannesson H 2006 *Phys. Rev. A* **73** 042320
[9] Coe J P, Sudbery A and D'Amico I 2008 *Phys. Rev. B* **77** 205122

Impurity Scattering in Luttinger Liquid with Electron-Phonon Coupling

Alexey Galda, Igor V. Yurkevich and Igor V. Lerner
School of Physics and Astronomy, University of Birmingham, Birmingham B15 2TT, United Kingdom
E-mail: galda@theory.bham.ac.uk

Abstract

We study the influence of electron-phonon coupling on electron transport through a Luttinger liquid with an embedded weak scatterer or weak link. We derive the renormalization group (RG) equations which indicate that the directions of RG flows can change upon varying either the relative strength of the electron-electron and electron-phonon coupling or the ratio of Fermi to sound velocities. This results in the rich phase diagram with up to three fixed points: an unstable one with a finite value of conductance and two stable ones, corresponding to an ideal metal or insulator.

PACS numbers: 71.10.Pm, 71.10.Hf, 72.10.Fk, 73.20.Mf

Interacting electrons in one dimension are known to form a Luttinger liquid (LL) characterized by power-law correlation functions [1]. This characteristic feature of the LL has been established via conductance measurements and a scanning tunneling microscopy (STM) both in carbon nanotubes [2] and semiconductor quantum wires [3]. Embedding a potential impurity into the LL leads [4, 5] to a universal (i.e. impurity-independent) power-law (in temperature T) suppression of the transmission amplitude through the LL and suppression of the tunneling density of states (TDoS) near the impurity, with the latter fading away with the distance [6, 7].

The electron-phonon (el-ph) coupling in addition to the (Coulomb) electron-electron (el-el) repulsion in the LL is known to result in the formation of two polaron branches with different propagation velocities (see, for example, [8, 9]). In the present Letter we show that embedding a single scatterer into such an el-ph liquid results in a rather rich phase diagram: depending on the relative strength of the el-el and el-ph coupling and on the ratio of the Fermi to sound velocities, the system can be an ideal metal (with conductance $g = 1$ in the units of e^2/h as in the absence of impurities [10]) or an ideal insulator for any impurity strength, or to be in an intermediate state from which it can flow either to the metallic or insulating limit, depending on the impurity strength.

To show this we consider both the weak back-scatterer (WS) and weak link (WL) limit, the latter corresponding to a weak tunneling coupling between the two halves of the LL. In the phononless case the WS and WL amplitudes are known to scale at low energies as $\varepsilon^{\widetilde{\gamma}_-}$ and $\varepsilon^{\widetilde{\gamma}_+}$, respectively. For the repulsive el-el interaction the exponents $\widetilde{\gamma}_- \equiv K - 1 < 0$ and $\widetilde{\gamma}_+ = K^{-1} - 1 > 0$ where $K < 1$ is the Luttinger parameter. Therefore, the renormalization group (RG) flows always go in the direction of stronger scattering (weaker transmission) resulting at $T = 0$ in the insulating phase for any scatterer [4]. The el-ph coupling leads to $\widetilde{\gamma}_\pm \to \gamma_\pm$, with each of the exponents $\gamma_\pm$ changing sign at different values of bulk parameters. In the WS limit which has been previously considered [11], the change of sign of γ_- was reversing the RG flows indicating that the weak scatterer becomes irrelevant for the sufficiently strong el-ph interaction and the LL thus remains in the metallic state. But the considerations of both the WS and WL limits presented here show that the change of signs in the exponents $\gamma_\pm$ happens at different values of the parameters, resulting in the rich phase diagram described in detail later.

To get the results, we employ the functional bosonization formalism in form developed in [7]. This allows us first to include the el-ph interaction which leads to electrons dressing with phonons, i.e. the formation of polarons, and only after that to bosonize the action. Prior to considering the embedded scatterer we describe the polaron formation in a pure LL, reproducing the known results [8, 9] in form convenient for further considerations.

We consider a model of 1D acoustic phonons linearly coupled to the electron density. The phonon spectrum is assumed to be linear with a cutoff at the Debye frequency, $\omega_{\rm D} = cq_{\rm D}$ (a straightforward modification for the 3D or optical phonons will be described elsewhere). Integrating out the phonon field in the standard way results in substituting the dynamical coupling, $V(\xi) = V_0 + \mathcal{D}_0(\xi)$, for the screened Coulomb interaction V_0 in the LL action:

$$S_{\rm LL} = \sum_{\eta=\pm1} \int\! \mathrm{d}\xi\, \bar{\psi}_\eta(\xi)\, \mathrm{i}\partial_\eta \psi_\eta(\xi) - \frac{1}{2}\int\! \mathrm{d}\xi\, V(\xi)\, n^2(\xi)\,. \tag{1}$$

Here $\partial_\eta \equiv \partial_t + \eta v_{\rm F}\partial_x$, the (spinless) electron field is decoupled into the sum of left- ($\eta=-1$) and right- ($\eta=1$) moving terms, $\psi(\xi) = \psi_{\rm R}(\xi)\mathrm{e}^{\mathrm{i}p_{\rm F}x} + \psi_{\rm L}(\xi)\mathrm{e}^{-\mathrm{i}p_{\rm F}x}$ with $\xi \equiv (x,t)$, and $n \equiv (\bar{\psi}_{\rm R}\psi_{\rm R} + \bar{\psi}_{\rm L}\psi_{\rm L})$. We use the Keldysh formalism [12] implying the time integration along the Keldysh contour in Eq. (1) and below. The free phonon propagator $\mathcal{D}_0(\xi)$ in $V(\xi) = V_0 + \mathcal{D}_0(\xi)$, can be defined by the Fourier transform of its retarded component,

$$\mathcal{D}_0^r(\omega,\, q) = \frac{\nu_0^{-1}\alpha_{\rm ph}\,\omega_q^2}{\omega_+^2 - \omega_q^2}\,, \qquad \omega_q = cq\,, \qquad \omega_+ \equiv \omega + \mathrm{i}0\,, \tag{2}$$

where c is the sound velocity, $\alpha_{\rm ph}$ is the dimensionless el-ph coupling constant, and $\nu_0 = (\pi v_{\rm F})^{-1}$ is the free spinless electron DoS. The form of the LL action in Eq. (1) implies neglecting the electronic backscattering. This remains justified in the presence of the el-ph coupling for low temperatures, $T \ll \omega_{\rm D}$.

The next step is the Hubbard-Stratonovich transformation which decouples the n^2 term in the action (1) and results in the mixed fermionic-bosonic action in terms of the auxiliary bosonic field φ minimally coupled to ψ:

$$S_{\rm eff} = -\frac{1}{2}\int\! \mathrm{d}\xi\, \varphi\, V^{-1}\varphi + \mathrm{i}\!\int\! \mathrm{d}\xi\, \bar{\psi}_\eta\, (\partial_\eta - \varphi)\, \psi_\eta\,. \tag{3}$$

We gauge out the coupling term by the transformation

$$\psi_\eta(\xi) \to \psi_\eta(\xi)\, \mathrm{e}^{\mathrm{i}\theta_\eta(\xi)}\,, \qquad \mathrm{i}\partial_\eta\theta(\xi) = \varphi(\xi)\,. \tag{4}$$

The Jacobian of this transformation results [7] in substituting $V^{-1} + \Pi$ for V^{-1} in Eq. (3), where $\Pi = \Pi_{\rm R} + \Pi_{\rm L}$ is the one-loop electronic polarization operator (exact for the LL [13]), with $\Pi_\eta(\xi) = \mathrm{i}g_\eta(\xi)g_\eta(-\xi)$ and the free electron Green function defined via the Fourier transform of its retarded component as $g_\eta^r(\omega,\, q) = [\omega_+ - \eta v_{\rm F} q]^{-1}$.

The phase θ_η is related to the auxiliary field φ by [14]:

$$\theta_\eta(\xi) = \int\! \mathrm{d}\xi'\, g_\eta^B(\xi, \xi')\, \varphi(\xi')\,,$$

where g_η^B is the bosonic Green functions that resolves Eq. (4). Its retarded component coincides with the free fermionic g_η^r (while the Keldysh components are naturally different).

The Green function of the interacting polarons is not gauge-invariant with respect to the transformation (4) and depends on the correlation function $\mathrm{i}U_{\eta\eta'} = \langle\theta_\eta\theta_{\eta'}\rangle$:

$$G_\eta(\xi, \xi') = g_\eta(\xi, \xi')\, \mathrm{e}^{\mathrm{i}U_{\eta\eta}(\xi,\xi')}\,.$$

The retarded Fourier component of $U_{\eta,\eta'}$ is found as

$$U_{\eta\eta'}^r(\omega, q) = \frac{\omega_+ + \eta' v_{\rm F} q}{\omega_+ - \eta\, v_{\rm F} q}\, \frac{V_0(\omega_+^2 - \omega_q^2) + \nu_0^{-1}\alpha_{\rm ph}\omega_q^2}{\left(\omega_+^2 - v_+^2 q^2\right)\left(\omega_+^2 - v_-^2 q^2\right)}\,, \tag{5}$$

where $v_\pm$ are velocities of the composite bosonic modes:

$$v_\pm^2 = \frac{1}{2}\left[v^2 + c^2 \pm \sqrt{(v^2-c^2)^2 + 4\alpha\, v^2c^2}\,\right]. \tag{6}$$

Here v is the speed of plasmonic excitations in the phononless LL, $v = v_{\rm F}(1+\nu_0 V_0)^{1/2} \equiv v_{\rm F}K^{-1}$, where K is the standard Luttinger parameter. Without phonons ($c=0$), one has $v_- = 0$ and $v_+ = v$, so that in this case (as well as for $\omega > \omega_{\rm D}$ or $\alpha_{\rm ph} = 0$), $U_{\eta\eta'}$ reduces to the usual LL plasmonic propagator.

We assume the parameter $\alpha \equiv \alpha_{\rm ph}K^2$ in Eq. (6) obeying the inequality $\alpha < 1$ to avoid the Wentzel–Bardeen instability [15] corresponding to $v_-^2 < 0$ (with the threshold shifted from $\alpha_{\rm ph} = 1$ for a pure el-ph model to $\alpha_{\rm ph} = 1/K^2 > 1$). Then the velocities $v_\mp$ of the slow and fast composite bosonic modes in Eq. (6) obey the inequalities $v_- < c$, $v < v_+$. These modes mean that the LL in the presence of the el-ph coupling becomes two-component. It is the existence of these two modes that leads to a rich phase diagram when a scatterer is embedded.

Following Kane and Fisher [4] we consider two types of scatterers (assuming them pinned at $x=0$ and not involved in lattice vibrations): a weak backscatterer (WS) or a weak tunneling link (WL). The WS action is

$$S_{\rm ws} = \lambda_0 \int {\rm d}t\, \bar\psi_{\rm R}(t)\psi_{\rm L}(t) + {\rm c.c.}, \qquad \psi_\eta(t) \equiv \psi_\eta(x=0,t), \tag{7}$$

with λ_0 being a bare backscattering amplitude.

The WL action contains the tunneling term linking the two halves of the LL, labeled by 1 and 2:

$$S_{\rm wl} = t_0 \int {\rm d}t\, \bar\psi_1\psi_2 + {\rm c.c.}, \qquad \psi_a = \psi_{a\rm R} + \psi_{a\rm L}, \tag{8}$$

with t_0 being a bare tunneling amplitude. We begin with the WS case, noticing in advance that the WL case is dual to it as well as for the standard LL [4].

The gauge transform (4) replaces λ_0 in Eq. (7) with

$$\lambda(t) = \lambda_0\, {\rm e}^{{\rm i}[\theta_L(t) - \theta_R(t)]}, \qquad \theta_\eta(t) \equiv \theta_\eta(x=0,t). \tag{9}$$

Integrating out all $x\neq 0$ fields results in the quadratic in $\theta_\eta(t)$ action with the scattering term (9) as in the phononless problem [4]. The difference is that the correlation function is governed by the composite bosonic modes, Eq. (5), and its retarded Fourier component is

$$\langle \theta_\eta(-\omega)\,\theta_{\eta'}(\omega)\rangle^r = {\rm i}\int \frac{{\rm d}q}{2\pi}\, U_{\eta\eta'}(\omega;q) \equiv \frac{\pi\gamma_{\eta\eta'}}{\omega_+}, \tag{10}$$

where the correlation matrix $\gamma_{\eta\eta'}$ is found from the straightforward integration above and can be expressed in terms of the exponents $\gamma_\pm$ as follows:

$$\gamma_\sigma = \gamma_{\rm LL} + \sigma\gamma_{\rm LR} = \kappa_\sigma K^{-\sigma} - 1, \qquad \sigma = \pm 1, \tag{11a}$$

$$\kappa_\sigma = \left\{ [1 - \alpha\delta_{\sigma,-1}]\left[1 + \frac{\alpha}{\left(\beta^\sigma + \sqrt{1-\alpha}\right)^2}\right]\right\}^{-\frac{1}{2}}, \tag{11b}$$

where $\gamma_{\rm LL} = \gamma_{\rm RR}$, $\gamma_{\rm LR} = \gamma_{\rm RL}$ and $\beta \equiv v/c$.

The only change in the $x=0$ action which is due to the el-ph coupling is the appearance of the dimensionless matrix $\gamma_{\eta\eta'}$ in Eq. (10). Thus, the RG analysis for the WS case is similar to that in [4] but the RG equation for the backscattering amplitude λ acquires a different prefactor γ_-, defined by

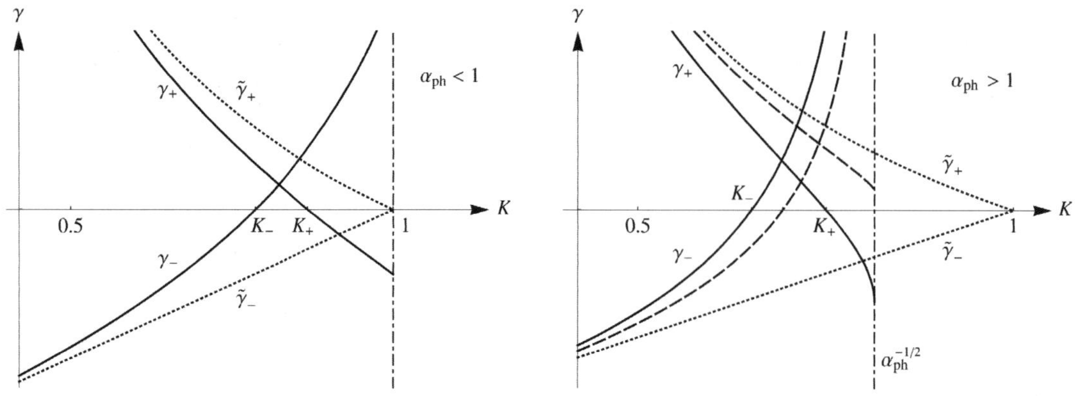

Figure 1: Weak tunneling (γ_+) and weak scattering (γ_-) exponents as functions of the Luttinger parameter $K<1$ for a weak (a) or strong (b) el-ph coupling ($\alpha_{\rm ph}=0.7$ and $\alpha_{\rm ph}=1.5$ were used). The dotted lines represent $\widetilde{\gamma}_\pm$ for the phononless LL, the solid and dashed lines correspond to $\beta_{\rm F}<\beta_{\rm F}^*$ and $\beta_{\rm F}>\beta_{\rm F}^*$, respectively (we used $\beta_{\rm F}=0.5$ and $\beta_{\rm F}=1.5$ respectively). The latter is shown only for the strong coupling where for $\beta_{\rm F}>\beta_{\rm F}^*$ the exponent γ_+ does not change sign within the stability region, $K<\alpha_{\rm ph}^{-1/2}$.

Eqs. (11):

$$\partial_l\,\lambda(E)=\begin{cases}(1-K)\lambda(E)\,, & E>\omega_{\rm D}\\ -\gamma_-\,\lambda(E)\,, & E<\omega_{\rm D}\end{cases}. \tag{12a}$$

Here E is a running cutoff and $l\equiv\ln E_0/E$ with $E_0\sim\varepsilon_{\rm F}$ being the bandwidth.

The WL case can be treated with the same $x=0$ action defined by Eq. (10). The gauge transform (4) changes the tunneling action (8) by replacing $\psi_a\to\psi_{a\rm R}e^{i\theta_a^-}+\psi_{a\rm L}e^{-i\theta_a^-}$ and $t_0\to t_0e^{i(\theta_2^+-\theta_1^+)}$, where $\theta_a^\pm=\frac{1}{2}(\theta_{a\rm R}\pm\theta_{a\rm L})$. The boundary conditions $\psi_{a\rm R}(x=0)=e^{i\chi}\,\psi_{a\rm L}(x=0)$ (where χ is an arbitrary phase shift acquired from edge scattering) demand that the fields $\theta_{a\eta}$ arising from the gauge transformation also obey the boundary conditions $\theta_{a\rm R}=\theta_{a\rm L}$, thus leaving ψ_a unchanged under the transformation. The only result, therefore, of the gauge transformation is an extra phase $\delta\theta^+(t)=\theta_2^+(x=0,t)-\theta_1^+(x=0,t)$ with correlations $\langle\delta\theta^+(-\omega)\delta\theta^+(\omega)\rangle^r=\pi\gamma_+/(\omega+i0)$. The RG analysis of the tunneling term yields the RG equation

$$\partial_l\,t(E)=\begin{cases}(1-K^{-1})\,t(E)\,, & E>\omega_{\rm D}\\ -\gamma_+\,t(E)\,, & E<\omega_{\rm D}\end{cases}, \tag{12b}$$

dual to Eq. (12a).

The RG equations (12) should be solved with the initial conditions $\lambda(E=E_0)=\lambda_0$ and $t(E=E_0)=t_0$. The solutions for $\varepsilon<\omega_{\rm D}$ are given by

$$\frac{\lambda(\varepsilon)}{\lambda_0}=\left(\frac{\omega_{\rm D}}{E_0}\right)^{\widetilde{\gamma}_-}\left(\frac{\varepsilon}{\omega_{\rm D}}\right)^{\gamma_-},\qquad \frac{t(\varepsilon)}{t_0}=\left(\frac{\omega_{\rm D}}{E_0}\right)^{\widetilde{\gamma}_+}\left(\frac{\varepsilon}{\omega_{\rm D}}\right)^{\gamma_+}. \tag{13}$$

The LL exponents $\widetilde{\gamma}_\pm$ are obtained with $\kappa_\pm=1$ in Eq. (11a). Hence, in the absence of the el-ph coupling Eqs. (13) reduce to the standard ones, $\lambda(\varepsilon)\sim(\varepsilon/E_0)^{K-1}$ and $t(\varepsilon)\sim(\varepsilon/E_0)^{K^{-1}-1}$, i.e. the backscattering amplitude increases (the WS case) and the tunneling amplitude decreases (the WL case) with $\varepsilon\to0$ for any $K<1$. This shows that embedding an arbitrary scatterer results in the LL becoming (at $T=0$) an ideal insulator.

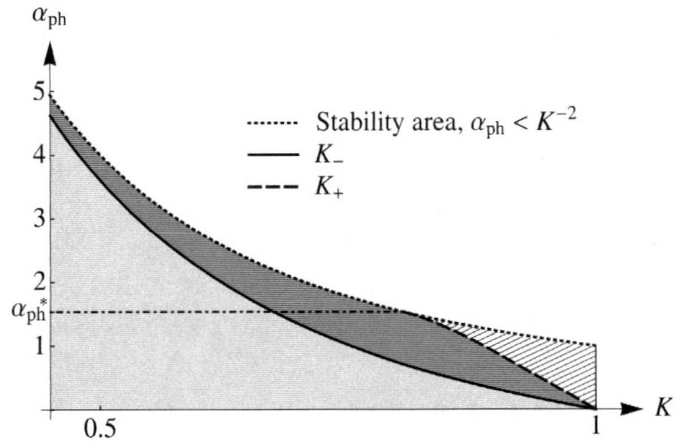

Figure 2: Phase diagram: The light-grey area represents the insulating regime. In the dark-grey area there exists a line of unstable fixed points separating the metallic state for a weak scatterer from the insulating phase for a strong one. The hatched area (reachable only for $\alpha_{ph} < \alpha^*_{ph}$) corresponds to the purely metallic phase. We used $\beta_F = 1.1$.

The el-ph coupling, however weak, changes drastically the above conclusion leading to the possibility of a metal-insulator transition with changing K.

Indeed, it follows from Eqs. (11)–(12) that RG flows can change directions depending on the values of K, α_{ph} and $\beta_F \equiv v_F/c$. It is easy to verify that $\kappa_+(K) \leq 1$ while $\kappa_-(K) \geq 1$, which results in $\gamma_\pm$ changing sign (see Fig. 1) when the interaction strength K equals $K_\pm$, with

$$\gamma_+(K_+) = 0\,, \qquad \gamma_-(K_-) = 0\,. \tag{14}$$

When the el-ph coupling is weak ($\alpha_{ph} < 1$), one can show that $K_- < K_+ < 1$, and K can take any values between 0 and 1 (repulsive el-el interaction of arbitrary strength), making three different regimes possible (Fig. 1a).

For $K_+ < K < 1$ both $\gamma_\pm$ change sign so that the WL amplitude $t(\varepsilon)$ increases while the WS amplitude $\lambda(\varepsilon)$ decreases with $\varepsilon \to 0$ in Eqs. (13), indicating [16] that adding a single scatterer does not change the metallic nature of the pure LL (where the dimensionless conductance $g = 1$) – in contrast to the phononless case.

Decreasing K (i.e. increasing the el-el interaction), we enter the region $K_- < K < K_+$, when the WL amplitude still decreases – as in the phononless case, but also the WS amplitude decreases due to γ_- changing its sign – opposite to the phononless case. This indicates that a strong scatterer (i.e. weak link) results in the insulating behavior (the $g = 0$ fixed point) while a weak scatterer leaves the LL in the metallic phase (the $g = 1$ fixed point). Thus there should exist an intermediate unstable fixed point with a finite g, separating the metallic and insulating regimes.

Finally, for a strong enough el-el interaction, $K < K_-$, the RG flows in Eqs. (13) remain for both the WL and WS amplitudes qualitatively the same as for the phononless case, so that any scatterer makes the LL insulating.

When the el-ph coupling is strong (which is the case for carbon nanotubes [17]) our considerations are bound by the stability requirement, $K \leq \alpha_{ph}^{-1/2}$, so that for $\alpha_{ph} > 1$ the Luttinger parameter is confined to the region $K \leq \alpha_{ph}^{-1/2}$. In this case there is an essential dependence on β_F and it is easy to see from Eqs. (11) and (14) that $K_+ > \alpha_{ph}^{-1/2}$ for large enough values of β_F:

$$\beta_F > \beta^*_F \equiv \left[\alpha_{ph}(\alpha_{ph} - 1)\right]^{-\frac{1}{2}}\,. \tag{15}$$

Hence, for such β_{F} the purely metallic regime is no longer accessible, as illustrated by the dashed lines in Fig. 1b. That is why on the phase diagram (Fig. 2) only two regimes exist for $\alpha_{\mathrm{ph}} > \alpha^*_{\mathrm{ph}} \equiv \frac{1}{2} + \frac{1}{2}\sqrt{1+4/\beta_{\mathrm{F}}^2}$ – which is an equivalent to (15) condition for a fixed β_{F}.

There are other properties of the LL strongly affected by the el-ph coupling. First, γ_+ is also the edge exponent of the TDoS at the boundary (edge) of the wire, $\nu_{\mathrm{edge}} \sim \varepsilon^{\gamma_+}$, the TDoS also experiences, depending on the parameters, a transition from vanishing to divergent with $\varepsilon \to 0$. Then, transport properties of the LL with resonant or antiresonant impurity [18], or of the disordered LL [19], or of the LL out of equilibrium [20] also experience qualitative changes with allowing for the el-ph interaction, as we will show elsewhere.

To summarize, we have shown that the el-ph coupling qualitatively changes the phase diagram of the LL with a single impurity. The change is not reducible to a redefinition of the Luttinger parameter K: the existence of slow and fast polaron modes with different weights in different regimes results in different RG flows for a weak scatter and a weak (tunneling) link. The resulting phase diagram (Fig. 2) has, depending on the parameters of the problem, regimes corresponding to purely metallic or purely insulating behavior and an intermediate regime with two stable fixed points (ideal metal and insulator) and one unstable, finite-conductance fixed point.

References

[1] S. Tomonaga, Prog. Theor. Phys., **5**, 544 (1950); J. M. Luttinger, J. Math. Phys., **4**, 1154 (1963); F. D. M. Haldane, J. Phys. C, **14**, 2585 (1981); J. von Delft and H. Schoeller, Ann. Phys., **7**, 225 (1998).

[2] M. Bockrath *et al.*, Nature, **397**, 598 (1999); Z. Yao *et al.*, *ibid.*, **402**, 273 (1999); H. Ishii *et al.*, *ibid.*, **426**, 540 (2003); J. Lee *et al.*, Phys. Rev. Lett., **93**, 166403 (2004).

[3] O. Auslaender *et al.*, Science, **295**, 825 (2002); E. Levy *et al.*, Phys. Rev. Lett., **97**, 196802 (2006).

[4] C. L. Kane and M. P. A. Fisher, Phys. Rev. Lett., **68**, 1220 (1992); Phys. Rev. B, **46**, 15233 (1992).

[5] K. A. Matveev, D. Yue, and L. I. Glazman, Phys. Rev. Lett., **71**, 3351 (1993).

[6] S. Eggert, Phys. Rev. Lett., **84**, 4413 (2000).

[7] A. Grishin, I. V. Yurkevich, and I. V. Lerner, Phys. Rev. B, **69**, 165108 (2004); I. V. Lerner and I. V. Yurkevich, in *Nanophysics: coherence and transport*, Les Houches Summer School Series, Vol. 81, edited by H. Bouchiat *et al.* (Elsevier, New York, 2005) p. 109.

[8] D. Loss and T. Martin, Phys. Rev. B, **50**, 12160 (1994).

[9] L. Mathey, D.-W. Wang, W. Hofstetter, M. D. Lukin, and E. Demler, Phys. Rev. Lett., **93**, 120404 (2004).

[10] D. L. Maslov and M. Stone, Phys. Rev. B, **52**, R5539 (1995).

[11] P. San-Jose, F. Guinea, and T. Martin, Phys. Rev. B, **72**, 165427 (2005).

[12] J. Rammer and H. Smith, Rev. Mod. Phys., **58**, 323 (1986); A. Kamenev and A. Levchenko, Adv. Phys., **58**, 197 (2009).

[13] I. E. Dzyaloshinskii and A. I. Larkin, Zh. Eksp. Teor. Phys., **65**, 411 (1973).

[14] After the gauge transformation (4) the remaining fermionic part can be bosonized with $\psi_\eta \propto \mathrm{e}^{\mathrm{i}\phi_\eta}$. This would allow one to rewrite the action in the standard, fully bosonized form in terms of $\Theta_\eta = \theta_\eta + \phi_\eta$.The mutually conjugated fields used in [4] correspond to $\Theta_{\mathrm{R}} \pm \Theta_{\mathrm{L}}$

[15] G. Wentzel, Phys. Rev., **83**, 168 (1951); J. Bardeen, Rev. Mod. Phys., **23**, 261 (1951).

[16] When either $t(\varepsilon)$ or $\lambda(\varepsilon)$ increases as $\varepsilon \to 0$, the one-loop RG gives only an indication of behavior in this limit.

[17] A. De Martino and R. Egger, Phys. Rev. B, **67**, 235418 (2003); S. Sapmaz, P. Jarillo-Herrero, Y. M. Blanter, and H. S. J. van der Zant, New J. Phys., **7**, 243 (2005).

[18] C. L. Kane and M. P. A. Fisher, Phys. Rev. B, **46**, 7268(R) (1992); D. G. Polyakov and I. V. Gornyi, **68**, 035421 (2003); A. Furusaki and K. A. Matveev, Phys. Rev. Lett., **88**, 226404 (2002); Y. V. Nazarov and L. I. Glazman, **91**, 126804 (2003); I. V. Lerner, V. I. Yudson, and I. V. Yurkevich, **100**, 256805 (2008); M. Goldstein and R. Berkovits, **104**, 106403 (2010).

[19] T. Giamarchi and H. J. Schulz, Phys. Rev. B, **37**, 325 (1988).

[20] D. B. Gutman, Y. Gefen, and A. D. Mirlin, Phys. Rev. Lett., **101**, 126802 (2008); Phys. Rev. B, **80**, 045106 (2009); **81**, 085436 (2010).

Journal of Physics: Conference Series **286** (2011) 012050 doi:10.1088/1742-6596/286/1/012050

Non-contact ultrasonic measurements of the elastic constants of magnetic materials

R.S. Edwards, R. Perry, D.J. Backhouse, I.J. Moore, D. Cleanthous, A.R. Clough and D. Stone

Department of Physics, University of Warwick, Coventry CV4 7AL, UK

E-mail: r.s.edwards@warwick.ac.uk

Abstract. Ultrasonic testing using contacting transducers such as quartz or PZT is well established. However, standard measurement techniques used require physical contact of the sample and ultrasonic transducer and some sort of couplant between the two. With this configuration there is a possibility of damaging the sample, transducer or bond during testing, thermal cycling, or removal of the transducer. We present results taken using recent advances in non-contact methods of ultrasound generation and detection using electromagnetic acoustic transducers (EMATs), which offer some significant benefits over contact ultrasonic techniques. Circumventing the need for couplant removes the possibility of contaminating the system, which is an issue for some material property measurements, and allows easier measurements over a wider range of temperatures. An automated data analysis system has been developed which allows the velocity of sound in the sample, and hence the elastic constants, to be determined to a high accuracy. This technique is illustrated using measurements of the alloy $Gd_{64}Sc_{36}$.

1. Introduction

Materials testing using contact transducers for generation and detection of ultrasonic waves is well used in areas ranging from medical physics, to non-destructive testing (NDT), through to fundamental physics experiments [1, 2, 3, 4, 5, 6]. Many of the future improvements in ultrasonic measurements are likely to be through developments in non-contact methods of generation and detection, as these offer significant benefits over contact ultrasonic techniques [2].

Standard ultrasonic measurements of single crystals use contact transducers, such as quartz or $LiNbO_3$, which require contact with the sample and couplant for transmission of the waves [6]. Repeated thermal cycling can affect this couplant, causing damage, dramatically reducing the coupling efficiency of the transducer to the sample, and ultimately leading to a complete loss of signal. Recent developments in NDT [2] have ensured that electromagnetic acoustic transducers (EMATs) are now a practical alternative to contact techniques.

Figure 1(b) shows two EMAT designs used for non-contact ultrasonic measurements, with lift-offs from the sample possible of up to several mm. The first is a typical EMAT used for NDT, which consists of a coil and a permanent magnet, with a current pulsed through the coil for generation of ultrasound [2]. These operate on electrically conducting samples (via the Lorentz force mechanism) and/or certain magnetic samples (via magnetoelastic mechanisms). For the Lorentz force mechanism the electrons in the mirror current in the sample experience a force and in turn 'drag' the atoms, with each current pulse generating an ultrasound pulse [2]. In magnetic materials, ultrasound is induced by the varying magnetic field from the current pulse.

Through careful consideration of the coil and magnet configuration various ultrasonic waves can be generated [7]. Detection is via a similar mechanism [2]. During measurements of the properties of a crystal the effects of temperature and magnetic field may be investigated [8, 9, 10]. In this case, the magnetic field for these experiments can be used as part of the EMAT in the place of the permanent magnet.

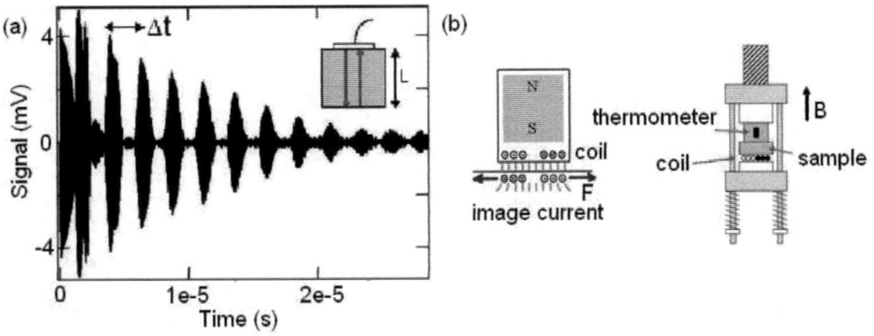

Figure 1. (a) shows a typical ultrasound signal from a single crystal, showing ultrasonic echoes with separation Δt. (b) shows EMATs developed for NDT (left) and single crystal measurements (right, after [8]).

EMATs have been used for some recent measurements of single crystals [8, 9, 11]; however, practically there are still significant issues. EMATs are, by their very nature, sensitive to electrical noise, and their signal to noise ratio when compared with standard techniques can be poor if the environment is electrically noisy. Filtering and lock-in techniques can be used to remove some of this noise, but must be used carefully to ensure that none of the information included within the signals is lost. However, the advantage of the no couplant requirement still makes EMATs an attractive prospect. The efficiency of non-contact ultrasound generation is significantly lower than that obtained when using piezoelectric transducers, and the signal to noise ratio can be an issue even without electrical noise. However, EMATs are highly sensitive to magnetic phase transitions; the exact generation mechanism is dependent on the magnetic state of the material, and hence the generation and/or detection efficiency will show a change at a magnetic phase change.

It is well known that the velocity of ultrasound propagation in a material is related to its elastic properties [3]. For a single crystal the velocity of sound is dependent on the type of sound wave and the direction of propagation. For example, in a hexagonal crystal such as Gd, for propagation of sound along the c-axis two elastic constants can be measured from the velocities v of longitudinal and shear waves; $C_{33} = \rho v_{long}^2$ and $C_{44} = \rho v_{shear}^2$, where ρ is the density of the material [3]. The elastic constants will change as the properties of the sample change [3].

Elastic constant measurements require accurate determination of the velocity [4, 5]. Typical measurements are set up as shown in the inset of Figure 1; a sample has an ultrasonic transducer fixed to it and reflections from the opposite face are detected, with a typical echo pulse train shown (the peaks in the figure show the envelope of the MHz-frequency oscillation of the transducer). The time between echoes, Δt, needs to be measured, with the velocity given by

$$\Delta t = \frac{2L}{v} + \frac{\phi}{2\pi} \cdot \frac{1}{f} \qquad (1)$$

where L is the sample thickness, f is the frequency of the ultrasonic signals (typically in the MHz range) and ϕ is the phase change on reflection. This phase correction is typically small [4].

It is difficult to define a suitable measurement point for each echo. In the 1960s pulse echo overlap (PEO) was developed to measure the time between echoes [5]. In this method, two pulses are generated and the time between them altered such that the echoes overlap, giving a phase velocity accuracy of 2 parts in 10^4, but requiring significant input from an operator. Recently, a digital equivalent using cross correlation has been developed [4], giving an accuracy of 1 part in 10^7.

2. Automating the velocity measurement

The availability of faster processors has meant that automation of data analysis, through recording all the ultrasonic echoes and digital signal analysis on a PC, is now possible. We have built real-time data acquisition and analysis routines using LabVIEW [12].

Ultrasonic echoes measured using EMATs are often noise dominated, and this adds to the problems of identifying the start time of an echo for measurement of Δt. Simply measuring the maximum amplitude point in each echo could lead to the wrong part of the echo being chosen due to the presence of noise, and hence automation is difficult. Cross correlation, however, enables a clearer measurement of the time between echoes [4]. We have implemented a routine using the cross-correlation VI in LabVIEW; the first measured echo is windowed and scanned across the series of echoes by varying an offset time. The maxima in the cross correlated signal therefore correspond to the offset times whereby the first and later echoes are in best agreement, and this gives the transit time within the material. At present, this method does not take into account phase changes on each reflection of the signals; these are assumed to be negligible and will be investigated later. However, initial results show this to be a suitable technique for velocity measurements. The LabVIEW routine for data analysis also measures echo amplitudes for the first few echoes, and an amplitude for the noise level between echoes. These are used to measure the signal attenuation, and to measure the signal to noise for the first echo, which gives a good measure of the EMAT efficiency.

3. Measurements of $Gd_{64}Sc_{36}$ using contact and non-contact ultrasonic techniques

Gd-rich alloys show a competition between ferromagnetic order (from Gd) and other magnetic phases from the alloy materials, such as Sc, which shows helical magnetic ordering [14]. $Gd_{64}Sc_{36}$ has been shown to be a simple helimagnet from the Néel point down to the lowest temperatures, with the turn angle locked in place by 30 K [13]. The single crystal sample used for these experiments was grown at the Centre for Materials Science, University of Birmingham, UK, and used for the work presented in [14]. It has dimensions of approximately 4 mm. This material was chosen as the proof of concept sample test, as measurements had previously been made using standard manual PEO techniques and quartz transducers, with results available for comparison.

Measurements were taken using a Matec 6600 pulse generator system, capable of generating narrowband signals between 1 and 350 MHz given suitable transducers. Temperature control and magnetic field were provided by an Oxford Instruments superconducting magnet and cryostat system, in combination with a heater fitted to the experimental probe. Data collection was through a Tektronix oscilloscope with 350 MHz bandwidth communicating with a PC running LabVIEW. Contact measurements used an X-cut quartz transducer, generating 15 MHz longitudinal waves with waves propagating along the c-axis and the magnetic field also applied along this axis. This was used in pulse-echo mode to both generate and detect ultrasonic signals. Non-contact measurements used several different designs of EMATs, either to just detect signals (with the quartz transducer used for generation) or to both generate and detect ultrasound. The coil designs investigated included a simple spiral (pancake) coil, and several designs of differential coil [15]. The latter design uses a pair of coils and was intended to reduce noise by providing a reference signal away from the sample. The reference coil was kept as close to the original site as possible to experience the same magnetic field.

Ultrasonic signals were recorded over a range of temperatures and analysed using cross-correlation methods [4], giving a measure of C_{33}. Figure 2(a) shows C_{33} as a function of magnetic field normalised to 1 for no applied field, measured using the differential coil EMAT to both generate and detect ultrasound, at a temperature of 90 K. Two main features can be seen; C_{33} shows a very clear dip at around 1.6 T, corresponding to the change from a fan to a ferromagnetic phase [14]. A smaller feature, showing as a change in gradient, is also present at around 0.6 T; this corresponds to the change from distorted helix to fan phase, and is clearer when the applied field is along the a-axis. Despite the noise problems inherent in an EMAT measurement, application of suitable frequency filtering and the use of cross-correlation techniques have allowed analysis of the data.

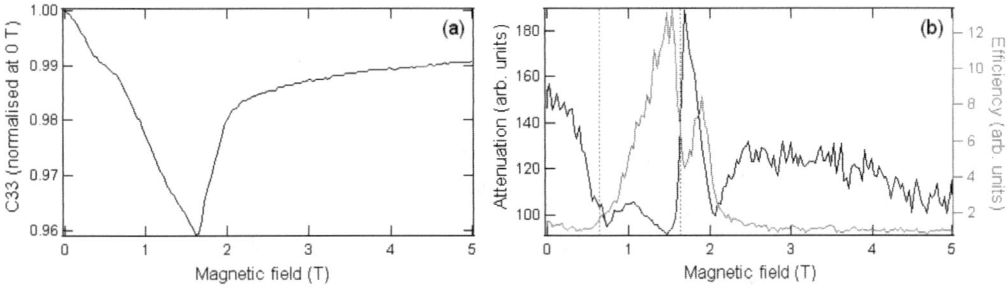

Figure 2. (a) Measurement of C_{33} using a differential coil for generation and detection, at a temperature of 90 K. (b) Attenuation (black) and EMAT efficiency (red) measurements.

The attenuation of the signal can also be measured, and is shown as the black trace in Figure 2(b). There is clearly a marked change in the signal attenuation around both phase transitions. With EMAT measurements a further measure of changes in the sample is possible through measuring the efficiency of generation and detection. As the exact generation and detection methods rely on the magnetic state of the material, it is to be expected that a change in this state will lead to a significant difference in the signal amplitudes being generated. For this measurement the efficiency is shown in red (right axis). In the region of phase changes the generation efficiency peaks, and it is clear that the fan phase is optimal for generation of signals. This will be investigated in more depth in a later publication.

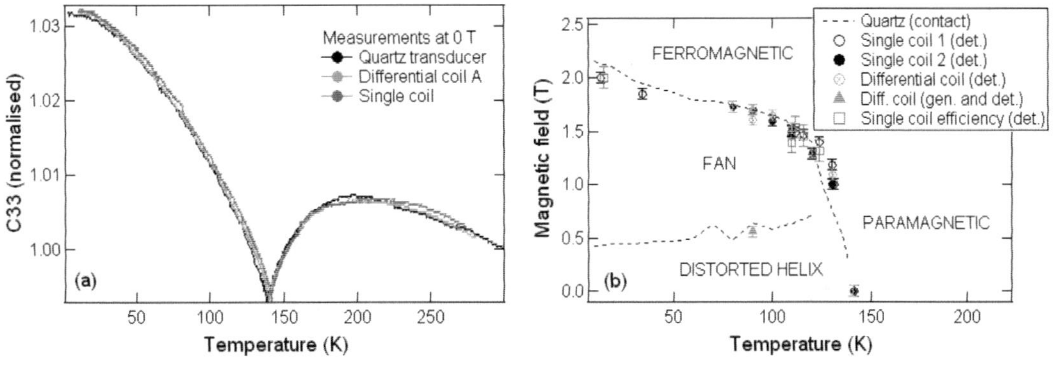

Figure 3. (a) C_{33} measurement using a quartz transducer to generate and either quartz (black), a single EMAT coil (red) or a differential coil (green) to detect ultrasound. (b) shows a phase diagram created using each set of data (after [14]).

Finally, we compare the results from several measurements using quartz transducers and/or EMATs with either a single coil for EMAT detection or a differential coil setup. Figure 3(a) shows normalised C_{33} values from each of these detection transducers for generation using quartz, measured for zero applied magnetic field and over a temperature range from room temperature down to 10 K. As can be seen, the results all show a similar trend, with a slight discrepancy in temperature possibly due to an observed degradation of the temperature sensor calibration.

Figure 3(b) shows the summary of results from each experiment on a phase diagram, with the quartz measurements used as a reference and shown as dashed lines. Points correspond to C_{33} measurements performed using different EMAT coils (for generation and detection) over a range of temperatures and magnetic fields. Red squares also show initial analysis of the EMAT efficiency. Again, very good agreement is shown between all methods.

4. Conclusions and future work

We have presented results from initial measurements using contact transducers, and using two different styles of EMAT coils for detection and generation of ultrasound. Results were compared for measurements of $Gd_{64}Sc_{36}$, and showed good agreement between the measurements. Results were analysed using a LabVIEW-based cross-correlation technique, which is a good step towards improving the problems caused by electrical noise in EMAT measurements.

Further improvement of the noise through the use of improved differential coils is anticipated. Also, other frequency-based techniques may be more appropriate for this echo measurement. Previous cross-correlation techniques have looked at the time difference between a reflection from a buffer rod and the first echo from a sample, and hence have fewer issues with phase changes on reflections. The fast Fourier transform of a string of echoes (as shown in Figure 1) gives a convolution of the frequency of the echo itself and the repeat pattern of the echoes [16], and is showing promise for analysing these measurements. We hope to extend these methods to further materials.

Acknowledgments

This work was funded by the ERC under grant 202735, NonContactUltrasonics. The authors would also like to thank the Nuffield Foundation for the Undergraduate Research Bursary (RP) and the University of Warwick for URSS project funding (DC). Finally, the authors would like to thank Bob Day and Dr Steve Dixon for their assistance.

References

[1] Edwards RS, Jian X, Fan Y and Dixon S 2005 *Appl. Phys. Lett.* 87 194104
[2] SB Palmer and S Dixon 2003 *Insight* **45** **(3)** 211-217
[3] Luthi B 2007 Physical Acoustics in the Solid State, Springer Press
[4] Pantea C, Rickel DG, Migliori A, Leisure RG, Zhang J, Zhao S, El-Khatib S and Li B 2005 *Review of Scientific Instruments* **76** 114902
[5] Papadakis EP 1967 *J. Acoust. Soc. Am.* /bf 42 (5) 1045-1051
[6] Truell R, Elbaum C and Chick BB 1969 Ultrasonic Methods in Solid State Physics, Academic Press
[7] Edwards RS, Dixon S, Fan Y and Jian X 2008 *Review of Progress in QNDE* **27A** 841-848
[8] Lim CM, Dixon S, Edwards C and Palmer SB 1998 *J. Phys. D: Appl. Phys.* **31** 1362-1367
[9] Dutton B, Gamboa-Perera F, Acosta M and Nahmad-Molinari Y 2006 *Review of Scientific Instruments* **77** 033903
[10] Melville RJ, Palmer SB, McIntyre GJ and Sousa JB 1988 *J. Appl. Phys.* **64** **(1)** 5889-5891
[11] Krstovska D, Galbova O and Sandev T 2008 *EPL* **81** 37006
[12] Details of LabVIEW available at http://www.ni.com
[13] Melville RJ, Bates S, McIntyre GJ, Sousa JB and Palmer SB 1988 *Europhys. Lett.* **6** 725
[14] Salgueiro da Silva M, Moreira JM, Pereira de Azevedo MM, Mendes JA, de Abreu CS, Sousa JB, Melville RJ and Palmer SB 1999 *J. Phys.: Cond. Matt.* **11** 71157124
[15] Shu L, Songling H and Wei Z 2007 *Sensors and Actuators A: Physical* **135** **(2)** 675-679
[16] Dixon S, Edwards C and Palmer SB 2001 *Ultrasonics* **39** 445-453

Journal of Physics: Conference Series **286** (2011) 012051 doi:10.1088/1742-6596/286/1/012051

Strictly localised triplet dimers on one- and two-dimensional lattices

S. Jackson and J.H.Samson

Department of Physics, Loughborough University, Loughborough, LE11 3TU, UK

E-mail: s.jackson3@lboro.ac.uk, j.h.samson@lboro.ac.uk

Abstract. Electrons may form inter-site pairs (dimers) by a number of mechanisms. For example, long-range (Fröhlich) electron-phonon interactions and strong on-site Hubbard U allow formation of small light bipolarons in some lattices. We identify circumstances under which triplet dimers are strictly localised by interference in certain one- and two-dimensional lattices. We assume a U-V Hamiltonian with nearest- and next-nearest-neighbour hopping integrals t and $t^{'}$, large positive U and attractive nearest- and next-nearest-neighbour interactions V and $V^{'}$. In the square ladder and some two-dimensional bilayers, if the dimer Hilbert space is restricted to nearest- and next-nearest-neighbour dimers, triplet dimers become strictly localised for certain values of these parameters. For example, in a square ladder with $t^{'} = t$ and $V^{'} = V$, all triplet bands become flat due to exact cancellation of hopping paths. We identify the localised eigenstates for all flat bands in each lattice. We show that many of the flat bands persist for arbitrary $t/t^{'}$ so long as other restrictions still apply.

Electrons may form inter-site pairs (dimers) by a number of mechanisms. For example, long-range (Fröhlich) electron-phonon interactions and strong on-site Hubbard U allow formation of small light bipolarons in some lattices. We identify circumstances under which triplet dimers are strictly localised by interference in certain one- and two-dimensional lattices. Our calculations are not specific to the pairing mechanism but our model can be obtained from the Coulomb-Fröhlich model [1, 2, 3, 4, 5, 6]. In this model electrons interact with lattice distortions on neighbouring sites. The electron together with the distortion forms a new quasiparticle – the polaron. An effective attraction between polarons can exist because two polarons can deform the lattice more effectively together than separately.

Applying the Lang-Firsov transformation [7] we obtain the UV model, valid for strong coupling [8]:

$$H = -\sum_{ij\sigma} t_{ij} c^{\dagger}_{i\sigma}\, c_{j\sigma} + U \sum_{i} n_{i\uparrow}\, n_{i\downarrow} + \frac{1}{2} \sum_{i} \sum_{j \neq i} \sum_{\sigma\sigma'} V_{ij} c^{\dagger}_{i\sigma} c_{i\sigma} c^{\dagger}_{j\sigma'} c_{j\sigma'} \qquad (1)$$

where $c^{\dagger}_{i\sigma}$ and $c_{j\sigma}$ are electron creation and annihilation operators respectively, t_{ij} is a renormalized hopping, U is the onsite Coulomb repulsion and V_{ij} is the polaron-polaron potential containing the Coulomb repulsion between electrons on neighbouring sites i and j and the (non-retarded) effective attraction due to the Fröhlich electron-phonon interaction (EPI). We take the on-site repulsion U to be infinite so that basis states with two electrons on a single site are suppressed.

The Hilbert space of two particles on on a d-dimensional tight-binding lattice can be represented as a particle on a $2d$-dimensional lattice, the tensor product of the two Hilbert spaces. However, if the attraction between the particles is strong enough to bind them into a dimer over the whole Brillouin zone, the low-lying states will have large amplitude only near a d-dimensional subspace where the particles are in close proximity. Accordingly a truncation of the Hilbert space of dimers to a small number of bond lengths will capture the essential physics. We call the truncated Hilbert space of the dimers the *dimer lattice*. Let

$$D_i = \{j : 0 < r_{ij} \leq L_{\max}\} \tag{2}$$

be the set of sites j whose distance r_{ij} from a site i is no more than $L_{\max}$. We shall call D_i the neighbours of i. A dimer will have (spin-independent) diagonal potential V_{ij}. We write V for the nearest-neighbour potential and V' for the next-nearest-neighbour potential.

If a lattice Λ has N sites and the mean number of neighbours $|D_i|$ is ν then the single-electron Hilbert space is $2N$-dimensional, the two-electron Hilbert space is $N(2N-1)$-dimensional and the dimer Hilbert space is $4\frac{\nu}{2}N$-dimensional. We can further reduce to one singlet and three triplet spaces, each of dimensionality $\nu N/2$. The dimer spaces for $S_z = 0$ are

$$\mathcal{S} = \text{span}\left\{\frac{1}{\sqrt{2}}\left(|i\uparrow j\downarrow\rangle + |j\uparrow i\downarrow\rangle\right) : i \in \Lambda, j \in D_i\right\} \tag{3}$$

$$\mathcal{T}_0 = \text{span}\left\{\frac{1}{\sqrt{2}}\left(|i\uparrow j\downarrow\rangle - |j\uparrow i\downarrow\rangle\right) : i \in \Lambda, j \in D_i\right\} \tag{4}$$

for singlets and triplets respectively. In each case basis states are double-counted in the span. We note that triplets are spatially antisymmetric and singlets are symmetric; if we represent the above dimer basis states as arrows pointing from i to j, then $|i \longrightarrow j\rangle = \pm|j \longrightarrow i\rangle$ with $+$ for singlets. The above formalism enables us to write the Hamiltonian of the dimers in each sector as a tight-binding Hamiltonian on a dimer lattice constructed by placing a node on the line joining each site i to each point $j \in D_i$. If $j \in D_i$ and $k \in D_i$, and $t_{jk} \neq 0$, then the dimer can hop from ij to ik. A dimer hopping vector is then drawn between the two nodes on the dimer lattice with hopping integral t_{jk}.

We consider a square ladder with nearest- and next-nearest-neighbour dimers, nearest and next-nearest-neighbour hopping t and t' respectively and nearest- and next-nearest-neighbour interactions V and V' respectively. Each sector of the truncated Hilbert space may be reduced to the five basis states indicated in figure 1(a), where n is the index of the unit cell. The dimer lattice comprises a chain of corner-sharing octahedra as indicated in figure 1(b). We find that some or all of the bands are flat depending on the ratios V'/V and t'/t. For $V' = V$ we obtain the triplet band structure:

$$E(k) - V = \pm 2t \tag{5}$$

$$E(k) - V = 0 \tag{6}$$

$$E(k) - V = \pm\sqrt{8t^2\cos^2\left(\frac{ka}{2}\right) + 8t'^2\sin^2\left(\frac{ka}{2}\right)} \tag{7}$$

For $V' = V$ and $t' = t$ we obtain the triplet band structure $E(k) - V = 0$, $E(k) - V = \pm 2t$, $E(k) - V = \pm 2\sqrt{2}t$. Since all bands are flat we infer that triplets are strictly localised for $V' = V$ and $t' = t$. This is because the points $|A, n\rangle$ in the dimer lattice (see figure 1(b)) are bottlenecks meaning that a dimer propagating along the ladder must repeatedly pass through these points. Any path from $|A, n\rangle$ to a neighbouring bottleneck $|A, n \pm 1\rangle$ can be replaced by a path of opposite sign (and equal magnitude if $t' = t$) by interchanging $|B, n\rangle$ with $|C, n\rangle$ and $|D, n\rangle$ with

$|E,n\rangle$. For example $|A,n\rangle \rightarrow |B,n\rangle \rightarrow |A,n+1\rangle$ is replaced with $|A,n\rangle \rightarrow |C,n\rangle \rightarrow -|A,n+1\rangle$. Since each path can be replaced by another path of opposite sign, and equal magnitude if $t^{'} = t$, the paths cancel each other for $t^{'} = t$ and triplet dimers become strictly localised. We have identified five linearly-independent localised eigenstates of the Hamiltonian.

We also consider square and honeycomb bilayers with restricted dimer-lengths and find that if some additional restrictions apply all triplet bands are flat for $V^{'} = V$ and $t^{'} = t$. As in the square ladder this is due to the existence of bottlenecks and the exact cancellation of paths connecting neighbouring bottlenecks. The band structure and localised eigenstates for these bilayers are essentially the same as for the square ladder.

We conclude that triplets are less mobile than singlets on a square ladder and that for $t^{'} = t$ and $V^{'} = V$, if dimer-lengths are restricted, triplets become strictly localised. This is also true for square and honeycomb bilayers.

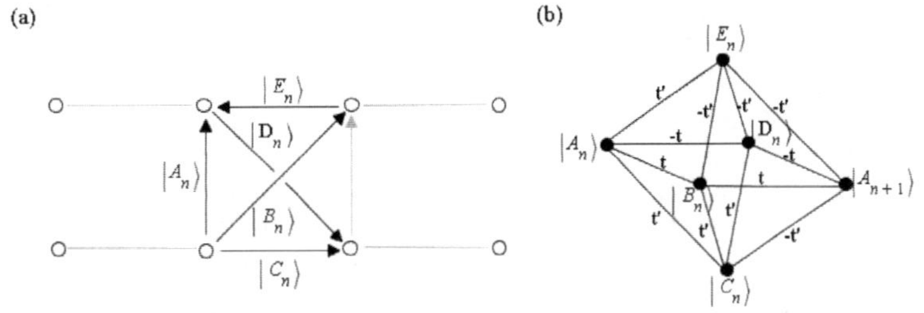

Figure 1. (a) Triplet basis states in the square ladder for nearest- and next-nearest-neighbour dimers. (b) Unit cell of the triplet dimer lattice. Adjacent cells are connected by the points $|A,n\rangle$ which form bottlenecks.

[1] A. S. Alexandrov 1996 *Phys. Rev. B* **53** 2863
[2] A. S. Alexandrov and P. E. Kornilovitch 2002 *J.Phys: Condensed matter* **14** 5337
[3] A. S. Alexandrov and P. E. Kornilovitch 1999 *Phys. Rev. Lett.* **82**, 807
[4] P. E. Spencer, J. H. Samson, P. E. Kornilovitch and A. S. Alexandrov 2005 *Phys. Rev. B* **71** 184310
[5] J. P. Hague, P. E. Kornilovitch, J. H. Samson and A. S. Alexandrov 2007 *J.Phys: Condensed matter* **19** 255214
[6] J. P. Hague, P. E. Kornilovitch, J. H. Samson, and A. S. Alexandrov 2007 *Physical Review Letters* **98** 037002
[7] I. G. Lang and Y. A. Firsov 1962 *Zh. Eksp. Teor. Fiz.* **43** 1843
I. G. Lang and Y. A. Firsov 1963 *Sov. Phys.-JETP* **16** 1301 (Engl. Transl.)
[8] A. S. Alexandrov and N. F. Mott 1994 *Rep. Prog. Phys.* **57** 1197

Study of magnetic anisotropy in Co-doped Mn_2Sb

Pallavi Kushwaha[1], **A. Thamizhavel**[2] **and R. Rawat**[1]

[1] UGC-DAE Consortium for Scientific Research, University Campus, Khandwa Road, Indore-452001, M.P, India

[2] Department of Condensed Matter Physics and Materials Science, Tata Institute for Fundamental Research, Colaba, Mumbai, India

E-mail: pallavi.bebu@gmail.com

Abstract.
We have grown single crystals of $Mn_{2-x}Co_xSb$ with x =0.20 and 0.15 . Magnetization studies show a first order ferrimagnetic to antiferromagnetic transition with lowering temperature at $T_N \approx 160\ K$ and ≈ 117 K for x = 0.20 and 0.15, respectively. T_N is found to be independent of applied magnetic field (1000 Oe) direction. Besides T_N, signature of a spin reorientation transition (T_{SR}) has also been observed near room temperature for x=0.15. Below spin reorientation transition, magnetization behavior is highly anisotropic and easy direction of magnetization are different for x= 0.20 and 0.15.

1. Introduction

The compound Mn_2Sb crystallizes in Cu_2Sb type tetragonal structure and orders ferrimagnetically (FRI) below 550 K [1]. Doping at either of Mn or Sb site results in first order ferrimagnetic (FRI) to antiferromagnetic (AFM) transition with lowering temperature (T_N) [2]. The magnetic structure of Mn_2Sb consist of triple layer like Mn_1-Mn_2-Mn_1 arrangement where Mn_1 (2.1 μ_B) and Mn_2 (3.9 μ_B) are two different crystallographic position with antiparallel spin alignment. In the ferrimagnetic state net moment of adjacent triple layers are parallel whereas in AFM state they are anti-parallel [4- 9]. With the application of magnetic field as well as pressure, T_N of doped Mn_2Sb can be shifted over a wide temperature range. Below T_N, application of magnetic field can induce the first order AFM-FRI transition and shows interesting magnetic behavior [7, 10]. In our studies on polycrystalline $Mn_{1.85}Co_{0.15}Sb$ we had addressed some of these interesting feature at low temperature and high magnetic field [11, 12]. In these studies irreversibilities arising due to slow dynamics of the transition (kinetic arrest) are distinguished from that arising due to supercooling/superheating across a first order transition. The critically slow dynamics of the phase transition at low temperature results in glass like high temperature FRI phase coexisting with equilibrium AFM phase. In continuation to these studies we have prepared single crystals of $Mn_{2-x}Co_xSb$ with x =0.15 and 0.20. Here we report our magnetization studies on these single crystals.

2. Experimental Details

The high purity constituent element with starting composition $Mn_{1.80}Co_{0.20}Sb$ and $Mn_{1.85}Co_{0.15}Sb$ were melted under high vacuum in a electrical furnace with a temperature gradient. The heat treatment was divided into few steps to reduce the evaporation of the

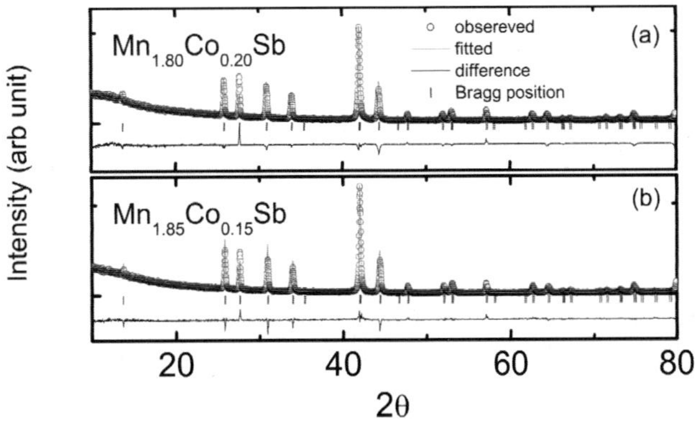

Figure 1. Rietveld refinement of powder x-ray diffraction pattern for $Mn_{2-x}Co_xSb$ (a) $x = 0.20$ and (b) $x = 0.15$.

elements and to obtain good homogeneity. Laue and powder x-ray diffraction techniques were used to orient and characterize the samples. Low temperature and high magnetic field magnetization measurements were performed using Superconducting Quantum Interference Device (SQUID).

3. Result and discussion

Figure 1 shows room temperature powder x-ray diffraction which has been taken after grinding a small piece from the single crystalline ingot. Rietveld refinement of xrd pattern shows that crystal are single phase in nature and the lattice parameters are in close agreement with polycrystalline sample reported earlier [10].

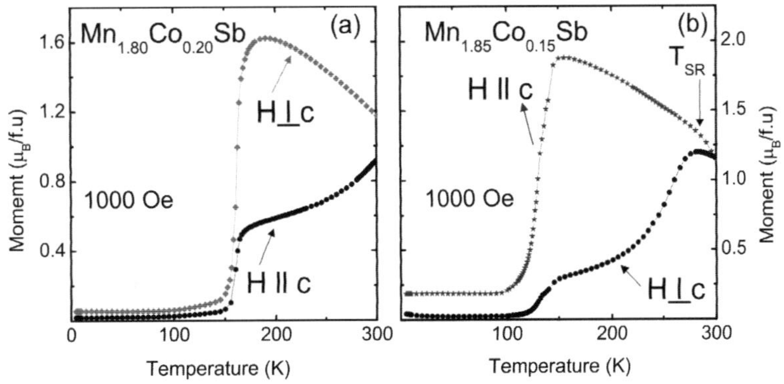

Figure 2. Magnetization variation with temperature for $Mn_{2-x}Co_xSb$ for (a) $x = 0.20$ and (b) 0.15 in the presence of 1000 Oe magnetic field.

Figure 2 (a) and (b) shows the temperature variation of magnetization for two different crystallographic direction in an applied magnetic field of 1000 Oe for $x = 0.20$ and 0.15 respectively. In the case of $x = 0.20$ magnetization increases up to 190 K with decreasing temperature for magnetic field perpendicular to c-axis. With further decrease in temperature, a FRI to AFM transition appears as a sharp decrease in magnetization around 160 K. This type of magnetization behavior is in agreement with the earlier studies [7]. However when magnetic field is applied parallel to c-axis magnetization decreases with decreasing temperature in the entire temperature range along with FRI to AFM transition. This FRI to AFM transition temperature is independent of the applied magnetic field (1000 Oe) direction. In case of sample with $x = 0.15$, magnetization behavior is similar to $x = 0.20$ sample but for exchange of crystallographic direction with respect to applied magnetic field. The FRI to AFM transition (T_N) appears at $\approx 117\ K$ for $x = 0.15$. However magnetic moment of 0.2 μ_B at 5 K is noticed when magnetic field is applied parallel to c axis. From earlier neutron studies, it was found that with decreasing temperature at spin reorientation transition (T_{SR}) spin direction flip from c-axis to the basal plane [1]. Spin reorientation is clearly seen near room temperature for $x = 0.15$ as a slope change in magnetization curve. Above spin reorientation transition magnetization curve matches for both the direction. In case of sample with $x = 0.20$, there was no signature of T_{SR} up to 300 K, however extrapolation of these curve suggest T_{SR} will be around 318 K.

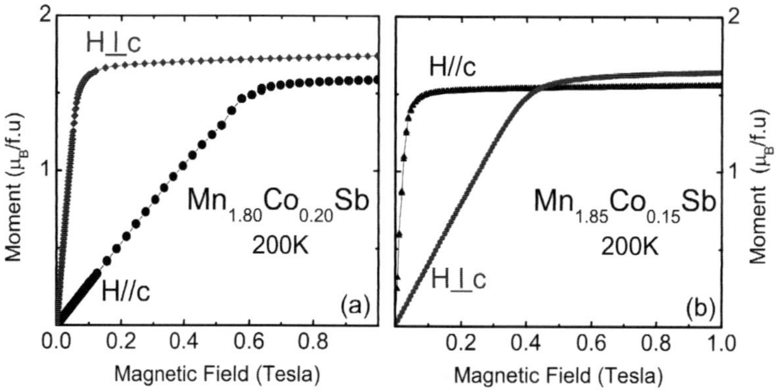

Figure 3. Isothermal Magnetization at 200 K (i.e below spin reorientation transition) for $Mn_{2-x}Co_xSb$ (a) $x = 0.20$ and (b) 0.15 after zero field cooling from room temperature.

Figure 3 shows the isothermal magnetization at 200 K (i.e below spin reorientation transition but in FRI state for both the samples) after zero field cooling from 300 K. The easy axis of magnetization has been found to be perpendicular to c-axis for $x = 0.20$. It is consistent with earlier reports [7]. However in case of $x = 0.15$ easy axis is parallel to c-axis which is a new observation. The saturation magnetic field along easy axis is almost same for both the sample however along hard axis it is 0.70 Tesla and 0.50 Tesla for $x = 0.20$ and 0.15, respectively.

4. Conclusion

We have grown single crystals of $Mn_{2-x}Co_xSb$ with $x = 0.20$ and 0.15 and characterized using magnetization measurement. Temperature variation of magnetization shows a first order FRI to AFM transition at $\approx 160\ K$ and $\approx 117\ K$ for $x = 0.20$ and 0.15, respectively. For $x = 0.15$ spin reorientation transition appears near room temperature. Below spin reorientation transition these systems are highly anisotropic in whole temperature range. The easy axis of magnetization is found to be in the basal plane for $x = 0.20$ and along c-axis for $x = 0.15$.

5. Acknowledgments

We thank Mrs. R. Kulkarni, Mr. N. Goyal and Mr. P. Das for their help during crystal growth and orientation. Mr. Ganesh for magnetization measurements. Dr. P. Chaddah and Prof A. Gupta for encouragement and support. Pallavi Kushwaha acknowledges CSIR, India for financial support.

References

[1] Wilkinson M K, Gingrich N S and Shull C G 1957 *J. Phys.Chem. Solids* B **2** 289
[2] Kanomata T and Ido H 1984 *J. Appl. Phys.* **55** 2039
[3] Swoboda T J, Cloud W H, Bither T A, Sadler M S and Jarrett H S 1960 *Phys. Rev. Lett.* **4** 509
[4] Bither T A, Walter P H L, Cloud W H, Swoboda T J and Bierstedt P E 1962 *J. Appl. Phys.* **33** (Suppl.) 134
[5] Darnell F J, Cloud W H and Jarrett H S 1963 *Phys. Rev.* **130** 647
[6] Wijngaard J H, Haas C and de Groot R A 1992 *Phys. Rev. B* **45** 5395
[7] Bartashevich M I, Goto T, Tomita T, Baranov N V, Zemlyanski S V, Hilscher G and Michor H 2002 *Physica B* **318** 198
[8] Zhang Y Q and Zhang Z D 2003 *Phys. Rev. B* **67** 132405
[9] Austin A E, Adelson E and Cloud W H 1963 *Phys. Rev.* **131** 1511
[10] Bartashevich M I, Goto T, Baranov N V and Gaviko V S 2004 *Physica B* **351** 71
[11] Kushwaha Pallavi, Rawat R and Chaddah P 2008 *J. Phys.: Condens. Matter* **20** 022204
[12] Kushwaha Pallavi, Lakhani Archana, Rawat R,Banerjee A and Chaddah P 2009 *Phys. Rev. B* **79** 132402

Fractional scaling of quantum walks on percolation lattices

Viv Kendon, Godfrey Leung[1], Joe Bailey[2] and Paul Knott

School of Physics and Astronomy, University of Leeds, Leeds, LS2 9JT, UK

E-mail: `V.Kendon@leeds.ac.uk`

Abstract. Quantum walks can be used to model processes such as transport in spin chains and bio-molecules. The enhanced spreading and mixing properties of quantum walks compared with their classical counterparts have been well-studied on regular structures and also shown to be sensitive to defects and imperfections. Using numerical simulation, we study the spreading properties of quantum walks on percolation lattices for both bond and site percolation. The randomly missing edges or sites provide a controlled amount of disorder in the regular Cartesian lattice. In one dimension (the line) we introduce a simple model of quantum tunneling to allow the walk to proceed past the missing edges or sites. This allows the quantum walk to spread faster than a classical random walk for short times, but at longer times the disorder localises the quantum walk. In two dimensions, we observe fractional scaling of the spreading with the number of steps of the walk. For percolation above the 85% level, we obtain faster spreading than classical random walks on the full lattice.

1. Introduction

Quantum versions of random walks have been used as simple models of quantum transport phenomena, applicable to systems as diverse as spin chains [1] and bio-molecules [2]. Here we investigate the properties of quantum walks on percolation lattices, disordered structures appropriate for modelling biological and experimentally realistic systems.

Quantum walks have continuous-time [3] and discrete-time (coined) versions [4, 5], in this work, we use only the coined quantum walk. We consider the effects of missing edges or sites for both the line and the 2D Cartesian lattice. In one dimension, a missing edge or site halts the progress of both quantum and classical random walks. Variants that allow progress to be made have been studied previously, e.g., dynamic broken links that change location at each time step [6]. In this work we introduce a simple model of quantum tunnelling [7] that also allows the quantum walker to make progress beyond the gaps. On 2D percolation lattices, there is in general more than one possible path between two sites, allowing the quantum walk to proceed around the missing edges without any tunnelling.

[1] Current address: School of Physics and Astronomy, University of Nottingham, University Park, Nottingham, NG7 2RD, United Kingdom. Contact: `ppxgl@nottingham.ac.uk`

[2] Current address: Centre for Mathematics and Physics in the Life Sciences and Experimental Biology (CoMPLEX), University College London, Gower Street London WC1E 6BT, United Kingdom. Contact: `joe.bailey.09@ucl.ac.uk`

2. Quantum walks

Quantum walks are quantum versions of classical random walks and are defined analogously. For a coined quantum walk, the quantum walker carries a quantum coin of dimension d, the maximum number of edges at any site. For the walk on the line, the coin has dimension $d = 2$ and the most common choice of coin operator to "toss" the coin is the Hadamard operator,

$$\mathbf{C}_2^{(\mathrm{Had})} = \frac{1}{\sqrt{2}} \begin{pmatrix} 1 & 1 \\ 1 & -1 \end{pmatrix}. \tag{1}$$

We write the basis states of the quantum walker as $|c, x\rangle$, where c is the state of the coin (one of the d directions, for the line $c = \pm 1$) and x is the position on the line. After "tossing" the coin with the coin operator, the particle moves to adjacent positions according the the coin state; this is expressed mathematically as a conditional shift operator,

$$\begin{aligned} \mathbf{S}|-1, x\rangle &= |-1, x-1\rangle \\ \mathbf{S}|+1, x\rangle &= |+1, x+1\rangle. \end{aligned} \tag{2}$$

One step of the quantum walk is produced by the unitary operator $\mathbf{U} = \mathbf{S}(\mathbf{C} \otimes \mathbb{1}_x)$. A quantum walk of t steps starting from an initial state $|\psi_0\rangle$ is thus

$$|\psi(x, t)\rangle = U^t |\psi_0\rangle. \tag{3}$$

Each vertex in a two-dimensional lattice has four edges connected to it, so the quantum walker needs a four-dimensional coin. We label the four directions $|L\rangle$, $|R\rangle$ corresponding to $|-1\rangle$ and $|+1\rangle$ on the x axis and $|D\rangle$, $|U\rangle$ for $|-1\rangle$ and $|+1\rangle$ on the y axis. The operator for "tossing" the coin can be any 4×4 unitary. The most common choice is based on Grover's diffusion operator,

$$\mathbf{C}_4 = \frac{1}{2} \begin{pmatrix} -1 & 1 & 1 & 1 \\ 1 & -1 & 1 & 1 \\ 1 & 1 & -1 & 1 \\ 1 & 1 & 1 & -1 \end{pmatrix}. \tag{4}$$

To complement this choice of coin operator, the shift operator $\mathbf{S}_4$ must also invert the coin states,

$$\begin{aligned} \mathbf{S}_4|L, x, y\rangle &= |R, x-1, y\rangle \\ \mathbf{S}_4|R, x, y\rangle &= |L, x+1, y\rangle \\ \mathbf{S}_4|D, x, y\rangle &= |U, x, y-1\rangle \\ \mathbf{S}_4|U, x, y\rangle &= |D, x, y+1\rangle. \end{aligned} \tag{5}$$

This ensures that the coin operator is then applied to a coin state that labels the edge from which the walker arrived, which is the one that receives the minus sign from the Grover coin operator. This preserves the symmetry of the walk.

The quantum walk on both the line and on the full Cartesian lattice spread at a rate proportional to t, a quadratic speed up over the classical $\sqrt{t}$. This can be quantified by calculating the root mean square distance of the walker from the starting point. On the line this is just $x_{rms}(t) = \sqrt{(\sum_{c,x} (x - x_0)^2 |\psi(x,t)|^2)}$ for a walk starting at x_0. On the 2D lattice,

$$r_{rms}(t) = \sqrt{(\sum_{c,x} \left[(x - x_0)^2 + (y - y_0)^2\right] |\psi(x,t)|^2)}. \tag{6}$$

3. Percolation lattices

Both bond (edge) and site percolation have similar definitions: with independent randomly chosen probability p the bond or site is present in the lattice. In two and higher dimensions, percolation lattices exhibit a phase transition from a set of small disconnected regions to a more highly connected structure ("one giant cluster"). On 2D Cartesian lattices, the critical probability $p_c = 0.5$ (bond) and $p_c = 0.5927...$ (site[8, 9]).

A typical percolation lattice will have sites with four, three, two, one and no edges. A suitable coin operator is needed for each, we used the Grover coin appropriate for the number of edges, padded with (minus) the identity for the missing edges (details in [7]). In one dimension (the line) any missing edge will stop the progress of the walker, effectively $p_c = 1$.

4. Numerical method

Numerical simulation of a quantum walk is a straightforward evaluation of equation (3) for chosen parameters. In high level computational environments such as Matlab, it takes only a few lines of code. However, we wrote our own routines in C and C++ to ensure they were efficient enough to to run up to 140 time steps for the walk on a two-dimensional lattice (the memory required scales linearly in the number of lattice sites). It was also necessary to repeat the simulations over a sufficient number of random percolation lattices to ensure reasonable statistics were obtained. The largest of our simulations took over a week on a 3GHz processor to complete this averaging for 5,000 random lattices. This was in fact largely due to the time required to generate the random numbers for specifying the percolation lattice (one per walk). We used routines from Park and Miller [10] freely available online for generating the random numbers, although the quality of the random numbers is not particularly important for this type of simulation.

5. Tunnelling on a broken line

Tunnelling is a generic property of quantum systems whereby a quantum particle has a small probability of passing through a barrier that prevents the passage of a classical particle. We model this using a biased Hadamard operator,

$$\mathbf{C}_2^{(\eta)} = \begin{pmatrix} \sqrt{\eta} & \sqrt{1-\eta} \\ \sqrt{1-\eta} & -\sqrt{\eta} \end{pmatrix}. \tag{7}$$

where η gives the probability of passing to the next site while $1-\eta$ is the probability of being reflected back. For a line with a random subset of edges missing, the quantum walk proceeds by using the standard Hadamard where both edges are present at a site, while the biased Hadamard is used at sites with one or both edges missing.

After 100 steps, the effect of the missing edges with tunnelling is not significantly different from the walk on the full line. However, by 1000 steps the spreading has significantly slowed, and after 10,000 steps the spreading rate has dipped below the classical $\sqrt{t}$ scaling. This is shown in figure 1. The random missing edges and consequent random use of different coin operators causes the walk to localise. Note that $p = 0$ (no edges) returns to quantum spreading, because there is no randomness to disrupt the quantum coherences if the tunnelling coin is used everywhere. Regular patterns of varying coin bias have been studied by Linden and Sharam [11], who find quantum spreading in most cases, but with widely varying spreading rates. Generalisation of this model by Shikano and Katsura [12] suggests that localization vs spreading depends on whether the variation in the coin is commensurate with the lattice spacing. Since our randomly missing edges do not have any periodicity, we expect to find localisation in all cases, with the range explored by the quantum walk governed by both the tunnelling probability η and the percolation probability p. Comparison of figures 1 and 2 show these effects clearly. Reducing η

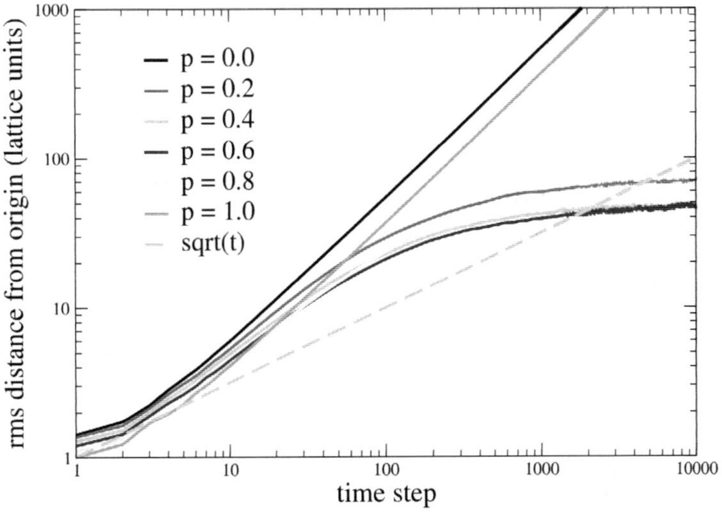

Figure 1. Spreading of a tunnelling quantum walk on a line with randomly missing edges at a density of p as given in key and tunnelling rate $\eta = 0.25$. Log-log plot with $\sqrt{t}$ shown for comparison.

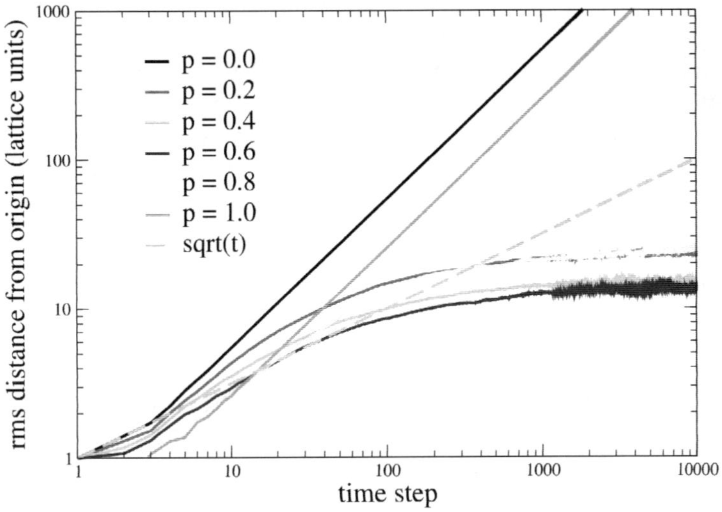

Figure 2. Spreading of a tunnelling quantum walk on a line with randomly missing edges at a density of p as given in key and tunnelling rate $\eta = 0.125$. Log-log plot with $\sqrt{t}$ shown for comparison.

by a factor of two (figure 2) reduces the range by roughly a factor of three, from around 70 to 23 for $p = 0.2$ or 0.8.

6. Spreading on 2D percolation lattices

On a 2D lattice there is no need for tunnelling, so we studied the basic quantum walk only. Below p_c, the quantum walk will not be able to spread as there are only small connected structures. Approaching p_c from above, the spreading slows down completely, as the number of long-distance connected paths reduces to zero. For $p = 1$, the lattice is fully connected, and the standard quantum walk spreading applies (linear in t). In between, we find the quantum walks show fractional scaling of the spreading, i.e., proportional to t^α for $(0.5 < \alpha < 1)$. The numerically determined scaling exponents α are shown in figure 3.

Our (numerical) results are skewed by finite size effects: the increase in α from zero begins slightly before $p = p_c$. The rise tails off at $\alpha = 0.5$, then rises again steeply to $\alpha = 1$ for the full lattice ($p = 1$). At this stage, we do not have enough data to predict the large t behaviour, but one possibility is that the steep rise to the quantum value of $\alpha = 1$ at $p = 1$ will becomes a "step" function at $p = 1$ as $t \to \infty$. The randomness in the percolation lattice would thus act as

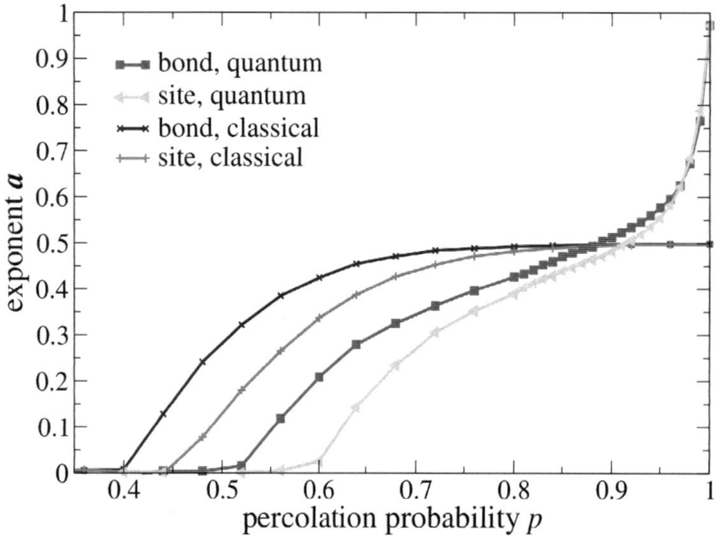

Figure 3. Fractional scaling exponent for 2D percolation lattices derived from data for $t = 80$ to $t = 140$ steps. Classical random walk exponents for the same processing and time steps shown.

decoherence in the large t limit, reducing the behaviour to that of a classical random walk. The classical behaviour is shown for numerical simulation using the same parameters. This data has much larger finite size effects. This is due to the classical spread of $\sqrt{t}$, which is only around 12 lattice units for 140 time steps. A classical random walk with equivalent spread (around 30 lattice units would require 900 steps.

7. Summary

We have found localisation on a line for quantum walks tunnelling through missing edges. However, the short time behaviour is still faster than classical spreading. In two dimensions we have found faster than classical scaling with fractional exponents. Although we conjecture this will disappear in the long time limit, again this would only be approached for quite large values of t, and from the point of view of models for disordered systems on smaller scales (tens to hundreds of sites), the faster-than-classical fractional scaling is very much the dominant feature.

Acknowledgments

We thank Jiajun Tan for assistance in preparing the references for this work. VK is funded by a Royal Society University Research Fellowship. PK was funded by a Nuffield Undergraduate Summer Bursary. JB was funded by a Royal Society Summer Bursary. GLMC was funded by the University of Leeds School of Physics and Astronomy Summer Bursary scheme.

References

[1] S Bose 2007 Quantum communication through spin chain dynamics: an introductory overview. *Contemp. Phys.*, **48**(13–30) p 13 DOI: 10.1103/PhysRevLett.91.207901
[2] Mohseni, P Rebentrost, S Lloyd, and A Aspuru-Guzik 2008 Environment-assisted quantum walks in photosynthetic energy transfer. *J. Chem. Phys.*, **129** 174106 DOI: 10.1063/1.3002335
[3] E Farhi and S Gutmann 1998 Quantum computation and decison trees. *Phys. Rev. A*, **58** pp 915–928 DOI: 10.1103/PhysRevA.58.915 (ArXiv:quant-ph/9706062)
[4] D Aharonov, A Ambainis, J Kempe, and U Vazirani 2001 Quantum walks on graphs. In *Proc. 33rd Annual ACM STOC* (ACM, NY) pp 50–59 DOI: 10.1145/380752.380758 (ArXiv:quant-ph/0012090)
[5] Andris Ambainis, Eric Bach, Ashwin Nayak, Ashvin Vishwanath, and John Watrous 2001 One-dimensional quantum walks. In *Proc. 33rd Annual ACM STOC* (ACM, NY) pp 60–69 DOI: 10.1145/380752.380757
[6] A Romanelli, R Siri, G Abal, A Auyuanet, and R Donangelo 2003 Decoherence in the quantum walk on the line. *Physica A*, **347** 137–152 DOI: 10.1016/j.physa.2004.08.070 (ArXiv:quant-ph/0403192)

[7] G. M. C. Leung, P. Knott, J. Bailey, and V. M. Kendon 2010 Quantum walks on percolation lattices. DOI: 10.1088/1367-2630/5/1/383 *New J. Phys.* Dec (forthcoming) (ArXiv:1006.1283)

[8] H Eugene Stanley Zorica V Djordjevic and Alla Margolina 1982 Site percolation threshold for honeycomb and square lattices. *J. Phys. A: Math. Gen.*, **15** pp L405–L412 DOI: 10.1088/0305-4470/15/8/006

[9] T. Gebele 1984 Site percolation threshold for square lattice. *J. Phys. A: Math. Gen.*, **17** pp L51–L54 DOI: 10.1088/0305-4470/17/2/005

[10] Steve Park and Keith Miller 1998 Random number generators: Good ones are hard to find. *Communications of the ACM*, **31** pp 1192–1201 DOI: 10.1145/63039.63042

[11] Noah Linden and James Sharam 2009 Inhomogeneous quantum walks. *Phys. Rev. A* **80**(5) 052327 DOI: 10.1103/PhysRevA.80.052327 (ArXiv:0906.3692v1)

[12] Yutaka Shikano and Hosho Katsura 2010 Localization and fractality in inhomogeneous quantum walks with self-duality. (ArXiv:1004.5394v1)

Journal of Physics: Conference Series **286** (2011) 012054 doi:10.1088/1742-6596/286/1/012054

Magnetoplasmons and SU(4) symmetry in graphene

Andrea M Fischer[1], Rudolf A Römer[1] and Alexander B Dzyubenko[2,3]

[1] Department of Physics and Centre for Scientific Computing, University of Warwick, Coventry, CV4 7AL, UK

[2] Department of Physics, California State University - Bakersfield, Bakersfield, California 93311, USA

[3] General Physics Institute, Russian Academy of Sciences, Moscow 119991, Russia

E-mail: `A.M.Fischer@warwick.ac.uk`

Abstract. We study magnetoplasmons or neutral collective excitations of graphene in a strong perpendicular magnetic field, which can be modelled as bound electron-hole pairs. The SU(4) symmetry of graphene arising from spin and valley pseudospin degrees of freedom is explored using Young diagrams to correctly predict the degeneracies of these excitations. The multiplet structure of the states is identical to that of mesons composed of first and second generation quarks.

1. Introduction

Graphene is a single layer of graphite or set of carbon atoms sp^2 bonded together to form a honeycomb lattice. The electron in the remaining p_z orbital is delocalised and responsible for graphene being an excellent electrical conductor [1]. Graphene is associated with numerous other superlatives such as strongest material [2] and of course thinnest material; it is the first example of an atomically thick membrane. Such properties make it an excellent candidate for industrial applications, although in many cases there are obstacles to be overcome before this can be put into practice [3]. Aside from its potential usefulness, graphene is scientifically interesting in its own right. Unlike other materials, including the 2D electron gas (2DEG), electrons in graphene have a linear dispersion relation close to the Fermi energy and in this sense behave quasirelativistically [4]. Effects such as Klein tunnelling, which were previously the dominion of high-energy particle physics, have now been observed in graphene [5]. For an undoped system, the Fermi energy is located at zero energy. The Fermi surface consists of six points which coincide with the six corners of the hexagonal Brillouin zone. Only two of these points are inequivalent and are commonly termed the $\mathbf{K}$ and $\mathbf{K}'$ valleys. They may also be referred to as pseudospins, since they result in the single particle wavefunctions having a spinor form [1].

In these proceedings we examine a monolayer of graphene in the presence of a strong perpendicular magnetic field of magnitude B. The Landau level (LL) energies for graphene are, $\epsilon_n = \mathrm{sign}(n)\hbar v_{\mathrm{F}}\sqrt{2|n|}/\ell_B$, where $v_{\mathrm{F}} \sim 10^6\mathrm{ms}^{-1}$ is the Fermi velocity, $\ell_B = \sqrt{\hbar c/eB}$ is the magnetic length and $\mathrm{sign}(0) = 0$. There is also the usual LL degeneracy resulting from the different possible guiding centres for electronic cyclotron orbits. This is described by the oscillator quantum number, m, in the symmetric gauge. We consider magnetoplasmons (MPs), which are essentially excitons created from a ground state described by the filling factor $\nu = 2$, where all states with LL indices $n \leq 0$ are occupied and all other states are empty. An exciton is formed when one of the electrons in this ground state is excited to a state in a higher lying LL and becomes bound to the hole it leaves behind. We focus on excitons with the lowest possible excitation energies, namely those with the hole in the $n = 0$ LL and the electron in the $n = 1$ LL. Taking into account the spin/pseudospin states of the electron and hole, this yields 16 resonant excitations. These are collective excitations on account of the LL degeneracy and also due to the fact that excitons with different spin/pseudospin characters may be mixed by the two-body Coulomb interaction. In the following, we study the SU(4) symmetry in graphene, which arises from the two spin and two pseudospin degrees of freedom, denoted by $\uparrow, \downarrow$ and $\Uparrow, \Downarrow$ respectively. This leads to a direct analogy between MPs in graphene and mesons composed of either first or second generation quarks. We then review the dispersion relation for MPs in pristine graphene and finally examine the states which become localised in the presence of a single Coulomb impurity. In both cases we observe the degeneracies predicted by the SU(4) symmetry and Young diagram techniques.

2. SU(4) symmetry in graphene

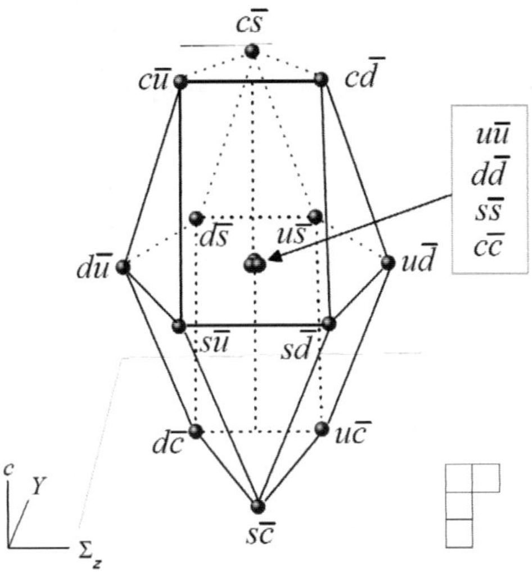

Figure 1. SU(4) [15]-plet describing the excitonic states in terms of flavour isospin (Σ_z), hypercharge (Y) and charm (c).

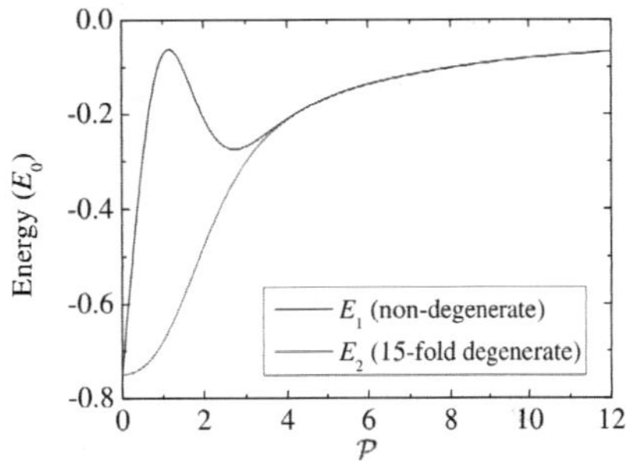

Figure 2. Dispersion relations for excitons with the hole in $n = 0$ LL and electron in $n = 1$ LL. $\mathcal{P}$ is the quasimomentum in units of $\hbar/\ell_B$.

An electron (or hole) in graphene in a strong perpendicular magnetic field is described by four quantum numbers: the LL index n; the oscillator quantum number m; the spin projection s_z and the pseudospin projection τ_z. We denote the fermionic creation operator for an electron by $c^\dagger_{nm\tau_z s_z}$ and that for a hole by $d^\dagger_{nm\tau_z s_z} \equiv c_{nm\tau_z s_z}$. The first two orbital quantum numbers are associated with the cyclotron orbits. The last two, when considered together, give four possible spin/pseudospin "flavours", leading to the SU(4) symmetry. We identify these with the four quark flavours up (u), down (d), charm (c) and strange (s) according to $\{\Downarrow\downarrow, \Uparrow\downarrow, \Downarrow\uparrow, \Uparrow\uparrow\} \equiv \{d, u, s, c\}$. The generators of SU(4) can be constructed using bilinear combinations of the fermionic creation and annihilation operators conserving the number of electrons and holes. We have

$$\hat{C}_{ij} = \sum_{nm} c^\dagger_{nmi} c_{nmj} - \sum_{nm} d^\dagger_{nmj} d_{nmi}, \tag{1}$$

where $i, j \in \{d, u, s, c\}$. The generators satisfy the commutation relations

$$[\hat{C}_{ij}, \hat{C}_{kl}] = \delta_{jk}\hat{C}_{il} - \delta_{il}\hat{C}_{kj}, \tag{2}$$

so are closed under commutation, confirming that they form a Lie algebra. The operators associated with spin and pseudospin may of course be expressed in terms of the generators. For example, $S_z = \frac{\hbar}{2}\left(\hat{C}_{ss} + \hat{C}_{cc} - \hat{C}_{uu} - \hat{C}_{dd}\right)$ and $S_+ = S_x + iS_y = \hbar\left(\hat{C}_{cu} + \hat{C}_{sd}\right)$.

In particle physics, mesons, which are a bound quark-antiquark pair, are classified using Young diagrams [6]. We use the same techniques to determine the multiplet structure for electron-hole (e-h) complexes in graphene. Young diagrams describe SU(n) multiplets and consist of n or fewer left-justified rows of boxes, such that each row is as least as long as the row beneath it. Roughly speaking, each box represents a particle. Each SU(n) multiplet is uniquely labelled by a list of $n-1$ non-negative integers, $(a_1, a_2, \ldots, a_{n-1})$. The integer in the ith position corresponds to the number of boxes in the ith row minus the number of boxes in the $(i+1)^{\text{th}}$ row i.e. the overhang of the ith row. By the definition of a multiplet label, adding columns of length n to the left of a Young diagram is redundant. Young diagrams can be used to calculate the multiplicity, $[N(n)]$, of an SU(n) multiplet. For the $n = 4$ case of current interest,

$$[N(4)] = \frac{(a_1+1)}{1}\frac{(a_2+1)}{1}\frac{(a_3+1)}{1}\frac{(a_1+a_2+2)}{2}\frac{(a_2+a_3+2)}{2}\frac{(a_1+a_2+a_3+3)}{3}. \tag{3}$$

There are two fundamental representations of SU(n): $[n]$, with Young diagram $\square$ and label $(1, 0, \ldots 0)$ and $[\bar{n}]$ with Young diagram composed of $n-1$ vertically stacked boxes and label $(0 \ldots, 0, 1)$. The latter represents an antiparticle; this makes sense if we identify n vertically stacked boxes with the vacuum. We shall also use the symbol, $\underline{1}$ to denote the vacuum. For the current purpose, the hole in a MP behaves as an antiparticle. Thus the electron has Young diagram $\square$ and the hole Young diagram . To determine the SU(4) multiplet structure for the MPs, we must combine these two diagrams. The rules for combining Young diagrams [6] yield

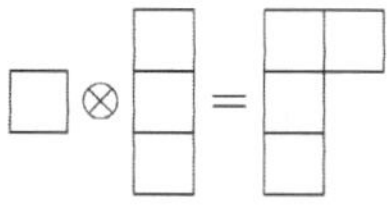
$\oplus \underline{1}$. (4)

Eq. (4) is equivalent to $(1,0,0) \otimes (0,0,1) = (1,0,1) \oplus (0,0,0)$ or $[4] \otimes [\bar{4}] = [15] \oplus [1]$. Now the generators, $\hat{C}_{ij}$, commute with the interacting Hamiltonian describing the electron system up to some small symmetry breaking terms [7]. In addition, the application of a generator to a state in a particular multiplet always returns a state in that same multiplet with the same Coulomb interaction energy. This ensures that states within a given multiplet are degenerate in their energies. Thus from this simple calculation, we have shown that the MPs in graphene will either have degeneracy 1 or 15. The [15]-plet is shown in Fig. 1 and is analogous to the meson multiplet of particle physics. The states are plotted as a function of: the flavour isospin projection $\hat{\Sigma}_z = \frac{1}{2}\left(\hat{C}_{uu} - \hat{C}_{dd}\right)$, the hypercharge $\hat{Y} = \frac{1}{3}\left(\hat{C}_{uu} + \hat{C}_{dd} + \hat{C}_{cc}\right) - \frac{2}{3}\hat{C}_{ss}$ and the charm $\hat{c} = \hat{C}_{cc}$. One may deduce the symmetry of states under exchange of quantum numbers for different particles from their Young diagrams [6].

3. Dispersion relation for magnetoplasmons in pristine graphene

As a precursor to our results for localised MPs in dirty graphene, we discuss the dispersion relation for extended MPs in a clean system. The calculation is explained in Ref. [8] in the same spirit as that for the 2DEG established by Kallin and Halperin [9], but with the additional

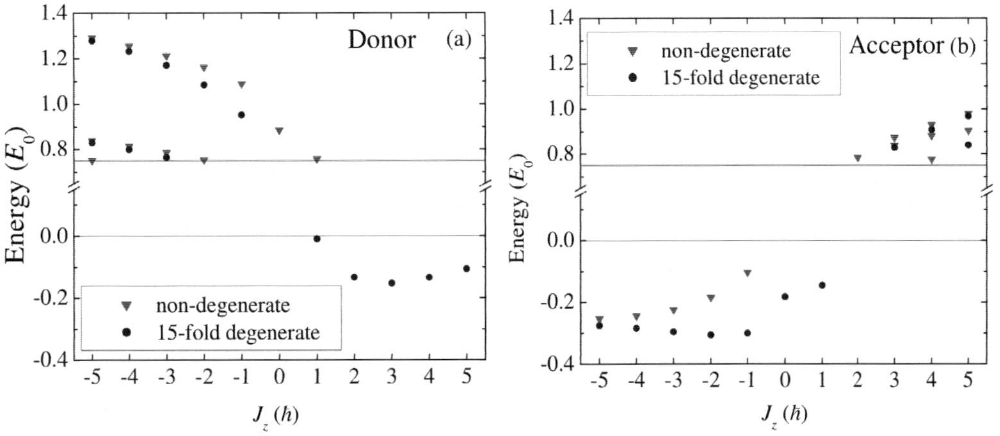

Figure 3. MPs from the ground state $\nu = 2$ localised on (a) a donor impurity and (b) an acceptor impurity. The hatched area of width $0.75E_0$ represents the continuum of extended MPs. Energies are given relative to the lower continuum edge in units of E_0 as a function of the generalised angular momentum projection, J_z.

complication of the valley pseudospin degree of freedom. The centre of mass quasimomentum for the exciton, **P**, can be shown to be a good quantum number. One can think of this as being due to the repulsive Lorentz forces and attractive Coulomb forces, which act on the electron and hole, cancelling out when they have a certain separation, enabling the exciton's centre of mass to move in a straight line. The Hamiltonian (in the symmetric gauge) for MPs is [10]:

$$\begin{aligned}\hat{H} &= \sum_{\mathcal{N},m} \tilde{\epsilon}_n c^\dagger_{\mathcal{N}m} c_{\mathcal{N}m} - \sum_{\mathcal{N},m} \tilde{\epsilon}_n d^\dagger_{\mathcal{N}m} d_{\mathcal{N}m} \\ &+ \sum_{\substack{\mathcal{N}_1,\mathcal{N}_2 \\ \mathcal{N}'_1,\mathcal{N}'_2}} \sum_{\substack{m_1,m_2 \\ m'_1,m'_2}} \mathcal{W}^{\mathcal{N}'_1 m'_1 \mathcal{N}'_2 m'_2}_{\mathcal{N}_1 m_1 \mathcal{N}_2 m_2} c^\dagger_{\mathcal{N}'_1 m'_1} d^\dagger_{\mathcal{N}'_2 m'_2} d_{\mathcal{N}_2 m_2} c_{\mathcal{N}_1 m_1},\end{aligned} \tag{5}$$

where we define the collective index $\mathcal{N} = \{n\tau_z s_z\}$. The first two terms give the single particle contribution. The $\tilde{\epsilon}_n$ are the LL energies renormalised by a self energy correction due to exchange interactions. The last term describes the *e*-*h* interaction; it has direct and exchange components. This Hamiltonian mixes the four excitons for which there are no spin or pseudospin flips and leaves the remaining twelve unmixed. The Coulomb interaction is treated exactly in the span of given LLs and the dispersion relations are found by diagonalising the Coulomb interaction Hamiltonian in the basis of two-particle electron-hole states compatible with magnetic translations (having definite quasimomentum P). The Coulomb interaction is treated as a perturbation and the dispersion relation found by diagonalising the Coulomb interaction matrix in the basis of these non-interacting two-particle wavefunctions. For the fully filled LL considered here, there are only two possible dispersion relations. The four mixed excitons give a non-degenerate dispersion branch [8]

$$E_1 = -\frac{E_0}{8} e^{-\frac{\mathcal{P}^2}{4}} \left[\left(6 + \mathcal{P}^2\right) I_0 \left(\frac{\mathcal{P}^2}{4}\right) - \mathcal{P}^2 I_1 \left(\frac{\mathcal{P}^2}{4}\right) \right] \tag{6}$$

and a triply degenerate dispersion branch

$$E_2 = E_1 + E_0 \mathcal{P} e^{-\frac{\mathcal{P}^2}{2}}, \tag{7}$$

where $\mathcal{P} = P\ell_B/\hbar$ is the dimensionless centre of mass quasimomentum, P, for the exciton and I_0 and I_1 are modified Bessel functions. The dispersion, E_2, is the same as for tthe 12 unmixed excitons, making the total degeneracy 15, as predicted in the previous section. These dispersion relations are plotted in Fig. 2. Energies are given in units of $E_0 = \sqrt{\frac{\pi}{2}}\frac{e^2}{\varepsilon\ell_B}$, the characteristic Coulomb interaction energy, where ε is an effective dielectric constant.

4. Localised magnetoplasmons in dirty graphene

We now consider how MPs may become localised in the presence of a single Coulomb impurity with potential, $V(r) = \frac{Ze^2}{\varepsilon r}$, where Z is the impurity charge in units of e [10]. The centre of mass quasimomentum is no longer well defined for a MP. However, the axial symmetry remains and the orbital angular momentum projection, defined in terms of the electron and hole orbital quantum numbers as, $J_z = |n_e| - |n_h| - m_e + m_h$, is a good quantum number. In the presence of an impurity, localised MPs break away from the continuum of extended MPs as discrete states. These localised MPs are shown for a donor ($Z = 1$) and acceptor ($Z = -1$) impurity in Fig. 3, plotted as a function of J_z. The yellow shaded area represents the continuum; it has width $0.75E_0$, as can be seen in Fig. 2. Note that for both impurity cases, there are non-degenerate branches and branches with 15-fold degeneracy. The predictions based on SU(4) symmetry hold here, because the Coulomb impurity, being long ranged, does not introduce an inequivalency between the valleys. This would not be the case for a short-ranged impurity located on one of the carbon sites [11].

5. Conclusions

In conclusion, we have studied neutral collective excitations of graphene in the presence of a strong perpendicular magnetic field. Electrons and holes in graphene can exist in four possible spin/pseudospin states, so that electron-hole complexes may be likened to mesons composed of quarks with four possible flavours. This analogy enabled us to use techniques established for treating mesons to obtain the multiplet structure of the neutral collective excitations for both extended excitations and those localised on a Coulomb impurity. An extension of this idea to charged collective excitations, which can be thought of as three particle complexes, can be found in Ref. [12]. These are analogous to baryons formed of three quarks.

Acknowledgments

We acknowledge funding by EPSRC (AMF and RAR). AMF is grateful for hospitality at CSU Bakersfield. ABD acknowledges Cottrell Research Corporation and the Scholarship of KITP, UC Santa Barbara.

References

[1] Peres N M R 2010 *Rev. Mod. Phys.* **82** 2673
[2] Lee C, Wei X, Kysar J W and Hone J 2008 *Science* **321** 385
[3] Geim A K 2009 *Science* **324** 1530
[4] Castro Neto A H, Guinea F, Peres N M R, Novoselov K S and Geim A K 2009 *Rev. Mod. Phys.* **81** 109
[5] Beenakker C W J 2008 *Rev. Mod. Phys.* **80** 1337
[6] Greiner W and Mueller B 1997 *Quantum Mechanics: Symmetries* (Berlin: Springer)
[7] Goerbig M O, Moessner R and Doucot B 2006 *Phys. Rev. B* **74** 161407
[8] Iyengar A, Wang J, Fertig H A and Brey L 2007 *Phys. Rev. B* **75** 125430
[9] Kallin C and Halperin B I 1984 *Phys. Rev. B* **30** 5655
[10] Fischer A M, Dzyubenko A B and Römer R A 2009 *Phys. Rev.B* **80** 165410
[11] Fischer A M, Dzyubenko A B and Römer R A 2009 *Ann. Phys. (Berlin)* **18** 944
[12] Fischer A M, Römer R A and Dzyubenko A B 2010 *EPL* **92** 37003

Generalising spin-ice: the magnetic ground-state of gadolinium titanate

M I Brammall, A K R Briffa[1] and M W Long

School of Physics, Birmingham University, Edgbaston, Birmingham, B15 2TT, UK

E-mail: [1] briffa@theory.bham.ac.uk

Abstract. We investigate the complex low-temperature magnetic ordering of the antiferromagnetic pyrochlore $Gd_2Ti_2O_7$. Mössbauer experiments indicate that the spins have equal-magnitude magnetic moments, which are restricted to lie in planes perpendicular to the local crystallographic directions. In addition neutron diffraction experiments show a magnetic scattering vector of (1/2,1/2,1/2) which is consistent with thirty-two atoms per magnetic unit cell. These restrictions are compatible with only two distinct magnetically ordered states.

This talk summarises the work presented in Ref. 1. Gadolinium titanate($Gd_2Ti_2O_7$) and the extensively studied 'spin-ices'($Ho_2Ti_2O_7$ and $Dy_2Ti_2O_7$[2]) both have the same pyrochlore structure: a three-dimensional lattice of corner-sharing tetrahedra. In spin-ice there is a large low-energy degeneracy. The ground-state manifold can be described using the 'ice-rules'. The strong crystal-field interactions mean that the spins have an Ising character: they are constrained to point along the local tetrahedral axes. The degenerate ground-state configurations consist of all tetrahedra containing two spins pointing inwards and the remaining two outwards (and, therefore, each tetrahedron has a ferromagnetic component). In contrast, in gadolinium titanate the antiferromagnetic nearest neighbour Heisenberg interaction is dominant, which is extremely frustrated on the pyrochlore lattice. This gives rise to a similar but much larger degeneracy than is seen in spin ice. It is this residual degeneracy that we wish to investigate.

Gadolinium is a spherically symmetric ion as a result of having a half-filled f-shell. It can therefore be modeled as just a fairly large classical spin. The value of the magnetic moment obtained from Curie-Weiss experiments is close to the expected value for the free ion[3]. A measured Curie temperature $\theta_{CW} = -9.6K$[3] indicates that the antiferromagnetic Heisenberg interaction dominates. Specific heat data shows two transitions: at $1.045K$ and 0.75[4]. These are associated with the magnetic ordering of the gadolinium ions. The transitions occur at temperatures of an order of magnitude below the Curie-Weiss temperature. This is consistent with a dominant frustrated antiferromagnetic Heisenberg interaction being lifted by other weaker interactions. Neutron diffraction measurements of the low temperature phase show that it is magnetically ordered with a scattering vector of $\mathbf{k} = \left(\frac{1}{2}, \frac{1}{2}, \frac{1}{2}\right)$[5].

To study the low temperature magnetic ordering, in addition to the dominant nearest neighbour Heisenberg interaction discussed above, we enforce two other experimentally motivated constraints. Firstly, the data from Mössbauer experiments[4] show an excellent fit when the spins are modeled using equal magnetic moment magnitudes. In addition the Mössbauer data indicates that the spins lie in planes which are perpendicular to the local crystallographic directions. These constraints are sufficient to reduce the degeneracy to a manageable level. The XY-like anisotropy can also be seen to be a result of partially minimising the direct dipolar interactions. As there are other competing interactions at a similar energy scale (for example longer-range Heisenberg), however, the XY-like anisotropy should be considered as an experimentally observed restriction rather than an energetic constraint. It is possible that small deviations from these planes could give rise to lower energy spin states which are not captured by our model. Using our assumptions we solve for a single tetrahedron to find all consistent spin states. This generates a more complex analogue of the ice-rules for gadolinium titanate.

At low temperatures we require that our rules are also consistent with the magnetic scattering vector $\mathbf{k} = \left(\frac{1}{2}, \frac{1}{2}, \frac{1}{2}\right)$. This turns out to lead to only two distinct magnetic states. They are unrelated to any previously proposed states. The energetics are complicated. The residual degeneracy appears to be lifted by a compromise between longer range Heisenberg and direct dipolar interactions of a similar energy scale. Our proposed states, however, do not appear to fit the neutron scattering Bragg peak intensities at all well. We therefore believe that it is possible to find still lower energy $\mathbf{k} = \left(\frac{1}{2}, \frac{1}{2}, \frac{1}{2}\right)$ states, of equal magnitude spins, by relaxing the restriction that the spins must lie exactly within the planes indicated by the Mössbauer data.

Acknowledgments
We wish to acknowledge useful discussions with A.J. Schofield.

References
[1] Brammall M I, Briffa A K R and Long M W 2010 *Preprint* arXiv:1008.1203
[2] Bramwell S T and Gingras M J P 2001 *Science.* **294** 1495
[3] Raju N P, Dion M, Gingras M J P, Mason T E and Greedan J E, 1999 *Phys. Rev. B* **59** 14489
[4] Bonville P, Hodges J A, Ocio M, Sanchez J P, Vulliet P, Sosin S and Braithwaite D 2003 *J. Phys.: Condens. Matter* **15** 7777
[5] Stewart J R, Ehlers G, Wills A S, Bramwell S T and Gardner JS 2004 *J. Phys.: Condens. Matter* **16** L321

Fluxoid formation: size effects and non-equilibrium universality

David J. Weir and Ray J. Rivers
Blackett Laboratory, Imperial College
London SW7 2AZ, U.K.

E-mail: david.weir03@imperial.ac.uk, r.rivers@imperial.ac.uk

Abstract. Simple causal arguments put forward by Kibble and Zurek suggest that the scaling behaviour of condensed matter at continuous transitions is related to the familiar universality classes of the systems at quasi-equilibrium. Although proposed 25 years ago or more, it is only in the last few years that it has been possible to devise experiments from which scaling exponents can be determined and in which this scenario can be tested.

In Ref. [1], an unusually high Kibble-Zurek scaling exponent was reported for spontaneous fluxoid production in a single isolated superconducting Nb loop, albeit with low density. Using analytic approximations backed up by Langevin simulations, we argue that densities as small as these are too low to be attributable to scaling, and are conditioned by the small size of the loop. We also reflect on the physical differences between slow quenches and small rings, and derive some criteria for these differences, noting that recent work on slow quenches does not adequately explain the anomalous behaviour seen here.

1. Overview

Causality imposes strong constraints on the non-equilibrium behaviour of condensed matter systems, particularly in the vicinity of continuous transitions where, adiabatically, correlation lengths become arbitrarily large. Since, in reality, disorder-order transitions take place in a finite time, correlation lengths must remain finite; causality imposes a maximum speed at which the system can order itself. The greater the transition rate, the more constrained the correlation lengths will be. Should the ordered system be frustrated, this frustration will be relieved by the spontaneous creation of topological defects, whose separation will reflect this finiteness.

The suggestion that causality alone could lead to the spontaneous creation of topological defects was originally made by Kibble [2, 3], not for condensed matter systems, but for the Grand Unified Theories (GUTs) present in the very early universe, which possess a cascade of transitions, almost all of which lead to the creation of cosmic strings (field vortices). Zurek, independently, had arrived at the same conclusions for the spontaneous formation of defects for condensed matter systems [4, 5, 6], and this mechanism is known as the Kibble-Zurek (KZ) scenario. This has led to an active collaboration between cosmologists and condensed matter physicists over many years, most notably over the classification of defects and the similarities in the symmetry-breaking patterns and defects of liquid ^3He and GUTs [7]. However, the more important aspect of the KZ scenario lies in the prediction of defect densities after a transition, since this should be amenable to experimental confirmation in condensed matter physics. Unfortunately, we cannot yet say the same for the defects of the early universe.

The simplest transitions are for systems cooled from one phase to another. Consider a temperature quench with quench time $\tau_Q = T_c(dT/dt)^{-1}_{T_c}$ in the vicinity of a continuous thermal transition with critical temperature T_c. It is predicted that, for a uniform quench applied to a homogeneous and isotropic system, the initial separation $\bar{\xi}$ between defects takes the scaling behaviour

$$\bar{\xi} \approx \xi_0(\tau_Q/\tau_0)^{\sigma} \tag{1}$$

where ξ_0 and τ_0 are fiducial distance and relaxation time scales, which depend upon the system. The important point is that the exponent σ is derived from the adiabatic critical exponents and therefore shows universality.

There are several ways to derive (1) from causal bounds [3, 6]. Most simply, we assume that the true correlation length tracks the adiabatic correlation length until the latter's rate of change is forced to try to exceed the relevant causal speed, whereupon the system freezes in, to unfreeze when the correlation length is again changing sufficiently slowly to be causally valid. This behaviour has been observed in a variety of simulations (e.g. see Refs. [8, 9]).

There have been many experimental tests of (1) that give results commensurate with it without being capable of determining σ, although not necessarily for temperature quenches. However, some experiments have been able to show scaling behaviour with quench time, in particular experiments on Rayleigh-Bernard convection patterns [10], charge density waves in $TbTe_3$ [11] and fluxons in annular Josephson junctions [12, 13]. Several proposals for future experiments have been made, such as experiments for the formation of kink defects in linear ionic crystals [14] and quantum transitions in cold gases (e.g. Refs. [15, 16, 17]).

Although this was the original motive, attempts to see these results as reflections of transitions in the early universe have to be taken with extreme caution. The early universe is lacking in boundaries and homogeneous and isotropic, to which (1) applies. Yet none of these attributes are possessed by laboratory condensed matter systems. In particular, inhomogeneity goes hand in hand with finite size. Inhomogeneity comes in two kinds. Firstly, that due directly to the finiteness of the system e.g. a trapped condensate has variable density [18]. The second is concerned with the inhomogeneous application of the quench. It is possible to incorporate such behaviours within scaling laws, but the end results are complicated [18, 19].

One way to try to avoid the worst of these problems is to use annular systems, originally proposed by Zurek [6]. Such annuli are achievable in superconductors, Josephson junctions and in annular traps for cold bosonic gases [17]. If the circumference of the ring can be taken to be larger than its other dimensions for the purposes of domain formation, then the periodicity of the ring means we can have a homogeneous system without boundaries. Further, if the ring can be kept small, the problem of quench inhomogeneity becomes less pressing. The disadvantage of finding less than one defect, on average, is mitigated by ease of observation and the absence of defect annihilation. There is yet a further reason for choosing small rings of superconductors. The KZ mechanism is not the only reason for the spontaneous generation of magnetic flux in superconductors. The long wavelength modes of the magnetic field can freeze in spontaneously, almost without regard to the superconductor itself. This mechanism, the Hindmarsh-Rajantie [20, 21] mechanism gives contributions to the spontaneous flux which have a totally different behaviour from (1). The effect can be ignored for small superconductors. It was with all these in mind that the experiments on fluxon formation in annular Josephson junctions were performed [12, 13], which showed remarkable allometric behaviour of the form given in (1).

The anticipated scaling exponent for annular Josephson tunnel junctions was unclear, because of ambiguities in the fabrication that may or may not have allowed for proximity effects [22]. However, single superconducting annuli should be less ambiguous and a recent experiment on low-temperature superconducting annuli showed a higher than expected scaling exponent [1]. This experiment is being revisited to look for possible systematic errors, but it is timely to see what effects due to size we might expect in superconducting systems that are smaller than the correlation length at the time of defect formation. We have some understanding of this for the quantum transitions of Bose gases [16, 17], but these are controlled by the Gross-Pitaevskii equation whereas for superconductors phase ordering is associated with diffusion.

The purpose of this paper is to advance numerical and analytical arguments that the behaviour of an annular system changes from KZ scaling to exponential suppression for defect densities as small as those reported in Ref. [1]. Such densities have not been widely studied in the literature, and certainly have not been widely explored for annular systems where the qualitative behaviour might be expected to be rather different from that for Bose-Einstein condensation experiments.

In Section 2 we estimate the effect of small size by making a Gaussian approximation for a diffusive system on a one-dimensional ring. It suggests an interpolation between the canonical scaling behaviour for large rings and exponential damping. This is followed in Section 3 by a demonstration that this is confirmed by a Langevin simulation for a simple complex field on the ring. More realistic simulations are proposed in Section 4. Our conclusions can be found in Section 5.

2. Gaussian approximation

Although causality does impose bounds, the KZ argument as presented above is too simple. It makes more sense to think of defect formation as due to the growth of long-wavelength unstable modes, where the growth of these is causally constrained. Surprisingly, treating the order parameter field as a Gaussian variable satisfying (short-time) linearised equations, which embodies this picture, recreates the scaling laws (1) [23, 24] and some key aspects of the nature of the transition [25]. It is borne out not only by the one-dimensional numerical simulations presented here, but also by more realistic simulations including the gauge field [26].

With due caution, we look to the effect of finite size on the winding number of a complex field $\Phi(x) = \chi(x)\, e^{i\theta(x)}$ confined to a loop parameterised by the coordinate x. We see this as a proxy for a Cooper pair field confined to a small superconducting annulus of circumference C. On quenching the system from the normal to the superconducting phase, we can define a winding number density $n(x) = \partial_x\theta(x)/2\pi$, where $0 \leq x < C$ is the distance along the annulus. The total winding number n around the loop is zero on average, but will have non-zero variance $(\Delta n)^2 = \langle n^2\rangle$, and it is this that we measure. The relevant quantity is the two-point correlation function $\langle n\,(x)\, n\,(y)\rangle$. We assume that the complex order parameter field $\Phi = (\phi_1 + i\phi_2)/\sqrt{2}$ is a Gaussian variable, whereby ϕ_1 and ϕ_2 are, independently, Gaussian fields.

The correlation function for the winding number density is now determined by the correlation function $G(x)$ for the field components, defined by

$$\langle \phi_a\,(x)\,\phi_b\,(y)\rangle = \delta_{ab}G\,(|x-y|)\,. \tag{2}$$

A simple approximation gives

$$\langle n\,(x)\,n\,(y)\rangle \;\approx\; \frac{2}{(2\pi)^2}\left[f'\,(x)^2 - f''\,(x)\,f\,(x)\right] \tag{3}$$

where $f\,(x) = G(x)/G(0)$ and $\langle \phi_a\,(x)\,\phi_b\,(y)\rangle = \delta_{ab}G\,(|x-y|)$.

For strongly coupled dissipative systems the linearised time-dependent Landau-Ginzburg equations suggest that, for *long rings* ($C/\bar{\xi} \gg 1$) $f\,(r)$ can be approximated well by the Gaussian form $f\,(r) \approx \exp(-r^2/2\xi_r^2)$ where $\xi_r \approx \bar{\xi}$. Although this expression does not show periodicity, for long rings this should not matter. Substituting in (3) gives $\langle n^2\rangle = \mathcal{O}(C/\xi_r)$ in the case where C is very large and the periodicity in the field can be ignored. This linear dependence of $\langle n^2\rangle$ on C is just as we would expect from a random walk in phase along the ring and is assumed as input in the KZ picture. Linearising the equations of motion for fast quenches then gives (1) with $\sigma = 1/4$, in agreement with the KZ prediction.

On the other hand, for short rings ($C/\bar{\xi} \ll 1$) we must have $f(x)$ periodic in x (mod C). To implement this we need to replace $f(x)$ by its Jacobi ϑ function generalisation

$$f(x)_{per} = \frac{\vartheta_3(\pi x/C|2\pi i\xi_r^2/C^2)}{\vartheta_3(0|2\pi i\xi_r^2/C^2)} \approx \; 1 - 4\sin^2(\pi x/C)\; e^{-2\pi^2\xi_r^2/C^2}, \tag{4}$$

whence $\langle n^2\rangle = \mathcal{O}(e^{-4\pi^2\xi_r^2/C^2})$. Thus, rather than the power falloff for large loops we have exponential damping

$$\ln\langle n^2\rangle \;\approx\; -4\pi^2\xi_r^2/C^2 + const. \tag{5}$$

This is in contrast to the proposal made in [1] that the slope be doubled, which is more likely for a disc. We should not push this approximation too far but it suggests exponential damping in the quench time whenever $\langle n^2\rangle \ll 1$, whether this be for very slow quenches in large rings or faster quenches in small rings.

3. Numerical simulations

Our Langevin simulations are for the U(1) scalar field theory described above in 1+1 dimensions with periodic boundary conditions, on a ring of circumference C. The potential is

$$V(|\Phi|^2) = \frac{1}{2}a(t)|\Phi|^2 + \frac{1}{4}|\Phi|^4, \tag{6}$$

where $a(t)$ is decreased linearly in time as $a(t) = -t/\tau_Q \;\; (t < \tau_Q)$ and $a(t) = -1 \;\; (t > \tau_Q)$ to model a slow quench. We start at $t = -2\tau_Q$ and continue until $t = 4\tau_Q$, by which time the defects have frozen out.

To study this time-dependent system with dissipation and coupled to a heat bath, we consider the second-order Langevin equation

$$\partial_t^2\phi_a - \partial_x^2\phi_a + \eta\partial_t\phi_a + \frac{\partial V}{\partial\phi_a} = \zeta_a, \tag{7}$$

with Gaussian noise

$$\big\langle\zeta_a(x',t')\zeta_b(x,t)\big\rangle = 2\eta T\delta(x'-x)\delta(t'-t)\delta_{ab}, \qquad \langle\zeta_a(x,t)\rangle = 0. \tag{8}$$

We take $\eta = 1$ (so that the system is heavily damped) and $T = 0.01$. It can be shown that such a system satisfies a fluctuation-dissipation theorem. We use a stochastic leapfrog method to evolve the equations of motion consistently [27].

In this model, there is no dynamical gauge field, a reasonable approximation for the present system; we are interested in small rings or slow quenches when relatively few defects form so $\langle n^2\rangle \ll 1$. Recent work [28] suggests that the KZ scaling result (1) should be modified for long times after slow quenches where defect annihilation is of importance. However, as we are dealing with flux trapping at very low densities, annihilation dynamics should be unimportant.

Applying periodic boundary conditions, we the phase angle is obtained through $\theta(x) = \mathrm{Pr}\arg(\phi_2(x) + i\phi_1(x))$, with a discretised winding number density $n(x) = (\theta(x+\delta x) - \theta(x))/\delta x$, mod 2π. The main bias in statistics of the winding number obtained from simulations will be a slight undercounting due to discretisation effects. One can account for this by checking that the results are robust to changes in lattice spacing.

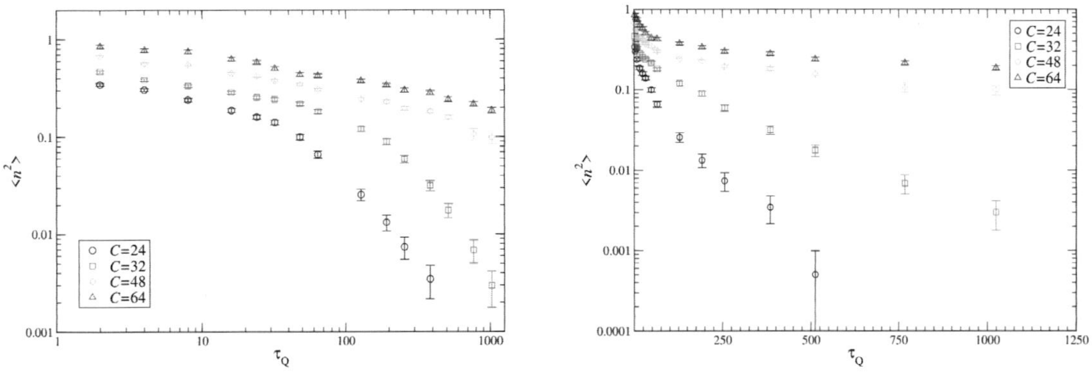

Figure 1. Results of simulations, showing change between KZ and exponential suppression regimes. The doubly logarithmic plot at left shows KZ scaling with $\sigma \approx 0.25$ giving way to an exponential suppression regime below $\langle n^2 \rangle \approx 0.2$. The same data are plotted again on log-linear axes in the plot at right.

Our numerical results are ensemble averages of 2000 separate simulations for each combination of parameters. Typically, the lattice spacing is $\delta x = 0.125$ and the timestep $\delta t = 0.0125$. The numbers of lattice sites used is in the region of $192-512$, giving circumferences C between 24 and 256 in appropriate units. Qualitative checks of the system's sensitivity to discretisation effects were carried out, and decreasing the lattice spacing or the timestep by a factor of 2 did not affect the results presented here. The Gaussian noise was generated using the 'ranlux' generator with Box-Müller transformations. In simulating the system, we need to quench the system rather slowly (potentially leading to growth of errors from the discretisation of the timestep), while keeping it fairly small (motivating a reduction in the lattice spacing). We think that our choice of parameters offers a reasonable trade-off.

In Figure 1 we see behaviour similar to that believed to have been seen in experiment [1]: the KZ scaling with $\sigma \approx 0.25$ gives way at small defect densities, $\langle n^2 \rangle \approx 0.2$, to exponential suppression. This is consistent with (5). It is easy to see how, for a limited set of data in the region of $\langle n^2 \rangle \approx 0.2$, finite-volume effects might be mistaken for a doubling of the scaling exponent.

The behaviour we see in Figure 1 does not particularly support the modified scaling behaviour at late times that was recently proposed [28]. In any case, it be unlikely that we would see defect interactions substantially alter a texture-forming quench of this kind at such low values of $\langle n^2 \rangle$.

While in experimental settings it is only possible to measure $\langle n^2 \rangle$, for the numerical simulation we can measure the probability of trapping exactly q fluxons or antifluxons $\bar{f}_q$. At very small trapping probabilities where $\bar{f}_q \approx 0$ for $q \geq 2$, then $\langle n^2 \rangle \approx \bar{f}_1$. Figure 2 demonstrates the effect of system size on $\bar{f}_q$, and also on $\langle n^2 \rangle$. The results demonstrate that keeping the quench time relatively short and simply varying the circumference also supports the Gaussian approximation. In other words, slow quenches and small systems show the same exponential suppression of flux trapping, as follows from (5).

4. More realistic simulations: the system in its environment

More realistic simulations require the incorporation of a dynamical gauge field. It is not clear how valid the arguments advanced in the previous sections will be for the case of a thick ring, or

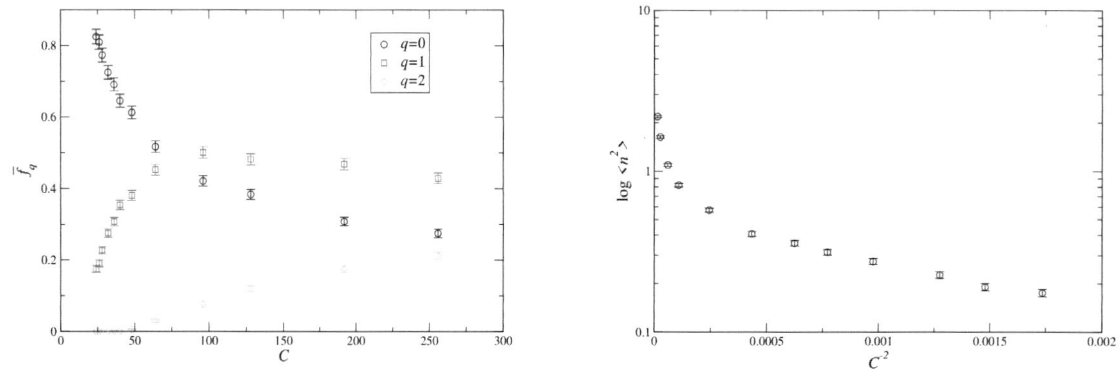

Figure 2. At left is a plot of the probability $\bar{f}_q$ of finding a given number of fluxons with system size with $\tau_Q = 24$; at right this has been combined into a measurement of $\langle n^2 \rangle$. The exponential scaling with C^{-2} is clear for small volumes. Our simulations suggest that small volumes and slow quenches have the same effect.

where the system is still annular but is no longer circular. Genuine experiments also typically have stray magnetic fields that cannot be truly eliminated. None of these effects can be studied in $1+1$-dimensional simulation as there is no dynamical magnetic field. Simulations of the noncompact Abelian Higgs model in $2+1$ dimensions may be little better [20]. For these reasons we favour treating the system as a thin ring in a larger simulational box containing only the U(1) field of vacuum electrodynamics [26].

There are no great technical obstacles to such a simulation. Performing a Langevin simulation while maintaining the Gauss constraint (having imposed temporal gauge $A_0 = 0$) is straightforward and has been used in previous Langevin simulations of similar systems, albeit with greater symmetry [29].

Indeed, the proposed simulation is not particularly elegant. In checking how external fields bias the trapping of flux, the penetration depth introduces a length scale which would be ignored for the simple case of a thin film. It is neatest to use periodic boundary conditions, in which we leave a gap between the edge of the annulus and the sides of the simulational box; the net flux through the system must vanish and so trapped flux must return around the edges of the ring.

We can measure flux trapping in such a simulation by calculating the magnetic flux through the ring or by measuring the scalar field winding around the inner boundary of the system. The results from both methods are in agreement based on preliminary results [26].

Perhaps, if we wish to examine finite-size issues in effective models of planar (non-annular) superconducting films, the most elegant way may be to create a film in the same way [30]. Then a direct parallel can be drawn with the experiments carried out in Bose-Einstein condensates [16, 17]. Periodic boundary conditions may be employed without hesitation so long as a gap is left at the edges of the film.

5. Conclusions

We have performed numerical simulations for small annular systems with broken $U(1)$ symmetry, an idealisation of superconducting rings. We have demonstrated that the simple power behaviour of (1) breaks down when the probability of finding a single defect falls below about 0.2, more or less independently of the microscopic properties of the model. As we anticipated in a simple

Gaussian approximation the likelihood of finding defects is then seen to become exponentially suppressed with quench rate or system size. This behaviour is very different from that of defect production in small Bose-Einstein condensates.

We have also discussed a potentially more valid simulational protocol for more realistic (planar) annular superconductors, incorporating the three dimensional electromagnetic field (and an external applied field) in a consistent way, and explained some of the limitations of such an approach.

Acknowledgments
RJR thanks R. Monaco for helpful discussions and both RJR and DJW are grateful to J. Mygind and A. Gordeeva at Danish Technical University for their kind hospitality and helpful discussions. DJW also thanks A. Rajantie for useful discussions. DJW is funded by the Science and Technology Facilities Council. This work made use of the Imperial College High Performance Computing Service.

References
[1] Monaco R, Mygind J, Rivers R J, Koshelets V P 2009 *Phys. Rev.* B **80** 180501
[2] Kibble T W B 1976 *J. Phys.* A **9** 1387
[3] Kibble T W B 1980 *Phys. Rep.* **67** 183
[4] Zurek W H 1985 *Nature* **317** 505
[5] Zurek W H 1993 *Acta Phys. Pol.* B **24** 1301
[6] Zurek W H 1996 *Phys. Rep.* **276** 177
[7] Volovik G E 2001 *Phys. Rep.* **351** 195
[8] Laguna P and Zurek W H 1998 *Phys. Rev.* D **58** 085021
[9] Antunes N D, Bettencourt L M A and Zurek W H 1999 *Phys. Rev. Lett.* **82** 2824
[10] Casado S, González-Viñas W, Mancini H and Boccaletti S 2001 *Phys. Rev.* E **63** 057301
[11] Yusupov R, Mertelj T, Kabanov V V, Brazovskii S, Kusar P, Chu J, Fisher I R and Mihailovic D 2010 *Nature Phys.* **6** 681
[12] Monaco R, Mygind J and Rivers R J 2003 *Phys. Rev.* B **67** 104506
[13] Monaco R, Mygind J, Aaroe M, Rivers R J and Koshelets V P 2006 *Phys. Rev. Lett.* **96** 180604
[14] del Campo A, De Chiara G, Morigi G, Plenio M B, Retzker A 2010 *Phys. Rev. Lett.* **105** 075701
[15] Damski B and Zurek W H 2007 *Phys. Rev. Lett.* **99** 130402
[16] Saito H, Kawaguchi Y and Ueda M 2007 *Phys. Rev.* A **76** 043613
[17] Dziarmaga J, Meisner J and Zurek W H 2008 *Phys. Rev. Lett.* **101** 115701
[18] Zurek W H 2009 *Phys. Rev. Lett.* **102** 105702
[19] del Campo A, Retzker A and Plenio M B 2010 Heterogeneous Kibble-Zurek mechanism: vortex nucleation during Bose-Einstein condensation *Preprint* 1010.6190
[20] Hindmarsh M and Rajantie A 2000 *Phys. Rev. Lett.* **85** 4660
[21] Rajantie A 2001 *J. Low Temp. Phys.* **124** 5
[22] Golubov A A, Houwman E P, Gijsbertsen J G, Krasnov V M, Flokstra J and Rogalla H 1995 *Phys. Rev.* B **51** 1073
[23] Karra G and Rivers R J 1998 *Phys. Rev. Lett.* **81** 3707
[24] Bowick M and Momen A 1998 *Phys. Rev.* D **58** 085014
[25] Moro E and Lythe G 1999 *Phys. Rev.* E **59** R1303
[26] Rajantie A, Rivers R J and Weir D J *In preparation*
[27] Borrill J and Gleiser M 1997 *Nucl. Phys.* B **483** 416
[28] Biroli G, Cugliandolo L F, Sicilia A 2010 *Phys. Rev.* E **81** 050101
[29] Stephens G J, Bettencourt L M A and Zurek W H 2002 *Phys. Rev. Lett.* **88** 137004
[30] Donaire M, Kibble T W B, Rajantie A 2007 *New J. Phys.* **9** 148

Journal of Physics: Conference Series **286** (2011) 012057 doi:10.1088/1742-6596/286/1/012057

Dynamics near a periodically forced robust heteroclinic cycle

T L Tsai[1] and J H P Dawes[2]

[1] Department of Mathematics, National Changhua University of Education, Jin-De Campus, Changhua, Taiwan 500, ROC

[2] Department of Mathematical Sciences, University of Bath, Claverton Down, Bath BA2 7AY, UK

E-mail: ttlong23@cc.ncue.edu.tw and J.H.P.Dawes@bath.ac.uk

Abstract. In this article we discuss, with a combination of analytical and numerical results, a canonical set of differential equations with a robust heteroclinic cycle, subjected to time-periodic forcing. We find that three distinct dynamical regimes exist, depending on the ratio of the contracting and expanding eigenvalues at the equilibria on the heteroclinic cycle which exists in the absence of forcing. By reducing the dynamics to that of a two dimensional map we show how frequency locking and complex dynamics arise.

1. Introduction

Nonlinear differential equations used to model competitive or cooperative interactions are widely recognised to be capable of generating complex dynamics. In many cases, such differential equations also contain subspaces of the solution phase space that are flow-invariant. Such invariant subspaces arise, for example, in the presence of symmetry or through modelling assumptions. The existence of invariant subspaces further produces generic, and 'natural' kinds of dynamics that are not at all generic in the absence of invariant subspaces. One of these is heteroclinic cycling, which arises robustly in the presence of invariant subspaces.

A *heteroclinic cycle* is a collection of flow-invariant sets $\{\xi_1, \dots, \xi_n\}$ and connecting orbits $\{\gamma_1(t), \dots, \gamma_n(t)\}$ whose α- and ω-limit sets satisfy $\alpha(\gamma_i) = \xi_i$ and $\omega(\gamma_i) = \xi_{(i \bmod n)+1}$. A heteroclinic cycle is said to be *robust* if, for every $1 \leq i \leq n$ there exists an invariant subspace P_i such that

- the connecting orbit γ_i is contained in P_i,
- $\xi_{(i \bmod n)+1}$ is a sink within P_i.

The flow-invariant sets $\xi_1, \dots, \xi_n$ may of course be periodic orbits or chaotic invariant sets. For simplicity we will here consider the ξ_i to be equilibria, and the invariant subspace P_i to be two-dimensional. The presence of invariant subspaces gives rise to a natural division of the eigenvalues of the Jacobian matrix at ξ_i into four classes: radial eigenvalues r_{ij} are those whose eigenvector lies in $L_i \equiv P_{i-1} \cap P_i$; expanding eigenvalues $e_{ij} > 0$ correspond to $P_i \setminus L_i$; contracting eigenvalues $-c_{ij} < 0$ correspond to $P_{i-1} \setminus L_i$. All other eigenvalues are denoted transverse (t_{ij}). Stability of the robust heteroclinic cycle (RHC) can be ensured when the local contraction dominates the expansion near each equilibrium, as given by the following result due to Krupa and Melbourne [1].

Theorem. Asymptotic stability is guaranteed if

$$t_i < 0 \;\; \forall \, i \qquad \text{and} \qquad \prod_{i=1}^{n} \min(c_i, e_i - t_i) > \prod_{i=1}^{n} e_i$$

where, at ξ_i, $t_i = \max_j\{Re\ t_{ij}\} < 0$ and $c_i = \min_j\{Re\ c_{ij}\} > 0$ are the real parts of the weakest transverse and contracting eigenvalues, and $e_i = \max_j\{Re\ e_{ij}\} > 0$ is the real part of the strongest expanding eigenvalue.

It should be noted that in many cases necessary and sufficient conditions for asymptotic stability differ from the above inequalities [2, 3]. Moreover, useful notions of stability that are weaker than asymptotic stability exist, for example 'essential asymptotic stability' [4, 5].

Turning to applications, there is a substantial literature on RHCs that has developed principally in two areas. The first of these applies methods developed in equivariant bifurcation theory [6, 7, 5] to fluid mechanical problems [8] and coupled oscillator networks [9, 10]. The second part of the literature is in evolutionary game theory and mathematical biology, including population dynamics [11, 12] and neuroscience [13]. The review article by M. Krupa [14] contains further details and many literature references.

In this article we examine the effect of an external time-periodic forcing on the dynamics near a RHC. While it is well-known that a constant symmetry-breaking perturbation typically breaks the heteroclinic cycle and generates a single periodic orbit which lies close to the remaining invariant subspaces, the effect of an external time-periodic forcing term has not previously been systematically studied. In this article we continue our earlier analysis [15] and identify and explain the existence of three distinct regimes as the eigenvalue ratio $\delta \equiv c/e > 1$ decreases towards unity. Our study is crucial to advancing the detailed understanding of the dynamics that arises when RHCs are coupled together [16, 17], since a second RHC acts similarly to an external periodic forcing. A full paper containing details of the results is currently in preparation [18].

The remainder of this article is laid out as follows. Section 2 introduces the model dynamical system. In section 2.1 we describe the systematic reduction of the dynamics to those of a two-dimensional map. Section 3 describes the dynamics at large δ where frequency locking does not occur (referred to as region I), and at intermediate δ where frequency locking appears (region II). In section 4 we reduce our return map to the well-known continuous-time ODEs for the forced damped pendulum and hence explain the bistability between frequency locking and quasiperiodic dynamics that arises when δ is very close to unity (region III).

2. Canonical example and ODE dynamics

Consider the vector field in $\mathbb{R}^3$ given by the differential equations

$$\dot{x} = x(1 - X - cy + ez) + \gamma(1 - x)\sin^2 \omega t, \tag{1}$$
$$\dot{y} = y(1 - X - cz + ex), \tag{2}$$
$$\dot{z} = z(1 - X - cx + ey), \tag{3}$$

where $X \equiv x + y + z$ and, for $\gamma = 0$, $c, e > 0$ are the absolute values of the contracting and expanding eigenvalues at the equilibrium points $\xi_1 \equiv (1,0,0)$, $\xi_2 \equiv (0,1,0)$ and $\xi_3 \equiv (0,0,1)$ which lie on a RHC. If $c > e$ (i.e. $\delta > 1$) then the above theorem due to Krupa and Melbourne implies that the RHC is asymptotically stable. System (1) - (3) with $\gamma = 0$ was analysed first by May and Leonard [19] as a model of competitive Lotka–Volterra type interactions between three populations. Independently [20], it was derived in the weakly nonlinear analysis of the Küppers–Lortz instability of convection rolls in rotating Rayleigh–Bénard convection. A proof

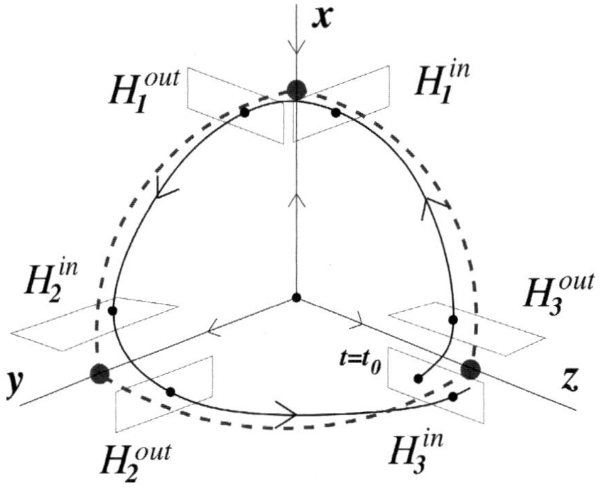

Figure 1. Robust heteroclinic cycle (thick dashed line) for (1) - (3) when $\gamma = 0$, and transverse planes H_i^{in} and H_i^{out} used in the return map construction for $\gamma > 0$.

of the existence of the RHC in (1) - (3) when $c, e > 0$ was given by Guckenheimer and Holmes [21]. Space does not permit a complete description of the dynamics of (1) - (3), save to remark that if $ce < 0$ then additional equilibria exist that destroy the connecting orbits on the RHC.

2.1. Return map

When $0 < \gamma \ll 1$ we analyse the dynamics by assuming that (after transients) trajectories continue to pass repeatedly through neighbourhoods of the three equilibria in turn, and converge to some (possibly chaotic) invariant set. The standard approach is to construct and analyse a Poincaré return map that gives the coordinates of successive intersections of a trajectory with a transverse surface, say H_3^{in}, near one of the equilibria, see figure 1. Between H_i^{in} and H_i^{out} we use the linearisation of (1) - (3) to integrate along trajectories and construct a 'local map'. Between H_i^{out} and $H^{in}_{(i \bmod 3)+1}$ we estimate the leading order behaviour of trajectories near the unstable manifold $W^u(\xi_i)$ and so construct a 'global map'. In both cases we consider the perturbation amplitude γ to be asymptotically small, and we keep only the leading order terms in γ. Composition of local and global maps results in a return map defined (at leading order) on two variables: the x-coordinate on H_3^{in} and the time t_n at which the trajectory hits H_3^{in}:

$$x_{n+1} = x_n^d + \gamma\mu_2 (1 - a_1 \cos 2\omega g_n - b_1 \sin 2\omega g_n) + \gamma f(x_n, t_n), \quad (4)$$

$$t_{n+1} = t_n + \mu_3 - \xi \log x_n - \gamma \frac{\xi}{2ex_n} (1 - a_2 \cos 2\omega t_n + b_2 \sin 2\omega t_n), \quad (5)$$

where we have kept terms up to $O(\gamma)$ only, $g_n \equiv t_n + \mu_3 - \xi_1 \log x_n$ and $d \equiv (c/e)^3$. We further define the parameters $\xi = (c^2 + ce + e^2)/e^3$, $a_1 = c^2/(c^2 + 4\omega^2)$, $a_2 = e^2/(e^2 + 4\omega^2)$, $b_1 = 2c\omega/(c^2 + 4\omega^2)$ and $b_2 = 2e\omega/(e^2 + 4\omega^2)$. The global maps cannot be computed analytically and lead to the introduction of the unknown function $f(x_n, t_n)$ into (4). For a suitable choice of $f(x_n, t_n)$ [15] the dynamics of the return map agree quantitatively extremely well with those of the ODEs (1) - (3) over a wide range of the parameters c, e and γ; this is illustrated in figure 2. Figure 2 shows a sequence of intervals in ω in which stable, frequency-locked periodic orbits exist, separated by intervals of more complicated (quasiperiodic or chaotic) dynamics. In figure 2(a), it is interesting to note that for $\omega < 0.05$, the periodic orbit undergoes a period-doubling bifurcation within each frequency-locking interval; this bifurcation behaviour occurs in

Journal of Physics: Conference Series **286** (2011) 012057 doi:10.1088/1742-6596/286/1/012057

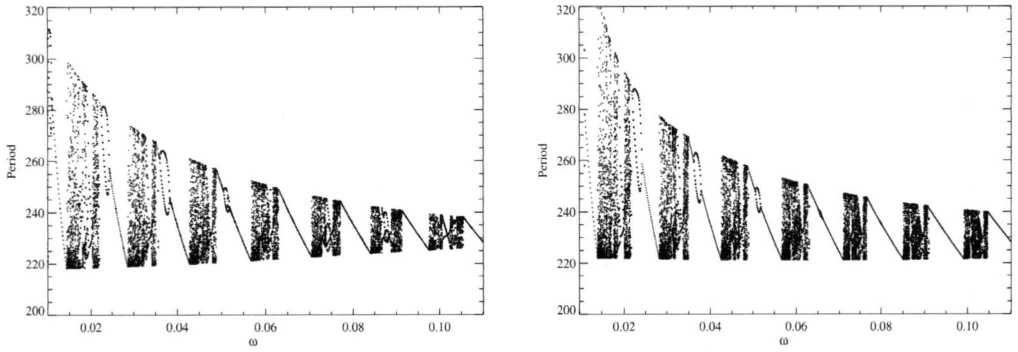

Figure 2. (a) Period T of trajectories of the ODEs (1) - (3) returning to H_3^{in}, as a function of driving frequency ω. (b) Dynamics of the return map (4) - (5), where the 'period' is $t_{n+1} - t_n$. Note the detailed agreement, including the regions of frequency locking in between complex dynamics (this corresponds to region II). $c = 0.25$, $e = 0.2$, $\gamma = 10^{-6}$.

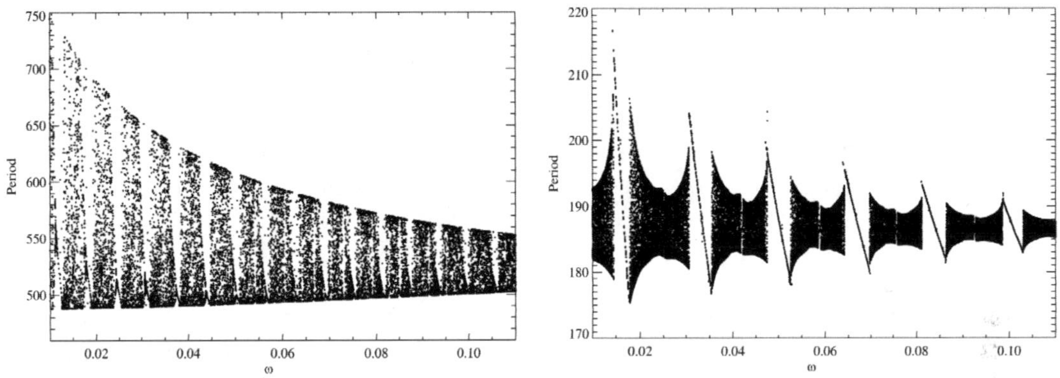

Figure 3. Period T of trajectories of the ODEs (1) - (3) returning to H_3^{in}, as a function of driving frequency ω. (a) $c = 0.5$ showing the disappearance of the intervals in ω of frequency locking at $\delta = 5/2$ (region I); (b) $c = 0.205$ showing the emergence of bistability between frequency-locking (slanted curve sections) and quasiperiodic dynamics (black regions), i.e. region III. Other parameters: $e = 0.2$, $\gamma = 10^{-6}$.

the map (4) - (5) as well [15]. It is also interesting to note that the position of the frequency-locked periodic orbit (which corresponds to a fixed point in the map) coincides exactly with one of the extrema of the ranges of periods for oscillations in the regions of complicated dynamics. Figure 3 illustrates the ODE dynamics for larger and smaller values of c (regions I and III respectively); the map dynamics at these parameter values also agree quantitatively with the ODE dynamics.

3. Regions I and II: complex dynamics and frequency locking

The system (1) - (3) was investigated by Rabinovich et al. [22] who noted the existence of intervals in ω containing frequency-locked periodic orbits, and presented an *ad hoc* construction of a circle map that was claimed to capture the system dynamics. In contrast, our results [15]

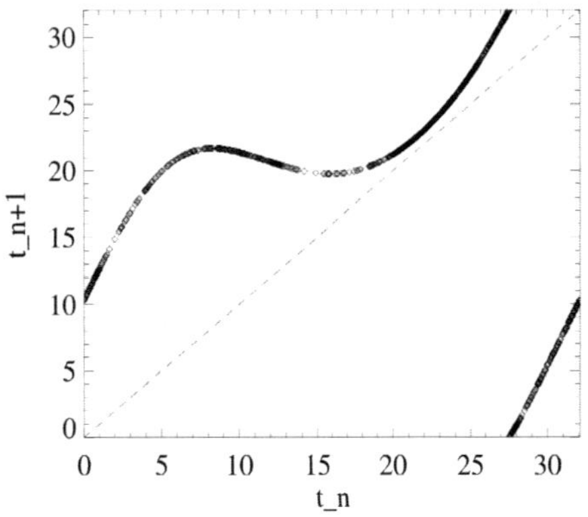

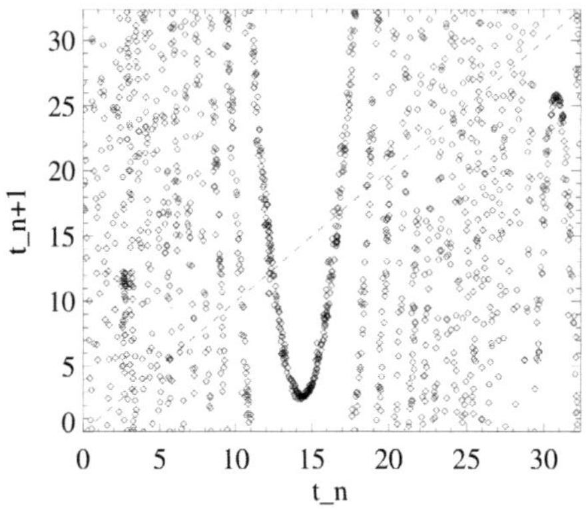

Figure 4. Plots of t_{n+1} mod π/ω against t_n mod π/ω for the ODEs (1) - (3). (a) One-dimensional circle-map-like dynamics for $c = 0.25$, $\omega = 0.098$ near the end of a frequency-locking interval in ω (corresponding to region II, same values of c and e as in figure 2); (b) complex dynamics for $c = 1.5$, $\omega = 0.097$ (corresponding to region I). Other parameters: $e = 0.2$, $\gamma = 10^{-6}$.

are based on a careful asymptotic reduction of the ODEs to the two-dimensional return map above. As discussed in [15], and illustrated in figure 4, numerical results obtained both for the ODEs (1) - (3) and for the map (4) - (5) indicate that the dynamics might be that of a circle map when δ is moderately close to unity (figure 4a), but is certainly not circle-map-like when δ is large (figure 4b). We distinguish between the region of complex dynamics that exists when $\delta \gg 1$ (region I) and the range of δ for which the dynamics contains substantial intervals of frequency-locking and therefore might be circle-map-like (referred to as region II). To determine whether or not the asymptotic dynamics is one-dimensional, we appeal to the 'Annulus Principle' due to Afraimovich et al. [23], reported also in Afraimovich and Hsu [24]:

Theorem (Annulus Principle). Suppose that the 2D map

$$x_{n+1} = F(x_n, t_n), \qquad t_{n+1} = t_n + G(x_n, t_n) \mod 2\pi,$$

maps an annulus $\mathcal{A} \equiv \{(x, t) : c_1 < x < c_2\}$ into itself, and also satisfies the four conditions

$$|(1 + G_t)^{-1}| < \infty, \qquad 1 - |(1 + G_t)^{-1}||F_x| > 2|(1 + G_t)^{-1}|\sqrt{|G_x||F_t|}, \tag{6}$$

$$|F_x| < 1, \qquad 1 + |(1 + G_t)^{-1}||F_x| < 2|(1 + G_t)^{-1}|, \tag{7}$$

then the maximal attractor in $\mathcal{A}$ is an invariant circle which is the graph of a smooth 2π–periodic function $x = h(t)$.

A direct analytic computation of the range of parameters c, e, ω under which the Annulus Principle holds for (4) - (5) becomes extremely messy. Instead we resort to a simplification of (4) - (5) that should be valid for δ near unity and $\omega \ll 1$ and which preserves the essential qualitative features of the dynamics. With these assumptions, we obtain the simplified map

$$x_{n+1} = x_n^d + \gamma\mu_2[1 - \sqrt{a_2}\cos(2\omega t_n)], \tag{8}$$

$$t_{n+1} = t_n + \mu_3 - \xi \log x_{n+1} \mod \pi/\omega. \tag{9}$$

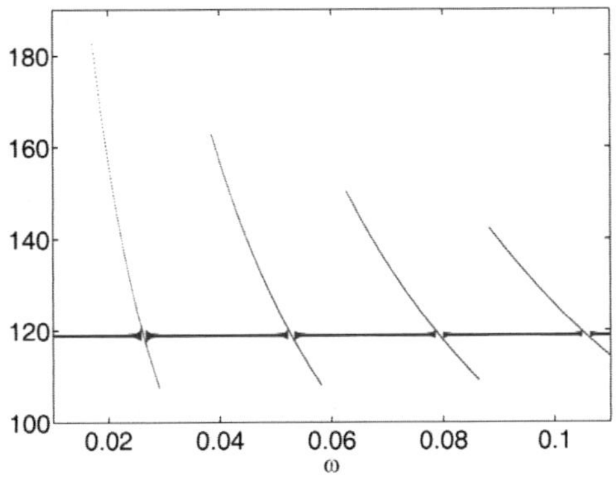

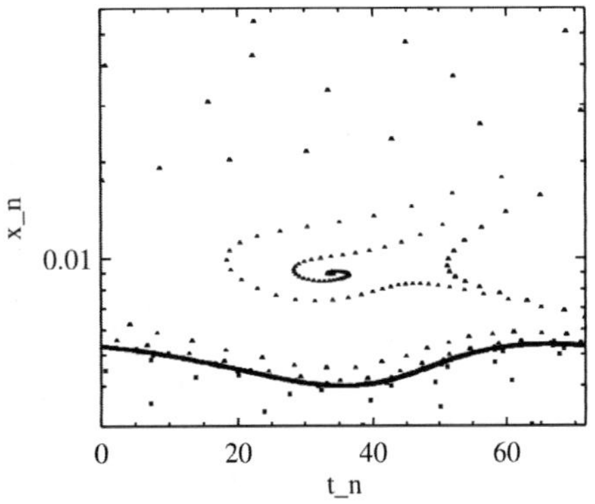

Figure 5. Dynamics in region III. (a) Period T of frequency-locked and quasiperiodic orbits in (8) - (9) returning to H_3^{in}, as a function of driving frequency ω. $c = 0.2001$, $e = 0.2$, $\gamma = 2 \times 10^{-5}$. (b) Dynamics of iterates of (4) - (5) in the (t, x) plane showing stable coexistence of frequency locked state (stable focus) and the invariant curve. $c = 0.20215$, $e = 0.2$, $\omega = 0.0439772$, $\gamma = 10^{-5}$.

Such a simplified map was discussed by Afraimovich, Hsu and Lin [25] as a model for the dynamics of a time-periodically forced system that is very similar to ours (the change of variable $y_n = x_n^d$ brings the simplified system (8) - (9) into their form). These authors refer to (8) - (9) as a 'dissipative separatrix map' since it corresponds closely to well-known and much-studied return maps near separatrices in perturbed Hamiltonian systems when $d = 1$. It is straightforward to check the inequalities in (6) - (7) for the dissipative separatrix map (8) - (9). Intriguingly, it turns out that the first of the four inequalities, (6a), is not satisfied in the limit $\delta \to 1^+$.

The qualitative nature of the conclusion is surprising; an attracting invariant curve $x = h(t)$ (and therefore one-dimensional dynamics) exists (numerically) for large enough $\delta > 1$, but we cannot guarantee that the dynamics remains one-dimensional as $\delta \to 1^+$. Further numerical results are consistent with this conclusion: for δ just above unity additional stable invariant sets exist - this region of bistability defines region III.

4. Region III: bistability between frequency locking and quasiperiodicity.

For δ close to unity we observe that there is bistability between frequency locking and quasiperiodic motion (region III), see figure 2(b). As $\delta \to 1^+$ this bistability becomes much more pronounced, see figure 5(a). The typical dynamics of iterates of (8) - (9) in the (t, x) phase space, near one of the frequencies ω where the frequency-locked state has a period close to that of the quasiperiodic dynamics (i.e. $T \approx 119$ in figure 5a) are shown in figure 5(b). Two fixed points exist away from the invariant curve; one saddle point and one stable focus. As usual, the stable manifold of the saddle provides the boundary dividing the basin of attraction of the invariant curve from that of the stable focus.

In the double limit in which $0 < \epsilon := \delta - 1 \ll 1$ and $\gamma \ll 1$ we find that (8) - (9) can be simplified to a second-order ODE for a rescaled time variable $s \sim t_{n+1} - t_n$; we obtain the canonical equation for a forced damped pendulum [26, 27]

$$\ddot{s} + \eta^{-1}\dot{s} - \sqrt{a_2}\cos s \;\; = \;\; \lambda,$$

where η and λ are combinations of the parameters in (8) - (9). If $|\lambda| > 1$ the only invariant set is a stable periodic orbit (the invariant curve corresponding to quasiperiodic dynamics here). If

$|\lambda| < 1$ and η^{-1} is large (i.e. at large $\epsilon > 0$) then only equilibria (frequency-locked solutions) exist; thus there are intervals in ω within which frequency locking occurs, but where the periodic orbit does not, see figure 2; this is region II. If $|\lambda| < 1$ and η^{-1} is small (i.e. at small ϵ) then equilibria coexist with the periodic orbit, as in figure 5; this bistability explains region III.

5. Summary

We have analysed the dynamics near a time-periodically forced robust heteroclinic cycle. There are three distinct regions for the dynamics; in region I the eigenvalue ratio $\delta \gg 1$ and frequency-locking does not occur. In region II, for intermediate values of δ, the 2D map can be reduced to a (non-invertible) circle map and frequency locking intervals exist. In region III where δ is close to unity, bistability exists and hence the dynamics in no longer that of a circle map: instead the 2D map can be reduced to the continuous time dynamics of a forced damped pendulum.

Since the forcing term in (1) - (3) has no particular nongeneric features we conjecture that the use of other forcing terms would lead to qualitatively similar dynamical regimes. Work in progress also indicates that similar dynamical regimes exist in systems of coupled RHCs.

Acknowledgments

TLT acknowledges past financial support from the Government of Taiwan and from the Cambridge Overseas Trust. JHPD is supported by the Royal Society through a University Research Fellowship.

References

[1] Krupa M and Melbourne I 1995 *Erg. Theory Dyn. Syst.* **15** 121
[2] Krupa M and Melbourne I 2004 *Proc. R. Soc. Edin.* A **134** 1177–97
[3] Postlethwaite C M and Dawes J H P 2010 *Nonlinearity* **23** 621–42
[4] Melbourne I 1991 *Nonlinearity* **4** 835–44
[5] Postlethwaite C M and Dawes J H P 2005 *Nonlinearity* **18** 1477-1509
[6] Ashwin P and Field M 1999 *Arch. Rat. Mech. Anal.* **148** 107–43
[7] Chossat P and Lauterbach R 2000 *Methods in Equivariant Bifurcations and Dynamical Systems* (Singapore: World Scientific)
[8] Holmes P, Lumley J L and Berkooz G 1996, *Turbulence, Coherent Structures, Dynamical Systems and Symmetry* (Cambridge, UK: Cambridge University Press)
[9] Ashwin P, Orosz G, Wordsworth J and Townley S 2007 *SIAM J. Appl. Dyn. Syst.* **6** 728–58
[10] Kuznetsov A S and Kurths J 2002 *Phys. Rev.* E **66** 026201
[11] Hofbauer J and Sigmund K 2002 *Evolutionary Games and Population Dynamics* (Cambridge, UK: Cambridge University Press)
[12] Szabó G and Fáth G 2007 *Phys. Rep.* **446** 97–216
[13] Rabinovich M I, Varona P, Selverston A I and Abarbanel H D I 2006 *Rev. Mod. Phys.* **78** 1213–65
[14] Krupa M 1997 *J. Nonlin. Sci.* **7** 129
[15] Dawes J H P and Tsai T L 2006 *Phys. Rev.* E **74** 055201(R)
[16] Tachikawa M 2003 *Prog. Theor. Phys. Suppl.* **150** 449–52
[17] Sato Y, Akiyama E & Crutchfield J P 2005 *Physica* D **210** 21–57
[18] Tsai T L and Dawes J H P 2010 Dynamics near a periodically-perturbed robust heteroclinic cycle *In preparation*
[19] May R M and Leonard W 1975 *SIAM J. Appl. Math.* **29** 243
[20] Busse F H and Heikes K 1980 *Science* **208** 173
[21] Guckenheimer J and Holmes P 1988 *Math. Proc. Camb. Philos. Soc.* **103** 189
[22] Rabinovich M I, Huerta R and Varona P 2006 *Phys. Rev. Lett.* **96** 014101
[23] Afraimovich V S, Gavrilov N N, Lukyanov V I and Shilnikov L P 1985 *Main Bifurcation of Dynamical Systems* (Gorky: Gorky University Press)
[24] Afraimovich V S and Hsu S-B 1998 *Lectures on Chaotic Dynamical Systems* Hsinchu, Taiwan: National Tsing-Hua University Press)
[25] Afraimovich V S, Hsu S-B and Lin H-E 2001 *Int. J. Bif. Chaos* **11** 435–47
[26] Andronov A A, Vitt A A and Khaikin S E 1966 *Theory of Oscillators* (New York: Pergamon)
[27] Coullet P, Gilli J M, Monticelli N and Vandenberghe N 2005 *Am. J. Phys.* **73** 1122–28

AUTHOR INDEX

AUTHOR INDEX